MW00843857

OXFORD LOGIC GUIDES: 32

General Editors

DOV GABBAY
ANGUS MACINTYRE
DANA SCOTT

OXFORD LOGIC GUIDES

---

1. Jane Bridge: *Beginning model theory: the completeness theorem and some consequences*
2. Michael Dummett: *Elements of intuitionism*
3. A. S. Troelstra: *Choice sequences: a chapter of intuitionistic mathematics*
4. J. L. Bell: *Boolean-valued models and independence proofs in set theory* (1st edition)
5. Krister Sederberg: *Classical propositional operators: an exercise in the foundation of logic.*
6. G. C. Smith: *The Boole–De Morgan correspondence 1842–1864*
7. Alec Fisher: *Formal number theory and compatibility: a work book*
8. Anand Pillay: *An introduction to stability theory*
9. H. E. Rose: *Subrecursion: functions and hierarchies*
10. Michael Hallett: *Cantorian set theory and limitation of size*
11. R. Mansfield and G. Weitkamp: *Recursive aspects of descriptive set theory*
12. J. L. Bell: *Boolean-valued models and independence proofs in set theory*
13. Melvin Fitting: *Computability theory: semantics and logic programming*
14. J. L. Bell: *Toposes and local set theories: an introduction*
15. R. Kaye; *Models of Peano arithmetic*
16. J. Chapman and F. Rowbottom: *Relative category theory and geometric morphisms: a logical approach*
17. Stewart Shapiro: *Foundations without foundationalism*
18. John P. Cleave: *A study of logics*
19. R. M. Smullyan: *Gödel's incompleteness theorem*
20. T. E. Forster: *Set theory with a universal set: exploring an untyped universe*
21. C. McLarty: *Elementary categories, elementary toposes*
22. R. M. Smullyan: *Recursion theory for metamathematics*
23. Peter Clote and Jan Krajíček: *Arithmetic, proof theory, and computational complexity*
24. A. Tarski: *Introduction to logic and to the methodology of deductive sciences*
25. G. Malinowksi: *Many valued logics*
26. Alexandre Borovik and Ali Nesin: *Groups of finite Morley rank*
27. R. M. Smullyan: *Diagonalization and self-reference*
28. Dov M. Gabbay, Ian Hodkinson, and Mark Reynolds: *Temporal logic: Mathematical foundations and computational aspects* (*Volume I*)
29. Saharon Shelah: *Cardinal arithmetic*
30. Erik Sandewall: *Features and fluents: Volume I: A systematic approach to the representation of knowledge about dynamical systems*
31. T. E. Forster: *Set theory with a universal set: exploring an untyped universe* (2nd edition)
32. Anand Pillay: *Geometric stability theory*
33. Dov M. Gabbay: *Labelled deductive systems*

Geometric Stability Theory

ANAND PILLAY

University of Notre Dame

University of Illinois at Urbana-Champaign

CLARENDON PRESS · OXFORD
1996

Oxford University Press, Walton Street, Oxford OX2 6DP

Oxford New York
Athens Auckland Bangkok Bombay
Calcutta Cape Town Dar es Salaam Delhi
Florence Hong Kong Istanbul Karachi
Kuala Lumpur Madras Madrid Melbourne
Mexico City Nairobi Paris Singapore
Taipei Tokyo Toronto

and associated companies in
Berlin Ibadan

Published in the United States
by Oxford University Press Inc., New York

A catalogue record for this book is available from the British Library

Library of Congress Cataloging in Publication Data
Pillay, Anand.
Geometric stability theory / Anand Pillay.
(Oxford logic guides; 32)
Includes bibliographical references and index.
1. Model theory. I. Title. II. Series.
QA9.7.P54 1996
511.3'3—dc20 96-17222 CIP

ISBN 0-19-853437-X

Typeset by Keyword Typesetting, Great Yarmouth Norfolk
Printed in Great Britain by Bookcraft Ltd. (Bath), Midsomer Norton, Avon

To my parents

Preface

The idea of writing the present book occurred to me in 1987. I began writing it at the end of 1990, and only now (the autumn of 1995) are the final revisions being made. I apologize for the delay in this work coming to fruition. In contrast to my earlier book, *An Introduction to Stability Theory*, the idea here is not only to describe technical machinery, but also to prove substantial and deep results (although Chapter 7 and 8 *are* largely concerned with the development of technology). As a consequence, the material is quite dense, and requires substantial work from the reader. The organization of the material bears the marks of my original conception: to develop first the 'finite rank' theory, which could be understood *without* having a firm grasp of general stability theory, and then generalize to stable theories, superstable theories, regular types, etc. As it turns out, I have added an introductory chapter on stability theory, and the language of general stability theory is used throughout.

Special thanks go to John Baldwin, Ambar Chowdhury, Ingo Kraus, Dave Marker, and Zeljko Sokolovic for their detailed and helpful comments on various portions of the book. I also wish to thank Gabriel Carlyle for his close reading of the entire final text and for catching and pointing out many errors and inaccuracies. Thanks are also due to the National Science Foundation for its financial support over many summers. Finally thanks are due to Margit Messmer for her support and encouragement.

South Bend A.P.
December 1995

Contents

Introduction

Geometric stability theory is a subject that has grown in an organic fashion out of categoricity theory and classification theory. I will try to describe, in the next few paragraphs, and with a minimum of technical language, both the subject and its development, insofar as it is treated in the main body of the book.

Morley proved in the early 1960s that if T is a complete countable first-order theory which has exactly one model (up to isomorphism) of some given uncountable cardinality κ, then T has exactly one model of cardinality λ for *all* uncountable cardinals λ. Such a theory is called *uncountably categorical*, or often just ω_1-categorical. The classical examples are the theory of algebraically closed fields (of a fixed characteristic) and the theory of vector spaces (over a fixed field). Shelah, in the late 1960s, then initiated a far-reaching programme of attempting, for an arbitrary first-order theory T, either to 'classify' the models of T (up to isomorphism) or to show such a classification to be impossible. This was 'classification theory', and at least for countable theories, it reached a successful conclusion in the early 1980s. 'Classifying' the models of T amounted to, from Shelah's viewpoint, describing models by certain nice trees of cardinal invariants. If this could be done (for a given theory T), the class of models of T was said to have a 'structure theorem'. (Uncountably categorical theories have the 'best' structure theorem.) The 'impossibility' of such a classification was usually taken to amount to showing that T has 2^λ models of cardinality λ for all large λ. This was called a 'non-structure theorem'. (There have been in the meantime refinements of the 'structure/non-structure' conceptual dichotomy, involving for example the question of whether models can be characterized by their theories in certain infinitary or generalized logics.

In order to implement his programme Shelah developed a series of 'fundamental dichotomies' (on the class of first-order theories). The most important of these was 'stability'. Roughly speaking, T is said to be *stable* if no model of T contains an infinite set of tuples on which some formula defines a linear ordering. Shelah proved that an unstable theory T has 2^λ models of cardinality λ for all $\lambda > |T|$, and thus, from the point of view of his programme, one could restrict attention to stable theories. (So uncountably categorical theories are stable, in fact satisfy a strong form of stability, called

ω-stability, and this fact was important on Morley's original work.) Under the assumption of stability, Shelah developed a sophisticated model-theoretic machinery (forking, orthogonality, regular types, etc.), with which to pursue this programme. Perhaps the most important of these was (non-)forking, a theory of independence in models of stable theories. Meaning was given to the expression: a is independent from b over A (a, b are tuples, and A some set in a model M of T). In the classical examples of ω_1-categorical theories mentioned above, this notion reduced to (field-theoretic) algebraic disjointness, and linear independence, respectively.

In the 1970s, Boris Zilber, with quite a different focus, developed a series of results and conjectures about uncountably categorical theories, the key (and bold) conjecture being the possibility of classifying such *theories*, up to 'bi-interpretability'. The *rough* idea was that if M is a model of an ω_1-categorical theory, then M *is* essentially either an algebraically closed field $(K, +, .)$ or a vector space $(V, +, 0, \lambda)_{\lambda \in F}$ (where λ are unary functions for scalar multiplication), or a 'degenerate' type of structure such as an infinite set with no relations or the integers equipped with the successor function. (This is a gross simplification, in many ways, and we will try to be more precise below.) The *foundational* impact of this point of view must be emphasized. It was being suggested that one could recover fundamental mathematical structures from assumptions of a purely 'logical' nature. Zilber recognized that such issues were intimately bound up with the 'geometry' of independence, and we will try to explain briefly how. First it was known that a model M of an ω_1-categorical theory contains certain 'irreducible' definable sets, called *strongly minimal.* The definable set D is *strongly minimal* if it is infinite, and every subset of D definable in M, with parameters is finite or cofinite. M is 'controlled' by D, in a general model-theoretic sense. Specifically M will be 'prime and minimal' over D. (All this had already been observed by Baldwin and Lachlan, whose work in a sense contains the germs of the geometric theory.) Inside a strongly minimal set, the dependence notion mentioned above is controlled by model-theoretic algebraicity: if $a \in D$, $A \subseteq B \subseteq D$, then a is dependent with B over A, if $a \notin \mathrm{acl}(A)$, but $a \in \mathrm{acl}(B)$. More generally, if $a_1, \ldots, a_n \in D$ and $A \subseteq B \subseteq D$, then $\{a_1, \ldots, a_n\}$ will be dependent with B over A if for some i,

$$a_i \notin \mathrm{acl}(A \cup \{a_1, \ldots, a_{i-1}, a_{i+1}, \ldots, a_n\})$$

but

$$a_i \in \mathrm{acl}(B \cup \{a_1, \ldots, a_{i-1}, a_{i+1}, \ldots, a_n\}).$$

So the nature of the dependence relation on D is reflected in the 'combinatorial pregeometry' $(D, \mathrm{acl}(\text{-}))$. Zilber's ideology was, more or less, that one should attempt to classify the possible pregeometries arising in this way, and as a consequence try to classify the possible strongly minimal sets (as sets equipped with all their definable relations), and then go on to classify uncountably categorical structures. A property which a strongly minimal set D might have is 'modularity': for any algebraically closed subsets $A, B \subseteq D$, A is independent from B over $A \cap B$. This property holds of any vector space $(V, +, 0, \lambda)_{\lambda \in F}$ over a division ring F (where λ denotes scalar multiplication by λ), where acl(-) is precisely linear closure. On the other hand Zilber proved that in a modular strongly minimal set D any definable group is abelian by finite. Modularity fails drastically in an algebraically closed field $(K, +, \ldots, 0, 1)$, where acl(-) is field-theoretic algebraic closure. Zilber's conjecture was (modified to put it in a more striking form) that if D is a strongly minimal set, then, after possibly naming some points, the pregeometry associated with D is degenerate ($\mathrm{acl}(A) = \cup\{\mathrm{acl}(a); a \in A\}$), or is the pregeometry associated to a vector space, or is the pregeometry associated to an algebraically closed field. Moreover, in the latter two cases, a structure of the appropriate kind is actually interpretable in D.

Zilber's conjectures and ground work represented the birth of geometric stability theory, although his activity was largely confined to the ω_1-categorical framework. It should also be mentioned that Zilber, in the early to mid-1970s, also initiated the study of groups definable in ω_1-categorical structures (later to become the theory of groups of finite Morley rank), and quite early on recognized the importance and relevance of such work for the understanding of uncountably categorical structures, in particular for the manner in which an uncountably categorical structure is built from a strongly minimal set.

Of course a breakthrough is often required before any new set of ideas enters the mainstream. In this case the notorious problem of the non-finite axiomatizability of totally categorical theories fitted the bill. (A complete countable theory T is said to be totally categorical if for *all* infinite cardinals λ, including ω, T has a unique model of cardinality λ.) This was solved by Zilber, and also by Cherlin, Harrington, and Lachlan in the early 1980s. Definable (automorphism) groups played an important role in Zilber's proof, and the classification of the geometries attached to certain totally categorical structures played an important role in that of Cherlin, Harrington, and Lachlan. In fact the classifcation of *ω-categorical* 'strictly minimal' sets (building blocks for totally categorical structures) was a fundamental result on its own right, and was proved by Cherlin (using the classification of finite simple groups) and Zilber (using some beautiful model theory). Such objects turned out to be either (infinite-dimensional) affine or projective spaces over finite fields, or (infinite) indiscernible sets. So under the additional assumption of ω-categoricity, the conjectured classification of strongly minimal sets was proved, the algebraically closed field case disappearing. Chapter 2 of this book is devoted to these 'classical' results of geometric stability theory.

My own interest in 'geometric' matters had developed, also around the early 1980s, from quite different perspectives. I had been interested in the (at present still unresolved) problem of whether a countable stable theory can have exactly n countable models, for some $1 < n < \omega$ (the original conjecture being *no*). Having proved this for theories of modules, I was looking for 'abstract' properties of a theory one could work with in a similar fashion, and come up with the notion of 'weak normality' (now called '1-basedness'). After conversations with Cherlin, it became clear that this was the key 'global' geometric property they had proved to hold of all totally categorical theories. Chapter 4 is devoted to the study of 1-based stable theories, and also to definable groups in such theories. The key result, as regards the latter, is that such groups are (with all their definable structure) essentially just modules, linking up the general theory with my original motivating example. Similar notions appear in the work of Palyutin and his circle, whose motivation was the understanding of quasivarieties, Horn theories, etc.

Around the same time, Buechler began trying to generalize the geometric theory outside the uncountably categorical or totally categorical framework, and proved his dichotomy theorem (a little technical to state at this point), which gave a kind of trade-off between model-theoretic complexity (non-ω-stability) and geometric complexity (non-modularity) in the context of 'U-rank 1 types'. (This also appears in Chapter 2.) This opened the door to using the geometric theory to prove results belonging to the problematic of (Shelah-style) classification theory.

In 1984–5, Ehud Hrushovski burst onto the scene, and has been the dominant force in the subject ever since. Both Zilber's work and that of Cherlin, Harrington, and Lachlan had, as a component, the deduction of 'global' properties of a totally or uncountably categorical theory (specifically the global geometric behaviour of the independence relation) from 'local' properties (such as the behaviour of independence on a strongly minimal set). Both myself and Buechler had tried (with some success) to generalize such results to superstable theories. In the same way as strongly minimal sets are the building blocks of uncountably categorical theories, regular types are the building blocks of superstable theories. A regular type is, roughly speaking, a type, on whose set of realizations, forking satisfies the Steinitz axioms for a dependence relation. In his doctoral thesis, Hrushovski classified the forking geometries of (locally) modular type, resurrected Shelah's theory of 'p-simple' types and 'p-weight' to give a satisfying generalization of the earlier 'local–global' results, proved that infinite definable groups exist in a number of contexts, lifted the theory of stable groups (group definable in models of stable theories) to new levels, and as a by-product solved an outstanding open problem in stability theory (the superstability of 'unidimensional' stable theories). We now had a systematic integration of the conceptual frameworks of the geometric and classical theories. Chapters 5 and 7 are devoted largely to the material in Hrushovski' thesis.

Together with Buechler's work mentioned above, Hrushovski's breakthroughs enabled Laskowski to solve a classification-theory-type problem concerning the models of an *uncountable* theory T which is $|T|^+$-categorical. They also enabled Buechler and Newelski to make enormous progress on Vaught's conjecture (the conjecture that a countable theory has *either* at most countably many *or* continuum many countable models), in the case of superstable theories of 'finite rank'. Chapter 6 is devoted to these applications to classification theory.

Although totally categorical theories had been shown not to be finitely axiomatizable, it was asked (by Lachlan) whether such theories could nevertheless be finitely axiomatized *modulo* the axiom schema saying there are infinitely many elements. This was answered positively by Ahlbrandt and Ziegler in the early to mid-1980s in some special cases, using the geometric theory together with some ingeneous combinatorial ideas. Hrushovski, in another *tour de force*, generalized their methods to prove the full conjecture (in fact, suitably rephrased, for the more general class of ω-stable, ω-categorical theories). This work is described in Chapter 3.

Superstability has been mentioned a few times in this introduction, and I now wish to say a little more about it. A stable theory is *superstable* if for any finite tuple a, and set A (in a model of T), there is a *finite* subset A_0 of A such that a is independent from A over A_0. Shelah had in fact proved that non-superstable theories satisfy a non-structure theorem: if T is such then T has 2^λ models of cardinality λ for all $\lambda > |T|$. Thus, from the point of view of Shelah's programme, superstability of T could be assumed. Shelah's 'Main Gap for countable theories' result can be roughly stated as: for countable superstable T, *either* T has 2^λ models of cardinality λ for all uncountable λ *or* for $\alpha \geq 0$, T has at most $\beth_{\omega_1}(|\alpha + \omega|)$ models of cardinality $\aleph_\alpha$. Although the result was somewhat more refined than this, it nevertheless left open exactly how many models in 'small' cardinalities T may have. Hrushovski, in as yet unpublished work, subsequently filled this gap, by arguments involving, among other things, the geometric theory. One of the tools was a generalization (by Hrushovski and Shelah) of Buechler's dichotomy theorem to regular types in superstable theories (under additional assumptions NDOP and NOTOP, properties which could be assumed, by the Shelah theory, to hold). We include the latter result, together with other related generalizations of the dichotomy theorem and additional material on superstability, in Chapter 8. There remain several open questions regarding classification theory, such as proving the Main Gap for *uncountable* theories, or proving a Main Gap result for the category of $|T|^+$-saturated models of a stable theory T. One expects the geometric theory to be of great use here.

The reader is referred to the 'Notes on chapters' for detailed references regarding the material discussed above.

Among the important developments that have *not* been included in this book is Hrushovski's construction of counterexample to Zilber's conjecture. Specifically he constructed strongly mininal sets which are not (locally) modular but in which no infinite group or field is definable. He used similar methods to build a *counteraxample* to a conjecture of Lachlan that a stable ω-categorical structure is superstable. These ingenious constructions and the general theory surrounding them deserve a book all to themselves. Another recent development (in the opposite direction) is work by Hrushovski and Zilber giving a *positive* answer to Zilber's conjecture, under additional 'topological' assumptions on the strongly minimal set. The latter result, as well as the material from Chapter 4 of this book, have played an important role in recent applications of geometric stability theory to diophantine geometry. In fact the present period is a very exciting one for model theory, in which there are both grand unifying trends within model theory (in which geometric stability theory has an important role) and new applications to and connections with core areas of mathematics. In fact, it is becoming increasingly clear that the scope of the ideas discussed above goes beyond stable theories, and that, in a sense, the prior concentration on stability was the result of the particular role of stability within Shelah's programme. But discussion and elaboration of these issues should be left for another book.

1
Stability theory

In this chapter, I survey the stability-theoretic results and notions required for the rest of the book. I include forking, canonical bases, ranks, orthogonality, and stable group theory. I give selective proofs. Hopefully this chapter will also provide a point of entry into general stability theory for the beginner, and thus I include a fairly thorough treatment of forking. I will be assuming knowledge of the basic ingredients of first-order model theory (languages, structures, elementary substructures, compactness, types, saturation, etc.). Such background can be found in Chang and Keisler's book [CK] for example, and also in Hodges recent book [Ho], but notation will be consistent with my earlier book [P1].

Strictly speaking, a full grasp of stability theory is not required for understanding the 'finite rank' part of the theory, covered in Chapters 2, 3, 5, 6, and part of Chapter 4. For much of this material, simply dimension and canonical bases in strongly minimal sets or minimal sets are all that the reader has to understand. So the reader who so wishes can begin at Chapter 2 and refer back to various sections of Chapter 1 (on minimal sets, ranks, etc.) when necessary. On the other hand, general stability provides a unifying conceptual framework and language for the various topics covered in the book, as well as being an essential feature in Chapter 4, 7, and 8.

1 Structures and imaginary elements

L will denote a first-order language, T will denote a complete L-theory with only infinite models, and $M, N, \ldots$, will denote models of T. For the moment the cardinality of L, denoted $|L|$, means the cardinality of the set of L-formulae. So $|T| = |L|$. We often identify the structure M with its universe (or underlying set). $|M|$ denotes the cardinality of the universe of M.

It is convenient (and common now in model theory) to fix a 'universal domain' which will contain as elementary substructures all models of T we are interested in. This universal domain, $\mathbf{C}$, which we will obtain in a moment, should have in addition certain 'homogeneity' features. We describe now these features. Recall that if κ is a cardinal, and M an L-structure, then M is said to be *κ-saturated* if whenever $\mathbf{x}$ is a finite tuple of variables, A is a subset of M of cardinality $<\kappa$, and $\Sigma(\mathbf{x})$ is a set of formulae with parameters in A which is consistent with $Th(M, a)_{a \in A}$, *then* $\Sigma(\mathbf{x})$ is realized in M. We say that M is *strongly* κ*-homogeneous*, if whenever A, B are subsets of M of cardinality $<\kappa$ and f is a bijection between A and B which is an elementary map in M, *then* f extends to an *automorphism* of M.

Fact. If κ is a regular cardinal $>|L|$, then there is a model M of T which is κ-saturated and strongly κ-homogeneous.

In fact in [Ho], Wilfred Hodges defines a notion 'κ-bigness' which implies κ-saturation and strong κ-homogeneity, and such that for any regular $\kappa > |L|$ there is a κ-big model of cardinality at most $|T| < \kappa$.

Let us fix a large regular cardinal κ_0, larger than any models of T we are interested in, and let $\mathbf{C}$ be a model of T given to us by the fact above. Thus any model of T of cardinality at most κ_0 is isomorphic to an elementary substructure of $\mathbf{C}$. By a *small model* we will thus mean an elementary substructure of $\mathbf{C}$ of cardinality $<\kappa_0$. M, N will denote such models. A, $B, \ldots$ will denote small subsets of $\mathbf{C}$, namely subsets of cardinality strictly less than κ_0. So the strong homogeneity condition on $\mathbf{C}$ reads: any partial elementary map between small subsets of $\mathbf{C}$ extends to an automorphism of $\mathbf{C}$. We will see later that under some stability assumption we may actually assume $\mathbf{C}$ to be of cardinality κ_0 (so saturated).

L_A is the language obtained from A by adjoining constants for the elements of A, which always have the natural interpretation. If σ is an L_A-sentence, then $\models \sigma$ means the same as $\mathbf{C} \models \sigma$, which thus means the same as $M \models \sigma$ for some (any) model M containing A. Whenever we mention (partial) elementary maps, we mean elementary in the sense of $\mathbf{C}$. Similarly if $\varphi(\mathbf{x})$, $\psi(\mathbf{x})$ are $L_{\mathbf{C}}$-formulae we will say φ is equivalent to ψ, if $\models (\forall \mathbf{x})(\varphi(\mathbf{x}) \leftrightarrow \psi(\mathbf{x}))$.

By a definable set, we mean (for now), a subset of $\mathbf{C}^n$ (for some n) which is defined by a formula $\varphi(\mathbf{x})$ of $L_{\mathbf{C}}$. A definable set X is said to be A-definable if X can be defined by a formula in L_A. If X is defined by $\varphi(\mathbf{x})$ and $\varphi(\mathbf{x})$ is over M, then by φ^M we mean $X \cap M^n$, or equivalently the set of n-tuples from M which satisfy φ in M (or in $\mathbf{C}$). Note that if $f\colon A \to \mathbf{C}$ is an elementary map, then f also acts on the collection of A-definable sets: if X is defined by $\varphi(\mathbf{x}, \mathbf{a})$ where $\varphi(\mathbf{x}, \mathbf{y})$ is an L-formula and $\mathbf{a}$ is from A, then $f(X)$ is the set defined by $\varphi(\mathbf{x}, f(\mathbf{a}))$. Moreover, this is *independent* of the particular formula $\varphi(\mathbf{x}, \mathbf{y})$ and particular tuple $\mathbf{a}$ from A chosen. The following lemma is proved by an elementary compactness argument.

LEMMA 1.1. *Let A, B be (small) subsets of* $\mathbf{C}$. *Let X be a B-definable set. The following are equivalent*:

(i) *X is A-definable.*

(ii) *Whenever* $f: A \cup B \to \mathbf{C}$ *is elementary and* $f \upharpoonright A$ *is the identity, then* $f(X) = X$.

(iii) *Whenever* f *is an automorphism of* $\mathbf{C}$ *and* $f \upharpoonright A = \text{identity}$ *then* $f(X) = X$.

We sometimes say that a definable set X is *over* A, if X is A-definable.

We will say that the definable set $X \subseteq \mathbf{C}^n$ is *almost* A*-definable* or *almost over* A, if there is an A-definable equivalence relation Y (on $\mathbf{C}^n$) which has finitely many classes, and X is a union of some of these classes, We then obtain:

LEMMA 1.2. *Suppose* X *is* B*-definable. The following are equivalent*:

(i) X *is almost* A*-definable.*

(ii) $\{f(X): f \upharpoonright A \cup B \to \mathbf{C}$ *is an elementary map which fixes* A *pointwise*$\}$ *is finite.*

(iii) $\{f(X): f$ *is an automorphism of* $\mathbf{C}$ *which fixes* A *pointwise*$\}$ *is finite.*

We now introduce a device which, in particular, enables definable sets to be viewed as objects of the same kind as elements of $\mathbf{C}$. Under the assumption of stability, this will also permit types (ultrafilters of definable sets) to be viewed as elements. In fact we will introduce a many-sorted language L^{eq}, an L^{eq}-theory T^{eq}, and natural 'expansion/extension' $\mathbf{C}^{\text{eq}}$ of $\mathbf{C}$ to a model of T^{eq}.

A few words should be said first about many-sorted languages and theories. A many-sorted language is a language which contains *sorts, relation symbols, function symbols* (we treat constant symbols as 0-ary function symbols), *and for each sort, a supply of variables of that sort.* To each n-ary relation symbol R will be associated an n-tuple $(S_1, \ldots, S_n)$ of sorts, and to each n-place function symbol f will be associated an n-tuple of sorts (for the domain of f) and also a target sort. Well-formed formulae are built up as usual, except we require that for $R(v_1, \ldots, v_n)$ to be a formula each variable v_i should be of sort S_i. Similarly for $f(v_1, \ldots, v_n) = v$ to be a formula we require the variables v_i and v to be of the right sorts. We make similar requirements for terms. (To be accurate we must also have a different equality symbol for each sort.) A structure M for such a language will then consist of a collection of disjoint domains S^M, corresponding to the various sorts S (so S^M will be the interpretation of the sort S), and relations and functions corresponding to the symbols of the language, with the stipulation that, for example, if the n-tuple $(S_1, \ldots, S_n)$ of sorts is associated to the n-ary relation symbol R, then the interpretation R^M of R in M is a subset of $S_1^M \times \cdots \times S_n^M$, and similarly for the interpretation of function symbols. Note that if v is a variable of sort S, then the quantifier $(\exists v)(\ldots)$ for example is interpreted as 'there is v in sort S such that...', that is all quantitifiers are 'relativized' to certain sorts.

All of predicate logic and model theory then goes through in this situation, in particular the notion of saturation. The thing to bear in mind is that compactness cannot be used to produce a structure with an element a which is not in any sort, the formal reason being that any variable must specifiy the sort to which it belongs.

We proceed to define the language L^{eq} (which depends on T). Let $ER(T)$ be the collection of L-formulae $E(\mathbf{x}, \mathbf{y})$ with $l(\mathbf{x}) = l(\mathbf{y})$, which define in $\mathbf{C}$ an equivalence relation on $\mathbf{C}^n$ (for $n = l(\mathbf{x})$). For each formula $E(\mathbf{x}, \mathbf{y}) \in ER(T)$, L^{eq} will contain a sort S_E. In particular there will be a sort $S_=$. For E as above, L^{eq} will also contain a new n-ary function symbol f_E, whose domain sort is $(S_=)^n$ and whose target sort is S_E. All the relation and function symbols of L will also be in L^{eq} but the associated tuples of sorts will all be of the form $(S_=, \ldots, S_=)$ (so we have a natural 'embedding' of L in L^{eq}). Note that $|L^{eq}| = |L|$.

We now enlarge $\mathbf{C}$ to a canonical L^{eq}-structure $\mathbf{C}^{eq}$. The interpretation of the sort $S_=$ in $\mathbf{C}^{eq}$ will be the set $\mathbf{C}$ itself, and those relation and function symbols of L^{eq} which are in L will be interpreted in $\mathbf{C}$ in the usual way. The interpretation of the sort S_E in $\mathbf{C}^{eq}$ will be the set $\{\mathbf{a}/E : \mathbf{a} \in \mathbf{C}^n\}$, and the interpretation of the function symbol f_E will be the function which takes $\mathbf{a}$ to $\mathbf{a}/E$. Similarly for any model M of T we produce an L^{eq}-structure M^{eq} such that the interpretation in M^{eq} of sort $S_=$ is exactly the structure M. By virtue of the natural embedding of L in L^{eq}, we can treat any L-formula $\varphi(\mathbf{x})$ as an L^{eq}-formula, by considering all the variables in φ (bound as well as free) as being of sort $S_=$ (or equivalently relativizing all quantifiers to $S_=$). We then clearly have:

Fact 1.3. For all models M of T, L-formulae $\varphi(\mathbf{x})$, and tuples $\mathbf{a}$ from M, $M \models \varphi(\mathbf{a})$ iff $M^{eq} \models \varphi(\mathbf{a})$.

The theory T^{eq} will be the L^{eq}-theory consisting of all the sentences of T together with axioms which assert that for each $E \in ER(T)$, f_E is a map from $(S_=)^n$ (for n = arity of f) *onto* S_E, and $f_E(\mathbf{x}) = f_E(\mathbf{y})$ iff $E(\mathbf{x}, \mathbf{y})$. Clearly M^{eq} is a model of T^{eq} whenever M is a model of T. Moreover, we have the following, whose proof is left as an exercise (see [Makk]).

LEMMA 1.4. (i) *Every model M^* of T^{eq} is of the form M^{eq} where M is a model of T, and $(S_=)^{M^*} = M$.*

(ii) *T^{eq} is complete.*

(iii) *Whenever* $E(1), \ldots, E(k) \in ER(T)$, *and* $\varphi(x_1, \ldots, x_k)$ *is an* L^{eq}-*formula with* x_i *a variable of sort* $S_{E(i)}$, *there is an L-formula* $\psi(\mathbf{y}_1, \ldots, \mathbf{y}_k)$, *such that* $T^{eq} \vdash (\forall \mathbf{y}_1 \ldots \mathbf{y}_k \textit{ from } S_=)(\psi(\mathbf{y}_1, \ldots, \mathbf{y}_k) \leftrightarrow \varphi(f_{E(1)}(\mathbf{y}_1), \ldots, f_{E(k)}(\mathbf{y}_k))$.

Note the following special case of 1.4(i): if $\varphi(x_1, \ldots, x_k)$ is a formula of L^{eq} and each x_i is a variable of sort $S_=$ then there is an L-formula $\psi(x_1, \ldots, x_k)$ which is equivalent to φ in T^{eq}.

To sum up: $\mathbf{C}$ is embedded naturally in $\mathbf{C}^{eq}$ as the elements of sort $S_=$; similarly for M in M^{eq}. We assumed M to be an elementary subtructure of $\mathbf{C}$, so M^{eq} is canonically a substructure of $\mathbf{C}^{eq}$. By Lemma 1.4(iii) we then obtain:

Lemma 1.5. (i) *For each model M,* M^{eq} *is an elementary substructure of* $\mathbf{C}^{eq}$.

(ii) $\mathbf{C}^{eq}$ *is in the definable closure (in the sense of* L^{eq}*) of* $\mathbf{C}$. *Similarly* $M^{eq} = \mathrm{dcl}^{eq}(M)$.

(iii) *Every subset of* $\mathbf{C}^n$ *which is definable (with parameters) in the structure* $\mathbf{C}^{eq}$ *is definable in* $\mathbf{C}$.

(iv) $\mathbf{C}^{eq}$ *is also* κ_0*-saturated and strongly* κ_0*-homogeneous.*

The following feature is then built into $\mathbf{C}^{eq}$: for any $E \in ER(T)$ and E-equivalence class $X \subseteq \mathbf{C}^n$ there is an element $a \in \mathbf{C}^{eq}$, and formula $\varphi(\mathbf{x}, y)$ (of L^{eq}), such that

(i) the formula $\varphi(\mathbf{x}, a)$ defines X; and

(ii) if $\varphi(\mathbf{x}, a')$ also defines X, then $a = a'$.

The element a is just $\mathbf{a}/E$ where $\mathbf{a} \in X$, and the formula $\varphi(\mathbf{x}, y)$ is simply $f_E(\mathbf{x}) = y$.

This suggests:

Definition 1.6. *We will say that a (possibly many-sorted) theory T has* elimination of imaginaries *if for any sufficiently saturated model M of T,* $\emptyset$*-definable equivalence relation E on* M^n, *and E-class X, there is some* tuple $\mathbf{a}$ *from M such that for some L-formula* $\varphi(\mathbf{x}, \mathbf{y})$, *X is defined by* $\varphi(\mathbf{x}, \mathbf{a})$ *and whenever* $\varphi(\mathbf{x}, \mathbf{a}')$ *is equivalent to* $\varphi(\mathbf{x}, \mathbf{a})$ *(i.e. also defines X), then* $\mathbf{a}' = \mathbf{a}$. *We call such* $\mathbf{a}$ a code for X.

An elementary compactness argument shows that $\mathbf{a}$ is a code for X if and only if for any automorphism f of $\mathbf{C}$, f fixes X setwise iff $f(\mathbf{a}) = \mathbf{a}$.

Remark 1.7. T^{eq} *has elimination of imaginaries.*

Proof. Let E be a $\emptyset$-definable equivalence relation in $\mathbf{C}^{\mathrm{eq}}$. Suppose the domain of E consists of n-tuples of sort $S_{E'}$, where $E' \in ER(T)$ defines an equivalence relation on $\mathbf{C}^k$. By Lemma 1.4(iii) there is an L-formula $\psi(\mathbf{y}_1, \ldots, \mathbf{y}_n, \mathbf{z}_1, \ldots, \mathbf{z}_n)$ such that for k-tuples $\mathbf{a}_1, \ldots, \mathbf{a}_n, \mathbf{b}_1, \ldots, \mathbf{b}_n$ from $\mathbf{C}$, $\mathbf{C} \models \psi(\mathbf{a}_1, \ldots, \mathbf{a}_n, \mathbf{b}_1, \ldots, \mathbf{b}_n)$ iff $\mathbf{C}^{\mathrm{eq}} \models E(\mathbf{a}_1/E', \ldots, \mathbf{a}_n/E', \mathbf{b}_1/E', \ldots, \mathbf{b}_n/E')$. But then the formula ψ defines an equivalence relation on $\mathbf{C}^{nk}$. Let X be an E-class. Let $(a_1, \ldots, a_n) \in X$, and let $\mathbf{a}_i$ be some k-tuple from $\mathbf{C}$ such that $\mathbf{a}_i/E' = a_i$. Then it is not difficult to see that the element $(\mathbf{a}_1, \mathbf{a}_2, \ldots, \mathbf{a}_n)/\psi$ of sort S_ψ is a code for X in $\mathbf{C}^{\mathrm{eq}}$, completing the proof. ■

Let us make some additional remarks connected with elimination of imaginaries:

Remark 1.8. (a) *If the tuple* $\mathbf{a}$ *is a code for the equivalence class* X *then* $\mathbf{a}$ *is* unique *up to interdefinability, in the sense that if* $\mathbf{b}$ *is another code for* X *then* $\mathbf{a} \in \mathrm{dcl}(\mathbf{b})$ *and* $\mathbf{b} \in \mathrm{dcl}(\mathbf{a})$.

(b) *Let* X *be any definable set, defined by the formula* $\varphi(\mathbf{x}, \mathbf{a})$ *say, where* $\varphi(\mathbf{x}, \mathbf{y})$ *is an* L*-formula. Let* $E(\mathbf{y}_1, \mathbf{y}_2)$ *be the* L*-formula* $(\forall \mathbf{x})(\varphi(\mathbf{x}, \mathbf{y}_1) \leftrightarrow \varphi(\mathbf{x}, \mathbf{y}_2))$. *So* $E \in ER(T)$. *Let* Y *be* E*-class of* $\mathbf{a}$ *(defined by the formula* $E(\mathbf{y}, \mathbf{a})$*). Then* X *and* Y *are 'interdefinable' in the sense that for any tuple* $\mathbf{b}$, X *is* $\mathbf{b}$*-definable iff* Y *is* $\mathbf{b}$*-definable. In particular some tuple* $\mathbf{b}$ *is a code for* X *(in the sense that* $\mathbf{b}$ *satisfies* (i) *and* (ii) *of Definition* 1.6 *for* X*) iff* $\mathbf{b}$ *is a code for* Y. *Thus if* T *has elimination of imaginaries then* every *definable set in* $\mathbf{C}$ *has some code which is again unique up to interdefinability.*

(c) *The property 'the tuple* $\mathbf{b}$ *is a code for the definable set* X*' can be characterized by: any automorphism* f *of* $\mathbf{C}$ *fixes* X *(as a set) iff it fixes* $\mathbf{b}$ *(as a tuple).*

(d) *Suppose* $\mathbf{a}$ *is a code for the definable set* X. *Let* A *be any set of parameters. Then* X *is almost* A*-definable iff* $\mathbf{a} \in \mathrm{acl}(A)$.

Note that passing to $\mathbf{C}^{\mathrm{eq}}$ yields a nice way of seeing Lemma 1.2. For by this means Lemma 1.2 amounts to:

(i) $a \in \mathrm{dcl}(A)$ iff any automorphism of $\mathbf{C}$ which fixes A pointwise fixes a;

(ii) $a \in \mathrm{acl}(A)$ iff a has only finitely many images under automorphisms of $\mathbf{C}$ which fix A pointwise.

The reader would be convinced now that passing from $\mathbf{C}$ to $\mathbf{C}^{\mathrm{eq}}$ is a convenient but harmless step. The objects of $\mathbf{C}^{\mathrm{eq}}$ are already present in $\mathbf{C}$, and the above lemmas and remarks serve solely to point out that these objects can be dealt with by the usual model-theoretic tools.

Typically we will work throughout in $\mathbf{C}^{eq}$, which should nevertheless be understood to be obtained from a one-sorted structure, in the manner outlined above. Once we start to work in $\mathbf{C}^{eq}$ the distinction between elements and tuples becomes somewhat blurred. For example, an n-tuple from $\mathbf{C}$ is an element of the sort $\mathbf{C}^n$. For this reason we will let a, b, c, etc., typically denote elements of or tuples from $\mathbf{C}^{eq}$, and similarly for variables x, y, z, etc. Of course there will be times when we are restricting our attention to a particular type $p(x)$ or definable set $D(x)$ (e.g. a strongly minimal set). In these situations we definitely *will* want to distinguish between, say, elements of D and tuples from D. In this situation we will revert to using $\mathbf{a}$, $\mathbf{b}$, etc., to denote tuples.

A helpful observation is:

REMARK 1.9. (*In* $\mathbf{C}^{eq}$) $\mathrm{tp}(a/\mathrm{acl}(A)) = \mathrm{tp}(b/\mathrm{acl}(A))$ *iff for each formula* $\varphi(x)$ *which is almost over* A *we have that* $\models \varphi(a)$ iff $\models \varphi(b)$.

By the *strong type* of a over A, we will mean $\mathrm{tp}(a/\mathrm{acl}(A))$, where acl(-) is in the sense of $\mathbf{C}^{eq}$.

2 Forking

Although only stable theories will be studied in this book, we will at times (specifically in Chapter 4) need to know something of the 'quantifier-free' theory. By the quantifier-free theory, I mean the theory which, given a set $\vartheta = \{\varphi_i(x, y_i) : i \in I\}$ of *stable formulae*, yields a smooth theory of forking for ϑ-types, namely types $p(x)$, the formulae of which are of the form $\varphi_i(x, a)$ or $\neg\varphi_i(x, a)$. For myself, this quantifier-free theory has certain *explanatory power* with regard to the general theory.

A formula $\varphi(\mathbf{x}, \mathbf{y})$ (which may have additional parameters) is said to be *stable* if there do not exist $\mathbf{a}_i$, $\mathbf{b}_i$ for $i < \omega$ such that $\models \varphi(\mathbf{a}_i, \mathbf{b}_j)$ iff $i \leq j$. Let us remark that when we write a formula φ as $\varphi(\mathbf{x}, \mathbf{y})$ we mean that the free variables in φ are among those in the sequence $(\mathbf{x}, \mathbf{y})$. The definition of stability of φ thus depends on the choice of $\mathbf{x}$ and $\mathbf{y}$. The theory T is said to be stable if each L-formula $\varphi(\mathbf{x}, \mathbf{y})$ (where $\mathbf{x}$, $\mathbf{y}$ are tuples of variables from sort $S_=$) is stable.

LEMMA 2.1. (i) *If* $\varphi(\mathbf{x}, \mathbf{y})$, $\psi(\mathbf{x}, \mathbf{z})$ *are stable then so are* $\neg\varphi(\mathbf{x}, \mathbf{y})$, $(\varphi \vee \psi)(\mathbf{x}, \mathbf{yz})$, *and* $(\varphi \vee \psi)(\mathbf{x}, \mathbf{yz})$.

(ii) *Let* $\psi(\mathbf{y}, \mathbf{x})$ *be* $\varphi(\mathbf{x}, \mathbf{y})$. *Then* $\varphi(\mathbf{x}, \mathbf{y})$ *is stable iff* $\psi(\mathbf{y}, \mathbf{x})$ *is stable.*

(iii) *If* T *is stable then every formula* $\varphi(\mathbf{x}, \mathbf{y})$ *of* L^{eq} *is also stable.*

(iv) $\varphi(\mathbf{x}, \mathbf{y})$ *is stable iff for some* $n < \omega$ *there do* not *exist* $\mathbf{a}_i$, $\mathbf{b}_i$, *for* $i \leq n$ such that $\models \varphi(\mathbf{a}_i, \mathbf{b}_j)$ iff $i \leq j$.

Let us now fix a formula $\delta(\mathbf{x}, \mathbf{y})$ without parameters. In the spirit of the remarks at the end of section 1, we write $\mathbf{x}$ as a single variable x, and $\mathbf{y}$ as y. A formula of the form $\delta(x, a)$ will be called an *instance* of δ. We want to fix our attention on types (consistent sets of formulae) consisting of instances and negations of instances of δ. Once we do this we may as well consider Boolean combinations of instances of δ, and also formulae $\varphi(x)$ equivalent to the latter. Thus we call a formula $\varphi(x)$ (with parameters) a *δ-formula* if $\varphi(x)$ is *equivalent to* some Boolean combination of instances and negations of instances of δ. (Note: by Boolean combination we always mean *finite* Boolean combinations.) By a complete δ-type over the set A, we mean a maximal consistent set of δ-formulae in L_A. We denote the set of complete δ-types over A by $S_\delta(A)$. (As usual we denote by $S(A)$ the set of complete types over A.) We often identify formulae with the sets they define (in the big model); in this context we define a δ-set to be a set defined by a δ-formula. Thus a complete δ-type over A is a maximal consistent collection of δ-sets which are over A. Let us remark that if M is a model and $p(x) \in S_\delta(M)$, then p is a consequence of, or is determined by, the set of instances and negations of instances of δ which are in p. This is because any δ-formula in L_M will be equivalent to a Boolean combination of formulae of the form $\delta(x, a)$, $\neg\delta(x, a)$ where a is in M.

If $\Sigma(x)$ is a collection of formulae, then $\Sigma \restriction \delta$ is the collection of those δ-formulae in Σ, and $\Sigma \restriction A$ is the collection of those formulas in Σ which are in L_A.

Let M be a model and $p(x)$ a complete type over M, or a complete δ-type over M. By a *δ-definition* of $p(x)$ we mean a formula $\psi(y)$ of L_M such that for all $b \in M$, $\delta(x, b) \in p(x)$ iff $\models \psi(b)$. $p(x)$ need not have a δ-definition, but if it does, this δ-definition is unique up to equivalence, and determines $p(x)$. (Two δ-definitions for p must define the same set in M, so as both these formulae are in L_M and M is a model, they define the same set in $\mathbf{C}$.) Almost all the combinatorial content of stability is contained in the following basic result:

Lemma 2.2. (Definability of types) *Suppose $\delta(x, y)$ is stable. Let M be a model and let $p(x) \in S(M)$ or $S_\delta(M)$. Then*

(i) *$p(x)$ has a δ-definition $\psi(y)$ which is a positive Boolean combination of formulae $\delta(c, y)$, $c \in M$.*

(ii) *Suppose A is a subset of M, and M is $|A|^+$-saturated. Then there are $c_1, c_2, \ldots$ in M such that c_{i+1} realizes $p \restriction (A \cup \{c_1, \ldots, c_i\})$, and the δ-definition of $p(x)$ is equivalent to some positive Boolean combination of formulae $\delta(c_i, y)$.*

Proof. We prove (i), and point out as we go along how the proof also yields (ii).

By Lemma 2.1, we can choose $N < \omega$ such that

(1) there do not exist c_i, a_i for $i \le N$ such that $\models \delta(c_i, a_j)$ iff $i < j$; and
(2) there do not exist c_i, b_i for $i \le N$ such that $\models \neg\delta(c_i, b_j)$ iff $i < j$.

Let c^* realize $p(x)$. We may assume that there exist d, $e \in M$, such that $\models \delta(c^*, d) \wedge \neg\delta(c^*, e)$.

We will define inductively elements $c_0, c_1, \ldots$ in M, sets $K(i)$, $L(i)$ of subsets of $\{0, \ldots, i\}$ for $i \ge -1$, and for each $i \ge -1$, elements a^s_{i+1} in M for $s \in K(i)$, and b^t_{i+1} in M for $t \in L(i)$, as follows.

Let c_0 be arbitrary and set $K(-1) = L(-1) = \{\emptyset\}$.

Suppose we have already defined $c_0, \ldots, c_n$, $K(-1), \ldots, K(n-1)$, $L(-1), \ldots, L(n-1)$, and a^s_i, b^t_i for $i = 0, \ldots, n$, $s \in K(i-1)$, $t \in L(i-1)$. We now define $K(n)$, $L(n)$ and the a^s_{n+1}, b^t_{n+1}:

(i) Let $K(n)$ be the set $\{W \subseteq \{0, \ldots, n\}$: there is $a \in M$ such that $\models \delta(c_j, a)$ for all $j \in W$, but $\models \neg\delta(c^*, a)\}$, and for each $W \in K(n)$, let a^W_{n+1} be an element of M which witnesses this (namely, $\models \delta(c_j, a^W_{n+1})$ for all $j \in W$, and $\models \neg\delta(c^*, a^W_{n+1})$).

Similarly let $L(n) = \{W \subseteq \{0, \ldots, n\}$: there is $b \in M$ such that $\models \neg\delta(c_j, b)$ for all $j \in W$, and $\models \delta(c^*, b)\}$. Again for $W \in L(n)$, choose b^W_{n+1} in M to witness this.

We now define c_{n+1}:

(ii) Let B_n be the set $\{a^s_{i+1}: i \le n, s \in K(i)\} \cup \{b^t_{i+1}: i \le n, t \in L(i)\}$. Then B is a finite set contained in M, so there is an element c in M such that for all $d \in B$, $\models \delta(c, d)$ iff $\models \delta(c^*, d)$. Let c_{n+1} be such an element c. In the case where M is an $|A|^+$-saturated model (for some fixed subset A of M), we may also choose c_{n+1} to realize $p(x) \upharpoonright (A \cup B \cup \{c_0, \ldots, c_n\})$.

We make some remarks on the construction.

(iii) Suppose $0 \le i \le n < \omega$, $s \in K(i-1)$ and $t \in L(i-1)$.

Then $\models \neg\delta(c_n, a^s_i) \wedge \delta(c_n, b^t_i)$. (This is clear from (i) and (ii). First by (i), $\models \neg\delta(c^*, a^s_i) \wedge \delta(c^*, b^t_i)$. Now a^s_i, b^t_i are in B_{n-1}, so by choice of c_n in (ii), we have what we want.)

(iv) Suppose that $i(0) < i(1) < \cdots < i(n) < \omega$, and that for some $a \in M$, $\models \delta(c_{i(0)}, a) \wedge \ldots \wedge \delta(c_{i(n)}, a) \wedge \neg\delta(c^*, a)$. *Then* there exist elements $d_0, \ldots, d_n$ in M such that for all j and r between 0 and n, $\models \delta(c_{i(j)}, d_r)$ iff $j < r$.

Proof of (iv). First note that $K(i(0)-1) \ne \emptyset$ (in fact $\emptyset \in K(i(0)-1)$, as $\models \neg\delta(c^*, a)$ and $a \in M$). Let d_0 be $a^s_{i(0)}$ for some $s \in K(i(0)-1)$, and by (iii) we have $\models \neg\delta(c_{i(j)}, d_0)$ for all $j = 0, \ldots, n$. Now let k be such that $0 \le k < n$.

Then from our assumption and the definition of $K(i(k))$, it must be the case that $W = \{(i(0), \ldots, i(k)\} \in K(i(k))$. Let d_{k+1} be $a^W_{i(k)+1}$. So $\models \delta(c_{i(0)}, d_{k+1}) \wedge \cdots \wedge \delta(c_{i(k)}, d_{k+1})$, and by (iii), $\models \neg\delta(c_{i(j)}, d_{k+1})$ for $j = k+1, \ldots, n$. This proves (iv). ■

(v) Suppose that $0 \le i(0) < i(1) < \cdots < i(n) < \omega$ and that for some $b \in M$, $\models \neg\delta(c_{i(j)}, b)$ for $j = 0, \ldots, n$, but $\models \delta(c^*, b)$. *Then* there are $e_0, \ldots, e_n$ in M such that $\models \neg\delta(c_{i(j)}, d_r)$ iff $j < r$ for all j and r between 0 and n.

Proof. Just like that of (iv), using the L in place of the K. ■

Now from (iv) and our assumption (1) on the number N, we conclude:

(I) if $a \in M$ and for some N-element subset $W$of $\{0, \ldots, 2N\}$, we have $\models \delta(c_i, a)$ for all $i \in W$, *then* $\models \delta(c^*, a)$ (namely, $\delta(x, a) \in p(x)$).

Similarly from (v) we conclude:

(II) if $b \in M$ and for some N-element subset W of $\{0, \ldots, 2N\}$ we have $\models \neg\delta(c_i, b)$ for all $i \in W$, *then* $\models \neg\delta(c^*, b)$ (namely, $\delta(x, b) \notin p(x)$).

From (I) and (II) we see that for any $a \in M$, $\delta(x, a) \in p(x)$ if and only if $\bigvee\{\bigwedge\{\delta(c_i, a) : i \in W\} : W$ an N-element subset of $\{0, \ldots, 2N\}\}$. This gives the required δ-definition of $p(x)$; note that the choice of the c_i in the case where M is $|A|^+$-saturated yields (ii). ■

Remark 2.3. *We obtain quite easily the usual 'number of types' characterization of stability.*

Let $\delta(x, y)$ be an L-formula. Then the following are equivalent:

(i) *$\delta(x, y)$ is stable*;

(ii) *for every model M and $p(x) \in S_\delta(M)$, $p(x)$ has a δ-definition*;

(*iii*) *for each cardinal $\lambda \ge |T|$, and set A of parameters with $|A| \le \lambda$, $|S_\delta(A)| \le \lambda$.*

(iv) *there is a cardinal $\lambda \ge |T|$ such that for every set A of parameters with $|A| \le \lambda$, $|S_\delta(A)| \le \lambda$.*

Proof. We proved (i) implies (ii) in 2.2.

(ii) → (iii). Let λ, A be as given. Let M be a model of cardinality λ containing A. As every $p(x) \in S_\delta(A)$ extends to some $q(x) \in S_\delta(M)$, it suffices to show that $|S_\delta(M)| \le \lambda$. This is immediate by (ii), as there are at most λ formulae in L_M and any $p(x) \in S_\delta(M)$ is determined by its δ-definition.

(iii) → (iv) is immediate.

(iv) $\rightarrow$ (i). Assume $\delta(x, y)$ to be unstable. Let λ be any cardinal $\geq |T|$. Let $\mu \leq \lambda$ be the smallest cardinal such that $2^\mu > \lambda$. Let ${}^\mu 2$ be the set of functions from μ to $\{0, 1\}$, and for $f, g \in {}^\mu 2$ define f to be 'less than' g if for some $\alpha < \mu$, $f \restriction \alpha = g \restriction \alpha$ and $f(\alpha) = 0$, $g(\alpha) = 1$. This is a linear ordering on ${}^\mu 2$. Let X be the subset of ${}^\mu 2$ consisting of eventually constant functions. Then X is dense in ${}^\mu 2$ and $|X| \leq \lambda$. By compactness, and the instability of $\delta(x, y)$, there are a_f, b_f for $f \in {}^\mu 2$, such that $\models \delta(a_f, b_g)$ iff f is less than g. Let M be a model of cardinality λ containing $\{b_g : g \in X\}$. Then if $f \neq g \in {}^\mu 2$, the complete δ-types of a_f and a_g over M are different, so $|S_\delta(M)| > \lambda$. ■

REMARK 2.4. *Suppose $p(x) \in S_\delta(M)$. Let A be a subset of M. We say that p is* definable over *A, if p has a δ-definition which is (equivalent to) an L_A-formula. Suppose M' is a model containing M, and $p(x) \in S_\delta(M)$ is definable over $A \subseteq M$. Let $\psi(y) \in L_A$ be some δ-definition of p. Then there is a* unique *$p'(x) \in S_\delta(M')$ such that $\psi(y)$ is a δ-definition of p'. In fact we only need to prove existence, which is rather easy as M is a model. Note that $p(x) \subseteq p'(x)$.*

Similarly if M' is a model which contains A and is contained in M, then $p \restriction M'$ is a complete δ-type over M' whose δ-definition is also ψ.

REMARK 2.5. *Let M be a model. Then $S_\delta(M)$ is not only a set but also a topological space: for each δ-formula $\varphi(x)$ over M, let $[\varphi] = \{p \in S_\delta(M)$: $\varphi \in p\}$, and let the set of such $[\varphi]$ be a basis for the topology. As such, $S_\delta(M)$ is compact, Hausdorff, and totally disconnected (i.e. has a basis of clopen sets). For any such topological space X one has the notion of the Cantor–Bendixon rank, $\mathrm{CB}_X(p)$ of a point $p \in X$, defined as follows: $\mathrm{CB}_X(p) \geq 0$ for all $p \in X$, and $\mathrm{CB}_X(p) = \alpha$* iff *$p$ is isolated in the space $\{q \in X$: $\mathrm{CB}_X(q) \geq \alpha\}$ (note that the latter set is a* closed *subspace of X). Note that if $\mathrm{CB}_X(p) < \infty$ for all $p \in X$, then $\{\mathrm{CB}_X(p)$: $p \in X\}$ has a maximum element α, and $\{p \in X$: $\mathrm{CB}_X(p) = \alpha\}$ is finite.*

LEMMA 2.6. *Suppose $\delta(x, y)$ is stable. Let M be a model and let X be the space $S_\delta(M)$. Then $\mathrm{CB}_X(p) < \infty$ for all $p \in$ X.*

Proof. Suppose by way of contradiction that for some $p \in X$, $\mathrm{CB}_X(p) = \infty$. So p is not isolated in $\{q \in X$: $\mathrm{CB}_X(q) = \infty\}$. So, given any δ-formula φ in p, there are $p_1, p_2 \in X$, $p_1 \neq p_2$, $\varphi \in p_1, p_2$. Thus we can find mutually contradictory δ-formulae $\varphi_1 \in p_1$, $\varphi_2 \in p_2$, both of which imply φ. Continuing this way we find a tree $\{\varphi_\mu$: $\mu \in {}^{\omega>}2\}$ of δ-formulae, such that for each $\lambda \in {}^\omega 2$, $\Sigma_\lambda = \{\varphi_{(\lambda \restriction n)}$: $n < \omega\}$ is consistent, and such that $\{\Sigma_\lambda : \lambda \in {}^\omega 2\}$ is pairwise contradictory. Let L_0 be a countable sublanguage of L containing $\delta(x, y)$

and M_0 a countable L_0-elementary substructure of M such that for each formula φ_μ is an $L_0(M_0)$-formula. Clearly $\delta(x, y)$ is stable in the theory $T_0 = Th(M_0)$. Each Σ_λ extends to some complete δ-type over M_0, whereby $|S_\delta(M_0)| = 2^\omega$. This contradicts Remark 2.3 (for the theory T_0). ■

LEMMA **2.7.** (Existence of 'non-forking extensions') *Let $\delta(x, y)$ be stable. Let $p(x) \in S(A)$, and let M be a model containing A. Then there is some $q(x) \in S_\delta(M)$ such that*

(i) *$p(x) \cup q(x)$ is consistent; and*

(ii) *the δ-definition* of $q(x)$ *is almost over A (i.e. $q(x)$ is definable over* acl(A)).

Proof. Remember we are working with imaginaries now, and so acl(-) is in the sense of $\mathbf{C}^{eq}$. Bearing in mind Remark 2.4 we may replace M by an $|A|^+$-saturated and strongly $|A|^+$-homogeneous elementary extension. Let $X = \{q \in S_\delta(M)\colon q(x) \cup p(x) \text{ is consistent}\}$. Then X (with the topology induced from $S_\delta(M)$) is compact Hausdorff and totally disconnected. By Lemma 2.6, $\mathrm{CB}_X(p) < \infty$, for all $p \in X$. By Remark 2.5 there is finite non-empty subset X_0 of X consisting of elements of maximum CB_X-rank. Note that X_0 is setwise invariant under $\mathrm{Aut}_A(M)$ (where $\mathrm{Aut}_A(M)$ denotes the group of automorphisms of M which fix A pointwise). Fix $q \in X_0$. By Lemma 2.2, q has a δ-definition $\psi(y) \in L(M)$. For any $f \in \mathrm{Aut}(M)$, $f(q) \in S_\delta(M)$ has δ-definition $f(\psi)$. As $q \in X_0$, $\{f(q)\colon f \in \mathrm{Aut}_A(M)\}$ is finite. Thus $\{f(\psi)\colon f \in \mathrm{Aut}_A(M)\}$ is finite. Our saturation and homogeneity assumptions on M, together with Lemma 1.2, imply that the formula $\psi(y)$ is almost over A. This completes the proof of the lemma. ■

From now on let $\delta(x, y)$ be a stable formula. Let $\varepsilon(y, x)$ be the formula $\delta(x, y)$. Then $\varepsilon(y, x)$ is also stable. Moreover, by Lemma 2.2, if $p(x) \in S_\delta(M)$, then the δ-definition of p is an $\varepsilon(y, x)$-formula (over M), and conversely if $q(y) \in S_\varepsilon(M)$ then the $\varepsilon(y, x)$-definition of q is a δ-formula over M.

LEMMA **2.8.** (Symmetry) *Suppose M is a model, $p(x) \in S_\delta(M)$ and $q(y) \in S_\varepsilon(M)$. Let $\psi(y)$ be the δ-definition of p, and $\chi(x)$ the ε-definition of q. Then $\chi(x) \in p(x)$ iff $\psi(y) \in q(y)$.*

Proof. Let M' be a $|M|^+$-saturated elementary extension of M, and let (by Remark 2.4) $p' \in S_\delta(M')$ have δ-definition $\psi(y)$, and $q'(y) \in S_\varepsilon(M')$ have ε-definition $\chi(x)$. Then $p \subseteq p'$ and $q \subseteq q'$. Suppose by way of contradiction that $\chi(x) \in p(x)$, but $\psi(y) \notin q(y)$. As $\psi(y)$ is an ε-formula over

M, then $\neg\psi(y) \in q(y)$. Define inductively a_i, b_j in M' such that a_i realizes $p \restriction (M \cup \{a_0, b_0, \ldots, a_{i-1}, b_{i-1}\})$, and b_i realizes $q' \restriction (M \cup \{a_0, b_0, \ldots, a_{i-1}, b_{i-1}, a_i\})$, Then $\models \chi(a_i) \wedge \psi(b_i)$ for all i. Thus, for $i \leq j$, $\models \varepsilon(b_j, a_i)$, but for $i > j$, $\models \neg\delta(a_i, b_j)$. Thus $\models \delta(a_i, b_j)$ iff $i \leq j$, contradicting the stability of δ. This proves the lemma. ■

COROLLARY **2.9.** (Uniqueness) *Let $p_i(x), p_2(x) \in S_\delta(M)$ be both definable over A where A is an* algebraically closed *subset of M^{eq}. Suppose also that $p_1 \restriction A = p_2 \restriction A$.* Then $p_1 = p_2$.

Proof. Let $\psi_1(y)$ be the δ-definition of p_1, and $\psi_2(y)$ the δ-definition of p_2. Both ψ_1, ψ_2 are over A and we must show that they are equivalent. Let, as above, $\varepsilon(y, x)$ be $\delta(x, y)$. Suppose b is in M and $\models \psi_1(b)$. Let $q(y) = \text{tp}(b/A)$ and by Lemma 2.7 let $q'(y) \in S_\delta(M)$ be such that $q(y) \cup q'(y)$ is consistent, and the ε-definition of q', say $\chi(x)$, is over A (and here we use $A = \text{acl}^{eq}(A)$). As $\psi_1(y)$ is an ε-formula over A, it follows that $\psi_1(y) \in q'$. By Lemma 2.8, $\chi(x) \in p_1$. But $\chi(x)$ is over A, so must also be in $p_2(x)$. By Lemma 2.8 again, $\psi_2(y) \in q'$. As ψ_2 is also over A, it must be the case that $\models \psi_2(b)$. Thus ψ_1 is equivalent to ψ_2, whereby $p_1 = p_2$, as required. ■

Now from Lemma 2.7 and Corollary 2.9, it follows that whenever A is an *algebraically closed* set and $p(x) \in S_\delta(A)$, there is formula $\psi(y)$ over A such that for any model M containing A, p has a unique extension to a complete δ-type $q(x) \in S_\delta(M)$ whose δ-definition is over A, and, moreover, $\psi(y)$ is precisely this δ-definition. Clearly $\psi(y)$ is unique up to equivalence. We say that $\psi(y)$ is the *δ-definition of p*. By Remark 1.8 (the set defined by) $\psi(y)$ has a *code*, say c, in $\mathbf{C}^{eq}$. By the *canonical base* of p, denoted $\text{Cb}(p)$, we mean $\text{dcl}(c)$ (in $\mathbf{C}^{eq}$).

REMARK **2.10.** *Let A, B be algebraically closed sets, $p(x) \in S_\delta(A)$, $q(x) \in S_\delta(B)$. Then the following are equivalent*:

(i) *The δ-definition of p is equivalent to the δ-definition of q.*

(ii) *For some model M containing $A \cup B$, there is $r(x) \in S_\delta(M)$ containing $p(x) \cup q(x)$ such that r is definable over A and also definable over B.*

Proof. Immediate. ■

If p, q satisfy the equivalent conditions of Remark 2.10, we will say that p and q are *parallel.* If in addition $B \supseteq A$, we say that q is *the non-forking extension* of p to a complete δ-type over B, or that q *does not fork* over A. We write $q = p|B$.

We also wish to define the notion of a non-forking extension of $p(x)$, when $p(x) \in S_\delta(A)$ where A is arbitrary (i.e. not necessarily algebraically closed). By a *finite equivalence relation*, we mean one which has only a finite number of classes.

LEMMA **2.11.** (Conjugacy and the finite equivalence relation theorem) *Let* $p(x) \in S_\delta(A)$. *Let* $B = \mathrm{acl}^{\mathrm{eq}}(A)$. *Let* $X = \{q(x) \in S_\delta(B)\colon q(x) \supseteq p(x)\}$. *Let* G *denote the group of elementary permutations of* B which fix A pointwise. (*Note that* G *acts on* X.) *Then*

(i) *G acts transitively on X*;

(ii) *there is an A-definable finite equivalence relation $E(x_1, x_2)$, such that each E-class is defined by some δ-formula, and such that for $q_1, q_2 \in X$, $q_1 = q_2$ iff $q_1(x_1) \cup q_2(x_2) \models E(x_1, x_2)$. In particular X is finite.*

Proof. First let $p_1(x)$ be any *complete type* over A extending $p(x)$.

CLAIM. *For each $q(x) \in X$, $q(x) \cup p_1(x)$ is consistent.*

Proof of claim. Otherwise there is, by compactness, some formula $\sigma(x) \in q(x)$ such that $\sigma(x) \cup p_1(x)$ is inconsistent. Thus each A-conjugate $\sigma'(x)$ of σ is inconsistent with $p_1(x)$. But $\sigma(x)$ is a δ-formula over $\mathrm{acl}(A)$, so in particular has only finitely many A-conjugates. Let $\psi(x)$ be the disjunction of these. So $\psi(x)$ is inconsistent with $p_1(x)$. But $\psi(x)$ is a δ-formula over A, implied by $\sigma(x)$, and thus in $p(x)$. This contradicts the fact $p_1(x)$ extends $p(x)$, proving the claim. ■

Now let $q_1, q_2 \in X$. By the claim, we can choose a realizing $q_1(x)$ and b realizing $q_2(x)$ such that $\mathrm{tp}(a/A) = \mathrm{tp}(b/A)$ $(=p_1(x))$. Let f be an A-elementary map such that $f(a) = b$. Extend f to an elementary map g whose domain contains B. Note that $g \restriction B$ is a permutation of B so $g|B \in G$. Clearly then $g(q_1) = q_2$. This proves (i).

Fix $q \in X$, and let $\psi(y)$ be the δ-definition of q. Then for $g \in G$ clearly $g(\psi)$ is the δ-definition of $g(q)$. (Let M be a strongly $|A|^+$-homogeneous model containing A. Then g extends to an automorphism of M.) But $\{g(\psi)\colon g \in G\}$ is clearly finite. By (i) this shows that X is finite.

Thus there is a finite set $\Phi(x)$ of δ-formulae over B, which separates the types in X. The set $\Phi(x)$ may be assumed to be closed under G. Let $E(x_1, x_2)$ be the formula $\wedge \{\phi(x_1) \leftrightarrow \phi(x_2)\ \phi \in \Phi\}$. This formula clearly works. ■

Definition 2.12. *Let* $A \subseteq B$, $p(x) \in S_\delta(A)$, $q(x) \in S_\delta(B)$. *We say that* q does not fork over A, or q is a non-forking extension *of* p, *if for some* $q_1(x) \in S_\delta(\mathrm{acl}(B))$ *extending* $q(x)$, q_1 *does not fork over* $\mathrm{acl}(A)$.

Remark 2.13. *With the set-up in Definition* 2.12, *the following are equivalent*:

(i) q *does not fork over* A;

(ii) *for* every $q_2(x) \in S_\delta(\mathrm{acl}(B))$ *extending* q, q_2 *does not fork over* A.

Proof. If $q_1, q_2 \in S_\delta(\mathrm{acl}(B))$ both extend q, then by Lemma 2.11(i), there is a B-elementary permutation f of $\mathrm{acl}(B)$ such that $f(q_1) = q_2$. Then f fixes A pointwise, so induces a perrmutation of $\mathrm{acl}(A)$. So if $\psi(y)$, the δ-definition of q_1, is over $\mathrm{acl}(A)$, then so is $f(\psi)$, which is the δ-definition of q_2. ■

Corollary 2.14. *Let* $A \subseteq B \subseteq C$, $p(x) \in S_\delta(A)$, $q(x) \in S_\delta(B)$, *and* $r(x) \in S_\delta(C)$, *with* $p(x) \subseteq q(x) \subseteq r(x)$. *Then* r *does not fork over* A *if and only if* r *does not fork over* B *and* q *does not fork over* A.

Remark 2.15. *Let* $M \subseteq N$ *be models,* $p(x) \in S_\delta(N)$. *Then* p *does not fork over* M *iff* p *is finitely satifiable in* M (*i.e. any formula* $\varphi(x) \in p$ *is satisified by an element of* M).

Proof. ($\Rightarrow$). Suppose p does not fork over M. So the δ-definition of p, say $\psi(y)$, is over M. Let $\varphi(x) \in p$. Without loss of generality $\varphi(x)$ is of the form $\delta(x, b_1) \wedge \cdots \wedge \delta(x, b_m) \wedge \neg\delta(x, c_1) \wedge \cdots \wedge \neg\delta(x, c_n)$, for some b_i, c_j in N. Let $\varepsilon(y, x)$ be $\delta(x, y)$. Let q_i be the complete ε-type of b_i over M, and r_j the complete ε-type of c_j over M. Let $\chi_i(x)$ be the ε-definition of q_i, and ξ_j the ε-definition of r_j. Let $p_0 = p|M$. So $\psi(y)$ is also the δ-definition of p_0. Now $\models \psi(b_i)$ for $i = 1, \ldots, m$, and $\models \neg\psi(c_j)$ for $j = 1, \ldots, m$. By Lemma 2.8, $\chi_i(x) \in p_0(x)$ for $i = 1, \ldots, m$, and $\neg\xi_j(x) \in p_0(x)$ for $j = 1, \ldots, n$. Thus (as p_0 is consistent) we can find $a \in M$ such that $\models \bigwedge_i \chi_i(a) \wedge \bigwedge_j \neg\xi_j(a)$. Then clearly $\models \varphi(a)$.

($\Leftarrow$). Suppose p is finitely satisfiable in M. Without loss N is $|A|^+$-saturated. Let $\psi(y)$ be the δ-definition of p. If ψ is not over M, then there are b, c in N such that $\mathrm{tp}(b/M) = \mathrm{tp}(c/M)$ and $\models \psi(b) \wedge \neg\psi(c)$. Thus the formula $\delta(x, b) \wedge \neg\delta(x, c)$ is in p. But this formula is clearly not satisfied in M, a contradiction. ■

The following result yields a convenient description/concept of a *formula* not forking over a set.

Lemma 2.16. *Let A be a set of parameters, and $\delta(x, a)$ an instance of $\delta(x, y)$. The following are then equivalent*:

(i) *There is $p(x) \in S_\delta(Aa)$ such that $\delta(x, a) \in p(x)$ and p does not fork over A.*

(ii) *Some positive Boolean combination of A-conjugates of $\delta(x, a)$ is A-definable and consistent.*

(iii) *Any set of* $\mathrm{acl}(A)$*-conjugates of $\delta(x, a)$ is consistent.*

(iv) *Whenever $(a_i\colon i < \omega)$ is an A-indiscernible sequence with* $\mathrm{tp}(a_i/A) = \mathrm{tp}(a/A)$, *then $\{\delta(x, a_i)\colon i < \omega\}$ is consistent.*

(v) *For* any *model $M_0 \supseteq A$* (so M_0 *need not contain a*), *$\delta(x, a)$ is satisfied by an element of M_0.*

Proof. We first make some constructions and observations. Let M be a model containing Aa and N an $|M|^+$-saturated and strongly $|M|^+$-homogeneous model containing M. Let $\varepsilon(y, x)$ be $\delta(x, y)$. Let $q(y) = \mathrm{tp}(a/\mathrm{acl}(A))$. Let (by 2.7) $q_1 \in S_\varepsilon(N)$ be such that $q_1(y) \cup q(y)$ is consistent, and $q_1(y)$ does not fork over $\mathrm{acl}(A)$. In particular $q_1(y)$ does not fork over M, so by (the proof of) Remark 2.15, $q_1(y) \cup q(y)$ is finitely satisfiable in M. We can easily extend $q_1(y) \cup q(y)$ to some complete type $q^*(y)$ over N such that $q^*(y)$ is finitely satisfiable in M.

Let $\sigma(x)$ be the ε-definition of $q^*(y)$, that is, the ε-definition of $q_1(y)$. By Lemma 2.2, there is a sequence $\{c_i\colon i < \omega\}$ in N, such that c_0 realizes $q^* \restriction M$ and c_{i+1} realizes $q^* \restriction (M \cup \{c_0, \ldots, c_i\})$, and $\sigma(x)$ is equivalent to a positive Boolean combination of the $\varepsilon(c_i, x)$, that is a positive Boolean combination of the $\delta(x, c_i)$. Note that

(a) the sequence $(c_i\colon i < \omega)$ is indiscernible over M (by the finite satisfiability in M of q^*); and

(b) $\mathrm{tp}(c_i/\mathrm{acl}(A)) = \mathrm{tp}(a/\mathrm{acl}(A))$ for each i.

We now set about proving the equivalences of (i)–(v).

(i) → (ii). Assuming (i) there is clearly some $p(x) \in S_\delta(N)$ which contains $\delta(x, a)$ and does not fork over A. Let $\psi(y)$ be the δ-definition of p. Then $\models \psi(a)$, and thus (as ψ is an ε-formula), $\psi(y) \in q_1(y)$. By Lemma 2.8, $\sigma(x) \in p(x)$, so is in particular consistent. Now as q_1 does not fork over $\mathrm{acl}(A)$, $\sigma(x)$ is over $\mathrm{acl}(A)$. Let $\sigma'(x)$ be the disjunction of the finitely many A-conjugates of σ. Then σ' is equivalent to a positive Boolean combination of conjugates of $\delta(x, a)$ (as, by (b), σ is), and is consistent and defined over A.

(ii) → (iii). Let $\tau(x)$ be some consistent positive Boolean combination of A-conjugates of $\delta(x, a)$ which is over A. Let $p_0(x) \in S_\delta(A)$ contain $\tau(x)$. Let

$p(x) \in S_\delta(N)$ be a non-forking extension of p_0. Thus $\tau(x) \in p(x)$. We may assume then, by saturation of N, that some A-conjugate, say $\delta(x, a')$ of $\delta(x, a)$, is in p. Let f be an A-automorphism of N taking a' to a and fixing A pointwise. Then $\delta(x, a) \in f(p) = p'$, and p' does not fork over A. In particular p' is definable over acl(A), so if a'' is in N and $\text{tp}(a''/\text{acl}(A)) = tp(a/\text{acl}(A))$, then $\delta(x, a'') \in p(x)$. By saturation of N, it follows that *any* set of acl(A)-conjugates of $\delta(x, a)$ is consistent.

(iii) → (iv). This is immediate, for if $\{a_i: i < \omega\}$ is an A-indiscernible sequence with $\text{tp}(a_i/A) = \text{tp}(a/A)$, then $\text{tp}(a_i/\text{acl}(A)) = \text{tp}(a_j/\text{acl}(A))$, for all $i, j < \omega$, so easily $\{\delta(x, a_i): i < \omega\}$ is consistent.

(iv) → (i). We revert to the notation and constructions given at the beginning of this proof. Assuming (iv) it follows that $\sigma(x)$ is consistent. Now $\sigma(x)$ is a δ-formula over acl(A). Let $p_0(x) \in S_\delta(\text{acl}(A))$ contain $\sigma(x)$, and let $p(x) \in S_\delta(M)$ be a non-forking extension of p_0. By Lemma 2.8, $\psi(y) \in q_1$, where $\psi(y)$ is the δ-definition of p. As q_1 is consistent with $\text{tp}(a/\text{acl}(A))$ and ψ is over acl(A), it follows that $\models \psi(a)$. Thus $\delta(x, a) \in p(x)$, and so (i) follows.

(i) → (v). Let $M_0 \supseteq A$ be any model. Let $N_0 \supseteq M_0 \cup \{a\}$. Assume $p(x) \in S_\delta(Aa)$ contains $\delta(x, a)$ and does not fork over A. Let $p'(x) \in S_\delta(N_0)$ be a non-forking extension of p. By Corollary 2.14, p' does not fork over M_0, so by 2.15 is finitely satisfiable in M_0. In particular $\delta(x, a)$ is satisfied in M_0.

(v) → (i). Assume (v). We again revert to the notation at the beginning of the proof. Let a^* realize $q_1(y) \cup q(y)$. So $\text{tp}(a^*/\text{acl}(A)) = \text{tp}(a/\text{acl}(A))$. Clearly then the formula $\delta(x, a^*)$ is realized in N, by, say, c. Thus $\models \sigma(c)$ (where $\sigma(x)$ is the ε-definition of q_1). By Lemma 2.8, $\psi(y) \in q_1(y)$, where ψ is the δ-definition of the complete δ-type $p_0(x)$ of c over acl(A). Thus $\models \psi(a^*)$. Thus also $\models \psi(a)$. In particular, $\delta(x, a)$ is in $p_0|\text{acl}(A)a$, proving (i). ■

Definition 2.17. *We will say that the formula* $\delta(x, a)$ does not fork over A, *if the equivalent conditions* (i)–(v) *of Lemma* 2.15 *are satisifed.*

Remark 2.17′. *Let* $\delta(x, a)$ *be an instance of* δ, *and let* A *be an* algebraically *set. By Lemma* 2.16 (*equivalence of* (ii) *and* (iii)) *there is a formula* $\sigma(x)$ *over* A *which is a positive Boolean combination of* A*-conjugates of* $\delta(x, a)$, *where the* consistency *of* $\sigma(x)$ *corresponds to* $\delta(x, a)$ *not forking over* A (*in the sense of Definition* 2.17). *We have* (*with this notation*):

'Open mapping theorem' (*for* A algebraically closed). *For any* $p(x) \in S_\delta(A)$, $\sigma(x) \in p(x)$ *iff* $\delta(x, a) \in p|(A, a)$.

Proof. This can be seen via 2.8 and noting how $\sigma(x)$ arose in the proof of 2.16, *or* can be seen as follows. Let $p(x) \in S_\delta(A)$, $M \supseteq Aa$ be $|A|^+$-saturated and strongly $|A|^+$-homogeneous, and $p'(x) \in S_\delta(M)$ the unique non-forking extension of p. We may assume that $\sigma(x)$ is a positive Boolean combination of conjugates of $\delta(x, a)$ under $\mathrm{Aut}_A(M)$. As p' is definable over A, p' is fixed by $\mathrm{Aut}_A(M)$. Thus $\delta(x, a) \in p'$ iff every conjugate of $\delta(x, a)$ under $\mathrm{Aut}_A(M)$ is in p' iff some conjugate of $\delta(x, a)$ under $\mathrm{Aut}_A(M)$ is in p'. So clearly $\delta(x, a) \in p'$ iff $\sigma(x) \in p'$ iff $\sigma(x) \in p$. ■

We deduce a similar result for arbitrary A:

'Open mapping theorem' (for arbitrary A). *Let $\delta(x, a)$ be an instance of δ. Then there is a δ-formula $\sigma_1(x)$ over A (which is even equivalent to a positive Boolean combination of A-conjugates of $\delta(x, a)$) such that for any $p(x) \in S_\delta(A)$, $\sigma_1(x) \in p(x)$ iff p has a non-forking extension $q(x) \in S_\delta(A, a)$ such that $\delta(x, a) \in q(x)$.*

Proof. Let $\sigma(x)$ be the formula over $\mathrm{acl}(A)$ given by the algebraically closed version above. Let $\sigma_1(x)$ be the disjunction of the A-conjugates of $\sigma(x)$. By 2.11, for $p(x) \in S_\delta(A)$, $\sigma_1(x) \in p$ iff $\sigma(x)$ is in some (necessarily non-forking) extension of p in $S_\delta(\mathrm{acl}(A))$. This is sufficient. ■

Finally in this section we wish to point out how the *global* theory of forking in stable theories follows smoothly from the local theory described above.

Let $\Delta(x)$ be a *finite* set $\{\delta_1(x, y_1), \ldots, \delta_n(x, y_n)\}$ of *stable* formulae. By a Δ-formula we simply mean a formula $\varphi(x)$ which is equivalent to a (finite) Boolean combination of δ_i-formulae for $i = 1, \ldots, n$. A complete Δ-type over A is simply a maximal consistent set of Δ-formulae in L_A. $S_\Delta(A)$ denotes the set of such types. By the Δ-definition of some $p(x) \in S_\Delta(M)$ we simply mean the sequence $(\psi_1(y_1), \ldots, \psi_n(y_n))$ where $\psi_i(y_i)$ is the δ_i-definition of $p \restriction \delta_i$ (or simply of p). Clearly such $p \in S_\Delta(M)$ is determined by its Δ-definition. We say that p is definable (almost) over A if each ψ_i is (almost) over A. All the above theory goes through for complete Δ-types in place of complete δ-types, in particular the notion 'p does not fork over A', where $p(x) \in S_\Delta(B)$, and $A \subseteq B$. However, the main point we wish to emphasize here is the analogue of Lemma 2.7.

Lemma 2.18. *Let $p(x) \in S(A)$, and let $M \supseteq A$. Then there is $q(x) \in S_\Delta(M)$ such that $q(x) \cup p(x)$ is consistent and $q(x)$ is definable over* $\mathrm{acl}(A)$.

Proof. As in the proof of 2.7 we assume M is $|A|^+$-saturated and strongly $|A|^+$-homogeneous. Let X be the *space* $\{q(x) \in S_\Delta(M)\colon q(x) \cup p(x)$ is consistent$\}$. All one has to know is that $\mathrm{CB}_X(q) < \infty$ for all $q \in X$. This follows exactly as in Lemma 2.6. (For countable L, over any countable model there are only countably many complete Δ-types, as each is definable and there are only countably many possible definitions, Δ being finite.) Given this, the proof follows that of Lemma 2.7. ■

Remark. *Lemma* 2.18 *implies in particular that* $\{q(x)$: *for some* $\delta \in \Delta$, $q \in S_\delta(M)$ *is the non-forking extension of* $p \upharpoonright \delta\}$ *is consistent, that is the non-forking extensions of the* $p \upharpoonright \delta$, $\delta \in \Delta$, *cohere to give a complete* Δ*-type over* M.

Suppose now T to be stable, that is every formula is stable. If $p(x)$ is a complete type over some set A, and $\delta(x, y)$ is an L-formula, then by $p \upharpoonright \delta$ we mean the set of δ-formulae in $p(x)$, so $p \upharpoonright \delta \in S_\delta(A)$.

Definition 2.19. *Let* $q(x)$ *be a complete type over* B, $A \subseteq B$ *and* $p(x) = q \upharpoonright A$. *We say* q does not fork over A, or q is a non-forking extension of p, *if for* each *formula* $\delta(x, y)$ of L, $q \upharpoonright \delta$ *does not fork over* A.

Note: *So if for example* M *is a model,* $A \subseteq M$, $q(x) \in S(M)$, *then* q *does not fork over* A, *iff for every* $\delta(x, y) \in L$, *the* δ*-definition of* $p(x)$ *is over* acl(A). *Also by definition if* $p(x) \in S(\text{acl}(A))$, *then* p *does not fork over* A.)

The basic properites now fall out of our previous local analysis.

Proposition 2.20. (i) (Existence of non-forking extensions) *Let* $p(x) \in S(A)$, *and* M *a model containing* A. *Then* $p(x)$ *has a non-forking extension* $q(x) \in S(M)$. *Moreover, if* $A = \text{acl}(A)$, *then* q is unique.

(ii) (Finite character of forking) *Let* $p(x) \in S(B)$. *Then* $p(x)$ *does not fork over* $A \subseteq B$ *iff for* every *formula* $\varphi(x) \in p(x)$, $\varphi(x)$ *does not fork over* A *(in the sense of Definition* 2.17).

(iii) (Transitivity) *Let* $p(x) \in S(C)$, *and* $A \subseteq B \subseteq C$. *Then* p *does not fork over* A *iff* p *does not fork over* B *and* $p \upharpoonright B$ *does not fork over* A.

(iv) (Conjugacy) *If* $p(x) \in S(A)$, *and* $B \supseteq A$, *then* $p(x)$ *has at most* $2^{|T|}$ *non-forking extensions in* $S(B)$. *If, moreover,* B *is a strongly* $(|T|^+ + |A|^+)$*-saturated model* M, *then* $\text{Aut}_A(M)$ *acts transitively on this set of non-forking extensions of* p. *Also if* $p_1, p_2 \in S(M)$ *are distinct non-forking extensions of* p, *then there is a finite equivalence relation* $E(x_1, x_2)$ *defined over* A *such that* $p_1(x_1) \cup p_2(x_2) \models \neg E(x_1, x_2)$.

(v) (Symmetry) *Suppose* $A \subseteq B$. *Then* tp(a/B) *does not fork over* A *iff for each tuple* b *from* B, tp(b/Aa) *does not fork over* A.

(vi) *If* $p(x) \in S(B)$ *is an* algebraic *type (i.e. has only finitely many realizations) and* p *does not fork over* $A \subseteq B$, *then* $p \upharpoonright A$ *is algebraic.*

(vii) *For any $A \subseteq B$ and formula $\varphi(x)$ over B there is a formula $\psi(x)$ over A such that for any $p(x) \in S(A)$, $\psi \in p$* iff *p has a non-forking extension $q(x) \in S(B)$ such that $\varphi(x) \in q$.*

Proof. (i) Let $p_1(x) \in S(\mathrm{acl}(A))$ be any extension of $p(x)$. From the definition of non-forking it suffices to prove that p_1 has a (unique) non-forking extension in $S(M)$. For each $\delta(x, y)$ in L, let $q_\delta(x) \in S_\delta(M)$ be the unique non-forking extension of $p_1 \restriction \delta$ (by 2.9). By Lemma 2.18, $\cup \{q_\delta(x): \delta \in L\}$ is consistent. (For each finite set Δ of such formulae $\delta(x, y)$, there is by 2.18 a non-forking extension $q_\Delta \in S_\Delta(M)$ of $p_1 \restriction \Delta$. In particular $q_\Delta \restriction \delta$ is a non-forking extension of $p_1 \restriction \delta$ for each $\delta \in \Delta$. By uniqueness (2.9), $q_\Delta \restriction \delta = q_\delta$ for each $\delta \in \Delta$.) So $q(x) = \cup \{q_\delta(x): \delta \in L\}$ is in $S(M)$, and is clearly the unique non-forking extension of $p_1(x)$.

(ii) The left to right direction is immediate from (i) of 2.16. For the right to left direction we may assume that A is algebracially closed and M is an $|A|^+$-saturated model. Let $p_0(x) = p \restriction A$. Let $p_1(x) \in S(M)$ be the unique non-forking extension of p_0. We show that $p_1 = p$. Let $\varphi(x) \in p$. By 2.16(ii), let $\chi(x)$ be some consistent formula over A which is equivalent to a positive Boolean combination of A-conjugates of $\varphi(x)$. Then $\chi(x) \in p_0(x)$, for otherwise $\varphi(x) \wedge \neg\chi(x) \in p(x)$, and some set of A-conjugates of it is inconsistent, contradicting our hypothesis and 2.16(iv). As $p_1(x)$ is invariant under A-automorphisms of M, it follows that $\varphi(x) \in p_1$. Thus $p(x) = p_1(x)$.

(iii) is immediate from the definitions and Corollary 2.14.

(iv) The bound on the number of non-forking extensions comes from the finite bound on the number of non-forking extensions of a complete δ-type.

Now clearly the group of elementary permutations of acl(A) which fix A pointwise acts transitively on the extensions of $p(x)$ in $S(\mathrm{acl}(A))$. So by part (i), $\mathrm{Aut}_A(M)$ acts transitively on the non-forking extensions of p in $S(M)$. The final part of (iv) comes from 2.11(ii).

(v) First we may assume that A is algebraically closed (in $\mathbf{C}^{\mathrm{eq}}$ as always). Second, by part (ii), we may assume the elements of A are named by constants of the language L. (As property (ii) in Lemma 2.15 is invariant under naming the elements of A.)

Suppose tp(a/B) does not fork over A. Let b be a tuple in B. Let x be a variable corresponding to a, y a tuple of variables corresponding to b, $\delta(x, y)$ an L-formula, and $\varepsilon(y, x)$ the formula $\delta(x, y)$. Let $\psi(y)$ be the δ-definition of $\mathrm{tp}(a/A) \restriction \delta$, and $\sigma_\varepsilon(x)$ the ε-definition of $\mathrm{tp}(b/A) \restriction \varepsilon$. By 2.8, $\models \sigma(a)$ iff $\models \psi(b)$. But as $\mathrm{tp}(a/B) \restriction \delta$ does not fork over A, $\models \psi(b)$ iff $\models \delta(a, b)$. Thus

$$(*) \qquad \models \varepsilon(b, a) \text{ iff} \models \sigma_\varepsilon(a).$$

Let M be a model containing Aa, and let $q(y) \in S(M)$ be the non-forking extension of $\mathrm{tp}(b/A)$. Then for each a' in M, and formula $\varepsilon'(y, x')$ of L, $\varepsilon'(y, a') \in q(y)$ iff $\models \sigma_{\varepsilon'}(a')$, where $\sigma_{\varepsilon'}$ is the ε'-definition of $\mathrm{tp}(b/A)$. In particular, by, $(*)$, $\mathrm{tp}(b/Aa) = q \restriction Aa$. Thus $\mathrm{tp}(b/Aa)$ does not fork over A, as required.

Now suppose $\mathrm{tp}(b/Aa)$ does not fork over A for all tuples b from B. As in the first part of the proof, we see that $\mathrm{tp}(a/Ab)$ does not fork over A for all b from B. By (ii), $\mathrm{tp}(a/B)$ does not fork over A, as required.

(vi) Suppose $p(x) \in S(B)$ is algebraic, and does not fork over A. We may assume B is algebracially closed, so $p(x) = \mathrm{tp}(b/B)$ for some $b \in B$. Thus the formula '$x = b$' is in $p(x)$. By (ii) the formula '$x = b$' does not fork over A. By property (iii) in Lemma 2.16, $b \in \mathrm{acl}(A)$. Thus $p \restriction A$ is also algebraic.

(vii) is left to the reader. (Use Remark 2.17′.) ■

The following lemma completes our picture of the connection between forking for complete (δ)-types and for formulae:

Lemma 2.21 (i) *If $\varphi(x)$ forks over A, and $\psi(x)$ forks over A, then $\varphi(x) \vee \psi(x)$ forks over A.*

(ii) *Let $\Delta(x)$ be finite, $p(x) \in S_\Delta(B)$, $A \subseteq B$. Then p does not fork over A (in the sense of the paragraph preceding Lemma* 2.18) *iff every formula in p does not fork over A.*

Proof. (i) We may assume A is algebraically closed. Write $\varphi(x)$ as $\varphi(x, a)$, where $\varphi(x, y) \in L$. Similarly write $\psi(x)$ as $\psi(x, b)$ where $\psi(x, z) \in L$. Let $\delta(x, y, z)$ be $\varphi(x, y) \vee \psi(x, z)$. Then $\delta(x, yz)$ is stable. Suppose that $\varphi(x, a) \vee \psi(x, b)$ does not fork over A, that is $\delta(x, a, b)$ does not fork over A. By definition, there is some complete δ-type $p(x)$ over Aab such that p does not fork over A. Let $p_0(x) = p \restriction A$. So $p_0 \in S_\delta(A)$. Let $\Delta(x) = \{\varphi(x, y), \psi(x, z), \delta(x, y, z)\}$, and let $p_1(x) \in S_\Delta(A)$ contain $p_0(x)$. Then $p_1(x)$ has, by 2.18, a unique non-forking extension $q(x) \in S_\Delta(Aab)$. Thus $q \restriction \delta = p$. So $\delta(x, a, b) \in q(x)$, so either $\varphi(x, a)$ or $\psi(x, b)$ is in $q(x)$. Thus, one of these formulae does not fork over A.

(ii) The proof is like (i). ■

We can, by the above, extend our forking terminology to partial types. We say the partial type $\Sigma(x)$ does not fork over A, if for every finite subset $\Sigma'(x)$ of $\Sigma(x)$, the formula $\wedge \Sigma'$ does not fork over A. This agrees with all previous terminology (i.e. when Σ is a complete type, complete δ-type, or complete Δ-type). Also if $\Sigma(x)$ does not fork over A and Σ is over A, then by Lemma

2.21(i), $\Sigma(x)$ extends to a complete type $p(x) \in S(A)$ which does not fork over A.

LEMMA 2.22. tp(ab/B) *does not fork over* $A \subseteq B$ *iff* tp(a/Bb) *does not fork over* Ab *and* tp(b/B) *does not fork over* A.

Proof. By (iii) and (v) in 2.20. ■

We will say that a is independent from B over C if tp$(a/B \cup C)$ does not fork over C. We say that A is independent from B over C if a is independent from B over C for all tuples a from A. Sometimes we may, by abuse of style, grammar, etc., use 'with' in place of 'from'. We may also say 'a forks with B over C' if tp$(a/B \cup C)$ forks over C. The various properties of forking above will clearly be true for infinite tuples or sets in place of finite tuples: for example, A is independent from B over C iff B is independent from A over C.

Suppose $A = \{A_i : i < \kappa\}$ is a family of sets. We say that A is B-independent iff A_i is independent from $\cup \{A_j : j < \kappa, j \neq i\}$ over B for all $i < k$.

LEMMA 2.23. *Suppose* $A = \{A_i : i < \kappa\}$, *and that* A_i *is independent from* $\cup \{A_j : j < i\}$ *over* B, *for all* $i < \kappa$. *Then* A *is* B*-independent.*

Proof. Exercise. ■

We end this section with some comments on stationary types, strong types, and canonical bases.

DEFINITION 2.24. (i) *The complete type* $p(x) \in S(A)$ *is said to be* stationary, *if* p *has a unique non-forking extension over any set* $B \supseteq A$.

(ii) *By* stp(a/A) (*the strong type of* a *over* A) *we mean* tp$(a/\mathrm{acl}(A))$. (acl$(-)$ *read in* $\mathbf{C}^{\mathrm{eq}}$ *as usual.*)

REMARK 2.25. (i) $p(x)$ *is stationary iff* p *has a unique extension to a complete type over* acl(A).

(ii) *Any strong type is stationary.*

Proof. (i) Any extension of p over acl(A) is a non-forking extension, so left to right is immediate.

On the other hand, if $M \supseteq \mathrm{acl}(A)$, then by Proposition 2.20(i), two non-forking extensions $p_1(x), p_2(x) \in S(M)$ of p are distinct iff $p_1 \upharpoonright \mathrm{acl}(A) \neq p_2 \upharpoonright \mathrm{acl}(A)$. The other direction follows.

(ii) Clear. ■

Suppose $p(x) \in S(A)$ to be stationary. Let $p_1(x) \in S(\mathrm{acl}(A))$ be its unique extension over $\mathrm{acl}(A)$. For each $\delta(x, y) \in L$, let $\psi_\delta(y)$ be the δ-definition of p_1 (i.e. the δ-definition of the non-forking extension of p_1 to a model). ψ_δ is over $\mathrm{acl}(A)$ and is unique up to equivalence. We define the canonical base of p, $\mathrm{Cb}(p)$, to be the definable closure of the set of codes for the various ψ_δ, as δ ranges over L. Equivalently $\mathrm{Cb}(p)$ is $\mathrm{dcl}(\cup\{\mathrm{Cb}(p_1|\delta): \delta(x, y) \in L\})$. If $p \in S(A)$ is a stationary type, and $B \supseteq A$, then by $p|B$ we will mean (when there is no ambiguity) the unique non-forking extension of p in $S(B)$. Note that if p, q are stationary parallel types (i.e. have a common non-forking extension) then $\mathrm{Cb}(p) = \mathrm{Cb}(q)$.

Remark 2.26. *Let $p(x) \in S(A)$ be stationary. Then*

(i) $\mathrm{Cb}(p) \subseteq \mathrm{dcl}(A)$.

(ii) *For any $B \subseteq A$, p does not fork over B iff* $\mathrm{Cb}(p) \subseteq \mathrm{acl}(B)$.

(iii) *For any $B \subseteq A$, p does not fork over B and $p \restriction B$ is stationary iff* $\mathrm{Cb}(p) \subseteq \mathrm{dcl}(B)$.

(iv) *Let M be a $|T|^+$-saturated and strongly $|T|^+$-homogeneous model, and suppose $p(x) \in S(M)$. Let f be an automorphism of M. Then $f(p) = p$ iff f fixes* $\mathrm{Cb}(p)$ *pointwise.*

(v) $\mathrm{Cb}(\mathrm{tp}(a/\mathrm{acl}(A))) \subseteq \mathrm{dcl}(A, a)$.

Proof. Left to the reader, except for (v) which we prove. Let for example M be a strongly $(|A|^+ + |T|^+)$-saturated model containing Aa. Let $p(x) \in S(M)$ be the non-forking extension of $\mathrm{tp}(a/\mathrm{acl}(A))$ over M. So $\mathrm{tp}(a/\mathrm{acl}(A))$ and p have the same canonical base, C say. Let f be an automorphism of M which fixes $A \cup \{a\}$ pointwise. Then clearly f fixes $\mathrm{tp}(a/\mathrm{acl}(A))$, and thus $f(p) = p$. So by (iv), f fixes C pointwise. The saturation of M implies that $C \subseteq \mathrm{dcl}(Aa)$ as required. ■

Definition 2.27. *Let $p(x) \in S(A)$ be stationary. By a Morley sequence in $p(x)$ of length κ, we mean a sequence $(a_i: j < \kappa)$ such that for each $i < \kappa$, a_i realizes $p|(A \cup \{a_j: j < i\})$. When we speak simply of a Morley sequence in p, we mean a Morley sequence of length ω.*

Lemma 2.28. *Let $p(x) \in S(A)$ be stationary. Let $(a_i: i < \omega)$ be a Morley sequence in p. Then*

(i) *$\{a_i: i < \omega\}$ is A-independent, and A-indiscernible, and if $(b_i: i < \omega)$ is another Morley sequence in p, then* $\mathrm{tp}((a_i)_{i<\omega}/A) = \mathrm{tp}((b_i)_{i<\omega}/A)$.

(ii) $\mathrm{Cb}(p) \subseteq \mathrm{dcl}(a_i: i < \omega)$.

Proof. (i) A-independence of $\{a_i: i < \omega\}$ is by Lemma 2.23. The rest follows from the stationarity of p, and is left to the reader.

(ii) Let M be a strongly $(\omega + |A|^+)$-saturated model containing $A \cup \{a_i: i < \omega\}$. Let $p_1(x) \in S(M)$ be the non-forking extension of p. Let $\delta(x, y)$ be an L-formula. Let ψ_δ be the δ-definition of p. By 2.26(i), ψ_δ is over A. By Lemma 2.2 there is a sequence $(c_i: i < \omega)$ in M such that c_i realizes $p_1 \upharpoonright (A \cup \{c_0, \ldots, c_{i-1}\})$ (the *restriction* of p_1 to $(A \cup \{c_0, \ldots, c_{i-1}\})$) and ψ_δ is over $\{c_i: i < \omega\}$. But then clearly $(c_i: i < \omega)$ is a Morley sequence in p. By part (i) and the homogeneity of M, there is an automorphism of M fixing A and taking $(c_i: i < \omega)$ to $(a_i: i < \omega)$. This automorphism fixes the formula ψ_δ, whereby ψ_δ is defined over $(a_i: i < \omega)$. Thus $\mathrm{Cb}(p)$ is in $\mathrm{dcl}(a_i: i < \omega)$, as required. ∎

Lemma 2.29. *Suppose $A \subseteq B$, and $\mathrm{tp}(a/B)$ forks over A. Let $\{a_i: i < \omega\}$ be a Morley sequence in $\mathrm{tp}(a/\mathrm{acl}(B))$. Then $\{a_i: i < \omega\}$ is* not *A-independent.*

Proof. Let a' realize the non-forking extension of $\mathrm{tp}(a/\mathrm{acl}(B))$ over $B \cup \{a_i: i < \omega\}$. If $\{a_i: i < \omega\}$ *were* A-independent, then (by indiscernibility of Morley sequences and the finite character of forking) $\{a_i: i < \omega\} \cup \{a'\}$ would be A-independent. In particular $\mathrm{tp}(a'/A \cup \{a_i: i < \omega\})$ would not fork over A. But by 2.28, $\mathrm{tp}(a'/B \cup \{a_i: i < \omega\})$ does not fork over $A \cup \{a_i: i < \omega\}$. So transitivity of forking would yield $\mathrm{tp}(a'/B)$ does not fork over A, a contradiction. ∎

Definition 2.30. *Let $p(x) \in S(A)$ be stationary. By $p^{(n)}$ we mean $\mathrm{tp}(a_1, \ldots, a_n \in /A)$ where $\{a_1, \ldots, a_n\}$ is a Morley sequence in p of length n. By 2.27, $p^{(n)}$ is well defined.*

For the remainder of this chapter (and book), T will denote a stable theory.

3 Ranks

Model-theoretic ranks of one kind or another appear throughout this book. In strongly minimal theories such ranks have a *geometric* or *dimension-theoretic* significance: for example, if V is a non-singular variety over the complex numbers then the Morley rank of V is the same as the dimension of V as a *complex* analytic manifold.

The ranks we deal with enter the picture in two ways: first in terms of the Cantor–Bendixon rank on a suitable space of types, and second in terms of forking.

We first recall the Cantor–Bendixon rank on a topological space (which has already appeared in section 2). Let X be compact Huasdorff topological space. For $p \in X$, $\mathrm{CB}(p)$ (or $\mathrm{CB}_X(p)$) is defined inductively by: $\mathrm{CB}(p) \geq 0$ for all $p \in X$, and $\mathrm{CB}(p) = \alpha$ iff p is isolated in the subspace $\{q \in X: \mathrm{CB}(q) \geq \alpha\}$. If $\mathrm{CB}(p) < \infty$ for all $p \in X$, then $\{\mathrm{CB}(p): p \in X\}$ has *greatest* element, α say, and $\{p \in X: \mathrm{CB}(p) = \alpha\}$ is finite, say of cardinality n. We say that α is the CB-rank of the space X, $\alpha = \mathrm{CB}(X)$, and n is the CB-multiplicity of the space X, $n = \text{CB-mult}(X)$. If $\mathrm{CB}(p) = \infty$ for some $p \in X$, we say $\mathrm{CB}(X) = \infty$.

Fix a variable x. Let $\Delta = \Delta(x)$ be some *finite* collection of formulae $\delta(x, y)$ (x is fixed, but y may vary). Although normally we consider only types over 'small' subsets of the universe, we here wish to consider also types over $\mathbf{C}$, namely *global* types. We denote such things, by $\mathbf{p}$, $\mathbf{q}$, etc. As usual the topology on $S_\Delta(\mathbf{C})$ is given by: the basic open sets are determined by Δ-formulae. That is if $\varphi(x)$ is a Δ-formula then $[\varphi] = \{\mathbf{p} \in S_\Delta(\mathbf{C}): \varphi(x) \in \mathbf{p}(x)\}$ is a basic open set, and all basic open sets are of this form.

Lemma 3.1. $\mathrm{CB}(S_\Delta(\mathbf{C}))$ *is finite* (*i.e ordinal valued and* $<\omega$).

Proof. Suppose not. Let X denote $S_\Delta(\mathbf{C})$ and let $\mathbf{p} \in X$ with $\mathrm{CB}(\mathbf{p}) \geq \omega$. Fix $n < \omega$. So $\mathrm{CB}(\mathbf{p}) \geq n + 1$. So $\mathbf{p}$ is not isolated among $\{\mathbf{q} \in X: \mathrm{CB}(\mathbf{q}) \geq n\}$, whereby there are $\mathbf{q}_1 \neq \mathbf{q}_2$ in X with $\mathrm{CB}(\mathbf{q}_i) \geq n$, $i = 1, 2$. Thus we can find $\varphi(x) \in \mathbf{q}_1$, $\neg\varphi(x) \in \mathbf{q}_2$, where $\varphi(x)$ is of the form $\delta(x, a)$ or $\neg\delta(x, a)$ for some $\delta(x, y) \in \Delta$, and $a \in \mathbf{C}$. Similarly, $\varphi(x)$ does not isolate $\mathbf{q}_1(x)$ in $\{\mathbf{q} \in X: \mathrm{CB}(\mathbf{q}) \geq n - 1\}$, so we can find distinct $\mathbf{r}_1, \mathbf{r}_2 \in X$, both containing $\varphi(x)$ and with $\mathrm{CB}(\mathbf{r}_i) \geq n - 1$. $\mathbf{r}_1$ is again distinguished from $\mathbf{r}_2$ by some formula $\delta'(x, b)$, where $b \in \mathbf{C}$ and $\delta' \in \Delta$. The same thing can be done for $\mathbf{q}_2$. The end result is to obtain a set $\{\varphi_\eta(x): \eta \in {}^{n\geq}2\}$ of formulae, such that (i) $\varphi_{\langle\rangle}(x)$ is $x = x$, (ii) for $0 \leq \mathrm{l}(\eta) < n$, $\varphi_{\eta^\wedge\langle 1\rangle}(x)$ is $\neg\varphi_{\eta^\wedge\langle 0\rangle}(x)$, and is of the form $\delta(x, a)$ for some $\delta \in \Delta$. (iii) for each $\mu \in {}^n2$, $\{\varphi_{\mu|i}(x): i \leq n\}$ is consistent. Now this can be done for any $n < \omega$. By compactness and the fact that Δ is finite, we can find Δ-formulae $\varphi_\eta(x)$ for $\eta \in {}^{\omega>}2$, such that $\varphi_{\eta^\wedge\langle 1\rangle}(x) = \neg\varphi_{\eta^\wedge\langle 0\rangle}(x)$ for all η, and such that for any $\mu \in {}^\omega 2$, $\{\varphi_{\mu\restriction n}(x): n < \omega\}$ is consistent. Choosing a countable sublanguage L_0 of L containing Δ, and a countable L_0-elementary substructure of the universe such that all φ_η are over M_0, we obtain clearly uncountably many complete Δ-types over M_0, contradicting Remark 2.3 (and the finiteness of Δ). This proves the lemma. ■

From Lemma 3.1 it follows that if Y is a (compact Hausdorff) subspace of $S_\Delta(\mathbf{C})$ (i.e. subset with the relative topology), then $\mathrm{CB}(Y)$ is also finite.

In the following, p, q, etc., denote complete types or complete Δ-types over small subsets of $\mathbf{C}$. $\Phi(x)$ denotes a set of formulae over a *small* subset of C.

Definition 3.2. *By the* Δ-*rank of* $\Phi(x)$, $R_\Delta(\Phi(x))$, *we mean the Cantor–*

Bendixon rank of the subspace $Y = \{\mathbf{q} \in S_\Delta(\mathbf{C})$: $q(x)$ *is consistent with* $\Phi(x)\}$ *of* $S_\Delta(\mathbf{C})$. *The* Δ*-multiplicity of* $\Phi(x)$, $\mathrm{mult}_\Delta(\Phi)$, *is* CB-mult$(Y)$. *If* $\phi(x)$ *is a formula then by* $R_\Delta(\phi(x))$ *we mean* $R_\Delta(\{\phi(x)\})$.

By the previous remarks $R_\Delta(\Phi(x)) < \omega$ for all Φ.

Note. The space Y in Definition 3.2 *is* compact and Hausdorff.

LEMMA 3.3. (i) *If* $\Phi(x) \subseteq \Psi(x)$ *(or equivalently* $\Psi(x) \models \Phi(x))$ *then* $R_\Delta(\Psi) \leq R_\Delta(\Phi)$.

(ii) $R_\Delta(\Phi(x)) = \min\{R_\Delta(\Phi'(x))$: Φ' *a finite subset of* $\Phi\}$.

(iii) $R_\Delta(\phi(x) \vee \psi(x)) = \max(R_\Delta(\phi(x)), R_\Delta(\psi(x)))$.

(iv) If $\Phi(x)$ *is a set of formulae over the small set* A, *then there is* $p(x) \in S_\Delta(A)$ *such that* $R_\Delta(\Phi(x) \cup p(x)) = R_\Delta(\Phi(x))$.

(v) *If* $\Phi(x)$ *is a set of* Δ*-formulae, then* $R_\Delta(\Phi) =$ $\max\{\mathrm{CB}_X(\mathbf{p})$: $\mathbf{p} \in X = S_\Delta(\mathbf{C})$ *and* $\Phi(x) \subseteq \mathbf{p}(x)\}$.

(vi) R_Δ *is preserved under automorphisms of* $\mathbf{C}$.

(vii) *If* $\varphi(x)$ *is a* Δ*-formula, then* $R_\Delta(\varphi) \geq n + 1$ *if there is an infinite set* $\{\varphi_i(x)$: $i < \omega\}$ *of pairwise contradictory* Δ*-formulae, each implying* $\varphi(x)$, *such that* $R_\Delta(\varphi_i) \geq n$ *for all* i.

(viii) *Suppose* $\varphi(x)$ *is a* Δ*-formula, with* $R_\Delta(\varphi) = n$. *Then* $\mathrm{mult}_\Delta(\varphi) =$ $\max\{k$: *there are pairwise contradictory* Δ*-formulae* $\varphi_1, \ldots, \varphi_k$, *each implying* φ, *and with* $R_\Delta(\varphi_i) = n$ *for all* $i\}$.

Proof. Left to the reader. ■

LEMMA 3.4. *Let* $A \subseteq B$, $q(x) \in S_\Delta(B)$, *and* $p(x) = q \restriction A (\in S_\Delta(A))$. *Then* q *does not fork over* A *if and only if* $R_\Delta(q) = R_\Delta(p)$.

Proof. Suppose q does not fork over A. Let $\varphi(x) \in q$ with $R_\Delta(\varphi) = R_\Delta(q)$ (by 3.3(ii)). By 2.21 $\varphi(x)$ does not fork over A. By 2.17′, some positive Boolean combination, say $\psi(x)$, of A-conjugates of $\varphi(x)$ is in $p(x)$. By 3.3(i) and (iii), $R_\Delta(\psi(x)) \leq R_\Delta(\varphi(x))$. Thus $R_\Delta(p) \leq R_\Delta(q)$, so we get equality by 3.3(i).

Conversely, suppose q forks over A. Let M be an $|A|^+$-saturated and strongly $|A|^+$-homogeneous model containing B, and $q'(x) \in S_\Delta(M)$ a non-forking extension of q. By the first part of the proof $R_\Delta(q) = R_\Delta(q')$. Also q' forks over A. So for some $\delta(x, y) \in \Delta$, the δ-definition of q', say $\psi(y)$, is not over $\mathrm{acl}(A)$. Thus clearly $q'(x)$ has infinitely many conjugates under A-automorphisms of M, say $\{q_i$: $i < \omega\}$. We can easily find an infinite subset S

of ω and formulae $\varphi_i(x) \in q_i$ for $i \in S$, which are mutually contradictory. It follows, using 3.3(vi) and (vii), that $R_\Delta(q') < R_\Delta(p)$, and thus $R_\Delta(q) < R_\Delta(p)$. ■

COROLLARY 3.5. *Let $A \subseteq B$, $q(x) \in S(B)$, $p(x) = q \restriction A$. Then q does not fork over A if and only if $R_\Delta(p \restriction \Delta) = R_\Delta(q \restriction \Delta)$ for all infinite sets $\Delta(x)$ of L-formulae.*

One shows in a similar fashion:

LEMMA 3.6. *With the hypotheses of 3.5, q does not fork over A iff $R_\Delta(p) = R_\Delta(q)$ for all finite sets $\Delta(x)$ of L-formulae.*

REMARK 3.7. *Let $p(x) \in S_\Delta(A)$. Then p is stationary (i.e. for any $B \supseteq A$ there is a unique non-forking extension $q(x) \in S_\Delta(B)$ of p) iff* $\text{mult}_\Delta(p) = 1$.

Proof. By Lemma 3.4 and 3.3(viii). ■

The global analogue of R_Δ is called Morley rank. Again fix a variable x. Let $S_x(\mathbf{C})$ be the space of complete types over M in the variable x. The basic open sets are determined by arbitrary formulae $\varphi(x)$ over M.

DEFINITION 3.8. (i) *T is called* totally transcendental *if for every variable x,* $CB(S_x(\mathbf{C})) < \infty$.

(ii) *Let $\Phi(x)$ be a set of formulae (over a small set). Let $Y = \{\mathbf{p} \in S(\mathbf{C}): \Phi(x) \subseteq \mathbf{p}(x)\}$, a (closed) subspace of $S_x(\mathbf{C})$. By the Morley rank of Φ, $RM(\Phi)$, we mean* $CB(Y)$ *(which may be ∞). If* $CB(Y) < \infty$, *then we say that $\Phi(x)$ is totally transcendental, and in this case $dM(\Phi)$, the Morley degree of Φ, is defined to the* CB-mult(Y).

So really $RM(\text{-})$ is simply $R_\Delta(\text{-})$ where Δ is the set of *all* formulae of L. The analogues of Lemmas 3.3 and 3.4 thus hold. This is again left to the reader.

LEMMA 3.9. (i) *RM has the following characteristic property: $RM(\phi(x)) \geq \alpha + 1$ iff there is an infinite set$\{\phi_i(x): i < \omega\}$ of pairwise contradictory formulae, each implying $\phi(x)$ such that $RM(\phi_i) \geq \alpha$ for all $i < \omega$.*

(ii) *$RM(\Phi(x)) = \max\{CB(\mathbf{p}): \mathbf{p} \in S_x(\mathbf{C})$ and $\Phi(x) \subseteq \mathbf{p}(x)\}$, where* CB(-) *here means Cantor–Bendixon rank in the space $S_x(\mathbf{C})$.*

(iii) *The Morley rank of a set $\Phi(x)$ of formulae is equal to the minimum* *of the Morley ranks of finite subsets of* $\Phi(x)$. *If* $RM(\Phi) < \infty$, *then* $dM(\Phi) = \min\{dM(\phi)$: ϕ *is a finite conjunction of formulae in* Φ *and* $RM(\phi) = RM(\Phi)\}$.

(iv) $RM(\phi(x)) = 0$ *iff* ϕ *is algebraic.*

(v) $RM(\phi \vee \varphi) = \max\{RM(\phi), RM(\varphi)\}$.

(vi) *If* $RM(\phi) = \alpha$ *then* $dM(\phi)$ *(the Morley degree of* ϕ*) is the greatest* k *such that there are mutually contradictory formulae* $\phi_1(x), \ldots, \phi_k(x)$, *each implying* $\phi(x)$ *and each with Morley rank* α.

(vii) $RM(\text{-})$ *is invariant under automorphisms of* **C**.

(viii) *Suppose* $p(x) \in S(A)$, $q(x) \in S(B)$, $A \subseteq B$, *and* $p(x) \subseteq q(x)$. *Suppose that at least one of* $RM(p)$, $RM(q)$ *is* $< \infty$. *Then* q *does not fork over* A *iff* $RM(p) = RM(q)$.

(ix) *For any set* $\Phi(x)$ *of formulae over a set* A, *there is a complete type* $p(x) \in S(A)$ *such that* $RM(\Phi) = RM(p)$.

We now pass to ∞-rank and U-rank.

We first define an ordering on formulae:

DEFINITION **3.10.** (i) *Let* $\varphi(x)$, $\psi(x)$ *be (consistent) formulae and* A *some set of parameters. We say* $\varphi <_A \psi$ *if* $\models \varphi(x) \to \psi(x)$, $\psi(x)$ *is over* A, *and* $\varphi(x)$ *forks over* A.

(ii) *We say* $\varphi < \psi$ *if* $\varphi <_A \psi$ *for some set* A.

REMARK **3.11.** (i) *Suppose* $\models \varphi_1(x) \to \varphi_2(x)$, *and* $\varphi_2(x) <_A \varphi_3(x)$. *Then* $\varphi_1 <_A \varphi_3$.

(ii) *The relation* $<$ *is transitive, and thus defines a partial ordering on formulae (up to equivalence).*

(iii) *Suppose* $\varphi(x) < \psi(x)$. *Then for* any *set* B *over which* ψ *is defined, there is a conjugate* $\varphi'(x)$ *of* φ *such that* $\varphi'(x) <_B \psi(x)$.

Proof. (i) One should only check that (under the given hypotheses), $\varphi_1(x)$ forks over A. But $\varphi_2(x)$ forks over A and thus (by 2.16(iii)) some set of acl(A)-conjugates of $\varphi_2(x)$ is inconsistent. The same is then true of $\varphi_1(x)$.

(ii) follows from (i).

(iii) Suppose that $\varphi(x) <_A \psi(x)$. Let c be some point over which φ is defined. Write φ as $\varphi(x, c)$. By hypothesis there are acl(A)-conjugates $c_1, \ldots, c_n$ of c such that $\{\varphi(x, c_i)$: $i = 1, \ldots, n\}$ is inconsistent. Let B be any other set over which ψ is defined. Let $(d_1, \ldots, d_n)$ realize the non-forking extension of tp$(c_1, \ldots, c_n/\text{acl}(A))$ over acl$(A \cup B)$. In particular $\{\varphi(x, d_i)$: $i = 1, \ldots, n\}$ is inconsistent. By 2.22, tp$(d_i/\text{acl}(AB))$ does not fork over acl(A).

Thus, by stationarity of $\mathrm{tp}(d_i/\mathrm{acl}(A)) = \mathrm{tp}(c/\mathrm{acl}(A))$, $\mathrm{tp}(d_i/\mathrm{acl}(AB)) = \mathrm{tp}(d_j/\mathrm{acl}(AB))$ for $i, j = 1, \ldots, n$. Thus the formulae $\varphi(x, d_i)$ are conjugate over $\mathrm{acl}(AB)$. This shows that $\varphi(x, d_1)$, say, forks over $A \cup B$, and thus over B. Note $\models \varphi(x, d_1) \to \psi(x)$ still. So $\varphi(x, d_1) <_B \psi(x)$. ■

Before the next definition, let us recall the *foundation* rank on a partial ordering $(P, <)$, defined inductively as follows: for $a \in P$, $\mathrm{rank}(a) \geq \alpha + 1$ if there is $b \in P$ with $b < a$ such that $\mathrm{rank}(b) \geq \alpha$, and for limit δ, $\mathrm{rank}(a) \geq \delta$ if $\mathrm{rank}(a) \geq \alpha$ for all $\alpha < \delta$. Of course we define $\mathrm{rank}(a) = \alpha$ if $\mathrm{rank}(a) \geq \alpha$ but *not* $\mathrm{rank}(a) \geq \alpha + 1$, and we put $\mathrm{rank}(a) = \infty$ if $\mathrm{rank}(a) \geq \alpha$ for all ordinals α. It is an easy and basic fact that rank(-) is ordinal valued on P iff $(P, <)$ is well founded, that is has no infinite descending chain $a_0 > a_1 > \ldots$.

DEFINITION **3.12.** (i) *T is said to be superstable if for any x there is no infinite descending chain of formulae $\varphi_0(x) > \varphi_1(x) > \ldots$.*

(ii) *The formula $\varphi(x)$ is said to be superstable if there is no such infinite descending chain with $\varphi_0 = \varphi$.*

(iii) *If $\varphi(x)$ is superstable we define the ∞-rank of φ, $R^\infty(\varphi)$ to be the foundation rank of φ with respect to the ordering $<$. If φ is not superstable, we put $R^\infty(\varphi) = \infty$.*

(iv) *If $\Phi(x)$ is any set of formulae, we put $R^\infty(\Phi) = \min\{R^\infty(\wedge \Phi')$: Φ a finite subset of $\Phi\}$.*

LEMMA **3.13.** (i) *If $\models \psi(x) \to \varphi(x)$, then $R^\infty(\psi) \leq R^\infty(\varphi)$.*

(ii) *$R^\infty(\varphi(x)) = 0$ iff $\varphi(x)$ is algebraic (i.e. has only finitely many solutions).*

(iii) *$R^\infty(\varphi(x) \vee \psi(x)) = \max\{R^\infty(\varphi), R^\infty(\psi)\}$.*

(iv) *R^∞(-) is invariant under automorphisms of* **C**.

(v) *For any set $\Phi(x)$ of formulae over a set A, there is complete type $p(x) \in S(A)$ such that $R^\infty(\Phi) = R^\infty(p)$.*

Proof. (i) is immediate from 3.11(iii).

(ii) If $\varphi(x)$ is algebraic and over A, and $\models \psi(x) \to \varphi(x)$, then the solutions of $\psi(x)$ are contained in $\mathrm{acl}(A)$, so by 2.16(iii), $\psi(x)$ could not fork over A. So $R^\infty(\varphi) = 0$.

Conversely, if $\varphi(x)$ is not algebraic, and is defined over A, let c satisfy $\varphi(x)$ with $c \notin \mathrm{acl}(A)$ (by compactness). Then again by 2.16(iii), the formula '$x = c$' forks over A. Thus $R^\infty(\varphi) > 0$.

(iii) One easily proves by induction that $R^\infty(\varphi \vee \psi) \geq \alpha$ implies $R^\infty(\varphi) \geq \alpha$ or $R^\infty(\psi) \geq \alpha$, which suffices by (i).

(iv) Immediate.

(v) follows from (iii). ■

LEMMA 3.14. *Suppose that $A \subseteq B$, $p(x) \in S(A)$, $q(x) \in S(B)$, and $p \subseteq q$. Then if q does not fork over A we have $R^\infty(q) = R^\infty(p)$. The converse is also true assuming that $R^\infty(q) < \infty$.*

Proof. Note first that by Lemma 3.13(i), $R^\infty(q) \leq R^\infty(p)$.

Suppose first that q does not fork over A. Without loss $B = \operatorname{acl}(B)$, then q does not fork over $\operatorname{acl}(A)$. Let $\varphi(x) \in q$ with $R^\infty(\varphi) = R^\infty(q)$ (which may be ∞). Then $\varphi(x)$ does not fork over $\operatorname{acl}(A)$. By 2.17′, there is a consistent formula $\psi(x)$ over $\operatorname{acl}(A)$, which is equivalent to a positive Boolean combination of $\operatorname{acl}(A)$-conjugates of $\varphi(x)$, and is in $q \restriction (\operatorname{acl}(A))$. By Lemma 3.13, and choice of φ, it follows that $R^\infty(\psi) = R^\infty(\varphi)$. Let $\psi'(x)$ be the disjunction of the finitely many A-conjugates of $\psi(x)$. Then $\psi(x) \in p(x)$, and by 3.13(iii), $R^\infty(\psi') = R^\infty(\psi)$. We have shown that $R^\infty(p) \leq R^\infty(q)$, which by the first remark in this proof yields equality.

Conversely, assume $R^\infty(q) = \alpha < \infty$, and that q forks over A. Let $\varphi(x)$ be any formula in p, and we clearly find, using 2.20(ii), some formula $\psi(x)$ of q such that ψ forks over A, $R^\infty(\psi) = \alpha$, and $\models \psi(x) \rightarrow \varphi(x)$. But then $\psi < \varphi$, so $R^\infty(\varphi) \geq \alpha + 1$. As $\varphi(x) \in p(x)$ was arbitrary this shows that $R^\infty(p) \geq \alpha + 1$. ■

LEMMA 3.15. *The following are equivalent*:

(i) *T is superstable.*

(ii) *There is no infinite chain $p_1(x) \subseteq p_2(x)\ldots$ of complete types $p_i(x) \in S(A_i)$, such that p_{i+1} is a forking extension of p_i for all $i < \omega$.*

(iii) *For each A and complete type $p(x) \in S(A)$, there is a finite subset A_0 of A such that p does not fork over A_0.*

Proof. The equivalence of (ii) and (iii) is immediate.

(i) → (ii) Suppose $p_i(x) \in S(A_i)$ for $i < \omega$, and p_{i+1} is a forking extension of p_i. Let (by Lemma 2.2(ii)) $\varphi_{i+1}(x) \in p_{i+1}(x)$ be a formula which forks over A_i. Let for $n < \omega$, $\psi_n(x)$ be the conjunction of the $\varphi_i(x)$ for $i \leq n$. Then clearly $\psi_1(x) > \psi_2(x) > \ldots$ so T is not superstable.

(ii) → (*i*) Suppose T is not superstable, so there is a chain $\varphi_1(x) > \varphi_2(x) > \cdots > \varphi_i(x) > \ldots$.

By 3.11(iii) and automorphisms, we can find formulae $\psi_i(x)$ and sets $A(i)$, for $i = 1, 2, \ldots$, such that $\psi_{i+1}(x) <_{A(i)} \psi_i(x)$ and $A(i) \subseteq A(i+1)$ for all i. Let $A = \cup\{A(i): i < \omega\}$, and let $p(x) \in S(A)$ contain $\cup(\psi_i(x): i < \omega\}$. Then $p|A_{i+1}$ forks over A_i for all i. ■

DEFINITION **3.16.** *For $p(x)$, $q(x)$ complete types in the variable x over sets $A \subseteq B$ respectively, $q < p$ means that q is a forking extension of p. The U-rank of a complete type $p(x)$, $U(p)$, is precisely the foundation rank of p in this partial ordering.*

CONCLUSION **3.17.** *$U(p) < \infty$ for all complete types $p(x)$ if and only if T is superstable.*

LEMMA **3.18.** (i) *$U(p) = 0$ if p is algebraic.*

(ii) *Let $p(x) \in S(A)$, $q(x) \in S(B)$, $p \subseteq q$. If q is a non-forking extension of p then $U(q) = U(p)$, and the converse holds if $U(q) < \infty$.*

(iii) *Suppose $b \in \operatorname{acl}(aA)$. Then $U(\operatorname{tp}(ab/A)) = U(\operatorname{tp}(a/A))$.*

Proof. (i) is like Lemma 3.13(ii).

(ii) Note first that $U(p) \geq U(q)$. The right to left direction of (ii) is immediate from the definitions. For the left to right direction first we may assume A is algebraically closed (why?), and we show by induction on α that $U(p) \geq \alpha$ implies $U(q) \geq \alpha$.

Suppose $U(p) \geq \alpha + 1$. Let (by definition of $U(\text{-})$), $p' \in S(C)$ be a forking extension of p with $U(p') \geq \alpha$. We may (as $U(\text{-})$ is clearly invariant under automorphisms) assume that C is independent from B over A (i.e. replace C by the realization of a non-forking extension of $\operatorname{tp}(C/A)$ over B). Let $p'' \in S(C \cup B)$ be a non-forking extension of p', realized by d say. By the induction hypothesis $U(p'') \geq \alpha$. Now by forking symmetry, $\operatorname{tp}(B/dC)$ does not fork over C, and so by transitivity of non-forking and choice of C, $\operatorname{tp}(B/dC)$ does not fork over A. Thus $\operatorname{tp}(B/dA)$ does not fork over A; hence by symmetry $\operatorname{tp}(d/B)$ does not fork over A. As p is stationary, $\operatorname{tp}(d/B) = q$. Clearly $\operatorname{tp}(d/B \cup C)$ forks over B, so $U(q) \geq \alpha + 1$, completing the proof.

(iii) is left to the reader. ■

We complete this section by giving some scattered but important facts about superstable and totally transcendental theories.

LEMMA **3.19.** *Suppose T is superstable, $p(x) \in S(A)$ is a stationary type. Then there is a* finite *set $A_0 \subseteq \operatorname{Cb}(p)$ such that $\operatorname{Cb}(p) \subseteq \operatorname{acl}(A_0)$. Moreover, if*

$\{a_i: i < \omega\}$ *is a Morley sequence in* p, *then* $\mathrm{Cb}(p) \subseteq \mathrm{dcl}(a_0, \ldots, a_n)$ *for some* $n < \omega$.

Proof. Let $A_1 = \mathrm{Cb}(p)$. By 2.26, $A_1 \subseteq \mathrm{dcl}(A)$, so we may assume $A_1 \subseteq A$ (as replacing A by $\mathrm{dcl}(A)$ (in $\mathbf{C}^{\mathrm{eq}}$) changes nothing). By 2.26 again p does not fork over A_1. By 3.15 there is a finite subset A_0 of A_1 such that p does not fork over A_0. By 2.26, $A_1 \subseteq \mathrm{acl}(A_0)$. This proves the first part.

Now for the second part. By 2.2.8, $A_0 \subseteq \mathrm{dcl}\{a_0, \ldots, a_{n-1}\}$ for some n. Thus $A_1 \subseteq \mathrm{acl}\{a_0, \ldots, a_{n-1}\}$. Note that $\mathrm{tp}(a_n/A \cup \{a_0, \ldots, a_{n-1}\})$ is (by definition of the Morley sequence $\{a_i\}$) the non-forking extension of p, and thus also has canonical base A_1. By Lemma 2.26, $\mathrm{tp}(a_n/A \cup \{a_0, \ldots, a_{n-1}\})$ does not fork over $\{a_0, \ldots, a_{n-1}\}$, and thus A_1 is the canonical base of $\mathrm{tp}\{a_n/\mathrm{acl}\{a_0, \ldots, a_{n-1}\}\}$). By 2.26(v), $A_1 \subseteq \mathrm{dcl}(a_0, a_1, \ldots, a_n)$ as required. ■

Lemma 3.20. *For any complete type* $p(x) \in S(A)$, $U(p) \leq R^\infty(p) \leq RM(p)$. *In particular* $R^\infty(\Phi(x)) \leq RM(\Phi(x))$ *for any set of formulae* $\Phi(x)$.

Proof. If $U(p) \geq \alpha + 1$, then by definition of $U(p)$, p has a a forking extension q such that $U(q) \geq \alpha$. By induction $R^\infty(q) \geq \alpha$, and by Lemma 3.14 $R^\infty(p) \geq \alpha + 1$.

For the remainder it suffices to prove that for any formula ϕ, if $R^\infty(\phi) \geq \alpha$ then $RM(\phi) \geq \alpha$. So suppose $R^\infty(\phi) \geq \alpha + 1$. Let ϕ be over A, and let $\phi_1(x)$ be a formula which implies ϕ and forks over A. By 2.16 there is a family $\{\phi_i(x): i < \omega\}$ of A-conjugates of ϕ, which is k-inconsistent (meaning that every k-element subset is inconsistent) for some $k < \omega$. Let B contain A and all the parameters in the ϕ_i. By 3.13(iii) $R^\infty(\varphi_i) \geq \alpha$ for all i; thus by 3.13(v) there is for each $i < \omega$ some $p_i(x) \in S(B)$ such that $\phi_i \in p_i$ and $R^\infty(p_i) \geq \alpha$. Among the p_i there must be infinitely many distinct types. Thus we can find an infinite subset J of ω and for $i \in J$, $\varphi_i(x) \in p_i$, such that $i \neq j$ implies φ_i and φ_j are contradictory. We may assume that $\models \varphi_i \to \phi$ for $i \in J$. Clearly $R^\infty(\varphi_i) \geq \alpha$ for each $i \in J$. By the induction hypothesis, $RM(\varphi_i) \geq \alpha$ for each i. Thus $RM(\phi) \geq \alpha + 1$ (by, say, 3.9(i)). ■

Lemma 3.21. *Suppose* T *is (countable and)* ω*-categorical, and* $p(x)$ *is a complete type over a finite set. Then whenever any two of the ranks* $R^\infty(p)$, $RM(p)$, $U(p)$ *are defined, then those two are equal (and similarly with* $R^\infty(\psi)$, $RM(\psi)$). *If also* T *is superstable then* T *is totally transcendental (so these ranks are defined and equal everywhere).*

Proof. Left to the reader. ■

Definition 3.22. *Let* $p(x) \in S(A)$. *The multiplicity of* p, $\mathrm{mult}(p)$, *is the number of extensions of* p *to complete types over* $\mathrm{acl}(A)$.

REMARK 3.23. *Clearly then, by* 2.20(i), mult(p) = *the number of non-forking extensions of* p *in* $S(M)$ *for* M *some (any) model containing* A. *Also it is rather easy to see that* mult(p) *is finite or* $\geq 2^{\omega}$. *By* 2.20(iv), $2^{|T|}$ *is an absolute bound for* mult(p).

LEMMA 3.24. *Suppose* $p(x) \in S(M)$ *and* $RM(p) < \infty$. *Then* mult(p) = $dM(p)$.

LEMMA 3.25. *The following are equivalent*:

(i) *T is totally transcendental.*

(ii) *T is superstable, for any variable x any complete type $p(x)$ has finite multiplicity,* and *for any countable sublanguage L_0 of L, there are only countably many complete L_0-types over $\varnothing$ (of any arity).*

(iii) *For any countable sublanguage L_0 of L, and any countable set of parameters A, there are at most countably many complete L_0-types over A (of any arity).*

Countable totally transcendental theories are called ω-stable, and we will see why in the next section. In the first pieces of work on stability theory (such as Morley's theorem), ω-stability played a crucial role. In Shelah's treatment, ω-stability and total transcendentality play a much less central role, and even appear as rather marginal notions compared with superstability) see our rather clumsy characterization in Lemma 3.25). With the onset of the geometric theory, ω-stability comes back to the fore, as many of the basic results consist of a dichotomy or trade-off between ω-stability (globally or 'locally') and geometric simplicity.

We now point out a nice 'additivity' property of U-rank, which falls out quite easily from the properties of forking. First, if α is an ordinal then α has a unique expression as a polynomial in ω, where the powers (or exponents) are ordinals and the coefficients are natural numbers: $\alpha = \Sigma\{\omega^{\gamma} n_{\gamma} : \gamma \in \text{ordinals}, n_{\gamma} \in \mathbf{N}\}$ where all but finitely many n_{γ} are 0. The *Cantor sum*, $\alpha \oplus \beta$, is obtained simply by addition of polynomials. Note that if $\alpha, \beta < \omega$, then $\alpha \oplus \beta = \alpha + \beta$. The following is very important. In fact we will prove a more general result in section 3 of Chapter 4.

LEMMA 3.26. *For any a, b, A.*

$$U(\text{tp}(a/bA)) + U(\text{tp}(b/A)) \leq U(\text{tp}(a, b/A)) \leq U(\text{tp}(a/bA)) \oplus U(tp(b/A)).$$

For other ranks we just need to note:

LEMMA 3.27. *Let $R(\text{-})$ be R^{∞} or RM. Then if $b \in \text{acl}(aA)$ we have $R(\text{tp}(ab/A)) = R(\text{tp}(a/A))$ and $R(\text{tp}(b/A)) \leq R(\text{tp}(a/A))$.*

Finally note:

LEMMA 3.28. *Fix some sort S. Let f be a definable permutation of* $S^{\mathbf{C}}$ *which induces an automorphism* $\mathbf{f}$ *of the* Boolean algebra *of definable subsets of* $S^{\mathbf{C}}$. *Then for any formula* $\phi(x)$ *(where x is of sort S),* $RM(\phi) = RM(\mathbf{f}(\phi))$, $dM(\phi) = dM(\mathbf{f}(\phi))$, *and* $R^{\infty}(\phi) = R^{\infty}(\mathbf{f}(\phi))$.

Proof. This is clear for RM, by 3.9(i) for example, or by the fact that $\mathbf{f}$ gives an automorphism of the topological space $S_x(\mathbf{C})$.

The definition of R^{∞}, however, is, on the face of it, given in terms of forking, rather than in terms of the Boolean algebra of definable sets. However, we should note the following characterization: suppose $\phi(x)$ is a formula over A; then $R^{\infty}(\phi) \geq \alpha + 1$ iff for arbitrary large cardinals λ, there are $k < \omega$, $\psi(x, y)$, and $\{a_i : i < \lambda\}$ such that $\models \psi(x, a_i) \rightarrow \phi(x)$ for all i, $R^{\infty}(\psi(x, a_i)) \geq \alpha$ for all i, and $\{\psi(x, a_i) : i < \lambda\}$ is k-inconsistent. In proving the right to left direction, the point is that by choosing λ sufficiently large, we can assume that all the a_i have the same type over acl(A), and thus that $\psi(x, a_i)$ forks over A.

Given the characterization, the Boolean algebra argument above works, once we note that if $\{X_i : i < \lambda\}$ is a uniformly definable family of definable sets, then so is $\{\mathbf{f}(X_i) : i < \lambda\}$ (as f is definable). ∎

4 Miscellaneous facts about stable theories

Traditionally infinite cardinal numbers play a big role in both the definitions and statements of theorems in stability theory. Although we do not emphasize this aspect so much, we neither try nor succeed to eliminate it. In any case, cardinals, and their role in defining suitable categories of models of T, figure a lot in the current section.

4.1 λ-stability and λ-saturation

The theory T is called λ-stable if for any set A of cardinality λ, $|S(A)| \leqslant \lambda$. We have decided to work in T^{eq}, where there are in general $|T|$-many sorts, and for each sort S, there are types of sort S. So in T^{eq}, λ-stability is only a non-vacuous notion if $\lambda \geq |T|$. (Note that the same problem arises with the notion of the *cardinality* of a structure.) We deal with this either by considering only $\lambda \geq |T|$, or by starting with the 1-sorted theory T, and defining T to be λ-stable if for any A of cardinality λ the set of complete types over A in some finite number of variables of the original sort has cardinality at most λ.

The reader can choose the convention that he or she prefers. In any case we have the following facts which the reader should be able to prove.

LEMMA **4.1.1.** (i) *T is stable iff T is λ-stable for some infinite cardinal λ, in which case T is also λ-stable whenever $\lambda = \lambda^{|T|}$.*

(ii) *T is superstable iff T is λ-stable for all $\lambda \geq 2^{|T|}$.*

(iii) *For countable T, T is ω-stable iff T is λ-stable for all λ iff T is totally transcendental.*

LEMMA **4.1.2.** *Suppose T is λ-stable. Then T has a λ-saturated model of cardinality λ.*

The proof of 4.1.2 (not given here) makes use of the invariant $\kappa(T)$ defined below. In any case, we see now that assuming T to be stable, the big model $\mathbf{C}$ can be choosen to be κ_0-saturated and of cardinality κ_0, for some suitably large cardinal κ_0.

4.2 $\kappa(T)$ and the category of *a*-models

We still assume T to be stable.

Shelah defined $\kappa(T)$ to be the smallest infinite cardinal, such that there is no chain $\{p_\alpha(x) \in S(A_\alpha): \alpha < \kappa\}$ of complete types such that for all $\alpha < \beta < \kappa$, $p_\beta(x)$ is a forking extension of $p_\alpha(x)$, where x is some n-tuple of variables in the original sort of T. In fact it is rather easy to check that we get the same value of $\kappa(T)$ if we instead allow x to be a variable from any sort in T^{eq}. $\kappa_r(T)$ denotes the first regular cardinal $\geq \kappa(T)$.

LEMMA **4.2.1.** (i) $\kappa(T) \leq |T|^+$.

(ii) *For any $p(x) \in S(A)$ there is $A_0 \subseteq A$ such that $|A_0| < \kappa(T)$ and p does not fork over A_0.*

(iii) *T is superstable iff $\kappa(T) = \omega$.*

(iv) *For countable T, $\kappa(T)$ is ω* or *ω_1.*

For totally transcendental theories T, the category of all models of T is nicely behaved, and one can, moreover, read off from this category certain fundamental properties of T. For example, $|T|^+$-categoricity is equivalent to no 'Vaughtian pairs'. Also the existence of 'prime models' over arbitrary sets is of great importance.

Shelah observed that in an arbitary stable theory, one has good behaviour (such as the existence of prime models) by working in a category of sufficiently saturated models. '*a*-saturation' is the weakest form of saturation under which a smooth theory can be developed in the general case. I have become psychologically accustomed to the use of a-models (or sufficiently saturated models) to help explicate orthogonality, domination, etc., but the technology is not essential.

In any case:

DEFINITION 4.2.2. *By an* a-model *we mean a model M of T such that for any subset A of M (or M^{eq}), with $|A| \leqslant \kappa_r(T)$, and any $p(x) \in S(acl(A))$, $p(x)$ is realized in M.*

REMARK 4.2.3. *Let M be an a-model, and $p(x) \in S(M)$. Then there is a countable set $A \subseteq M^{eq}$ such that p does not fork over A and the restriction of p to A is stationary.*

Proof. By Lemma 4.2.1, there is a set $A_0 \subseteq M$ of cardinality $< \kappa_r(T)$ such that p does not fork over A_0. Now as M is an a-model we can find a Morley sequence $\{a_i : i < \omega\}$ in $p|\mathrm{acl}(A)$ in M. Now use Lemma 2.27. ■

The category of a-models brings along with it suitable versions of the notions 'prime model', 'isolated type', etc. For example, we say that M is *a-prime* over A if M is an a-model, $M \supseteq A$, and whenever $N \supseteq A$ is an a-model, then there is an A-elementary embedding of M into N. We say that $\mathrm{tp}(b/B)$ is *a-isolated* if there is some $B_0 \subseteq B$ with $|B_0| < \kappa_r(T)$, such that $\mathrm{tp}(b/\mathrm{acl}(B_0)) \models \mathrm{tp}(b/B)$. (The reader should confirm that this definition depends only on the *type* of b over B, not on b.)

LEMMA 4.2.4. *For any set A, there is a model M which is a-prime over A. Such a model M is unique up to A-isomorphism and is also* a-atomic *over A (i.e. for every finite tuple b from M,* $\mathrm{tp}(b/A)$ *is a-isolated).*

In fact the a-prime model M over A is *a-constructible* over A, meaning that we can write M as $\{a_i : i < \gamma\}$ (some γ) such that for each $\alpha < \gamma$, $\mathrm{tp}(a_\alpha/A \cup \{a_\beta : \beta < \alpha\})$ is a-isolated. Moreover, this feature characterizes the a-prime model over A. From this we can deduce, for example, that if M is a-prime over A and b is a (finite) tuple from M, then M is also a-prime over $A \cup b$.

Of course in the case of totally transcendental theories we have the classical result: if T is totally transcendental, then over any set A there is a prime model M, which is (i) unique up to A-isomorphism, and (ii) atomic over A, Again M will be *constructible* over A.

4.3 Orthogonality and domination

If $p(x)$, $q(y)$ are both complete types over A, we say that p is *weakly* or *almost* orthogonal to q if whenever a realizes p, and b realizes q, then $\mathrm{tp}(a/Ab)$ does not fork over A. (This notion is symmetric in p, q by forking symmetry.) If $p(x) \in S(A)$, $q(y) \in S(B)$ are *stationary* types (over perhaps different sets), we say p is *orthogonal* to q if whenever $C \supseteq A \cup B$, then $p|C$ is weakly orthogonal to $q|C$ (where remember $p|C$ denotes non-forking extension of p over C etc.). Note that an algebraic type is orthogonal to every type.

Lemma 4.3.1. (i) *Orthogonality is a relation between* parallelism classes *of stationary types (or equivalently between strong types).*

(ii) *If $p(x)$, $q(y) \in S(M)$ where M is an a-model, then p is orthogonal to q iff p is weakly orthogonal to q.*

(iii) *Suppose $p(x)$, $q(y) \in S(A)$ are both stationary. Then p is orthogonal to q iff $p^{(n)}$ is weakly orthogonal to $q^{(n)}$ for all $n < \omega$.*

Proof. (i) and (ii) are routine applications of forking. To illustrate this standard technique we prove (ii). Suppose $p(x)$, $q(y) \in S(M)$ are not orthogonal. Thus there is $C \supseteq M$, a realizing $p|C$, and b realizing $q|C$ such that $\mathrm{tp}(a/bC)$ forks over C. Thus $\mathrm{tp}(a/bC)$ forks over M, whereby (2.20(ii)) there is a finite tuple c from C such that $\mathrm{tp}(a/bcM)$ forks over M. Let $A_0 \subseteq M$ be of cardinality $< \kappa(T)$ such that $\mathrm{tp}(a, b, c/M)$ does not fork over A_0. Then the properties of forking imply that (a) each of $\mathrm{tp}(a/M)$, $\mathrm{tp}(b/M)$ does not fork over A_0, (b) each of $\mathrm{tp}(a/A_0c)$, $\mathrm{tp}(b/A_0c)$ does not fork over A_0, and (c) $\mathrm{tp}(a/bcA_0)$ forks over cA_0. Choose c' in M realizing $\mathrm{tp}(c/\mathrm{acl}(A_0))$, and then choose b' somewhere such that $\mathrm{tp}(a, b', c'/\mathrm{acl}(A_0)) = \mathrm{tp}(a, b, c/\mathrm{acl}(A_0))$. Thus (a), (b), (c) above are true with (a, b', c') replacing (a, b, c). It follows that $\mathrm{tp}(a/A_0c')$ and $\mathrm{tp}(b'/A_0c')$ are not weakly orthogonal. Let (a'', b'') realize the non-forking extension of $\mathrm{tp}(a, b'/A_0c')$ over M. Then $p(x) = \mathrm{tp}(a''/M)$, $q(y) = \mathrm{tp}(b''/M)$, and the properties of forking imply that $\mathrm{tp}(a''/Mb'')$ forks over M. Thus p is not weakly orthogonal to q.

(iii) The left to right implication is easily seen to be true by induction on n.

Conversely, suppose $p(x)$, $q(y) \in S(A)$ are (stationary) and non-orthogonal. Let M be a κ-saturated model containing A, for some $\kappa > |A| + |T|$. Let $p'(x)$, $q'(y) \in S(M)$ be the non-forking extensions of p, q respectively. By part (ii), p' is non-orthogonal to q', so there are a realizing p', b realizing q' such that $\mathrm{tp}(a/bM)$ forks over M. So $\mathrm{tp}(ab/M)$ forks over A. Let $B \supseteq A$, $B \subseteq M$ be such that $|B| < \kappa$ and $\mathrm{tp}(ab/M)$ does not fork over B. By the κ-saturation of M we can find in M a Morley sequence $\{(a_i, b_i): i < \omega\}$ in $\mathrm{tp}(a, b/\mathrm{acl}(B))$. By 2.28(ii) and Remark 2.26, $\mathrm{tp}(a, b/M)$ does not fork over $\{a_i, b_i: i < \omega\}$, and thus $\mathrm{tp}(a, b/A \cup \{a_i, b_i: i < \omega\})$ forks over A. By the finite character of forking $\text{tp}(a, b/A \cup \{a_0, b_0, \ldots, a_n, b_n\})$ forks over A for some $n < \omega$. But clearly $(a_0, \ldots, a_n, a)$ realizes $p^{(n+2)}$, and $(b_0, \ldots, b_n, b)$ realizes $q^{(n+2)}$. One then deduces that $p^{(n+2)}$ is not weakly orthogonal to $q^{(n+2)}$, as required. ■

REMARK **4.3.2.** *If* $\{a_1, \ldots, a_n\}$ *is* A*-independent, and* $\text{tp}(a_i/\text{acl}(A))$ *is orthogonal to* q *for each* i*, then* $\text{tp}(a_1, \ldots, a_n/\text{acl}(A))$ *is orthogonal to* q.

We have discussed above the notion of orthogonality between types. There is also a notion of orthogonality between a type and a set (of parameters). Let $p(x) \in S(A)$ be a stationary type, and B some set of parameters. We say that *p is orthogonal to B*, if for every type $q(y)$ over $\text{acl}(B)$, p is orthogonal to q.

LEMMA **4.3.3.** *Let* $p(x) \in S(A)$ *be stationary, and let* B *be a set of parameters. Then the following are equivalent*:

(i) *p is orthogonal to B.*

(ii) *Whenever C is independent from A over B, then p is orthogonal to BC.*

(iii) *Whenever* $\text{tp}(A'/\text{acl}(B)) = \text{tp}(A/\text{acl}(B))$, *and A' is independent from A over B, then p is orthogonal to p', where p' is the 'copy' of p over A'.*

Proof. (i) → (ii). Suppose C is as in the hypothesis of (ii) and that p is non-orthogonal to q for some $q \in S(\text{acl}(BC))$. Let $D = \text{acl}(ABC)$. By 4.3.1, there are Morley sequences $(a_1, \ldots, a_n)$ in $p \restriction D$ and $(b_1, \ldots, b_n)$ in $q \restriction D$ such that $\text{tp}(a_1, \ldots, a_n/D \cup \{b_1, \ldots, b_n\})$ forks over D. Forking calculus implies that $\text{tp}(a_1, \ldots, a_n/C \cup \{b_1, \ldots, b_n\} \cup AB)$ forks over AB, and so $\text{tp}(a_1, \ldots, a_n/\{c, b_1, \ldots, b_n\} \cup AB)$ forks over AB for some tuple c from C. But it is easy to see that $\text{tp}(c, b_1, \ldots, b_n/AB)$ does not fork over B. So using Remark 4.3.2, we conclude that $p \restriction D$ is non-orthogonal to $\text{tp}(c, b_1, \ldots, b_n/B)$. Thus p is non-orthogonal to B.

(ii) → (iii) is immediate.

(iii) → (i) Suppose p is non-orthogonal to B. Let M be an a-model containing $A \cup B$ say, and let a realize $p|M$, b realize $q|M$ for some $q \in S(\text{acl}(B))$ such that $\text{tp}(a/bM)$ forks over M. Clearly then, $\text{tp}(aM/dB)$ forks over B. Let $aM, a_1M_1, \ldots$ be a Morley sequence in $\text{tp}(aM/\text{acl}(dB))$. Let $a_0 = a$, $M_0 = M$. By 2.29,

$(*)$ $\{a_0M_0, a_1M_1, a_2M_2, \ldots\}$ is not B-independent.

On the other hand, M_0 is independent from $\{a_1M_1, a_2M_2, \ldots\}$ over bB, so also

$(**)$ M_0 is independent from $\{a_1M_1, a_2M_2, \ldots\}$ over B.

Similarly for each i,

$(***)$ a_iM_i is independent from $\{M_j : j \neq i\}$ over B.

Let A_i be the 'copy' of A in M_i, and p_i the copy of p over A_i. Then by (***) we have

(****) for each i, a_i realizes $p_i|(M_0M_1, \ldots)$.

By (*) let n be least such that $\{a_0M_0, \ldots, a_nM_n\}$ is not B-independent. Then $\{a_1M_1, \ldots, a_nM_n\}$ is B-independent, whereby $\{a_1, \ldots, a_n\}$ is $M_1M_2\ldots M_n$-independent, so by (**) also $M_0M_1\ldots M_n$-independent. Using (**), it must be the case that $\mathrm{tp}(a_0/a_1a_2\ldots a_nM_0M_1\ldots M_n)$ forks over $M_0M_1\ldots M_n$. By 4.3.2 (and (****)), p_0 is non-orthogonal to p_i for some $i = 1, \ldots, n$. By (**) A_i is independent from A over $\mathrm{acl}(B)$ (and, moreover, has the same type over $\mathrm{acl}(B)$ as A). So (iii) is contradicted. ■

The 'opposite' notion to orthogonality is *domination*, We say that A dominates B over C, if for any set D, if D is independent from A over C, then D is independent from B over C (or equivalently for all d, if $\mathrm{tp}(d/AC)$ does not fork over C then $\mathrm{tp}(d/BC)$ does not fork over C). If p and q are stationary types over maybe different domains, we say that *p dominates q* (or, as sometimes is said, *p eventually dominates q*) if there is some set C containing the domains of p, q and a realizing $p|C$, b realizing $q|C$ such that a dominates b over C. For M an a-prime model, and A a set, let $M[A]$ denote the a-prime model over $M \cup A$. For $p \in S(M)$, let $M[p]$ denote $M[a]$ for a some realization of p.

Lemma 4.3.4. (i) *Suppose A dominates B over C, $C_0 \subseteq C$ and* $\mathrm{tp}(AB/C)$ *does not fork over C_0. Then A dominates B over C_0.*

(ii) *Suppose A dominates B over C, $C \subseteq C_1$ and* $\mathrm{tp}(A/C_1)$ *does not fork over C. Then A dominates B over C_1 (and B is independent from C_1 over C).*

(iii) *Let M be an a-model. Then* $\mathrm{tp}(c/Mb)$ *is a-isolated if and only if b dominates bc over M.*

(iv) *Let M be an a-model, and p, $q \in S(M)$. Then p dominates q if and only iff q is realized in $M[p]$ (iff q is realized in* every *a-model $N \supseteq M$ in which p is realized).*

Proof. (i) is simple forking calculus. Assume D is independent with A over C_0. We may assume that D is independent with C over ABC_0, from which by forking symmetry and transivity we deduce that (a) D is independent from A over C, and (b) D is independent from BC over BC_0. From (a) and our hypothesis we conclude that D is independent from B over C, and thus, by (b), D is independent from B over C_0.

(ii) Similar.

(iii) Suppose that $\text{stp}(c/Ab) \models \text{tp}(c/Mb)$ for some $A \subseteq M$ with $|A| < \kappa_r(T)$. This means that $\text{stp}(c/Ab)$ has a unique extension over Mb, which must therefore be a non-forking extension, and thus means

($*$) $\text{stp}(c/Ab)$ is weakly orthogonal to $\text{stp}(d/Ab)$ for any tuple d from M.

We may also assume that $\text{tp}(ab/M)$ does not fork over A.

We show that b dominates bc over A. Suppose d' is independent from b over A. As M is an a-model, $\text{stp}(d'/A)$ is realized in M by some d. Then d is also independent from b over A, whereby $\text{tp}(d/Ab) = \text{tp}(d'/Ab)$. By ($*$), $\text{tp}(c/Abd')$ does not fork over Ab. By Lemma 2.22, $\text{tp}(bc/Ad')$ does not fork over A.

This shows that b dominates bc over A. As b is independent from M over A, we conclude by (ii) that b dominates bc over M.

Conversely suppose b dominates bc over M. Let $A \subseteq M$ be such that $|A| < \kappa_r(T)$ and $\text{tp}(bc/M)$ does not fork over A (so also $\text{tp}(b/M)$ does not fork over A, and $\text{tp}(c/Mb)$ does not fork over Ab). By (i) b dominates bc over A. Let c' realize $\text{stp}(c/Ab)$. Then b dominates bc' over A. Now M is independent from b over A, so M is independent from bc' over A, so c' is independent from Mb over Ab. Thus $\text{tp}(c'/Mb) = \text{tp}(c/Mb)$.

(iv) Assuming that p dominates q, we may find b realizing p, c realizing q such that b dominates c over M. (This is by using (i), (ii), and the fact that M is an a-model, and is left to the reader.) Now choose $A \subseteq M$ with $|A| < \kappa_r(T)$ such that $\text{tp}(b, c/M)$ does not fork over A. By (i) b dominates c over A. Now $\text{stp}(c/bA)$ is realized in $M[b]$ by some c' (as $M[b]$ is by definition an a-model, and $|Ab| < \kappa_r(T)$). Clearly (by automorphism) b dominates c' over A. But b is independent with M over A; thus c' is independent with M over A, whereby c' realizes q.

The other direction of (iv) is given by (iii) and Lemma 4.2.4. ■

REMARK 4.3.5. *Suppose that* (i) *b dominates bc over A, and* (ii) *c dominates b over A. Then c dominates bc over A.*

Proof. Suppose d is independent from c over A. Then by (ii) d is independent from b over A, so by (i), d is independent with bc over A. ■

4.4 Weight

The *preweight* of a type $p(x) = \text{tp}(a/A)$, denoted $\text{prwt}(p)$ or $\text{prwt}(a/A)$, is the supremum of the set of cardinals κ for which there exists an A-independent set $\{b_i : i < \kappa\}$ such that $\text{tp}(a/b_iA)$ forks over A for all i. It is easy to see that if $q(x) \in S(B)$ is a non-forking extension of $p(x) \in S(A)$, then $\text{prwt}(q) \geq \text{prwt}(p)$. The *weight* of a type p, $\text{wt}(p)$, is the supremum of $\{\text{prwt}(q)$: q a non-forking extension of $p\}$.

It is easy to see that $\mathrm{prwt}(a/A) = \mathrm{prwt}(a/\mathrm{acl}(A))$, and also $\mathrm{wt}(a/A) = \mathrm{wt}(a/\mathrm{acl}(A))$; prwt and wt also make sense for infinitary types. So we speak sometimes of $\mathrm{prwt}(B/A)$ or $\mathrm{wt}(B/A)$.

Note first that always $\mathrm{prwt}(a/A) \leq \kappa(T)$. For suppose $\{b_i: i < \kappa\}$ is A-independent, and each b_i forks with a over A. Then properties of forking imply that $\mathrm{tp}(a/A \cup \{b_i: i < \alpha\})$ forks over $A \cup \{b_i: i < \beta\}$ whenever $\beta < \alpha < \kappa$. Thus $\kappa < \kappa(T)$. Thus also $\mathrm{wt}(p) \leq \kappa(T)$ for all types p.

Lemma **4.4.1.** (i) *Suppose* $\{a_1, \ldots, a_n\}$ *is B-independent. Then* $\mathrm{wt}(a_1, \ldots, a_n/B) = \sum \{\mathrm{wt}(a_i/B): i = 1, \ldots, n\}$.

(ii) *Suppose* $\mathrm{wt}(a/B) = n$, $\{a_1, \ldots, a_n\}$ *is B-independent, and* $\mathrm{tp}(a/a_iB)$ *forks over B for each i. Then* $(a_1, \ldots, a_n)$ *dominates a over B.*

(iii) *Suppose X is a set of elements and for all* $a \in X$, $\mathrm{wt}(a/B) = 1$. *Let I, J be both maximal B-independent subsets of X (i.e. I and J are each* bases *of X over B). Then* $|I| = |J|$.

(iv) *If M is an a-model, and* $p(x) \in S(M)$, *then* $\mathrm{prwt}(p) = \mathrm{wt}(p)$.

Proof. (i) is a simple application of forking calculus, making use of the easy fact: for any $a_1, \ldots, a_n$ and B, $\mathrm{wt}(a_1, \ldots, a_n/B) \leq \sum (\mathrm{wt}(a_i/Ba_i \ldots a_{i-1})$: $i = 1, \ldots, n\})$ (and likewise for prwt).

(ii) is immediate from the definitions.

(iii) The idea is to show the following: suppose $I = \{a_i: i < \kappa\}$ is a basis for X over B, and $b \in X$. Let $\alpha < \kappa$ be least such that $\mathrm{tp}(b/\{a_i: i \leq \alpha\} \cup B)$ forks over B. Let I' be the result of replacing a_α in I by b. *Then* I' is a basis for X over B.

This is shown by easy forking calculus together with (ii) (when $n = 1$). Having shown this, it follows, by repeatedly substituting elements of J for elements of I, that $|J| \leq |I|$. By symmetry we conclude $|I| = |J|$.

(iv) This is a standard argument involving pulling down parameters from the domain of a non-forking extension of p, into M. ■

Lemma **4.4.2.** (i) *Suppose* $p(x)$, $q(y)$ *are both stationary types of weight* 1. *Then p dominates q iff q dominates p iff p is non-orthogonal to q.*

(ii) *The relation of non-orthogonality between weight* 1 *stationary types is an equivalence relation.*

Proof. (i) it follows directly from 4.4.1(ii).

(ii) follows from the trivial fact that if a dominates b over A and b dominates c over A then a dominates c over A. ■

4.5 Regular types

A non-algebraic stationary type $p(x) \in S(A)$ is said to be *regular* if for any stationary forking extension q of p, p is orthogonal to q. Although we usually are concerned only with *stationary* regular types, we can define an arbitrary type $p(x) \in S(A)$ to be regular if some completion of p over $\mathrm{acl}(A)$ is regular.

It is easily checked that for *stationary* types, regularity is a parallelism invariant, and that for an arbitrary type $p(x) \in S(A)$, p is regular iff *every* completion of p over $\mathrm{acl}(A)$ is regular.

Regular types will be the topic of Chapter 7. But nevertheless we will here summarize some important properties.

LEMMA **4.5.1.** *Suppose $p(x) \in S(A)$ is regular and stationary. Suppose a realizes p, B, C are sets of realizations of p, for all $b \in B$, $\mathrm{tp}(b/CA)$ forks over A, and $\mathrm{tp}(a/BA)$ forks over A, Then $\mathrm{tp}(a/CA)$ forks over A.*

Proof. Otherwise $\mathrm{tp}(a/CA)$ is a non-forking extension of p. Thus (as p is regular) for any $b \in B$, a is independent from b over CA, that is $\mathrm{tp}(a/bCA)$ is a non-forking extension of p. Continuing, we see that $\mathrm{tp}(a/BC)$ is a non-forking extension of p, contradicting a hypothesis. ■

Lemma 4.5.1, together with forking symmetry, says that the relation 'a forks with B over A', for $\{a\} \cup B$ a set of realizations of p, satisfies the Steinitz axioms for a dependence relation (see the section on geometries in Chapter 2).

LEMMA **4.5.2.** *Suppose $p(x) \in S(A)$ is regular and stationary. Let $C \supseteq A$, let $(a_1, \ldots, a_n)$ realize $p^{(n)}|C$ (or equivalently $(p|C)^{(n)}$) and let $\mathrm{tp}(a/C)$ be a forking extension of p. Then $\mathrm{tp}(a_1, \ldots, a_n/aC)$ does not fork over C (and thus does not fork over A).*

Proof. For each $i < n$, $\mathrm{tp}(a_i/C \cup \{a_1, \ldots, a_{i-1}\})$ is a non-forking extension of p, but $\mathrm{tp}(a/C \cup \{a_1, \ldots, a_{i-1}\})$ is a forking extension of p. Thus by regularity of p, for each i, $\mathrm{tp}(a_i/C \cup \{a_1, \ldots, a_{i-1}\} \cup \{a\})$ does not fork over C. Lemma 2.22 yields the desired conclusion. ■

LEMMA **4.5.3.** *Any regular type has weight* 1.

Proof. Suppose $p(x) \in S(A)$ is regular. We may assume p stationary. It is enough to show that $\mathrm{prwt}(p) = 1$. Suppose not, so there is $a \models p$, and there are tuples b, c, such that $\{b, c\}$ is A-independent, but a is dependent with each of b, c (separately) over A. Let M be an $|A|^+ + |T|^+$-saturated model containing bA such that $\mathrm{tp}(ac/M)$ does not fork over bA. In particular

(i) $\text{tp}(a/M)$ does not fork over bA and (ii) $\text{tp}(c/M)$ does not fork over A. Also (iii) $\text{tp}(a/M)$ forks over A.

Let $\{a_i : i < \omega\}$ be a Morley sequence in $\text{tp}(a/\text{acl}(Ab))$ contained in M. By 2.27, $\text{tp}(a/M)$ does not fork over $\{a_i : i < \omega\}$. By (ii) and the finite character of forking, $\text{tp}(a/A \cup \{a_1, \ldots, a_n\})$ forks over A, for some $n < \omega$. Now $\text{tp}(a/A) = p = \text{tp}(a_i/A)$ for all i, so by Lemma 4.5.1, there is an A-independent subset of $\{a_i : i < n\}$, say A_0, such that

$$(*) \qquad \text{tp}(a/A_0A) \text{ forks over } A.$$

But $A_0 \subseteq M$, so by (ii), A_0 is an Ac-independent set of realizations of $p|Ac$, Also $\text{tp}(a/Ac)$ forks over A. By Lemma 4.5.2, $\text{tp}(A_0/Aca)$ does not fork over A, and thus $\text{tp}(A_0/Aa)$ does not fork over A. This contradicts $(*)$, proving the lemma. ■

If $p_1, \ldots, p_n$ are stationary types over a set A, then by $p_1 \otimes p_2 \otimes \cdots \otimes p_n$ we mean $\text{tp}(a_1, \ldots, a_n/A)$ where $\text{tp}(a_i/A) = p_i$ and $\{a_1, \ldots, a_n\}$ is A-independent. Clearly this type is well defined, and we call it the product of $p_1, \ldots, p_n$.

Lemma 4.5.4. *Let M be an a-model, $q \in S(M)$, and $p_1, \ldots, p_n \in S(M)$ regular types. Then the following are equivalent*:

(i) *q is domination equivalent to $p_1 \otimes \cdots \otimes p_n$.*

(ii) *There are realizations b of q and $(a_1, \ldots, a_n)$ of $p_1 \otimes \cdots \otimes p_n$ such that $M[b] = M[a_1, \ldots, a_n]$.*

Proof. (All we really use about the p_i is that they have weight 1.)

(ii) → (i) is a direct consequence of Lemma 4.3.4(iv) (and does not use anything about regularity).

(i) → (ii) Let $(a_1, \ldots, a_n)$ realize $p_1 \otimes \cdots \otimes p_n$. Then by 4.3.4(iv), q is realized in $N = M[a_1, \ldots, a_n]$ by, say, b. By 4.3.4, there is a realization $(c_1, \ldots, c_n)$ of $p_1 \otimes \cdots \otimes p_n$ in N such that b dominates $(c_1, \ldots, c_n)$ over M. In particular $\text{tp}(c_i/a_1, \ldots, a_nM)$ forks over M for all $i = 1, \ldots, n$. Applying 4.4.1(iii) to $X = \{a_1, \ldots, a_n, c_1, \ldots, c_n\}$ and using the fact that each p_i has weight 1, we conclude that $\text{tp}(a_i/c_1, \ldots, c_nM)$ forks over M for $i = 1, \ldots, n$.

By Lemma 4.4.1(i) and (ii), we conclude: $(c_1, \ldots, c_n)$ dominates $(a_1, \ldots, a_n)$ over M. Transititivity of domination implies that b dominates $(a_1, \ldots, a_n)$ over M. But $(a_1, \ldots, a_n)$ dominates $(a_1, \ldots, a_n, b)$ over M. So by Remark 4.3.5, b dominates $(a_1, \ldots, a_n, b)$ over M. By 4.3.4(iii), $\text{tp}(a_1, \ldots, a_n/bM)$ is a-isolated. It easily follows that N is a-prime over $M \cup b$ (using for example the remarks following Lemma 4.24). Namely, $N = M[b]$, as required. ■

Remark 4.5.5. *Let $p(x)$ be a stationary regular type, M an a-model, and suppose p is non-orthogonal to M. Then there is a regular type $r \in S(M)$ which is non-orthogonal to (and thus domination equivalent to) p.*

Proof. First we may assume that $p = \text{tp}(a/\text{acl}(A))$ for some A with $|A| < \kappa(T)$. Assuming p non-orthogonal to M there is clearly some set $B \subseteq M$ with $|B| < \kappa_r(T)$ such that p is non-orthogonal to B. By augmenting B we may assume that $\text{tp}(A/M)$ does not fork over B. As M is an a-model, we can find A' in M such that $\text{tp}(A/\text{acl}(B)) = \text{tp}(A'/\text{acl}(B))$. The $\text{acl}(B)$-elementary map taking A to A' extends to one taking $\text{acl}(A)$ to $\text{acl}(A')$. Let p' be the image of p under such a map. Then p' is regular, and by Lemma 4.3.3(iii), p' is non-orthogonal to p. Let $r \in S(M)$ be $p'|M$. ■

Finally we point out that for superstable T any type is domination equivalent to a *finite* product of regular types, and thus has finite weight. This result will be generalized in Chapter 4 beyond the superstable context.

Lemma 4.5.6. *Suppose T is superstable, $M \subseteq N$ a-models, and $b \in N - M$ such that $U(\text{tp}(b/M))$ is least possible. Then $\text{tp}(b/M)$ is regular.*

Proof. Let $p = \text{tp}(b/M)$. If p is not regular, there are $M \subseteq C$, b' realizing $p|C$, and c realizing a forking extension of p over C such that b' forks with c over C. We can find $A_0 \subseteq M$ of cardinality $< \kappa(T)$ and $C_0 \subseteq C$ of cardinality $< \kappa(T)$ such that $A_0 \subseteq C_0$, p (and thus also $\text{tp}(b/C)$) does not fork over A_0, $\text{tp}(c/C)$ does not fork over C_0, and b' forks with c over C_0.

In particular $U(c/C_0) = U(c/C) < U(p)$ (by Lemma 3.18).

As M is an a-model, we can find $D_0 \subseteq M$ such that $\text{tp}(D_0/\text{acl}(A_0)) = \text{tp}(C_0/\text{acl}(A_0))$. Then $\text{tp}(b', C_0) = \text{tp}(b, D_0)$. So there is c' such that $\text{tp}(b, c', D_0) = \text{tp}(b', c, C_0)$. As N is an a-model, we may choose such c' in N. Then

(i) $\text{tp}(b/c'D_0)$ forks over D_0, whereby $c' \notin M$; and
(ii) $U(c'/M) \leq U(c'/D_0) = U(c/C_0) < U(p)$.

(i) and (ii) contradict the minimal choice of $U(b/M)$. So $\text{tp}(b/M)$ is regular. ■

Corollary 4.5.7. *Suppose T is superstable. Then any type is domination equivalent to a finite product of regular types. Specifically, if $q \in S(M)$ where M is an a-model, then there are regular types $p_1, \ldots, p_n \in S(M)$ such that q is domination equivalent to $p_1 \otimes \cdots \otimes p_n$.*

Proof. Let $q \in S(M)$ where q is an a-model. Let b realize q, and let $N = M[b]$. Let $\{a_i : i < \kappa\}$ be a maximal M-independent subset of N such that $\text{tp}(a_i/M)$

is regular for each i. Then b forks with each a_i over M, and thus by superstability κ is finite, say $\kappa = n$. (Forking calculus yields that $\mathrm{tp}(b/A \cup \{a_i : i \leq \alpha\})$ forks over $A \cup \{a_i : i < \alpha\}$ for each $\alpha < \kappa$, so $\kappa < \kappa(T) = \omega$.) So b dominates $(a_1, \ldots, a_n)$ over M.

We will show that $(a_1, \ldots, a_n)$ dominates b over M.

In fact let $B \subseteq N$ be a *maximal* set containing $(a_1, \ldots, a_n)$ such that $(a_1, \ldots, a_n)$ dominates B over M. Transitivity of domination and 4.3.4 (iii) implies that B is an a-model M_1, say. If $M_1 = N$ then we are finished. Otherwise by Lemma 4.5.6 there is some $c \in N - M_1$ such that $\mathrm{tp}(c/M_1)$ is regular.

Case (*i*). $\mathrm{tp}(c/M_1)$ is orthogonal to M. Then clearly M_1 dominates $M_1 c$ over M, so M_1 was not maximal after all, a contradiction.

Case (*ii*). $\mathrm{tp}(c/M_1)$ is non-orthogonal to M. By Remark 4.5.5, there is a regular type $r \in S(M)$ which is non-orthogonal to $\mathrm{tp}(c/M_1)$. Let $r_1 = r|M_1$. Then by 4.3.1(ii) $\mathrm{tp}(c/M_1)$ is not weakly orthogonal to r. As r has weight 1, it follows from 4.4.1(ii) that $\mathrm{tp}(c/M_1)$ dominates r_1 whereby (by 4.3.4(iv)) r_1 is realized (by d, say) in N. But then d is independent from $\{a_1, \ldots, a_n\}$ over M and $\mathrm{tp}(d/M)$ is regular, contradicting maximal choice of $\{a_1, \ldots, a_n\}$. This final contradiction proves that $M_1 = N$. Thus $(a_1, \ldots, a_n)$ dominates b over M. So q is domination equivalent to $p_1 \otimes \cdots \otimes p_n$, where $p_i = \mathrm{tp}(a_i/M)$. ■

COROLLARY **4.5.8.** *If T is superstable then every type has finite weight.*

If $p \in S(A)$ is stationary and regular and M is a model containing A, then by $\dim(p, M)$ (the dimension of p in M) we mean the cardinality of a maximal independent (over A) subset of p^M. This is well defined by Lemma 4.4.1 and Lemma 4.5.3 for example (or directly using Lemma 4.5.1). We wish to point out a certain 'additivity' of dimension property. First note:

LEMMA **4.5.9.** (i) *Suppose M is a model, $p(x) \in S(M)$, and $\varphi(x)$ is a formula in p, with φ over $A \subseteq M$. Then $p \restriction (\varphi^M \cup A) \vDash p$.*

(ii) *Suppose $A \subseteq M$, M is an $|A|^+$-saturated model and $p(x) \in S(M)$. Let $X = p^M$. Then $p \restriction (X \cup A) \vDash p$.*

Proof. (i) Let $p_0 = p \restriction (\varphi^M \cup A)$. By Lemma 2.2 (or its proof) if $q(x) \in S(M)$ is *any* extension of p_0, then q is definable over φ^M, and so is the unique non-forking extension of p_0. This shows that $p_0 \vDash p$.

(ii) is similar using (ii) of Lemma 2.2. ■

LEMMA **4.5.10.** *Let $A \subseteq M \subseteq N$, where M is an a-model. Let $p(x) \in S(A)$ be a stationary regular type. Then* $\dim(p, N) = \mathit{dim}(p, M) + \dim(p|M, N)$.

Proof. Let I be a maximal A-independent subset of p^M. Let J be a maximal $A \cup I$-independent subset of $(p|A \cup I)^N$. Thus $\dim(p, N) = |I \cup J|$, $\dim(p, M) = |I|$, and it suffices to prove that

(*) J is an M-independent set of realizations of $p|M$.

Let $A_0 \subseteq A$ be such that $|A_0| < \kappa(T)$ and p does not fork over A_0. Let $p_0 = p \restriction \mathrm{acl}(A_0)$, and let $X = p_0^M$. Let $\mathbf{c} = (c_1, \ldots, c_n)$ be a tuple from J. By Lemma 4.5.9, $\mathbf{c}$ is independent from M over $A_0 \cup X$. So $\mathbf{c}$ is independent from M over $A \cup X$. Note that $I \subseteq X$ and that for any $a \in X$, $\mathrm{tp}(a/A \cup I)$ is a forking extension of p_0. But p_0 is regular, and $(c_1, \ldots, c_n)$ is an independent tuple of realizations of $p_0|A \cup I$. It easily follows that for each tuple $\mathbf{a}$ from X, $\mathbf{c}$ is independent from $\mathbf{a}$ over $A \cup I$. Thus $\mathrm{tp}(\mathbf{c}/A \cup X)$ does not fork over $A \cup I$. Thus $\mathrm{tp}(\mathbf{c}/M)$ does not fork over $A \cup I$. (*) follows. ■

Let $p(x) \in S(A)$ be stationary, and $\varphi(x)$ some formula in p. We say that (p, φ) is *strongly* regular if for any $B \supseteq A$ and stationary $q(x) \in S(B)$ which contains $\varphi(x)$, either $q = p|B$ or q is orthogonal to p. We will say that $p(x)$ is strongly regular if (p, φ) is strongly regular for some φ in p. It is quite easy to see that any strongly regular type is regular, and we have also:

Lemma 4.5.11. *Suppose $p(x) \in S(A)$ is strongly regular, $A \subseteq M \subseteq N$. Then* $\dim(p, N) = \dim(p, M) + \dim(p|M, N)$.

Proof. Like 4.5.10, using 4.5.9(i). ■

5 Weakly minimal sets and unidimensional theories

The terms weakly minimal, strongly minimal, minimal, are confusing and rather unfortunate. As these terms still have wide currency in the 'community' we will stick with them.

Definition 5.1. (i) *The* formula $\theta(x)$ *is said to be* strongly minimal *if* $RM(\theta) = 1$, $dM(\theta) = 1$.

(ii) *The* formula $\theta(x)$ *is said to be* weakly minimal *if* $R^\infty(\theta) = 1$.

(iii) *The* partial type $\Phi(x)$ *(over A, say) is said to be* minimal *if for any* $A \subseteq B$, $\Phi(x)$ *has a* unique *extension $p(x) \in S(B)$ which is not algebraic.*

Remark 5.2. (i) *We call a set X strongly minimal, weakly minimal, or minimal, according to whether X is the set of solutions (or realizations) in* **C** *of a strongly minimal formula, weakly minimal formula, or minimal partial type, respectively.*

(ii) *The definable set X is strongly minimal iff for any definable subset Y of X, either Y is finite or $X - Y$ is finite.*

(iii) *Similarly the set (of realizations of some partial type) X is minimal if for every 'definable' subset Y or X, Y is finite or $X - Y$ is finite. (By a definable subset of X, we here mean a set of the form $X \cap Z$ where Z is definable.) Thus $\theta(x)$ is strongly minimal iff $\{\theta(x)\}$ is minimal.*

(iv) *A non-algebraic formula $\theta(x)$ (over A, say) is weakly minimal iff for any $B \supseteq A$ there are at most $2^{|T|}$ complete non-algebraic types $p(x) \in S(B)$ which contain $\theta(x)$. Moreover, any non-algebraic stationary type containing $\theta(x)$ is minimal.*

(v) *A strongly minimal formula (or even a formula of Morley rank 1) is weakly minimal. Moreover, if $\theta(x)$ is a weakly minimal formula over a model M, and there are only finitely many complete non-algebraic types over M containing $\theta(x)$, then there are strongly minimal formulae $\varphi_1(x), \ldots, \varphi_n(x)$ such that $\models \theta(x) \leftrightarrow \bigvee \{\varphi_i(x)\colon i = 1, \ldots, n\}$.*

(vi) *A complete type $p(x) \in S(A)$ is minimal iff p is stationary and $U(p) = 1$. Moreover, a complete minimal type is regular.*

Lemma 5.3. *Suppose $\theta(x)$ is a weakly minimal formula, defined over a set A. Let D be $\theta^{\mathbf{C}}$.*

(*i*) *Let $b \in D$, and $B \supseteq A$. Then $\mathrm{tp}(b/B)$ forks over A iff $b \notin \mathrm{acl}(A)$ and $b \in \mathrm{acl}(B)$.*

(ii) *Let $\varphi(x, y)$ be a formula over A such that $\models \varphi(x, y) \to \theta(x)$. Then there is a formula $\delta(y)$ over A such that for any c, $\models \delta(c)$ iff $\varphi(x, c)^{\mathbf{C}}$ is infinite.*

Proof. (i) By 5.2, $\mathrm{tp}(b/\mathrm{acl}(A))$ is either algebraic or of U-rank 1. In the latter case the forking extensions of it have to be algebraic.

(ii) Let M be an $(|A|^+ + (2^{|T|})^+)$-saturated model containing A. Let $\{p_\alpha\colon \alpha < 2^{|T|}\}$ be the non-algebraic types over M containing $\theta(x)$ (by 5.2(iv)). Again by 5.2(iv), each p_α does not fork over A. Thus for each α there is a formula $\delta_\alpha(y)$ over $\mathrm{acl}(A)$ such that for any $c \in M$, $\models \delta_\alpha(c)$ iff $\varphi(x, c) \in p_\alpha$. Now for any c in M, clearly $\varphi(x, c)^{\mathbf{C}}$ is infinite iff $\varphi(x, c) \in p_\alpha$ for some α. Compactness yields a finite disjunction $\delta(y)$ of some of the δ_α which is equivalent to the infiniteness of $\varphi(x, y)$. $\delta(y)$ is clearly A-invariant, so over A. ■

Remark 5.4. *Of course a similar (and easier) result holds if D is a minimal set defined over A (i.e. D is the solution set of a minimal partial type $\Phi(x)$ over A).* (i) *has the same proof. In this context,* (ii) *of* 5.3 *should read: for any formula $\varphi(x, y)$ there is $\delta(y)$ over A, such that for any c, $\models \delta(c)$ iff*

$D \cap \varphi(x, c)^{\mathbf{C}}$ is infinite. $\delta(y)$ will be simply the $\varphi(x, y)$ definition of the unique non-algebraic type $p(x) \in S(\mathrm{acl}(A))$ which extends $\Phi(x)$.

It follows from 5.3 and 5.4 that if D is a weakly minimal or minimal set defined over A, then the relation '$b \in \mathrm{acl}(C, A)$' for $b \in D$ and $C \subseteq D$ satisfies the Steinitz axioms for a dependence relation: (i) finite character (clear); (ii) transitivity: if $b \in \mathrm{acl}(C, A)$ and for all $c \in C$, $c \in \mathrm{acl}(E, A)$, then $b \in \mathrm{acl}(E, A)$ (also clear); (iii) if $b \in \mathrm{acl}((C \cup \{c\}, A)$ and $b \notin \mathrm{acl}(C, A)$, then $c \in \mathrm{acl}(C \cup \{b\}, A)$ (by 5.3(i), 5.4(i) and forking symmetry).

Thus for $B \subseteq D$, $\dim(B/A)$ = 'the cardinality of a maximal subset B_0 of B such that for all $b \in B_0$, $b \notin \mathrm{acl}(B_0 - \{b\}, A)$' is well-defined.

Alternatively, we could use the fact that for each $b \in D - \mathrm{acl}(A)$, $\mathrm{tp}(b/\mathrm{acl}(A))$ is regular, so by section 4.5 has weight 1. Then for $B \subseteq D$, let $B_1 = \{b \in B: b \notin \mathrm{acl}(A)\}$, and we see easily that $\dim(B/A)$ = the cardinality of a maximal A-independent subset of B_1 (see 4.4.1(iii)). (If D is minimal then there is a unique stationary non-algebraic type $p(x) \in S(A)$, say, extending '$x \in D$'. The notion '$\dim(p, M)$' has been already introduced at the end of section 4.5, and we see that if M is a model containing A, then $\dim(D^M/A) = \dim(p, M)$.)

LEMMA 5.5. *Let D be a weakly minimal or minimal set defined over A. Let $\mathbf{b}$ be a finite tuple of elements from D. Then* $\dim(\mathbf{b}/A) = U(\mathrm{tp}(\mathbf{b}/A))$.

Proof. We know that for a single element $b \in B$, $\dim(b/A) = 0$ iff $U(\mathrm{tp}(b/A)) = 0$ iff $b \in \mathrm{acl}(A)$, and also $\dim(b/A) = 1$ iff $U(\mathrm{tp}(b/A)) = 1$ iff $b \notin \mathrm{acl}(A)$. On the other hand, if $\mathbf{b}$, $\mathbf{c}$ are finite tuples from D then clearly $\dim(\mathbf{b}, \mathbf{c}/A) = \dim(\mathbf{b}/\mathbf{c}A) + \dim(\mathbf{c}/A)$. But by 3.26, the same is true for U-rank in place of dim. This yields the lemma. ■

In fact it is not difficult to see that in the weakly minimal case, dim $(\mathbf{b}/A)$ also equals $R^\infty(\mathrm{tp}(\mathbf{b}/A))$. In the case where D is strongly minimal, it is also equal to $\mathrm{RM}(\mathrm{tp}(\mathbf{b}/A))$. We will in fact prove a rather stronger result, including also the 'definability' of the ranks R^∞ and RM in suitable context (using essentially 5.3(ii)).

DEFINITION 5.6. *The theory T is said to be* unidimensional *if any for two non-algebraic stationary types p, q, p is non-orthogonal to q.*

In Chapter 8 we will see that unidimensional theories are superstable (although we could have made this observation much earlier). But for now we wish to make some observations about the behaviour of ranks in superstable unidimensional theories. Note that in any superstable theory there is a formula of ∞-rank 1. Similarly in any totally transcendental theory there is a formula with Morley rank 1. We work now in a slightly more general context. The following assumption will be in effect up to and including Corollary 5.14.

ASSUMPTION **5.7.** *T is superstable, $\theta(x)$ is a formula over $\varnothing$ with $R^\infty(\theta) = 1$, and for every non-algebraic type $p(y)$, there is some $q(x)$ containing θ such that p is non-orthogonal to q.*

LEMMA **5.8.** *For any complete type p, $U(p) < \omega$.*

Proof. If not, then clearly there is some complete type p with $U(p) = \omega$. By our assumptions p is non-orthogonal to some r containing $\theta(x)$. So after replacing p, r by non-forking extensions over a suitable set A (which preserves the U-ranks of these types) we have realizations a of p and b of r such that $\mathrm{tp}(a/bA)$ forks over A. By 3.18(ii), $U(\mathrm{tp}(a/bA)) < \omega$. Also of course $U(\mathrm{tp}(b/A)) = 1$. By Lemma 3.24, $U(\mathrm{tp}(ab/A)) < \omega$. (In fact a direct and simple application of Lemma 2.22 suffices here, that is $U(\mathrm{tp}(ab/A)) \leq U(\mathrm{tp}(a/Ab)) + 1$.) Thus $U(\mathrm{tp}(a/A)) < \omega$, a contradiction. ■

DEFINITION **5.9.** (For superstable theories) *If $\varphi(y)$ is any formula, then by $U(\varphi(y))$ we mean* $\sup\{U(p(y))$: *p a complete type containing $\varphi\}$.*

For arbitrary superstable theories, U-rank may not be continuous; that is, given a complete type p, there may be no formula $\varphi \in p$ with $U(p) = U(\varphi)$. However, we will see that in the *unidimensional* case, U-rank *will be* continuous.

LEMMA **5.10.** *Suppose $p(y) \in S(A)$, and $U(p) = n$. Then there is an L-formula $\varphi(y, z)$ and some $c \in A$ such that* (i) *$\varphi(y, c) \in p$, and* (ii) *for any $c' \in C$, if $\varphi(y, c')$ is consistent then $U(\varphi(y, c')) = n$. Moreover, for any L-formula $\psi(y, u)$ and $d \in A$ such that $\psi(y, d) \in p$, φ can be chosen so that z includes u and $\models \varphi(y, z) \rightarrow \psi(y, u)$.*

Proof. We prove the lemma by induction on n. For $n = 0$, everything is clear. Suppose $U(p) = n + 1$. Let M be an a-model containing A, and $p' \in S(M)$ a non-forking extension of p. So $U(p') = n + 1$. We first find a suitable formula $\varphi(y, x)$ for p'.

Now p' is non-orthogonal to r for some $r(x) \in S(M)$ containing θ, and M is an a-model, so by Lemma 4.3.1 there are realizations a of $p'(y)$ and b of $r(x)$ such that $\mathrm{tp}(a/bM)$ forks over M. Thus $\mathrm{tp}(b/aM)$ forks over M, whereby $b \in \mathrm{acl}(Ma)$. Lemma 3.24 implies that $U(\mathrm{tp}(a/bM)) = n$. By the induction hypothesis, there is an L-formula $\chi(y, x, z)$ and $c \in M$ such that $\models \chi(a, b, c)$, $U(\chi(y, b, c)) = n$, and whenever $\chi((y, b', c')$ is consistent then also $U(\chi(y, b', c')) = n$. We may also assume that for all a', c', $\chi(a', x, c')$ is algebraic (i.e. has at most finitely many solutions). Let $\psi(y, z)$ be the formula $(\exists x)(\theta(x) \wedge \chi(y, x, z))$. Using Lemma 5.3(ii), let $\delta(z)$ be an L-formula expressing 'there are infinitely many x such that $\theta(x) \wedge (\exists y)(\chi(y, x, z))$'. Let $\varphi(y, z)$ be the formula $\psi(y, z) \wedge \delta(z)$. Note that $\models \varphi(a, c)$, so $\varphi(y, c) \in p'(y)$.

We claim that $\varphi(y, z)$ works (for p').

Suppose that $\varphi(y, c')$ is consistent, and let $q(y)$ be any complete type over some $C \supseteq c'$ containing $\varphi(x, c')$. Let a' realize q. Then there is b' satisfying $\theta(x)$ such that $\models \chi(a', b', c')$. Now $U(\mathrm{tp}(b'/C)) \leq 1$, and (by choice of χ) $U(\mathrm{tp}(a'/b'C)) \leq n$ and $U(\mathrm{tp}(b'/a')) = 0$. Thus by 3.24, $U(\mathrm{tp}(a'/C)) \leq n + 1$.

On the other hand, given any C containing c', we can (as $\delta(c')$ holds) find b' satisfying $\theta(x)$ with $b' \notin \mathrm{acl}(C)$, and $\chi(y, b', c')$ consistent. Again by choice of χ we can find a' satisfying $\chi(y, b', c')$ such that $U(\mathrm{tp}(a'/b'C)) = n$. This time we have $U(\mathrm{tp}(b'/C)) = 1$, and still $b' \in \mathrm{acl}(a'C)$, whereby 3.24 yields $U(\mathrm{tp}(a'/C)) = n + 1$.

We have shown that $U(\varphi(y, c')) = n$ whenever $\varphi(y, c')$ is consistent. To complete the proof we must find a suitable formula for p. Let (by 2.17′) $\varphi'(y)$ be a formula in $p(y)$ which is a positive Boolean combination of A-conjugates of $\varphi(y, c)$. Write $\varphi'(y)$ as $\varphi'(y, d)$ where $\varphi'(y, w) \in L$ and $d \in A$. So $\varphi'(y, d)$ is equivalent to $\bigwedge_i\{\bigvee_j\{\phi(y, c_{ij})\}\}$ for some finite set of i and j indices, and where $\mathrm{tp}(c_{ij}/A) = \mathrm{tp}(c/A)$ for all i, j. Moreover, we may clearly assume (by adjoining $\varphi'(y, d)$ to each $\varphi(y, c_{ij})$) that $\varphi(y, c_{ij})$ implies $\varphi'(y, d)$ for all i, j.

Let $\varepsilon(w)$ be the formula

$$(\exists z_{ij})_{ij}(\textstyle\bigwedge_{ij}\{(\exists y)\varphi(y, z_{ij})\} \wedge (\forall y)(\varphi'(y, w)$$
$$\leftrightarrow \textstyle\bigwedge_i\{\bigvee_j\{\varphi(y, z_{ij})\}\}) \wedge \bigwedge_{i,j}(\forall y(\varphi(y, z_{ij}) \rightarrow \varphi'(y, w)).$$

Thus the formula $\varphi'(y, d) \wedge \varepsilon(d) \in p(y)$. So all we have to show is that $U(\varphi'(y, d') \wedge \varepsilon(d')) = n + 1$ whenever $\varphi'(y, d') \wedge \varepsilon(d')$ is consistent. Suppose this formula to be consistent. Choose c'_{ij} given by ε. Let a' satisfy $\varphi'(y, d')$ and we may suppose a' is independent from all the c'_{ij} over d'. Then $\models \phi(a', c'_{ij})$ for some i, j, whereby $U\mathrm{tp}(a'/c_{ij})) \leq n + 1$ by the above. Thus $U(\mathrm{tp}(a'/d')) \leq n + 1$. On the other hand, choose c'_{ij} such that $\varphi(y, c'_{ij})$ is consistent, and by the above we can find a' satisfying $\varphi(y, c_{ij})$ with $U(\mathrm{tp}(a'/c'_{ij})) = n + 1$. We may assume a' is independent from d' over c'_{ij}, so $U(\mathrm{tp}((a'/d')) \geq n + 1$. By definition of ε, $\models \varphi(a', d')$. We have shown that $U(\varphi'(y, d')) = n + 1$ as required.

The lemma is proved (the moreover clause is clear). ■

Corollary **5.11.** (Boundedness and definability of U-rank) (i) *For any sort*

S (in T^{eq}) there is a finite bound to $\{U(p(y)): p(y)$ a complete type$\}$, where the variable y is of sort U.

(ii) *For any L-formula $\varphi(y, z)$ and $n < \omega$ there is an L-formula $\psi(z)$ such that for any $c \in \mathbf{C}$, $\models \psi(c)$ iff $U(\varphi(y, c)) = n$.*

Proof. (i) For each complete type $p(y)$ over $\varnothing$, there is by Lemma 5.10 a formula $\varphi(y) \in p$, such that $U(\varphi(y)) = U(p(y)) = n_\varphi$, say. By compactness there is a finite subset Φ of these formulae such that $\models S(y) \leftrightarrow \vee \{\varphi(y): \varphi \in \Phi\}$. Clearly then $\max\{n_\varphi: \varphi \in \Phi\}$ is the required bound.

(ii) For each complete type $p(y, z)$ over $\varnothing$ containing $\varphi(y, z)$, we can by Lemma 5.10 find a formula $\psi(p)(y, z)$ in p, and $n(p) \in \omega$, such that

(a) $\models \psi(y, z) \rightarrow \varphi(y, z)$;

(b) for some (any) c such that $p(y, c)$ is consistent, $U(p(y, c)) = n(p)$;

(c) for any c, if $\psi(y, c)$ is consistent then $U(\psi(y, c)) = n(p)$.

Again by compactness, $\varphi(y, z)$ is equivalent to the disjunction of a finite number of such formulae, say $\psi(p_1)(y, z), \ldots, \psi(p_m)(y, z)$. Then for any c, and $k < \omega$, $U(\varphi(y, c)) = k$ iff $k = \max\{n(p_i): (\exists y)(\psi(p_i)(y, c))\}$. ■

LEMMA **5.12.** (i) *For any complete type p, $U(p) = R^\infty(p)$.*

(ii) *If $RM(\theta(x)) = 1$, then T is totally transcendental and for all complete types p, $U(p) = RM(p)$.*

Proof. (i) We start by pointing out

Claim. (Which is true for *any* stable theory) For any formula $\varphi(y)$, if $U(\varphi(y)) = n < \omega$ then $R^\infty(\varphi(y)) = n$.

Proof of claim. By Lemma 3.20 we always have $U(\varphi) \leq R^\infty(\varphi)$. So it suffices to prove that $U(\varphi(y)) = n$ implies $R^\infty(\varphi(y)) \leq n$, which we do by induction on n. Suppose $U(\varphi(y)) = n + 1$, but that $R^\infty(\varphi(y)) > n + 1$. Thus there is a formula $\psi(y)$ such that $\models \psi(y) \rightarrow \varphi(y)$, $\psi(y)$ forks over A for some A containing the parameters in φ, and $R^\infty(\psi) > n$. By the induction hypothesis $U(\psi) > n$. Let B be a set containing A such that ψ is over B, and let $q(y) \in S(B)$ be a type which contains $\psi(y)$ and with $U(q) > n$. But q also contains $\varphi(y)$. Thus $U(q) = n + 1$, and also $U(q \restriction A) = n + 1$. So q does not fork over A, a contradiction.

(i) Now follows from the claim and Lemma 5.10 (i.e. the fact that for any complete $p(x)$ there is a formula φ in p such that $U(\varphi) = U(p) < \omega$).

(ii) Let us assume now that $RM(\theta(x)) = 1$.

We start with an easy exercise which is left to the reader.

Claim. (In any theory) Suppose $\theta(x)$, $\chi(y, x)$ are formulae (with parameters maybe) such that $RM(\theta) = 1$, $RM(\chi(y, b)) = n$, for all b satisfying θ, and $\chi(a, x)$ is algebraic (or inconsistent) for all a. Let $\varphi(y)$ be the formula $(\exists x)(\theta(x) \wedge \chi(y, x))$. Then $RM(\varphi(y)) = n + 1$.

Given the claim, we prove by induction on n that $U(p) = n$ implies $RM(p) = n$.

Suppose $U(p) = n + 1$. Without loss p is over an a-model M. As in the proof of Lemma 5.10, there is $b \notin M$ satisfying $\theta(x)$, and a realizing $p(y)$ such that $b \in \mathrm{acl}(aM)$ and $U(\mathrm{tp}(a/bM)) = n$. By Lemma 5.10 there is a formula $\chi(y, x)$ over M such that $\models \chi(a, b)$ and for all b', $\chi(y, b')$ has U-rank n, if consistent. By the induction hypothesis, $RM(\chi(y, b')) = n$, whenever $\chi(y, b')$ is consistent. We may assume that $\models \theta(x) \rightarrow (\exists y)(\chi(y, x))$. Let $\varphi(y)$ be $(\exists x)(\theta(x) \wedge \chi(y, x))$. By the claim, $RM(\varphi(y)) = n + 1$. Now $\varphi(y)$ could easily have been chosen to imply any given formula in p. Thus $RM(p) = n + 1$, completing the proof of (ii). ■

COROLLARY 5.13. *$R^\infty(\varphi) < \omega$ for all φ. Moreover, for any L-formula $\varphi(y, z)$ and $n < \omega$ there is an L-formula $\psi(z)$ such that for any c, $\models \psi(c)$ iff $R^\infty(\varphi(x, c)) = n$. Similarly for RM in place of R^∞, in the case where $RM(\theta(x)) = 1$.*

We now drop Assumption 5.7.

PROPOSITION 5.14. *Suppose T is superstable and unidimensional. Then*

(i) *For any complete type $U(p) = R^\infty(p) < \omega$ (so in particular $R^\infty(\varphi) < \omega$ for any formula φ).*

(ii) *For each formula $\varphi(y, z)$ of L and $n < \omega$ there is an L-formula $\psi(z)$ such that, for any $d \in \mathbf{C}$, $\models \psi(d)$ iff $R^\infty(\varphi(y, d)) = n$.*

(iii) *If for some non-algebraic formula $\theta(x)$ (possibly with parameters), $RM(\theta) < \infty$,* then *T is totally transcendental and* (i), (ii) *also hold with RM in place of R^∞.*

Proof. Let $\theta(x, c)$ be a weakly minimal formula. Then Assumption 5.7 is valid for T' where $T' = Th(\mathbf{C}, c)$ (i.e. after adding constants for c to the language). Thus 5.12 and 5.13 are true for T'. It is clear (by taking non-forking extensions over c) that 5.12 and the finiteness of R^∞ remain true for T. The only trouble is definability of rank *over* $\varnothing$. But this is dealt with easily as follows. Let $q = \mathrm{tp}(c/\mathrm{acl}(\varnothing))$. Let $\varphi(y, z)$ be any L-formula and $n < \omega$. By 5.13 let $\psi(z)$ be a formula over c such that for every d, $\models \psi(d)$ iff $R^\infty(\varphi(y, d)) = n$.

Write $\psi(z)$ as $\psi(z, c)$, with $\psi(z, w) \in L$. Let $\delta(z)$ be the $\psi(z, w)$-definition of $\mathrm{tp}(c/\mathrm{acl}(\emptyset))$. It is easy to check that for any d, $\models \delta(d)$ iff $R^\infty(\varphi(y, d)) = n$. Now $\delta(z)$ is over $\mathrm{acl}(\emptyset)$, but can be replaced by the disjunction of its conjugates over $\emptyset$.

The above discussion shows (i) and (ii) of the proposition, as well as (iii), if $RM(\theta(x, c)) = 1$. But then (iii) follows, for clearly if some formula has Morley rank $< \infty$, then some formula $\theta(x, c)$ has Morley rank 1. ∎

Definition 5.15. *Let D be some A-definable set. By $(D_A)^{\mathrm{eq}}$ we mean $\{c \in \mathbf{C}^{\mathrm{eq}}: c \in \mathrm{dcl}(D \cup A)\}$. If D is defined by the formula $\theta(x)$ over A, we may sometimes refer to $(D_A)^{\mathrm{eq}}$ as $(\theta_A)^{\mathrm{eq}}$.*

As essentially a special case of the previous analysis we have:

Proposition 5.16. *Suppose $\theta(x)$ is a weakly minimal formula over A, say (inside a stable but not necessarily superstable theory). Then*

(i) *For any $b \in (\theta_A)^{\mathrm{eq}}$ and $B \supseteq A$, $U(\mathrm{tp}(b/B)) = R^\infty(\mathrm{tp}(b/B)) < \omega$.*

(ii) *For any L_A-formula $\varphi(y, z)$ which implies '$y \in (\theta_A)^{\mathrm{eq}}$', and $n < \omega$, there is some L_A-formula $\psi(z)$ such that for any $d \in \mathbf{C}$, $\models \psi(d)$ iff $R^\infty(\varphi(y, d)) = n$.*

If $\theta(x)$ is of Morley rank 1, then (i) *and* (ii) *are true with RM in place of R^∞.*

Proof. This is not a purely formal consequence of 5.12, 5.13, for Assumption 5.7 is not necessarily true here. However, we *are* here concerned just with types and definable sets (of elements in $(\theta_A)^{\mathrm{eq}}$), all of whose extensions are non-orthogonal to some type containing θ, and which can be seen from the start to have finite U-rank. The proofs of results 5.8 to 5.13 apply in such a context. ∎

Remark 5.17. (i) *It is worth noting the* dimension-theoretic *significance of U-rank/R^∞ under the hypotheses of* 5.16. *Suppose $A = \emptyset$, for simplicity (so θ is weakly minimal, defined over $\emptyset$). Suppose $c \in \theta^{\mathrm{eq}}$ and B is some set of parameters. Then $c \in \mathrm{dcl}(a_1, \ldots, a_n)$ for some $a_1, \ldots, a_n$ satisfying θ. Choose a maximal subtuple $\mathbf{b}$ of $\{a_1, \ldots a_n\}$, such that $\mathrm{tp}(\mathbf{b}/cB)$ does not fork over B. Then $\mathrm{tp}(c/\mathbf{b}B)$ does not fork over B. Also (as every element of θ has U-rank* 0 *or* 1) *it should be clear that for every $a_i \notin \mathbf{b}$, $a_i \in \mathrm{acl}(c\mathbf{b}B)$. Thus $(a_1, \ldots, a_n) \in \mathrm{acl}(c\mathbf{b}B)$ and of course $c \in \mathrm{acl}(a_1, \ldots, a_n, \mathbf{b}B)$. So $U(\mathrm{tp}(c/B)) = U(\mathrm{tp}(c/\mathbf{b}B)) = U(\mathrm{tp}(a_1, \ldots, a_n)/\mathbf{b}B)) = \dim(a_1, \ldots, a_n/\mathbf{b}B)$.*

(ii) *Proposition* 5.16(ii) *has a much simpler proof in the case where y is a tuple of variables $(x_1, \ldots, x_n)$ and $\models \varphi(y, z) \to \theta(x_i)$ for $i = 1, \ldots, n$. The reader* *is invited to find such a proof, but we give the main point (corresponding to Lemma 5.10.) Assume θ is over $\emptyset$. Suppose $a_1, \ldots, a_n$ satisfy θ, $\dim(a_1, \ldots, a_k/B) = k$, and $a_{k+1}, \ldots, a_n \in \operatorname{acl}(a_1, \ldots, a_k, B)$. Thus $\dim(a_1, \ldots, a_n/B) = k$. We exhibit a formula witnessing this, uniformly in the parameters (as in Lemma 5.10).*

Let $\varphi(x_1, \ldots, x_n)$ be a formula over B true of $(a_1, \ldots, a_n)$ such that $\models \varphi(x_1, \ldots, x_n) \rightarrow x_i \in \operatorname{acl}(Bx_1, \ldots, x_k)$, for $i = k+1, \ldots, n$. Then clearly $\dim(\varphi(x_1, \ldots, x_n)) = k$. Suppose the formula φ has parameter $c \in B$. Rewrite φ as $\varphi(x_1, \ldots, x_n, c)$ where $c \in B$ and $\varphi(x_1, \ldots, x_n, y)$ is an L-formula. We may assume, by strengthening φ, that $\models \varphi(x_1, \ldots, x_n, y) \rightarrow x_i \in \operatorname{acl}(x_1, \ldots, x_k, y)$ for each $i = k+1, \ldots, n$. By Lemma 5.3(ii), the following condition on y, (for infinitely many x_1 (for infinitely many x_2(...(for infinitely many $x_k(\exists x_{k+1}, \ldots, x_n(\varphi(x_1, \ldots, x_n))))\ldots)))$, is expressible by a formula $\delta(y)$ over $\emptyset$. We see then that for any c', the formula $\varphi(x_1, \ldots, x_n, c') \wedge \delta(c')$ has dimension (or ∞-rank) k, if consistent.

If D is a *minimal* set (defined by a partial type over $\emptyset$), we can also talk about D^{eq}. All types here will have finite U-rank. Although ∞-rank and Morley rank may not apply, the theory (definability of rank, multiplicity, etc.) is in fact very similar to the case where D is a *strongly minimal* definable set, as long as one is willing to deal with 'relatively definable' subsets of D^n, or D^{eq}, rather than definable sets. Details will be given in Chapter 2.

We finally recall some classical results linking the above notions to *categoricity*. Remember that a theory is said to be λ-categorical if it has a unique model of cardinality λ, up to isomorphism.

THEOREM 5.18. *Suppose T is λ-categorical for some $\lambda > |T|$. Then T is superstable and unidimensional. If, moreover, T is countable, then T is also totally transcendental (ω-stable).*

Proof. We give an outline, although we need at some point to refer to a theorem of Shelah.

CLAIM I. *Suppose that T is an arbitrary (not necessarily stable) complete theory. Then for any $\lambda > |T|$, T has a model M such that*

(i) *$|M| = \lambda$; and*

(ii) *for any subset A of M of cardinality $|T|$, at most $|T|$ many complete types over A are realized in M.*

Sketch of proof. The idea is to expand T to a complete theory T' of the same cardinality as T which has Skolem functions, that is such that whenever

$\varphi(x, y)$ is a formula of $L' = L(T')$ then there is some function symbol $g(\mathbf{y})$ of L' such that $T' \models (\forall \mathbf{y})(\exists x)(\varphi(x, \mathbf{y}) \rightarrow \varphi(g(\mathbf{y}), \mathbf{y}))$. So for any subset A of a model M' of T', $\mathrm{dcl}(A)$ is the universe of an elementary substructure of M'. Now for any $\lambda > |T|$, let $(a_i : i < \lambda)$ be an indiscernible sequence in some model of T' (by compactness). Let M' be $\mathrm{dcl}(a_i : i < \lambda)$. Then M' has cardinality λ, and the reduct of M' to a model of T has the desired property, proving the claim. ■

Let us now assume that T is λ-categorical for some $\lambda > |T|$. It follows from Claim I that T must be $|T|$-stable. By Lemma 4.1.1 T is stable and, moreover, totally transcendental (and so superstable) of T is countable. If T is uncountable, we must refer to Chapter VIII of [Sh1] where it is shown that any non-superstable theory actually has 2^λ models of cardinality λ for any $\lambda > |T|$. (The argument is a little easier in the case where λ is regular.)

In any case we now have:

Claim II. *T is superstable, $|T|$-stable (and totally transcendental if T is countable).*

It follows in particular that for any finite set A there are at most $|T|$ many strong types over A. Thus (using $\kappa(T) = \omega$) we have:

Claim III. *Any model M of T is contained in an a-model M' with $|M'| = |M| + |T|$.*

We need finally to show that T is unidimensional. *Suppose not*, and let p, q be orthogonal types. By Claim III we may assume that both $p, q \in S(M)$ where M is an a-model of cardinality $|T|$. Fix $\lambda > |T|$. Let $(a_i : i < \lambda)$ be a Morley sequence in p of cardinality λ. Let N be the a-prime model over $M \cup \{a_i : i < \lambda\}$. Then $|N| = \lambda$ (by Claim III again). Let $b \in N$ be such that q does not fork over b. Let $q_0 = q \upharpoonright \mathrm{acl}(b)$.

Claim IV. *N does not contain any Morley sequence $B = (b_i : i < \lambda)$ in q_0.*

Proof. Now for each tuple c from M there is (by superstability) some finite $B_0 \subseteq B$ such that $\mathrm{tp}(c/bB)$ does not fork over bB_0. Thus as $|M| = |T| < \lambda$, there is some set $B_1 \subseteq B$ with $|B_1| < \lambda$ such that M is independent from bB over bB_1. In particular B is independent from M over B_1. But as $|B| = \lambda$ there is $b_\alpha \in B - B_1$ so clearly $\mathrm{tp}(b_\alpha/M) = q$. But $b_\alpha \in N$, so by Lemma 4.3.4, p dominates q, which clearly contradicts the orthogonality of p to q. This proves Claim IV. ■

Finally, it is easy to construct (using the remark before Claim III), for any $\lambda > |T|$, a model M_λ of T of cardinality λ, such that for any finite $c \in M$ and any type $p(x) \in S(\mathrm{acl}(c))$, M_λ contains a Morley sequence of length λ in p. Choosing λ such that T is λ-categorical, this gives a contradiction to Claim IV. This proves that T was unidimensional after all. ■

THEOREM 5.19. *Suppose T is countable and λ-categorical for some uncountable λ. Then*

(i) *T is λ-categorical for* every *uncountable λ; and*

(ii) *either T is ω-categorical or T has exactly ω countable models (up to isomorphism of course).*

Theorem 5.19(i) is Morley's theorem, and Theorem 5.19(ii) is the Baldwin–Lachlan theorem. Shelah generalized (i) to uncountable theories, and Laskowski, using the 'geometric' theory in an essential way, generalized (ii) to uncountable theories. Part of Chapter 6 will be devoted to an exposition of these latter results (given in a way that includes Theorem 5.19 as a special case).

6 Stable groups

By a *stable* group is usually meant a group definable in (a model of) a stable theory. It is convenient to study more generally groups defined by (partial) types rather than simply formulae. We call these '∞-definable groups'. Also there is not much additional work in developing the general theory for *homogeneous spaces.*

As usual we are working in the monster model $\mathbf{C}$ of a stable theory.

DEFINITION 6.1. *Let $\Phi(y)$, $\Psi(x)$, $\Theta(y_1, y_2, y_3)$, $\Xi(y_1, x_1, x_2)$ be partial types over some (small) set A.*

We say that $(\Phi^{\mathbf{C}}, \Theta^{\mathbf{C}}, \Psi^{\mathbf{C}}, \Xi^{\mathbf{C}})$ is an ∞-definable homogeneous space, defined over A, if

(i) *$\Theta^{\mathbf{C}}$ is the graph of a group operation on $\Phi^{\mathbf{C}}$; and*

(ii) *$\Xi^{\mathbf{C}}$ is the graph of a transitive group action of the group $\Phi^{\mathbf{C}}$ on the set $\Psi^{\mathbf{C}}$.*

If the partial types Φ, Θ, Ψ, Ξ are each (equivalent to) *formulae* φ, θ, ψ, ξ, over A, then we speak just of a *definable homogeneous space, defined over A* (or A-definable homogeneous space).

Similarly the data $(\Phi^{\mathbf{C}}, \Theta^{\mathbf{C}})$ is called an ∞-definable group, defined over

A, and simply an A-definable group if Φ, Θ are equivalent to formulae. This can (and will) be also considered as a (∞-definable) homogeneous space with respect to the action of G on itself by left translation.

REMARKS 6.2. (i) *Recall that a transitive action of a group G on a set S is a map from $G \times S$ to S taking (g, s) to $g.s$) such that $g(h(s)) = gh(s)$ for all $g, h \in G$ and $s \in S$, $1.s = s$ for all $s \in S$, and such that for all $s_1, s_2 \in S$ there is $g \in G$ such that $g.s_1 = s_2$.*

(ii) *We usually denote an ∞-definable homogeneous space above by (G, S) where $G = \Phi^{\mathbf{C}}$, $S = \Psi^{\mathbf{C}}$ (and Θ, Ξ are understood). Both the group operation on G and the action of G on S will de denoted $g.h$, $g.s$ if there is no ambiguity. In fact we may even sometimes speak of the (∞-definable) homogeneous space S (in which case the existence of a suitable G is presupposed). Similarly ∞-definable groups are denoted G, H, etc.*

(iii) *If $(\Phi^{\mathbf{C}}, \Theta^{\mathbf{C}}, \Psi^{\mathbf{C}}, \Xi^{\mathbf{C}})$ is a ∞-definable homogeneous space defined over A, then there are, by compactness, formulae $\theta(y_1, y_2, y_3)$ and $\xi(y_1, x_1, x_2)$ over A such that*

(a) $\models \forall y_1 y_2 \exists^{\leq 1} y_3(\theta(x_1, x_2, x_3))$, *and* $\models \forall y_1 x_1 \exists^{\leq 1} x_2(\xi(y_1, x_1, x_2))$; *and*

(b) $\Theta^{\mathbf{C}} = \theta^{\mathbf{C}} \cap (\Phi^{\mathbf{C}})^3$, *and* $\Xi^{\mathbf{C}} = \xi^{\mathbf{C}} \cap (\Psi^{\mathbf{C}})^3$.

(iv) *If the action of G on S is regular (i.e. for $a, b \in S$ there is unique $g \in G$ such that $g.a. = b$), then we call (G, S) an (∞-definable)* principal homogeneous space. *Note that the action of G on itself by left translation (and by right translation) makes G into a principal homogeneous space.*

We now fix some ∞-definable homogeneous space (G, S) as above, defined over $\varnothing$. We identify the expression $S(x)$ with the partial type $\Psi(x)$ which defines S, and similarly for $G(y)$. Let θ, ξ be as in Remark 6.2(iii). If $\theta(b_1, b_2, b_3)$ holds we will write $b_1.b_2 = b_3$, and say that $b_1.b_2$ is defined. Similarly for ξ. By compactness there is some formula $\varphi_0(y)$ in $\Phi(y)$ and there is some formula $\psi_0(x)$ in $\Psi(x)$ such that whenever y_1, y_2, y_3 satisfy φ_0, and x satisfies ψ_0, then $y_1.(y_2.y_3) = (y_1.y_2).y_3$, $y_1.1 = 1.y_1 = y_1$, $y_1.(y_2.x) = (y_1.y_2).x$, and $1.x = x$.

We are interested in types $p(x)$ extending $S(x)$, and the effect of the group action on forking. If $\varphi(x, z)$ is a formula, then by $\varphi(y.x, z)$ we really mean the formula '$\varphi_0(y) \wedge \psi_0(x) \wedge \varphi(y.x, z)$'. Similarly for a formula $\varphi(y_1.y, z)$.

By a *relatively* definable subset of S, we mean a set of the following kind: $\{a \in S : \models \psi(a)\}$ where ψ is some formula.

DEFINITION 6.3. (i) *A relatively definable subset X of S is said to be* generic *(with respect to the given transitive action of G on S) if there are $g_1, \ldots, g_n$ in G such that $g_1.X \cup \cdots \cup g_n.X = S$.*

(ii) *If* $\mathbf{p}(x) \in S(\mathbf{C})$ *extends* $S(x)$ *then we say that* $\mathbf{p}$ *is generic iff every relatively definable subset of S in* $\mathbf{p}$ *is generic.*

LEMMA **6.4.** *Let X be a relatively definable subset of S. Then*

(i) *X or* $S - X$ *is generic;*

(ii) *X is non-generic iff there is an indiscernible (over the parameters defining X) set* $\{g_i : i < \omega\}$ *in G such that* $\{g_i.X : i < \omega\}$ *is inconsistent.*

Proof. We introduce an auxiliary structure $\mathbf{C}_0 = (S, G, R)$, where R is the binary relation defined as $R(x, y)$ if $x \in S$, $y \in G$, and $x \in y.X$. It is clear that the formula $R(x, y)$ is stable in $\mathbf{C}_0$. Let $T_0 = Th(\mathbf{C}_0)$. Moreover, for any $g \in G$, the map which takes $s \in S$ to $g.s$ and $h \in G$ to $g.h.$ is clearly an *automorphism* of $\mathbf{C}_0$, whereby there is a *unique* type over $\varnothing$ realized in S (in the structure $\mathbf{C}_0$). Let *1* denote the identity element of G. So $R(x, 1)$ defines the set X.

CLAIM. *X is generic iff the formula* $R(x, 1)$ *does not fork over* $\varnothing$ *in* (T_0).

Proof of claim. By Lemma 2.16(ii) applied to the theory T_0 and formula $R(x, y)$ we see (by the above remarks) that $R(x, 1)$ does not fork over $\varnothing$ iff some positive Boolean combination of formulae of the form $R(x, g)$ is consistent and $\varnothing$-definable in $\mathbf{C}_0$. But any consistent $\varnothing$-definable subset of S (in $\mathbf{C}_0$) must be S itself. So $R(x, 1)$ does not fork over $\varnothing$ iff $R(x, g_1) \vee R(x, g_2) \vee \cdots \vee R(x, g_n)$ is equivalent to $S(x)$ for some $g_1, \ldots, g_n$ in G. The latter condition means precisely that X is generic, proving the claim. ■

Clearly the same proof shows that $S - X$ is generic iff $\neg R(x, 1)$ does not fork over $\varnothing$ in T_0. But either $R(x, 1)$ or $\neg R(x, 1)$ forks over $\varnothing$ (in T_0). Thus part (i) of the lemma is proved.

Using the claim and part (iv) of Lemma 2.16, we see that X is non-generic iff in some elementary extension $\mathbf{C}_0'$ of $\mathbf{C}_0$, there is an indiscernible sequence $(g_i : i < \omega)$ such that $\{R(x, g_i) : i < \omega\}$ is inconsistent. As $\mathbf{C}$ is a saturated model of T, this is clearly equivalent to there being a sequence $(g_i : i < \omega)$ in G which is indiscernible over the parameters defining X, such that $\{g_i.X : i < \omega\}$ is inconsistent. This proves (ii). ■

It is convenient for the next result to introduce another auxiliary structure $\mathbf{C}_1$ which we define as follows: first for each formula $\varphi(x, z)$, let Γ_φ be the following set $\{S \cap \varphi(g.x, b)^{\mathbf{C}} : g \in G, b \in \mathbf{C}\}$ of relatively definable subsets of S, and let ε_φ denote the membership relation between S and Γ_φ. Let $\mathbf{C}_1$ be the many-sorted structure $(S, \{\Gamma_\varphi, \varepsilon_\varphi : \varphi(x, z)$ an L-formula$\})$. So the sorts of

$\mathbf{C}_1$ are S and the Γ_φ, and the relations are the ε_φ. Let T_1 be $Th(\mathbf{C}_1)$. T_1 need not be stable (although it will be if both G, S are defined by *formulae* in T). Let us note that there is a natural identification ($\mathbf{p} \leftrightarrow \mathbf{p}'$) of the set of complete *global* types over $\mathbf{C}$ which extend $S(x)$ with the set of complete quantifier-free types over $\mathbf{C}_1$ containing $S(x)$. That is, if $\mathbf{p}(x) \in S(\mathbf{C})$ and $\mathbf{p}(x) \vDash S(x)$, then the corresponding type $\mathbf{p}'(x) \in S(\mathbf{C}_1)$ contains the formula $x\varepsilon_\varphi X$ (for $X \in \Gamma_\varphi$) whenever $\mathbf{p}(x) \vDash$ '$x \in X$'.

REMARK 6.5. (i) *Each formula ε_φ is stable for the theory T_1.*

(ii) *For each $g \in G$, the map which takes $x \to g.x$ for $x \in S$, and $X \to g.X$ for X in any Γ_φ, is an automorphism of $\mathbf{C}_1$.*

(iii) *In the structure $\mathbf{C}_1$, all elements in S have the same type over $\varnothing$.*

Proof. (i) is clear as each formula $\varphi(y.x, z)$ $(=\varphi'(x; z, y))$ is stable in $\mathbf{C}$.

(ii) is simply a matter of checking.

(iii) follows from (ii) as G acts transitively on X. ■

We restrict our attention to types $\mathbf{p}(x) \in S(\mathbf{C})$ extending $S(x)$ (and we also consider the associated types $\mathbf{p}'(x) \in S(\mathbf{C}_1)$). Note that G acts on the set of such types; this can be seen in one of two ways. We define $g.p$ to contain exactly those relatively definable subsets of S of the form $g.X$, where X is some relatively definable subset of S in p. Alternatively, let the element a realize $p(x)$ in some elementary extension $\mathbf{C}_1$ of $\mathbf{C}$ and define $g.p$ to be $\operatorname{tp}(g.a/\mathbf{C})$.

PROPOSITION 6.6. *Let $Y = \{\mathbf{p}(x) \in S(\mathbf{C})$: $\mathbf{p}$ is generic$\}$. Then*

(i) *Y is non-empty.*

(ii) *G acts transitively on Y.*

(iii) *$\mathbf{p}(x) \in Y$ iff for all $g \in G$, $g.\mathbf{p}$ does not fork over $\varnothing$.*

(iv) *Let G^0 (*the connected component of G*) be the intersection of all relatively $\varnothing$-definable subgroups of G of finite index. Then for any $g \in G^0$ and $\mathbf{p} \in Y$, $g.\mathbf{p} = \mathbf{p}$.*

Proof. Let $Y_1 = \{\mathbf{p}'(x) \in S(\mathbf{C}_1)$: for each $\varphi(x, z)$, $\mathbf{p}'(x) \restriction \varepsilon_\varphi$ does not fork over $\varnothing$ (in T_1)$\}$. By Corollary 2.9 and Lemma 2.18, Y_1 is non-empty. By 6.5(iii) there is a unique 1-type over $\varnothing$ containing $S(x)$ (in $\mathbf{C}_1$), say p_0'. For each $\varphi(x, z)$, let E_φ be the finite equivalence relation given by Lemma 2.11(ii) (for the formula ε_φ and the type $p_0 \restriction \varepsilon_\varphi$). Then E_φ is $\varnothing$-definable in $\mathbf{C}_1$ (so G-invariant, by Remark 6.5(ii)) and also relatively $\varnothing$-definable in $\mathbf{C}$. Thus (as G acts transitively on S), G acts transitively on the E_φ-classes, and thus (by Lemma 2.11(ii)), G acts transitively on $\{\mathbf{p} \restriction \varepsilon_\varphi : \mathbf{p}' \in Y_1\}$. Similarly if E_1 is the intersection of finitely many of the equivalence relations E_φ, G acts transitively on the E_1-classes. By compactness (applied in $\mathbf{C}$) and the fact that each E_φ is relatively definable in $\mathbf{C}$, we see that G acts transitively on the E-classes, where E is the intersection of all the E_φ. Thus G acts transitively on Y_1. So, at least to prove (i) and (ii) it suffices to show that:

Claim. $Y_1 = \{\mathbf{p}'(x) \in S(\mathbf{C}_1) : \mathbf{p}(x) \in Y\}$.

Proof. We must show that a relatively definable subset X of S is generic iff X is contained in some $\mathbf{p}' \in Y_1$. Suppose first X to be generic. As finitely many G-translates of X cover S, some $g.X$ must be contained in some $\mathbf{p}' \in Y_1$. But then $g^{-1}.g.X \in g^{-1}.\mathbf{p}'$, and the latter type is in $\mathbf{Y}_1$. On the other hand, suppose $X \in \mathbf{p}'$, and $\mathbf{p}' \in Y_1$. Suppose $X \in \Gamma_\varphi$. We have already seen that $\{\mathbf{p}' \restriction \varepsilon_\varphi : \mathbf{p}' \in Y_1\}$ is finite, and that G acts transitively on this set. Thus some finite union of G-translates, say X', of X is in every $\mathbf{p}'$ in Y_1. By what we have just proved the complement of X' could not be generic. By Lemma 6.4, X' is generic, whereby clearly so is X. This proves the claim, and also (i) and (ii) of the proposition.

(iii) Suppose $\mathbf{p}(x)$ is generic. Then by the claim $\mathbf{p}'(x) \in Y_1$ so does not fork over $\varnothing$ in T_1. By 2.20(iv), $\mathbf{p}'$ has at most $2^{|T|}$ images under automorphisms of $\mathbf{C}_1$. But any automorphism of $\mathbf{C}$ induces an automorphism of $\mathbf{C}_1$. Thus $\mathbf{p}$ has at most $2^{|T|}$ images under automorphisms of $\mathbf{C}$. By 2.20(vii), $\mathbf{p}$ does not fork over $\varnothing$. For the same reason $g.\mathbf{p}$ does not fork over $\varnothing$ for all $g \in G$.

Conversely if $\mathbf{p}$ is not generic, then some relatively definable subset X of S in $\mathbf{p}$ is non-generic. By Lemma 6.4(ii) we can find a sequence $(g_i : i < \kappa = (2^{|T|})^+)$ from G such that $\{g_i.X : i < \kappa\}$ is k-inconsistent (for some $k < \omega$). Thus among $\{g_i.\mathbf{p} : i < \kappa\}$ there are κ distinct types. By 2.16, at least one of these forks over $\varnothing$. This concludes the proof of the proposition.

(iv) For any L-formula $\varphi(x, z)$ let E_φ be the finite equivalence relation on S defined at the beginning of this proof. As remarked there, E_φ is relatively $\varnothing$-definable and G-invariant. Let $G^0_\varphi = \mathrm{Fix}(S/E_\varphi)$. Then clearly G^0_φ is a (normal) relatively $\varnothing$-definable subgroup of G of finite index. If $g \in G^0$ then $g \in G^0_\varphi$ for all φ, so by the above remarks (and 2.11), $g.\mathbf{p} = \mathbf{p}$ for every $\mathbf{p} \in Y$. ∎

Remark 6.7. (i) *The proof above shows that* (a) *a relatively definable subset X of S is generic if and only if for all $g \in G$, $g.X$ does not fork over $\varnothing$, and* (b) *if X, Y are relatively definable subsets of S*) *and $X \cup Y$ is generic, then one of X, Y is generic.*

(ii) *Although Proposition 6.6 was stated for complete types over* $\mathbf{C}$, *it is clearly valid for complete types* $p(x) \in S(M)$ *containing* $S(x)$ *where* M *is any* $|T|^+$*-saturated model.*

(iii) *By looking at the definition of generic formulae we see that the notion of a generic type* $\mathbf{p}(x) \in S(\mathbf{C})$ *of* S *is invariant under naming parameters in* $\mathbf{C}$. *To be more specific, let* A *be some (small) set of parameters and let* M *be some* $(|A|^+ + |T|^+)$*-saturated model. Let* $p(x) \in S(M)$ *extend* $S(x)$. *Then* p *is a generic type of* S *for* T *iff* T *is a generic type of* S *for* $Th(\mathbf{C}, a)_{a \in A}$.

DEFINITION **6.8.** *Let* $p(x) \in S(A)$ *and* $p(x) \models S(x)$. *We call* $p(x)$ *generic if every relatively definable subset of* S *in* p *is generic.*

LEMMA **6.9.** *Let* $p(x) = \mathrm{tp}(a/A)$ *where* $a \in S$. *The following are equivalent.*

(i) $p(x)$ *is generic;*

(ii) *some non-forking extention of* $p(x)$ *to a saturated model* M *is generic;*

(iii) *every non-forking extension of* $p(x)$ *to a saturated model is generic;*

(iv) *whenever* $g \in G$ *and* a *is independent from* g *over* A, *then* $g.a$ *is independent from* $A \cup \{g\}$ *over* A.

Proof. (i) $\rightarrow$ (ii). By Remark 6.7(i), $p(x)$ has an extension in $S(M)$ which contains only generic relatively definable subsets of S, and thus is generic.

(ii) $\rightarrow$ (iii). All non-forking extensions of p over M are conjugate under $\mathrm{Aut}_A(M)$, and this clearly preserves genericity.

(iii) $\rightarrow$ (iv). Without loss $g \in M$ where M is a saturated model containing A and a is independent with M over A. By (iii) $\mathrm{tp}(a/M)$ is generic, and thus $\mathrm{tp}(g.a/M)$ does not fork over $\emptyset$. In particular $\mathrm{tp}(g.a/A \cup \{g\})$ does not fork over A.

(iv) $\rightarrow$ (i). Let M be an $|A|^+ + |T|^+$-saturated model containing A such that a is independent from M over A. Let $g \in G^M$. Then $\mathrm{tp}(a/M)$ does not fork over $A \cup \{g\}$, so $\mathrm{tp}(g.a/M)$ does not fork over $A \cup \{g\}$. So by hypothesis $\mathrm{tp}(g.a/M)$ does not fork over A. This shows that $\mathrm{tp}(a/M)$ is generic, after naming parameters from A. By Remark 6.7(iii), $\mathrm{tp}(a/M)$ is generic, and thus so is $\mathrm{tp}(a/A)$. ■

By convention, a generic type of the ∞-definable group G is a generic type for the action of G on itself by left multiplication. Namely, a relatively definable subset X of G is said to be *(left) generic* if $g_1.X \cup \cdots \cup g_n.X = G$ for some $g_1, \ldots, g_n \in G$, and a type $p(y)$ extending $G(y)$ is said to be (left) generic of it contains only left generic subsets of G. Considering the action of G on itself by *right multiplication*, we obtain the notion of a *right generic* type of G. In particular a relatively definable subset X of G is called *right generic* if finitely many *right translates* of X cover G. It turns out that 'left generic = right generic'. The following amusing proof shows this to be essentially an issue of forking symmetry.

Lemma 6.10. *Let X be a relatively definable subset of G. Then X is left generic iff X is right generic.*

Proof. Let $\mathbf{C}_0$ be the structure (G, R) where $R(x, y)$ iff $x \in y.X$. As in the proof of 6.4, $R(x, y)$ is a stable formula for $T_0 = Th(\mathbf{C}_0)$. Suppose X is left generic. As in the claim in 6.4, $R(x, 1)$ does not fork over $\varnothing$ in T_0. It is then easy to deduce from Lemma 2.8 that $R(g, y)$ does not fork over $\varnothing$ (in T_0) for some $g \in G$. As in the claim of 6.4 (using the fact that a unique 1-type over $\varnothing$ is realized in $\mathbf{C}_0$), it follows that $R(g_1, y) \cup \cdots \cup R(g_n, y)$ is equivalent to $y = y$ (in $\mathbf{C}_0$) for some $g_1, \ldots, g_n \in G$. But then clearly $G = X.g_1^{-1} \cup \cdots \cup X.g_n^{-1}$. So X is right generic.

We have shown that the left generic implies right generic. The other direction follows by symmetry. ■

It follows from 6.10 that if $\mathrm{tp}(g/A)$ is a generic type of G, then so is $\mathrm{tp}(g^{-1}/A)$.

We will say 'a is generic in S over A, if $a \in S$ and $\mathrm{tp}(a/A)$ is a generic type of S (for the action of G on S). Similarly we say 'g is generic in G over A, if $g \in G$ and $\mathrm{tp}(g/A)$ is a (left or right) generic type of G.

Lemma 6.11. *Suppose $a \in S$ and g is generic in G over $\{a\}$. Then $g.a$ is generic in S over $\varnothing$.*

Proof. Let $h \in G$ such that h is independent from $g.a$ over $\varnothing$. We will show that $h.(g.a)$ is independent from h over $\varnothing$ (which will prove the lemma by 6.9). We may assume that h is independent from $\{g.a\}$ over $\varnothing$. In particular g is independent from h over a. By 6.9, $h.g$ is independent from h over a. But $h.g$ is independent from a over $\varnothing$. Thus h is independent from $\{h.g, a\}$ over $\varnothing$, so h is independent from $h.(g.a)$ over $\varnothing$, as required. ■

Lemma 6.12. (i) *Suppose that T is superstable*: (a) *if $p(x)$ is a complete type containing $S(x)$ then p is generic iff $R^\infty(p) = R^\infty(S)$*; (b) *among the complete types $p(x)$ containing $S(x)$ there are types of maximum U-rank. Moreover, these are precisely the generic types of S.*

(ii) *If T is totally transcendental then* (i)(a) *holds with RM in place of R^∞.*

Proof. (i) Let M be a saturated model. Then all generic types of S over M

are conjugate under the action of G^M (by 6.6). Thus they all have the same U-rank and R^∞-rank (by 3.28).

On the other hand if $p(x) \in S(\emptyset)$ is any type containing $S(x)$, and a realizes p, then by 6.11, there is $g \in G$, independent from a over $\emptyset$ such that $\mathrm{tp}(g.a/\emptyset)$ is generic. But $U(\mathrm{tp}(a/\emptyset)) = U(\mathrm{tp}(a/g)) = U(\mathrm{tp}(g.a/g)) \leq U(\mathrm{tp}(g.a/\emptyset))$ (and similarly for R^∞). This proves (a) and (b).

(ii) is proved similarly. ∎

LEMMA **6.13.** *Let $\varphi(y, z)$ be some formula of L. Let $\{c_i : i \in I\}$ be a collection of parameters such that for each i, $G_i = \{g \in G : \models \varphi(g, c_i)\}$ is a subgroup of G. Then there is finite $J \subseteq I$ such that $\cap_{i \in I} G_i = \cap_{i \in J} G_i$.*

Proof. Consider the structure $\mathbf{C}_2 = (G, \mathbf{C}, R(y, z))$ where $R(y, z)$ holds iff $y \in G$ and $z \in \mathbf{C}$. Then clearly R is stable in $Th(\mathbf{C}_2)$ (and $\mathbf{C}_2$ has a countable language). If the lemma fails then we can clearly find a countable elementary substructure of $\mathbf{C}_2$ over which there are continuum many complete R-types, contradicting Remark 2.3. ∎

Recall from 6.6(iv) that the connected component G^0 of G is the intersection of all $\emptyset$-relatively definable subgroups of finite index in G. We say that G is *connected* if $G = G^0$, or equivalently if G contains no relatively definable subgroup of finite index.

COROLLARY **6.14.** *G^0 is contained in every relatively definable subgroup of finite index of G, and hence is the intersection of all such subgroups. In particular G^0 is a normal subgroup of G.*

Proof. For each formula $\varphi(y, z)$ of L, let G^0_φ be the intersection of all subgroups of finite index of G of the form $\varphi(y, c)^{\mathbf{C}} \cap G$ for some c in M. By Lemma 6.13, G^0_φ is the intersection of finitely many such subgroups and is thus clearly relatively $\emptyset$-definable, and hence contains G^0. ∎

DEFINITION **6.15.** (i) *Let $\mathbf{p}(x) \in S(\mathbf{C})$ extend $S(x)$. By* Stab($\mathbf{p}$) *(the stabilizer of* $\mathbf{p}$*) we mean $\{g \in G : g.\mathbf{p} = \mathbf{p}\}$.*

(ii) *If $p(x) = \mathrm{stp}(a/A)$ where $a \in S$, then by* Stab(p) *we mean* Stab($\mathbf{p}'$) *where $\mathbf{p}' \in S(\mathbf{C})$ is the non-forking extension of p.*

We assemble some facts about stabilizers, and some additional facts about generic types.

LEMMA **6.16.** (i) *If $\mathbf{p}(x) \in S(\mathbf{C})$ extends $S(x)$, then* Stab($\mathbf{p}$) *is the intersection* *of at most* $|T|$ *relatively definable subgroups of G. Moreover,* Stab(**p**) *is* ∞*-definable, defined over* Cb(**p**).

(ii) *If* $p(x) = \text{stp}(a/A)$ *where* $a \in S$, *then* $g \in \text{Stab}(p)$ *iff whenever* a' *realizes* $p|(A \cup \{g\})$ *then* $g.a'$ *realizes* $p|(A \cup \{g\})$.

(iii) *Let* $\mathbf{p}(x) \in S(\mathbf{C})$ *extend* $S(x)$. *Then* **p** *is a generic type of S iff* $\text{Stab}(\mathbf{p}) \supseteq G^0$.

(iv) *Suppose that* (G, S) *is a principal homogeneous space. Let* $p(x) \in S(\mathbf{C})$ *extend* $S(x)$. *Then* **p** *is a generic type of S iff* $\text{Stab}(\mathbf{p}) = G^0$. *Moreover, for each orbit* $G^0.a$ *of* G^0 *on S, there is a unique generic type* $\mathbf{p}(x) \in S(\mathbf{C})$ *of S extending '*$x \in G^0.a$*'.*

(v) *For each formula* $\varphi(x, z)$ *of L,* $\{\mathbf{p} \upharpoonright \varphi \colon \mathbf{p}(x) \in S(\mathbf{C})$ *a generic type of* $S\}$ *is finite.*

Proof. (i) Let $\varphi(x, z)$ be any formula. Let $\varphi'(x; y, z)$ be the formula $\varphi(y.x, z)$. Let $\delta(y, z)$ be the φ'-definition of $\mathbf{p}(x)$. Then $\delta(y, z)$ is Cb(**p**)-definable. Define $\text{Stab}_{\varphi'}(\mathbf{p}) = \{g \in G\colon (\forall y, z)\delta(y.g, z) \leftrightarrow \delta(y, z))\}$. So $\text{Stab}_{\varphi'}(\mathbf{p})$ is a relatively definable subgroup of G (defined over Cb(p)) and clearly $\text{Stab}(\mathbf{p}) = \cap_{\varphi'} \text{Stab}(\mathbf{p})$.

(ii) is easy.

(iii) We have already seen in Proposition 6.6(iv) that if $\mathbf{p}(x) \in S(\mathbf{C})$ (extending $S(x)$) is generic then $\text{Stab}(\mathbf{p}) \supseteq G^0$. On the other hand, if Stab(**p**) does not contain G^0 then by (i) and 6.14, there is a relatively definable subgroup H of G which has infinite index in G and contains Stab(**p**). Thus the index of H in G is $> 2^{|T|}$. If g_1 is not congruent to g_2 mod H then $g_1.\mathbf{p} \neq g_2.\mathbf{p}$. So clearly $g.\mathbf{p}$ forks over $\varnothing$ for some $g \in G$, whereby **p** is not generic.

(iv) To prove the first assertion, it suffices by (iii) to prove that for every $\mathbf{p}(x) \in S(\mathbf{C})$, $\text{Stab}(\mathbf{p}) \subseteq G^0$. Fix **p**. Now for each relatively $\varnothing$-definable subgroup H of G with finite index in G, H has only finitely many orbits on S, so for some $a \in S$, $\mathbf{p}(x)$ contains the 'formula' $x \in H.a$. By compactness there is $a \in S$ such that $\mathbf{p}(x)$ contains the partial type '$x \in G^0.a$'. Suppose $g \in \text{Stab}(\mathbf{p})$. Then '$x \in g.G^0.a$' is contained in **p**, and thus $g.G^0.a \cap G^0.a \neq \varnothing$, so for some $g_1, g_2 \in G^0$, $g.g_1.a = g_2.a$. As G acts regularly on S, $g.g_1 = g_2$, whence $g \in G^0$, as required.

For the second assertion, let first $\mathbf{p}(x) \in S(\mathbf{C})$ be some generic type of S. As in the proof of (iv), there is $a \in S$ such that '$x \in G^0.a$' is contained in **p**. By 6.6, the set of generic types of $S = \{g.\mathbf{p}\colon g \in G\}$. Given $b \in G$, let $g \in G$ be such that $g.a = b$. Then (by normality of G^0), $g.G^0.a = G^0.g.a = G^0.b$. Thus '$x \in G^0.b$' $\in g.\mathbf{p}$. This shows that every orbit of G^0 on S is contained in some generic type of S. On the other hand **p** must be the unique generic type of S containing $G^0.a$: if **q** were another, then $\mathbf{q} = g.\mathbf{p}$ for some $g \in G$. By normality of G^0, $g.G^0.a = G^0.g.a$ which must therefore equal $G^0.a$, whereby $g \in G_0$, so $\mathbf{q} = \mathbf{p}$.

(v) As in the proof of (i) above, let $\varphi'(x, y, z)$ be the formula $\varphi(y.x, z)$. Let $p(x) \in S(\mathbf{C})$ be a generic type of S, and let $H = \text{Stab}_{\varphi'}(\mathbf{p})$ as defined in the proof of (i). By (iii), the relatively definable subgroup H of G has finite index in G. As G acts transitively on the generic types of S it follows that the set of φ'-types of generic types of S is in one-to-one correspondence with G/H. Thus there are finitely many such φ'-types, so also φ-types. ■

Note that (iv) in 6.16 applies to the case where G acts on itself by left (or right) multiplication.

Remark 6.17. *The above theory can be developed 'locally' for homogeneous spaces and δ-types, where $\delta(x, z)$ is an 'equivariant' stable formula, as the proofs of* 6.4 *and* 6.6 *show.*

The next result, although not difficult, is quite fundamental.

Lemma 6.18. *Let G be an ∞-definable group, defined over the set A, say. Then there is a* definable *group H defined over A such that*

(i) G is a subgroup of H; and

(ii) $G = \cap_i H_i$, for some set $\{H_i : i < |T|\}$ of A-definable subgroups of H.

Proof. Assume A to be $\varnothing$ for notational convenience. Assume G to be the solution set of the partial type $\Phi(x)$. As in the beginning of this section, let $\varphi_0(x)$ be a formula in Φ such that whenever a, b, c satisfy φ_0 then $(a.b).c$ and $a.(b.c)$ are defined and equal, and $a.1 = 1.a = a$. For each formula $\varphi(x)$ in $\Phi(x)$, let $\delta_\varphi(x, y)$ be the formula $\varphi(y.x) \wedge \varphi_0(y) \wedge \varphi_0(x)$. By 6.16(v), let $\varepsilon_\varphi(y)$ be the conjunction of the δ_φ-definitions of the genetic types of S (of which there are finitely many by 6.16(v)). (So for any b. $\models \varepsilon_\varphi(b)$ iff whenever $a \in G$ is generic over b, then $\models \delta_\varphi(a, b)$.) Clearly $\varepsilon_\varphi(y)$ is over $\varnothing$. We assume that Φ is closed under finite conjunctions. Note that if $\models \varphi(x) \rightarrow \psi(x)$ then $\models \varepsilon_\varphi(y) \rightarrow \varepsilon_\psi(y)$.

Claim 1. *$\Phi(x)$ is equivalent to $\{\varphi_0(x)\} \cup \{\varepsilon_\varphi(x) : \varphi \in \Phi\}$.*

Proof of claim. Suppose b satisfies the right-hand side. Let $a \in G$ be generic over b. So $\models \delta_\varphi(a, b)$ for all $\varphi \in \Phi$. As $\models \varphi_0(b) \wedge \varphi_0(a)$, it follows that $\models \varphi(b.a)$ for all $\varphi \in \Phi$, that is $b.a \in G$. Thus $b \in G$ (i.e. b satisfies Φ). The other direction is immediate. The claim is proved. ■

Using the claim we can find $\varphi(x) \in \Phi$ such that whenever a, b satisfy $\varphi_0(x) \wedge \varepsilon_\varphi(x)$ then $a.b$ satisifies $\varphi_0(x)$.

CLAIM 2. *If b satisfies $\varphi_0(x) \wedge \varepsilon_\varphi(x)$, and $a \in G$, $b.a$ satisfies $\varphi_0(x) \wedge \varepsilon_\varphi(x)$.*

Proof. Choose $c \in G$ generic over $\{a, b\}$. In particular (by Lemma 6.9) $a.c$ is generic over b. As b satisfies $\varphi_0 \wedge \varepsilon_\varphi$, it follows that $b.(a.c)$ satisfies φ, and thus $(b.a).c$ satisifes φ. By definition of ε_φ, we conclude that $\models \varepsilon_\varphi(b.a)$. Also by choice of φ, we have $\models \varphi_0(b.a)$. ■

Let H_0 be the set defined by $\varphi_0(x) \wedge \varepsilon_\varphi(x)$. Let $H_1 = \{a \in H_0$: for all $b \in H_0, b.a \in H_0\}$. Then H_1 is $(\varnothing)$-definable and by Claim 1 contains G. Note also that H_1 is closed under multiplication. Let $H = \{a \in H_1$: there is $b \in H_1$, $a.b = b.a = 1\}$, that is, H is the set of 'invertible' elements of H_1. Then $H \supseteq H_1$, and is a group under the operation.

So we have succeeded in finding the definable group H, proving (i). For any formula $\psi(x)$ in Φ whose solution set is contained in H, we can, by replacing φ_0 by ψ in the above proof, find a $\varnothing$-definable subgroup H_ψ of H containing G, and contained in ψ^H. Clearly then G is the intersection of the $\varnothing$-definable groups H_ψ. Lemma 6.13 shows that G is the intersection of at most $|T|$ such definable subgroups. ■

The reader is invited to prove the analogous result for homogeneous spaces (and also principal homogeneous spaces).

LEMMA **6.19.** *Suppose (G, S) is an ∞-definable (principal) homogeneous space, defined over A. Then there is a* definable *(principal) homogeneous space (H, P) defined over A, with G a subgroup of H, S a subset of P, and there are* definable *subgroups H_i of H for $i < |T|$, and definable subsets P_i of P for $i < |T|$ such that $G = \cap H_i$, $S = \cap P_i$, H_i acts (strictly) transitively on P_i, and everything in sight commutes.*

REMARK **6.20.** *Suppose G is a definable group, defined over $\varnothing$. Let H be an ∞-definable subgroup of G. Then H has a 'canonical base', that is there is a set A of elements (from* $\mathbf{C}^{\text{eq}}$*) such that* (i) *H can be defined by a partial type over A, and* (ii) *any automorphism which fixes H setwise fixes A pointwise.*

Proof. We have seen in 6.18 that H is the intersection of some set $\{H_i: i \in I\}$ of definable subgroups of G. For each i. let H_i' be the intersection of all images (under automorphisms of $\mathbf{C}$) of H_i which contain H. By Lemma 6.13, H_i' is *definable* (being a finite intersection of definable subgroups). H_i' is now something canonical, in the sense that any automorphism which fixes H

setwise fixes H'_i setwise. Note that $H = \cap_i H'_i$. Let a_i be a canonical parameter for H'_i (i.e. H'_i viewed as an element in $\mathbf{C}^{eq}$). Then clearly $A = \{a_i: i \in I\}$ is the required set. ∎

Remark 6.21. (i) *If T is totally transcendental, then there is no infinite strictly descending chain of definable groups $G_0 > G_1 > \dots$. In particular any ∞-definable group is definable.*

(ii) *If T is superstable then there is no infinite descending chain of groups $G_0 > G_1 > \dots$, where G_{i+1} has infinite index in G_i for all i. In particular for any ∞-definable group G there is a* definable *supergroup H such that H contains the connected component of G, whence $U(G) = U(H)$, and $R^\infty(G) = R^\infty(H)$.*

Proof. (i) Assume T totally transcendental. By 3.28, if H is a definable subgroup of the definable group G, then for any $a \in G$, $RM(H) = RM(a.H)$, and similarly for dM(-). Thus if H has *infinite* index in G, then $RM(H) < RM(G)$, and if H has finite index (say n) in G, then $RM(H) = RM(G)$ and $n.dM(H) = dM(G)$. This shows that any descending chain of definable groups must stabilize. By Lemma 6.18 we deduce that any ∞-definable group is definable.

(ii) Suppose T is superstable. By 3.28 again, if H is a definable subgroup of the definable group G, then for each $g \in G$, $R^\infty(H) = R^\infty(g.H)$. If also H has infinite index in G, then we can clearly find a sequence $(g_i: i < \omega)$ of elements of G which is indiscernible over A (where H, G are defined over A), and such that $g_i.H \neq g_j.H$ for $i \neq j$. By Lemma 2.16, the formula '$x \in g_0.H$' forks over A. Thus $R^\infty(G) > R^\infty(g_0.H) = R^\infty(H)$. Thus there can be no infinite descending chain of definable groups, each of infinite index in the preceding one. Note that by 3.28 again (and 3.13(iii)) if H has finite index in G then $R^\infty(H) = R^\infty(G)$. Now suppose G is an ∞-definable group. By Lemma 6.18, $G = \cap\{G_i: i < |T|\}$, where each G_i is definable. We may assume the family of G_i is closed under finite intersections. By what we have just seen, choose $i(0)$ such that $G_i \cap G_{i(0)}$ has infinite index in $G_{i(0)}$ for all $i < |T|$. So $R^\infty(G) = \min\{R^\infty(G_i): i < |T|\} = R^\infty(G_{i(0)})$.

Let $\mathbf{p}(x) \in S(\mathbf{C})$ be a generic type of $G_{i(0)}$. By 6.12(b) $U(\mathbf{p}) = U(G_{i(0)})$. By compactness there is $g \in G_{i(0)}$ such that $p(x) \models x \in g.G$. Then $g^{-1}.\mathbf{p}$ is also a generic type of $G_{i(0)}$, but also a generic type of G (by the charcterization in 6.16(iv) for example). Thus $U(G_{i(0)}) = U(g^{-1}.\mathbf{p}) = U(G)$ (by 6.12(b)). ∎

Finally in this section we state without proof some results which will be used later.

Fact 6.22. Let G be a connected ∞-definable group whose generic type is regular. Then G is commutative.

Fact 6.23. If K is an ∞-definable infinite field then K is connected. If, moreover, $U(K) < \infty$, then K is algebraically closed. If T is superstable then K is definable.

Fact 6.24. Any connected definable group G which has Morley rank at most 2 and is definable in an ω-categorical (stable) structure is nilpotent.

REMARK. *In fact one can deduce just from finite Morley rank and ω-categoricity that G is abelian-by-finite, as proved by Baur–Cherlin–Macintyre. Alternatively, Cherlin proves that if G is any connected non-nilpotent group of Morley rank 2, then an infinite field is definable in G, and the latter clearly contradicts ω-categoricity.*

Fact 6.25. Suppose (G, S) is an ∞-definable homogeneous space, defined over A, where S is minimal (in the sense of Remark 5.2(i)). Then one of the following holds:

(i) G is minimal (and thus commutative) and the action of G on S is *regular*.

(ii) $U(G) = 2$ and there is an A-definable (algebraically closed) field structure $(K, +,.)$ on S such that G is precisely the group of transformations $\{x \to a.x + b \colon a, b \in K\}$.

(iii) $U(G) = 3$, S has (definably over A) the structure of $P^1(K)$ for some definable) field $(K, +,.)$, and G is the group $PSL_2(K)$ of linear fractional transformations $x \to (a.x. + b)/(c.x + d)$ $(a, b \in K)$.

2

The classical finite rank theory

In this chapter I develop what I call the 'classical' part of the geometric theory, consisting essentially of results of Zilber and Cherlin–Harrington–Lachlan from the 1970s and early 1980s. Early work for Buechler falls naturally into this framework.

The main new idea (compared with the viewpoint of Shelah) is to study the 'geometry' of forking, in particular the behaviour of algebraic closure in strongly minimal sets, both for its own sake as well as an aid in attacking other problems (such as the non-finite axiomatizability of totally categorical theories).

The word 'geometry' here refers to 'combinatorial' geometries or matroids, rather than geometry in the usual sense (curves, surfaces, etc.). However, in our study of strongly minimal sets and ω_1-categorical structures, geometric intuitions of a more classical nature are also present.

1 Geometries and pregeometries

Definition 1.1. *By a (combinatorial)* pregeometry, *we mean a set S together with a certain closure operation* cl(-) *from P(S), the power set of S, to P(S), such that the following hold (X, Y denote subsets of S, and a, b denote elements of S):*

(i) $X \subseteq \mathrm{cl}(X)$;

(ii) $\mathrm{cl}(\mathrm{cl}(X)) = \mathrm{cl}(X)$;

(iii) *if* $a \in \mathrm{cl}(X \cup \{b\}) - \mathrm{cl}(X)$ *then* $b \in \mathrm{cl}(X \cup \{a\})$;

(iv) *if* $a \in \mathrm{cl}(X)$ *then* $a \in \mathrm{cl}(Y)$ *for some finite subset Y of X.*

The pregeometry (S, cl) *is said to be a* geometry, *if* $\mathrm{cl}(\emptyset) = \emptyset$ *and* $\mathrm{cl}\{a\}) = \{a\}$ *for all* $a \in S$.

The pregeometry (S, cl) *is said to be* homogeneous *if for any closed subset X of S and* $a, b \in S - X$, *there is an* automorphism *of S (namely, a* cl(-)*-preserving permutation of S) which fixes X pointwise and takes a to b.*

Remark 1.2. (i) *To any pregeometry* (S, cl) *we associate a canonical geometry*

(S', cl'): $S' = \{\mathrm{cl}(\{a\}): a \in S - \mathrm{cl}(\varnothing)\}$ *and for* $X \subseteq S$, $\mathrm{cl}'(\{\mathrm{cl}\{a\}: a \in X\}) = \{\mathrm{cl}(\{b\}): b \in \mathrm{cl}(X)\}$. *If* (S, cl) *is homogeneous so is* (S', cl').

(ii) *If* (S, cl) *is a pregeometry and A is a subset of S we can* localize *this pregeometry at A to obtain the pregeometry* (S, cl_A) *where for any* $X \subseteq S$, $\mathrm{cl}_A(X)$ *is defined to be* $\mathrm{cl}(A \cup X)$.

(iii) *By virtue of the 'Steinitz exchange' axiom* (iii) *in the definition of a pregeometry* (S, cl) *we can talk about 'independence' and 'dimension': if A and B are subsets of S, then we say A* is independent over *B if for any* $a \in A$, $a \notin \mathrm{cl}((A - \{a\}) \cup B)$. *We say that* $A_0 \subseteq A$ *is a basis for A over B if* $A \subseteq \mathrm{cl}(A_0 \cup B)$ *and* A_0 *is independent over B. All bases for A over B will have the same cardinality which we call* $\dim(A/B)$. *We will say that A is independent from B over C if for all finite subsets A′ of A*, $\dim(A'/B \cup C) = \dim(A'/C)$. *It follows that A is independent from B over C iff B is independent from A over C.*

Definition 1.3. *Let* (S, cl) *be a pregeometry.*

(i) (S, cl) *is said to be* trivial *(or degenerate) if for every* $X \subseteq S$, $\mathrm{cl}(X) = \cup \{\mathrm{cl}(\{a\}): a \in X\}$.

(ii) (S, cl) *is said to be* modular *if for any closed sets* $X, Y \subseteq S$, *X is independent from Y over* $X \cap Y$. *Equivalently, for any finite-dimensional closed sets, X, Y*, $\dim(X) + \dim(Y) - \dim(X \cap Y) = \dim(X \cup Y)$.

(iii) (S, cl) *is said to be* locally modular *if for some* $a \in S$, $(S, \mathrm{cl}_{\{a\}})$ *(the localization of S at* $\{a\}$*) is modular.*

(iv) (S, cl) *is said to be* projective *if it is non-trivial and modular, and* locally projective *if non-trivial and locally modular.*

(v) (S, cl) *is* locally finite *if for any finite* $A \subseteq S$, $\mathrm{cl}(A)$ *is finite.*

Note that triviality implies modularity.

Lemma 1.4. (i) *Each of the properties, triviality, modularity, local modularity, passes to the associated geometry, and is preserved under localization.*

(ii) *If* $(S \cup \mathrm{cl})$ *is locally modular and homogeneous, then* $(S, \mathrm{cl}_{\{b\}})$ *is modular for any* $b \in S - \mathrm{cl}(\varnothing)$.

(iii) (S, cl) *is modular if and only if whenever* $a, b \in S$, $A \subseteq S$, $\dim(\{a, b\}) = 2$, *and* $\dim(\{a, b\}/A) \leq 1$ *then there is* $c \in \mathrm{cl}(\{a, b\}) \cap \mathrm{cl}(A)$, $c \notin \mathrm{cl}(\varnothing)$.

Proof. Exercise. ■

We now give some examples. We leave the verification of certain assertions to the reader.

Example 1.5. Let S be any set and define $\mathrm{cl}(A) = A$ for all subsets A of S. Then (S, cl) is a (pre)geometry which is homogeneous and trivial.

Example 1.6. Let F be a division ring. Let V be a κ-dimensional vector space over F. For $A \subseteq V$ define $\mathrm{cl}(A)$ to be the F-linear span of A, that is the F-subspace of V generated by A. Then (V, cl) is a homogeneous pregeometry which is modular. Also $\dim(V/\varnothing) = \kappa$. The associated geometry is precisely the collection of 1-dimensional subspaces of V with the induced closure operation. This latter geometry is usually called $(\kappa - 1)$-dimensional projective geometry over F, denoted $P^{(\kappa-1)}(F)$ (although its dimension as defined above is κ). For example, the projective plane over F is the set of 1-dimensional subspaces of a vector space V of dimension 3 over F.

Example 1.7. Let F, V, and κ be as in 1.6. By an F-affine subspace of V we mean a subset of V of the form $W + a$ where $a \in V$ and W is an F-linear subspace. If we now define $\mathrm{affcl}(A)$ to be the smallest F-affine subspace of V containing A, we obtain again a homogeneous pregeometry (V, affcl) of dimension $\kappa + 1$ in the sense of 1.2(iii). This pregeometry happens to be a geometry (note that for any element $a \in V$, $\{a\}$ is an affine subspace). This geometry is called κ-dimensional affine geometry over F. The geometry is locally modular: localize at 0 to obtain $(V, \mathrm{affcl}_{\{0\}})$. Clearly for any $A \subseteq V$, $\mathrm{affcl}_{\{0\}}(A)$ is the smallest F-linear subspace of V containing A so we are in the situation of Example 1.6.

The geometry (V, affcl) is *not* modular. Let a, b, c be non-collinear points in V. Then $\dim\{a, b, c\} = 3$. Let $d = a - b + c$. Then $d \in \mathrm{affcl}(\{a, b, c\})$. So $\dim\{a, b, c, d\} = 3$. Clearly $\dim\{a, b\} = \dim\{c, d\}) = 2$. However, $\mathrm{affcl}(\{a, b\}) \cap \mathrm{affcl}(\{c, d\}) = \varnothing$, as $\mathrm{affcl}(\{a, b\})$ and $\mathrm{affcl}(\{c, d\})$ are affine lines which are 'parallel'.

Example 1.8. Let K be an algebraically closed field of infinite transcendence degree over its prime subfield. Let cl(-) be (field-theoretic) algebraic closure. Then (K, cl) is a homogeneous pregeometry. Local modularity fails quite drastically. For let $c_0, \ldots, c_n, a \in K$ be algebraically independent. Let $b = c_n a^n + \cdots + c_1 a + c_0$. Then

$$\dim\{a, b, c_{n-1}, c_n\}/\{c_0, \ldots, c_{n-2}\}) = 3,\ \dim(\{a, b\}/\{c_0, \ldots, c_{n-2}\}) = \dim(\{c_{n-1}, c_n\}/\{c_0, \ldots, c_{n-2}\}) = 2.$$

But it can be shown that

$$\mathrm{cl}(\{c_0, \ldots, c_{n-2}, a, b\} \cap \mathrm{cl}(\{c_0, \ldots, c_{n-2}, c_{n-1}, c_n\}) = \mathrm{cl}(\{c_0, \ldots, c_{n-2}\}).$$

Thus the localization of (K, cl) at $\{c_0, \ldots, c_{n-2}\}$ is not modular. (A more complete proof of the non-local modularity of (K, cl) will appear in the next section.)

Example 1.9. Let D be a minimal set defined over a set A in a saturated model $\mathbf{C}$ of cardinality $\kappa > |A|$ of a stable theory T. Let cl(-) on D be defined by: $\mathrm{cl}(B) = \mathrm{acl}(B \cup A) \cap D$. Let $\mathbf{D}_A$ be (D, cl). Then from section 5 of Chapter 1 (see the remarks following 1.5.4) we see that $\mathbf{D}_A$ is a pregeometry. In fact $\mathbf{D}_A$ is homogeneous. For if $B \subseteq D$, $B = \mathrm{acl}(B \cup A) \cap D$, and $a, b \in D - B$, then $\mathrm{tp}(a/B \cup A) = \mathrm{tp}(b/B \cup A)$, and there will be an automorphism f of $\mathbf{C}$ which fixes $B \cup A$ pointwise and takes a to b. (One should be a little careful here as B may be of cardinality κ. But such an automorphism nevertheless exists, as the reader is asked to check.) Of course Example 1.8 is a special case of this. (There the structure $(K, +, .)$ is strongly minimal.)

Note that $D - \mathrm{acl}(A) = D'$ is also minimal and, moreover, that $\mathbf{D}_A$ and $\mathbf{D}'_A$ have the same associated geometries. But now D' is the set of realizations of a complete stationary type p, say (of U-rank 1). We will say that p is trivial, modular, or locally modular, according to whether the geometry $\mathbf{D}'_A$ is.

If D is a weakly minimal set defined over A, then $\mathbf{D}_A$, as defined in the first paragraph, will also be a pregeometry, but not necessarily homogeneous.

Remark 1.10. *The question arises of the distinction between a structure D as in Example* 1.9 *and the pregeometry* $(\mathbf{D}_A, \mathrm{cl})$. *In some cases these are essentially identical. For example, suppose F is a finite field. Let* (P, cl) *be an infinite-dimensional projective geometry over F. For each n, let* R_n *be an* $n + 1$*-ary relation on P such that* $(a, b_1, \ldots, b_n) \in R_n$ *iff* $a \in \mathrm{cl}(\{b_1, \ldots, b_n\})$. *Then the structure* $M = (P, R_n)_{n<\omega}$ *is* ω*-categorical and strongly minimal (i.e. the theory of this structure is* ω*-categorical, and the formula* $x = x$ *is strongly minimal). In particular M is saturated in its own cardinality. For* $A \subseteq P$, $a \in P$, $a \in \mathrm{acl}(A)$ *in the sense of the structure M iff* $a \in \mathrm{cl}(A)$ *in the sense of the geometry. Similarly if V is an infinite-dimensional vector space over the finite field F, then let* R_n *be the* $n + 1$*-ary relation on V*: $R_n(a, b_1, \ldots, b_n)$ *iff* $a \in \mathrm{affcl}(\{b_1, \ldots, b_n\})$. *Then again* $(V, R_n)_{n<\omega}$ *is* ω*-categorical and strongly minimal and algebraic closure in this structure is precisely F-affine closure.*

More will be said on this topic in Chapter 3.

In Chapter 5 *it will be pointed out that there is no strongly minimal structure M whose natural pregeometry is a projective* geometry *over an* infinite *field.*

It is convenient in this chapter to make use of the following facts:

Fact 1.11. Let (S, cl) be a locally projective, locally finite, infinite homogeneous geometry. Then (S, cl) is isomorphic to some affine or projective geometry over a finite field.

It is also worth mentioning:

Fact 1.12. If (S, cl) is a projective geometry of dimension at least 4, such that all 2-dimensional closed sets have at least three elements, then (S, cl) is a projective geometry over some division ring.

2 Minimal types and linearity

We have seen in section 1 that a minimal set D in a stable structure has a canonical homogeneous pregeometry. The objective of this section is to give a characterization of when this pregeometry is locally modular, in terms of families of 'plane curves' over D.

In this section D will be a minimal set defined over a small set A in a saturated model $\mathbf{C}$ of a stable theory.

We first set notation. Let $\mathbf{a}$ be a finite tuple over D. We have, a priori, three notions of the 'rank' or 'dimension' of $\mathbf{a}$ over A. One is $\dim(\mathbf{a})$ in the sense of the pregeometry $\mathbf{D}_A$. The second is $\dim(\mathbf{a}/A)$ in the sense of Lemma 1.5.5. The third is $U(\mathbf{a}/A)$, which, recall, means $U(\mathrm{tp}(\mathbf{a}/A))$. By common sense, together with Lemma 1.5.5, these are all the same. We will use '$U(\mathbf{a}/A)$'. Recall (as this is finite) that for $\mathbf{a}$, $\mathbf{b}$ tuples from D, $U(\mathbf{a}, \mathbf{b}/A) = U(\mathbf{a}/\mathbf{b}A) + U(\mathbf{b}/A)$.

By $(D_A)^{\mathrm{eq}}$ we will mean $\mathrm{dcl}(D \cup A) \cap \mathbf{C}^{\mathrm{eq}}$. So clearly for any $c \in (D_A)^{\mathrm{eq}}$, also $U(c/A)$ is finite.

The main point we want to make here is that if one restricts one's attention to types of tuples from D or D^{eq} then U-rank behaves like Morley rank and types have a finite multiplicity like Morley degree. We summarize this now.

By a *relatively definable subset* of D^n we mean a set of the form $X \cap D^n$ where X is definable. As usual by an ∞-definable subset of D^n we mean a subset of D^n which is the solution set of a partial type. If X is an ∞-definable subset of D^n, defined over $B \supseteq A$, then by $U(X)$ we mean $\max\{U(\mathbf{a}/B)\colon \mathbf{a} \in X\}$, which is clearly at most n. Similarly we talk about relatively definable and ∞-definable sets in $(D_A)^{\mathrm{eq}}$. For example, X will be called a relatively definable subset of $(D_A)^{\mathrm{eq}}$ if X can be obtained as follows: for some A-definable function f whose domain contains D^n for some n, and whose range is contained in the sort S of $\mathbf{C}^{\mathrm{eq}}$, and for some definable subset Y of S, X is precisely $f(D^n) \cap Y$.

The following can be proved easily by modifying remarks from section 5 of Chapter 1.

LEMMA 2.1. (i) *Let $\varphi(x_1, \ldots, x_n, y)$ be a formula over A. Then for any $m < \omega$, $\{c\colon U(\varphi(\mathbf{x}, c)^{\mathbf{C}} \cap D^n) = m\}$ is defined by some formula over A.*

(ii) *Let X be an ∞-definable subset of D^n, defined over $B \supseteq A$. Then* $\mathrm{mult}(X) =_{\mathrm{def}} |\{p(\mathbf{x}) \in S(\mathrm{acl}(B))\colon U(p) = U(X)\}|$ *is finite.*

(iii) *Let* $\mathbf{a} \in D^n$, $B \supseteq A$. *Let* $U(\mathbf{a}/B) = m$, *and* $\text{mult}(\text{tp}(\mathbf{a}/B)) = d$. *Then there is a relatively definable subset* X *of* D^n, *defined over* B, *such that* $U(X) = m$ *and* $\text{mult}(X) = d$.

(iv) *The above also holds for tuples in* $(D_A)^{\text{eq}}$ *and relatively (or* $\infty-$*) definable subsets of* $(D_A)^{\text{eq}}$.

It is worth noting that canonical bases of types of tuples from D (and even D^{eq}) are essentially elements, which, moreover, can be chosen also in $(D_A)^{\text{eq}}$.

Specifically, let $a_1, \ldots, a_n \in D$, and let B be an algebraically closed set (in $\mathbf{C}^{\text{eq}}$) containing A.

Claim 2.2. $\text{Cb}(\text{stp}(a_1, \ldots, a_n/B))$ *is (the definable closure of) an* element *in* $(D_A)^{\text{eq}}$.

Proof. Suppose $U(a_1, \ldots, a_n/B) = k$. We may suppose that $a_1, \ldots, a_k$ are algebraically independent over B, and so $a_{k+1}, \ldots, a_n \in \text{acl}(a_1, \ldots, a_k, B)$. Let $\varphi(x_1, \ldots, x_n)$ be a formula over B which is satisfied by $(a_1, \ldots, a_n)$ and such that $\varphi(a_1, \ldots, a_k, x_{k+1}, \ldots, x_n)$ isolates $\text{tp}(a_{k+1}, \ldots, a_n/a_1, \ldots, a_k, B)$. By stability (specifically Lemma 1.4.5.9) there is an m-tuple $\mathbf{c}$ from D such that $\varphi(x_1, \ldots, x_n)$ is equivalent on D to $\psi(x_1, \ldots, x_n, \mathbf{c})$ for some ψ. Consider the equivalence relation: $\mathbf{c}\, E\, \mathbf{d}$ iff $U((\psi(\mathbf{x}, \mathbf{c}) \wedge \psi(\mathbf{x}, \mathbf{d}))^{\mathbf{C}} \cap D^m) = k$. By 2.1, this is A-definable. Thus $\mathbf{c}/E \in (D_A)^{\text{eq}}$. It is easily checked that $\text{Cb}(\text{stp}(a_1, \ldots, a_n/B)) = \text{dcl}(\mathbf{c}/E)$. ■

2.3 Plane curves

If K is an algebraically closed field, then an (irreducible) plane curve C is simply an irreducible one-dimensional algebraic subvariety of $K \times K$. If C' is the image of C under an automorphism of K then either $C = C'$ or $C \cap C'$ is finite. So, if $b \in K^{\text{eq}}$ is a 'code' for C in the sense of 1.1.6 , then the family of conjugates of C is parametrized by the family of conjugates of c, and the 'size' of this family is measured by $\dim(c)$. Note that if C is the curve defined by $y = c_n x^n + \ldots + c_1 x + c_0$, and if $c_0, \ldots, c_n$ are algebraically independent, then $(c_0, \ldots, c_n)$ is a code for C, and thus the family of conjugates of C has dimension $n + 1$.

If we work in an arbitrary minimal set D, then there is no notion of 'Zariski-closed' subsets of $D \times D$, but we still want to measure the 'size' of the family of conjugates of a minimal subset of $D \times D$. This can be done by means of canonical bases. Let X be a minimal subset of $D \times D$. If X' is the image of X under an A-automorphism of $\mathbf{C}$, then define X to be equivalent to X', $X \approx X'$, if $X \cap X'$ is infinite. We want to measure the size of the family of equivalence classes $Y/\approx$, where Y is an A-conjugate of X. Let X be B-definable, let $(a_1, a_2) \in X\text{-acl}(B)$. Then $q(x_1, x_2) = \text{stp}(a_1, a_2/B)$ is uniquely determined. By 2.2 $\text{Cb}(q) = c \in (D_A)^{\text{eq}}$. Then $U(c/A)$ is the required number: note that for any A-automorphism f of $\mathbf{C}$, $f(X) \approx X$ iff $f(c) = c$.

DEFINITION 2.4. *We will say that the minimal set D (defined over A) is* linear, *if for any $a_1, a_2 \in D$, and set B of parameters containing A such that $U(a_1, a_2/B) = 1$, then if $c = \text{Cb}(\text{stp}(a_1, a_2/B))$, we have $U(c/A) \leq 1$.*

So linearity over D means intuitively that every 'definable' family of plane curves over D is at most a 1-parameter family.

LEMMA 2.5. *Let $a_1, a_2 \in D$, and let $B \supseteq A$ be such that $U(a_1, a_2/B) = 1$. Let $c = \text{Cb}(\text{stp}(a_1, a_2/B))$. Then $U(c/A) \leq 1$ iff $c \in \text{acl}(a_1, a_2, A)$.*

Proof. Note that $U(a_1, a_2/cA) = 1$ and $U(a_1, a_2/A) \leq 2$. If $U(a_1, a_2/A) = 1$, then by 1.2.26, $c \in \text{acl}(A)$ and there is nothing to do.

So we may suppose $U(a_1, a_2/A) = 2$.

We have: $U(a_1, a_2, c/A) = U(a_1, a_2/cA) + U(c/A) = U(c/a_1, a_2, A) + U(a_1, a_2/A)$. Clearly then $U(c/A) = 1$ iff $U(c/a_1, a_2, A) = 0$, as required. ∎

The reader should check that the linearity of D does not depend on the particular set A. In fact this will follow from the next proposition, which also implies that local modularity of $\mathbf{D}_A$ depends only on D, not on the choice of A.

PROPOSITION 2.6. *The following are equivalent*:

(i) *for some (small) set $B \supseteq A$, $\mathbf{D}_B$ is modular;*

(ii) *D is linear;*

(iii) *$\mathbf{D}_A$ is locally modular.*

Proof. For convenience we assume that $A = \emptyset$.

(i) $\rightarrow$ (ii). So assume that for some B, $\mathbf{D}_B$ is modular.

Let $q = \text{tp}(a_1, a_2/E)$ be minimal with $a_1, a_2 \in D$. Let $c = \text{Cb}(q)$.

We may assume (by automorphism) that $a_1 a_2 E$ is independent from B over $\emptyset$, and thus also $c = \text{Cb}(\text{stp}(a_1, a_2/E \cup B))$. We may also assume (by a Morley sequence argument, or as in the proof of 2.2) that E is a finite subset of D.

We may also assume that $U(\mathrm{tp}(a_1, a_2/\emptyset)) = 2$, for otherwise $\mathrm{tp}(a_1, a_2/E)$ does not fork over $\emptyset$, whereby (by 1.2.26), $c \in \mathrm{acl}(\emptyset)$. Thus $\dim(a_1, a_2/B) = 2$, but $\dim(a_1, a_2/E \cup B) = 1$. By modularity of $\mathbf{D}_B$, there is $d \in \mathrm{acl}(a_1, a_2\ B) \cap \mathrm{acl}(E \cup B) \cap D$, $d \notin \mathrm{acl}(B)$.

Forking symmetry (or Steinitz exchange in $\mathbf{D}_B$) implies that $\dim(a_1, a_2/Bd) = 1$. That is, $U(a_1, a_2/Bd) = 1$, whereby $\mathrm{tp}(a_1, a_2/E \cup B)$ does not fork over Bd. By 1.2.26, $c \in \mathrm{acl}(Bd)$. But $U(d/B) = 1$. Thus $U(c/B) \leq 1$. But c is independent from B over $\emptyset$ (as $c \in \mathrm{acl}(E)$); thus $U(c/\emptyset) \leq 1$, as required.

(ii) → (iii). Assume D to be linear. Let $e \in D - \mathrm{acl}(\emptyset)$. We must show that the localization of $\mathbf{D}$ at e is modular. Let $a_1, a_2 \in D$, and let B be a finite subset of D such that $U(a_1, a_2/e) = 2$ and $U(a_1, a_2/Be) = 1$.

(*) We must find some $d \in \mathrm{acl}(a_1, a_2, e) \cap \mathrm{acl}(B, e)$, with $d \notin \mathrm{acl}(e)$.

Let $c = \mathrm{Cb}(\mathrm{stp}(a_1, a_2/Be))$. So $c \in \mathrm{acl}(Be)$. By linearity and 2.5, also $U(c) = 1$ and $c \in \mathrm{acl}(a_1, a_2)$. If c forks with a_1 over $\emptyset$ then $a_1 \in \mathrm{acl}(c)$, and we can take $d = a_1$ in (*). So we may assume c to be independent from a_1 (and from a_2) over $\emptyset$.

As $\{a_1, a_2\}$ is independent from e over $\emptyset$, and $c \in \mathrm{acl}(a_1, a_2)$, c is independent from e over $\emptyset$. Thus (as $\mathrm{tp}(a_1/\emptyset)$ is stationary), $\mathrm{tp}(a_1/c) = \mathrm{tp}(e/c)$. Thus there is d such that $\mathrm{tp}(a_1, a_2/c) = \mathrm{tp}(e, d/c)$. In particular $d \in \mathrm{acl}(c, e)$ and $d \notin \mathrm{acl}(e)$. So d satisfies (*).

(iii) → (i) is immediate. ■

Returning to the example of algebraically closed fields, we see from the discussion in 2.3 that for any $n < \omega$ there are types $p = \mathrm{tp}(a_1, a_2/B)$ (plane curves) with $U(\mathrm{Cb}(p)) = n$. Thus an algebraically closed field is non-linear (so non-locally modular).

3 The dichotomy theorem

The aim of this section is to prove that a weakly minimal stationary type (i.e. a stationary type with ∞-rank 1) is locally modular or has Morley rank 1. This is a dichotomy between geometric simplicity and model-theoretic simplicity. This result is fundamental for the understanding of superstable theories of finite rank, in particular for questions such as Vaught's conjecture. Generalizations of this dichotomy to regular types in superstable theories will appear in Chapter 8.

In this section T is assumed to be a stable theory. We begin with an easy but important observation.

Lemma 3.1. *Suppose T is superstable and $p(x) \in S(A)$ is a minimal type (i.e. a complete stationary type of U-rank 1). Suppose p is not trivial. Then $R^\infty(p) = 1$.*

Proof. Non-triviality of p means that there are realizations $a_1, \ldots, a_n, b$ of p, such that $b \in \operatorname{acl}(a_1, \ldots, a_n, A)$ but $b \notin \operatorname{acl}(a_i, A)$ for all i. Minimize n. So $\{a_1, \ldots, a_n\}$ is A-independent. Let $B = A \cup \{a_1, \ldots, a_{n-2}\}$. Let p_1 be the non-forking extension of p over B. We will show that $R^\infty(p_1) = 1$. Now a_{n-1}, a_n, b each realize p_1, they are pairwise independent over B, but each is algebraic over B together with the other two. Rename a_{n-1}, a_n, b as a, b, c. Let $\varphi(x, y, z)$ be a formula over B such that $\models \varphi(a, b, c)$ and for all a', b', c', if $\models \varphi(a', b', c')$ then $a' \in \operatorname{acl}(b', c', B)$ and $b' \in \operatorname{acl}(a', c', B)$. As T is superstable we can find a formula $\theta(x)$ in p_1 such that $R^\infty(p_1) = R^\infty(\theta) = \alpha < \infty$, say.

Let $\chi(x, y)$ be the formula $(\exists z)(\varphi(x, y, z) \wedge \theta(z))$. Let $\psi(y)$ be the $\chi(x, y)$ definition of the stationary type p_1. Then $\psi(y)$ is over B. Note that $\models \psi(b)$ (as a realizes the non-forking extension of p_1 over Bb), and $\models \chi(a, b)$.

Thus $\psi(y) \in p_1(y)$.

Claim. $R^\infty(\psi(y)) = 1$.

Proof of claim. Clearly $\psi(y)$ is not algebraic, so $R^\infty(\psi) \geq 1$.

To show that $R^\infty(\psi) = 1$, we must show that whenever b' satisfies ψ, then any forking extension of $\operatorname{stp}(b'/B)$ is algebraic, that is $U(b'/B) \leq 1$. So let b' realize ψ. Let a' realize the non-forking extension of p_1 over Bb'. By choice of ψ, there is c' such that $\models \varphi(a', b', c') \wedge \theta(c')$. By choice of φ, $a' \in \operatorname{acl}(b', c', B)$. Thus by 1.3.27,

$$\alpha = R^\infty(a'/b', B) \leq R^\infty(c'/b', B) \leq R^\infty(c'/B) \leq \alpha.$$

Thus we have equality in the line above. By 1.3.14, c′ is independent from b' over B. So b' is independent from c' over B. Thus $U(b'/B) = U(b'/c', B)$. On the other hand $b' \in \operatorname{acl}(a', c', B)$, so by 1.3.18(iii), $U(b'/c', B) \leq U(a'/c'B) \leq 1$. This shows that $U(b'/B) \leq 1$. The claim is proved. ■

The claim shows that $R^\infty(p_1) = 1$. As p_1 is a non-forking extension of p, we have (by 1.3.14) that $R^\infty(p) = 1$, completing the proof of the proposition. ■

Proposition 3.2. *Let $p \in S(A)$ be a stationary type with $R^\infty(p) = 1$. Assume that p is not locally modular. Then $RM(p) = 1$.*

Proof. For convenience we assume that $A = \varnothing$. Let $\theta(x) \in p(x)$ be a weakly minimal formula. We will work entirely in θ^{eq}. So we will be able to make use of the results following Assumption 5.7 in Chapter 1.

By 2.6, p is not linear. Thus there are a_1, a_2 realizing p, and some c

such that $U(a_1, a_2/c) = 1$, $c = \mathrm{Cb}(\mathrm{stp}(a_1, a_2/c))$ and $U(c/\varnothing) = n \geq 2$. It follows that $U(a_1, a_2/\varnothing) = 2$, and $U(a_1/c) = U(a_2/c) = 1$. Let $q(x_1, x_2, y)$ be $\mathrm{tp}(a_1, a_2, c/\mathrm{acl}(\varnothing))$ and let $r(y) = \mathrm{tp}(c/\mathrm{acl}(\varnothing))$. Now $U(a_1, a_2, c) = U(a_1, a_2/c) + U(c/\varnothing) = U(c/a_1, a_2) + U(a_1, a_2/\varnothing)$. Thus

(i) $U(c/a_1, a_2) = n - 1$ (and we know that $U(a_1, a_2/c) = 1$).

(ii) If c_1, c_2 realize r and $c_1 \neq c_2$, then $q(x_1, x_2, c_1)$ and $q(x_1, x_2, c_2)$ have only finitely many realizations in common.

Using Lemma 1.5.10, and applying compactness to (i) and (ii), we find a formula $\sigma(x_1, x_2, y) \in q(x_1, x_2, y)$ such that

(iii) If $(\mathbf{a}', c')$ satisfies $\sigma(\mathbf{x}, y)$ then $U(\mathbf{a}'/\varnothing) \leq 2$, $U(c'/\varnothing) \leq n$, $U(\mathbf{a}'/c') \leq 1$, and $U(c'/\mathbf{a}') \leq n - 1$. Also

(iv) For any $c' \neq c''$, $\sigma(\mathbf{x}, c') \wedge \sigma(\mathbf{x}, c'')$ is algebraic (or inconsistent).

We now aim to show that the formula $\sigma(x_1, x_2, c)$ has Morley rank 1. As (by (iii)) $\sigma(x_1, x_2, c)$ has U-rank 1, it clearly suffices to show that there are only finitely many non-algebraic strong types over c containing $\sigma(\mathbf{x}, c)$.

Claim 1. *Suppose* $\models \sigma(\mathbf{b}, c)$, *and* $\mathbf{b} \notin \mathrm{acl}(c)$. *Then* $U(\mathbf{b}/\varnothing) = 2$ *and* $U(c/\mathbf{b}) = n - 1$.

Proof. By (iii) $U(c/\mathbf{b}) \leq n - 1$. So c forks with $\mathbf{b}$ over $\varnothing$. But by (iii) again (and the assumption that $\mathbf{b} \notin \mathrm{acl}(c)$) we must have $U(\mathbf{b}/c) = 1$. So by 1.3.18, $U(\mathbf{b}) \geq 2$. By (iii) $U(\mathbf{b}) = 2$, whence $U(c/\mathbf{b}) = n - 1$. ■

Claim 2. *Suppose that* $\models \sigma(\mathbf{b}, c)$ *and* $\mathbf{b} \notin \mathrm{acl}(c)$. *Let* c' *realize the non-forking extension of* $\mathrm{stp}(c/\mathbf{b})$ *over* $\mathbf{b}c$. *Then* c *is independent from* c' *over* $\varnothing$ *(and note that* $\models \sigma(\mathbf{b}, c')$ *too).*

Proof. By Claim 1, $U(\mathbf{b}, c, c') = U(\mathbf{b}) + U(c/\mathbf{b}) + U(c'/c\mathbf{b}) = 2 + (n - 1) + (n - 1) = 2n$. Note that $c \neq c'$ (as by Claim 1, $U(c/\mathbf{b}) = n - 1 \geq 1$). So by (iv) $b \in \mathrm{acl}(c, c')$. Thus (using $U(\mathbf{b}, c, c') = U(\mathbf{b}/c, c') + U(c, c')$) we see that $U(c, c') = 2n$. As both c and c' realize the type r which has U-rank n, it follows that $U(c'/c) = n$, that is c is independent from c' over $\varnothing$. ■

Let $P = \{\mathrm{tp}(\mathbf{b}/\mathrm{acl}(c))$: for some c_1 realizing r, with c_1 independent from c over $\varnothing$, $\models \sigma(\mathbf{b}, c) \wedge \sigma(\mathbf{b}, c_1)\}$. Then P is finite: fix c_1 realizing r, with c_1 independent from c over $\varnothing$. Then by (iv) $\sigma(\mathbf{x}, c) \wedge \sigma(\mathbf{x}, c_1)$ has only finitely many solutions. Let c_2 be some other realization of r, independent with c over $\varnothing$. Then (as r is stationary) there is an automorphism which fixes $\mathrm{acl}(c)$ and takes c_1 to c_2.

By Claim 2, any non-algebraic complete extension of $\sigma(\mathbf{x}, c)$ over acl(c) is in P.

We have shown that $\sigma(\mathbf{x}, c)$ has Morley rank 1. It follows that the formula $(\exists x_2)(\sigma(x_1, x_2, c))$ has Morley rank at most 1. This formula is in $\text{tp}(a_1/c)$. So $RM(a_1/c) \leq 1$. But $\text{tp}(a_1/c)$ is a non-forking extension of the non-algebraic type p, so (by 1.3.9(viii)) $RM(p) = 1$. The proof of 3.2 is complete. ■

COROLLARY **3.3.** *Let $p \in S(A)$ be a minimal type in a superstable theory. Then either p is locally modular or $RM(p) = 1$.*

Proof. By 3.1 and 3.2, noting that triviality implies local modularity. ■

4 Zilber's theorem

Here we prove a fundamental result due to Zilber (and independently Cherlin, Mills, Neumann) that an ω-categorical strongly minimal set is locally modular (and thus, by 1.11, its associated geometry, if non-trivial, is affine or projective over some finite field). The proof we give of this result is due to Hrushovski and yields more (as we shall see in Chapter 5): a 'unimodular' minimal set is locally modular.

We will introduce the notion of a unimodular mininal set, the word 'unimodular' coming from the theory of Haar measure on locally compact groups. It will be seen easily that a strongly minimal ω-categorical structure (or set) is unimodular. We will prove here that a unimodular minimal set D is '2-pseudolinear'. This means that the canonical base of a plane curve over D has U-rank at most 2. Specializing back to the strongly minimal ω-categorical case we will show that 2-pseudolinearity implies local modularity. In Chapter 5 we will show that any pseudolinear minimal set is locally modular.

The proof below depends on assigning rational numbers to types/definable sets in D^{eq}. For example, if X is a definable set of U-rank n, with a unique stationary extension of U-rank n, and 'generically' X is in some definable k to 1 relation with D^n, then the rational number assigned to X will be k. So this number (the 'Zilber degree') is analogous to the degree of an algebraic curve. The important thing is that it will be well defined (using unimodularity) and will depend only on 'generic' behaviour. In Proposition 4.14, under an hypothesis which shall be eventually ruled out, we will prove a Bezout-type theorem, expressing the cardinality of the intersection of two 1-dimensional sets as the product of the Zilber degrees of those sets. From this 2-pseudolinearity will be deduced in 4.15.

Recall that a complete 1-sorted theory T in a countable language L is

ω-categorical if for any n, there are only finitely many formulae $\varphi(x_1, \ldots, x_n)$ of L up to T-equivalence. Equivalently there are, for each n, only finitely many complete n-types of T. This is a classical result in model theory, following from the omitting types theorem.

This notion can be generalized as follows:

Definition 4.1. *Let* $\mathbf{C}$ *be a big model of a theory* T. *Let* X *be a definable set in* $\mathbf{C}$, *defined by a formula* $\varphi(x, a)$ *where* a *is the canonical parameter for* X. *We say that* X *(or the formula* $\varphi(x, a)$*) is* ω*-categorical, if for any* n, *there are only finitely many* a*-definable subsets of* X^n.

We now pass to unimodularity. Let us fix, as in section 2, a *minimal* set D defined in a big model $\mathbf{C}$ of a stable theory.

We assume for simplicity that D *is defined over* $\varnothing$ (i.e. D is the solution set of a partial type over $\varnothing$). We recall some notation: if $a \in \mathrm{acl}(A)$, then $\mathrm{mult}(a/A)$ denotes the number of elements having the same type over A as a. This agrees with the notation in Lemma 2.1 in case a is a tuple from D.

Definition 4.2. *We say that* D *is* unimodular *if whenever* $d_1, \ldots, d_n$, $d'_1, \ldots, d'_n \in D$ *with* $\dim(\mathbf{d}/\varnothing) = \dim(\mathbf{d}'/\varnothing) = n$, *and* $\mathrm{acl}(\mathbf{d}) = \mathrm{acl}(\mathbf{d}')$, *then* $\mathrm{mult}(\mathbf{d}/\mathbf{d}') = \mathrm{mult}(\mathbf{d}'/\mathbf{d})$.

Remark 4.3. *Assume* D *is strongly minimal and* ω*-categorical. Then* D *is unimodular.*

Proof. Let $\mathbf{d}$, $\mathbf{e}$ be tuples from D such that $\mathrm{tp}(\mathbf{e}/\varnothing) = \mathrm{tp}(\mathbf{d}/\varnothing) = r$, say, and $\mathrm{acl}(\mathbf{d}) = \mathrm{acl}(\mathbf{e})$. Let $X = \mathrm{acl}(\mathbf{d}) \cap D$. Then X is finite, by ω-categoricitiy of D. Let Y be the set of realizations of r, all of whose coordinates are in X. Similarly let W be the set of realizations of $\mathrm{tp}(\mathbf{d}, \mathbf{e}/\varnothing)$ in X. Then Y and W are both finite and non-empty. Clearly $|W| = |Y|.\mathrm{mult}(\mathbf{e}/\mathbf{d}) = |Y|.\mathrm{mult}(\mathbf{d}/\mathbf{e})$. So $\mathrm{mult}(\mathbf{d}/\mathbf{e}) = \mathrm{mult}(\mathbf{e}/\mathbf{d})$. In particular this shows that D is unimodular. ■

Let us return to the general case of minimal D.

Definition 4.4. *Let* $c \in D^{\mathrm{eq}}$. *Let* $d_1, \ldots, d_n$ *be algebraically independent elements of* D *such that* $c \in \mathrm{acl}(d_1, \ldots, d_n)$. *Suppose that* $\{d_1, \ldots, d_k\}$ *is independent from* c *over* $\varnothing$ *and that* $\mathbf{d} \in \mathrm{acl}(c, d_1, \ldots, d_k)$. *(So* c *is inter-algebraic with* $(d_{k+1}, \ldots, d_n)$ *over* $\{d_1, \ldots, d_k\}$.*) We define* $Z_{\mathbf{d}}(c)$ *to be* $\mathrm{mult}(c/\mathbf{d})/\mathrm{mult}(\mathbf{d}/(c, d_1, \ldots, d_k))$.

We will show in a moment that if D is unimodular then the value of $Z_{\mathbf{d}}(c)$ depends only on c. Before doing this let us remark how $Z_{\mathbf{d}}(c)$ can be computed in terms of indices of automorphism groups: with c, $\mathbf{d}$ as in Definition 4.4, let $X = \operatorname{acl}(\mathbf{d}) \cap D$. Let X^{eq} denote $\operatorname{dcl}(X) \cap D^{\mathrm{eq}}$. Let $G(X)$ denote the group of elementary permutations of X, or equivalently the permutations of X which extend to automorphisms of $\mathbf{C}$. Note that $G(X)$ acts also on X^{eq}. For any subset A of X^{eq} let $G(X, A)$ denote the subgroup of $G(X)$ consisting of maps which fix A pointwise. If $\mathbf{a}$ is an enumeration of A we also write $G(X, \mathbf{a})$ for $G(X, A)$. Then we have:

REMARK 4.5. $Z_{\mathbf{d}}(c) = [G(X, \mathbf{d})\colon G(X, (c, \mathbf{d})]/[G(X, (c, d_1, \ldots, d_k))\colon G(X, (c, \mathbf{d}))]$.

Proof. All that has to be remarked is that if $a, b \in X^{\mathrm{eq}}$ and $X \subseteq \operatorname{acl}(a)$, then $\operatorname{mult}(b/a) = [G(X, a)\colon G(X, (a, b))]$. ■

LEMMA 4.6. *If D is unimodular then $Z_{\mathbf{d}}(c)$ does not depend on the choice of* $\mathbf{d}$.

Proof. Let $e_1, \ldots, e_r, e_{r+1}, \ldots, e_m$ be algebraically independent elements of D such that $c \in \operatorname{acl}(\mathbf{e})$, $\{e_1, \ldots, e_r\}$ is independent from c over $\varnothing$, and $\mathbf{e} \in \operatorname{acl}(c, e_1, \ldots, e_r)$. We must show that $Z_{\mathbf{e}}(c) = Z_{\mathbf{d}}(c)$.

Let us write $(d_1, \ldots, d_k)$ as $\mathbf{d}^1$, $(d_{k+1}, \ldots, d_n)$ as $\mathbf{d}^2$, $(e_1, \ldots, e_r)$ as $\mathbf{e}^1$, and $(e_{r+1}, \ldots, e_m)$ as $\mathbf{e}^2$.

We can assume that $\operatorname{tp}(\mathbf{e}/\mathbf{d}c)$ does not fork over c. It follows easily that each of $(\mathbf{d}^1, \mathbf{e}^1, \mathbf{d}^2)$, $(\mathbf{d}^1, \mathbf{e}^1, \mathbf{e}^2)$ is an algebraically independent tuple of elements of D, and moreover, $\operatorname{acl}(\mathbf{d}^1, \mathbf{e}^1, \mathbf{d}^2) = \operatorname{acl}(\mathbf{d}^1, \mathbf{e}^1, \mathbf{e}^2)$. Also $(\mathbf{d}^1, \mathbf{e}^1)$ is independent from c over $\varnothing$ and $\operatorname{acl}(\mathbf{d}^1, \mathbf{e}^1, \mathbf{d}^2) = \operatorname{acl}(\mathbf{d}^1, \mathbf{e}^1, c) = \operatorname{acl}(\mathbf{d}^1, \mathbf{e}^1, \mathbf{e}^2)$.

CLAIM 1. $Z_{\mathbf{d}}(c) = Z_{(\mathbf{d}^1, \mathbf{e}^1, \mathbf{d}^2)}(c)$.

Proof of claim. Note that $\mathbf{e}^1$ is independent from $\{\mathbf{d}, c\}$ over $\varnothing$ and $\operatorname{tp}(\mathbf{e}^1/\varnothing)$ is stationary. Thus $\operatorname{mult}(c/\mathbf{d}) = \operatorname{mult}(c/\mathbf{d}, \mathbf{e}^1)$ (i.e any $\mathbf{d}$-automorphisc image of c is a $(\mathbf{d}, \mathbf{e}^1)$-automorphic image). Similarly $\operatorname{mult}(\mathbf{d}/c, \mathbf{d}^1) = \operatorname{mult}(\mathbf{d}/c, \mathbf{d}^1, \mathbf{e}^1)$. This is enough. ■

Similarly:

CLAIM 2. $Z_{\mathbf{e}}(c) = Z_{(\mathbf{d}^1, \mathbf{e}^1, \mathbf{d}^2)}(c)$.

Let $Y = \operatorname{acl}(\mathbf{d}, \mathbf{e}) \cap D$. By Claims 1 and 2 and Remark 4.5 we have to show

$$(*)\ [G(Y, (\mathbf{d}, \mathbf{e}^1))\colon [G(Y, (c, \mathbf{d}, \mathbf{e}^1))]/[G(Y, (c, \mathbf{d}^1, \mathbf{e}^1))\colon G(Y, (c, \mathbf{d}, \mathbf{e}^1))]$$
$$= [G(Y, (\mathbf{e}, \mathbf{d}^1))\colon [G(Y, (c, \mathbf{e}, \mathbf{d}^1))]/[G(Y, (c, \mathbf{e}^1, \mathbf{d}^1))\colon [G(Y, (c, \mathbf{e}, \mathbf{d}^1))].$$

CLAIM 3. $[G(Y, (\mathbf{d}, \mathbf{e}^1)): G(Y, (c, \mathbf{d}, \mathbf{e}))] = [G(Y, (\mathbf{e}, \mathbf{d}^1)): G(Y, (c, \mathbf{d}, \mathbf{e}))]$ ($=n$, say).

Proof. By the definition of unimodularity, applied to $(\mathbf{d}^1, \mathbf{e}^1, \mathbf{d}^2)$ and $(\mathbf{d}^1, \mathbf{e}^1, \mathbf{e}^2)$ and 4.5, we have:

$$[G(Y, (\mathbf{d}, \mathbf{e}^1)): G(Y, (\mathbf{d}, \mathbf{e}))] = [G(Y, (\mathbf{e}, \mathbf{d}^1)): G(Y, (\mathbf{d}, \mathbf{e}))].$$

The claim follows. ■

Let now
$$u = [G(Y, (c, \mathbf{d}^1, \mathbf{e}^1)): [G(Y, (c, \mathbf{d}, \mathbf{e}^1))]$$
$$v = [G(Y, (\mathbf{d}, \mathbf{e}^1)): [G(Y, (c, \mathbf{d}, \mathbf{e}^1))]$$
$$w = [G(Y, (c, \mathbf{d}^1, \mathbf{e}^1)): G(Y, (c, \mathbf{e}, \mathbf{d}^1))]$$
$$x = [G(Y, (\mathbf{e}, \mathbf{d}^1)): G(Y, (c, \mathbf{e}, \mathbf{d}^1))].$$

Also let
$$y = [G(Y, (c, \mathbf{d}, \mathbf{e}^1)): G(Y, (c, \mathbf{d}, \mathbf{e}))]$$
$$z = [G(Y, (c, \mathbf{e}, \mathbf{d}^1)): G(Y, c, \mathbf{d}, \mathbf{e}))]$$

Then

(4) $\quad u.y = w.z \; (=[G(Y, (c, \mathbf{d}^1, \mathbf{e}^1)): G(Y, (c, \mathbf{d}, \mathbf{e}))])$

(5) $\quad v.y = x.z = n$ (by Claim 3).

From (4) and (5) we obtain

(6) $\quad v/u = x/w$, which is exactly (*).

The lemma is proved. ■

For the remainder of this section we assume D to be unimodular.

DEFINITION 4.7. *Let $a, b, c \in D^{\mathrm{eq}}$.*

(i) *We define $Z(c)$ to be $Z_{\mathbf{d}}(c)$ for some algebraically independent tuple $\mathbf{d} = (d_1, \ldots, d_n)$ of elements of D such that $c \in \mathrm{acl}(\mathbf{d})$. (This is well defined by Lemma 4.6.)*

(ii) *We define $Z(a/b)$ to be $Z(ab)/Z(b)$.*

If B is a finite set in D^{eq}, then by $Z(a/B)$ we mean $Z(a/\mathbf{b})$ where $\mathbf{b}$ is some (any) enumeration of B.

LEMMA 4.8. (i) *For $a, b, c \in D^{\mathrm{eq}}$, $Z(ab/c) = Z(a/bc).Z(b/c)$.*

(ii) *If $\mathrm{tp}(ab) = \mathrm{tp}(a'b')$ $(a, b \in D^{\mathrm{eq}})$ then $Z(a/b) = Z(a'/b')$.*

(iii) *If* $d_1, \ldots, d_n \in D$ *and are algebraically independent over* b *then* $Z(d_1, \ldots, d_n/b) = 1$.

(iv) *If* $a \in \mathrm{acl}(b)$ $(a, b \in D^{\mathrm{eq}})$ *then* $Z(a/b) = \mathrm{mult}(a/b)$.

Proof. (i) and (ii) are immediate.

(iii) By (i) it is enough to prove it for $n = 1$.

So suppose $d \in D$ and $d \notin \mathrm{acl}(b)$. Let $d_1, \ldots, d_n \in D$ be algebraically independent such that $b \in \mathrm{acl}(d_1, \ldots, d_n)$ We may assume that $\{d_1, \ldots, d_n, d\}$ is also algebraically independent. We may assume that b is independent from $\{d_1, \ldots, d_k\}$ over $\varnothing$ (some $k \leq n$) and that b is interalgebraic with $(d_1, \ldots, d_n)$ over $\{d_1, \ldots, d_k\}$. Let $\mathbf{d}$ denote $(d_1, \ldots, d_n)$ and $\mathbf{d}^1$ denote $\{d_1, \ldots, d_k\}$. Then $Z(b) = \mathrm{mult}(b/\mathbf{d})/\mathrm{mult}(\mathbf{d}/b\mathbf{d}^1)$. Clearly also $Z(db) = \mathrm{mult}(db/\mathbf{d}d)/\mathrm{mult}(\mathbf{d}d/db\mathbf{d}^1)$. But clearly

$$\mathrm{mult}(b/\mathbf{d}) = \mathrm{mult}(bd/\mathbf{d}d) \quad \text{and} \quad \mathrm{mult}(\mathbf{d}/b\mathbf{d}^1) = \mathrm{mult}(\mathbf{d}d/db\mathbf{d}^1).$$

So $Z(db) = Z(b)$, and $Z(d/b) = 1$.

(iv) Choose again $d_1, \ldots, d_k, \ldots, d_n \in D$ algebraically independent such that b is independent from $\{d_1, \ldots, d_k\}$ over $\varnothing$ and b is interalgebraic with $\mathbf{d}$ over $\{d_1, \ldots, d_k\}$. The same is true for ab in place of b.

So

$$Z(a/b) = (\mathrm{mult}(ab/\mathbf{d})/\mathrm{mult}(\mathbf{d}/abd_1, \ldots, d_k))/(\mathrm{mult}(b/\mathbf{d})/\mathrm{mult}(\mathbf{d}/bd_1, \ldots, d_k)).$$

This is easily seen to be equal to $\mathrm{mult}(a/bd_1, \ldots, d_k)$ which equals $\mathrm{mult}(a/b)$ as $\mathrm{tp}(d_1, \ldots, d_k/\varnothing)$ is stationary and $(d_1, \ldots, d_k)$ is independent from ab over $\varnothing$. ■

LEMMA **4.9.** (*All elements are assumed to be from* D^{eq}.)

(i) *Let* $c = \mathrm{Cb}(\mathrm{stp}(a/b))$. *Then* $Z(a/b) = \mathrm{mult}(\mathrm{tp}(a/b)).Z(a/bc)$.

(ii) *Let* $\mathrm{tp}(a/b)$ *be stationary, and suppose* a *is independent from* c *over* b. *Then* $Z(a/b) = Z(a/bc)$.

Proof. (i) Note first (by 1.2.26) that $c \in \mathrm{acl}(b)$. Moreover, it is not difficult to see that $\mathrm{mult}(\mathrm{tp}(a/b)) =_{\mathrm{def}}$ number of non-forking extensions of $\mathrm{tp}(a/b)$ over $\mathrm{acl}(b)$, is precisely $\mathrm{mult}(c/b)$ which equals $Z(c/b)$ by 4.8(iv). (These non-forking extensions are, by Chapter 1, conjugate over b, and determined by the image of c.) Also by 1.2.26 again $c \in \mathrm{dcl}(b, a)$.

Now $Z(abc) = Z(a/bc).Z(c/b).Z(b) = Z(c/ab).Z(a/b).Z(b)$ Thus $Z(a/bc)$. $\mathrm{mult}(\mathrm{tp}(a/b).Z(b) = 1.Z(a/b).Z(b)$, which gives what we want.

(ii) Clearly we can find $d_1, \ldots, d_n \in D$, such that $c \in \mathrm{dcl}(d_1, \ldots, d_n)$ and a is independent from $\{d_1, \ldots, d_n, b\}$ over b. We first prove

$(_*)$ for each $i = 0, \ldots, n-1$, $Z(a/bd_1 \ldots . d_i) = Z(a/bd_1 \ldots d_{i+1})$.

Proof of (*). It is enough to do it for $i = 0$. Suppose first that $d_1 \in \mathrm{acl}(b)$. As above $Z(d_1/ab).Z(a/b) = Z(a/d_1b).Z(d_1/b)$. But as $\mathrm{tp}(a/b)$ is stationary and a is independent from d_1 over b, $Z(d_1/ab) = \mathrm{mult}(d_1/ab) = \mathrm{mult}(d_1/b) = Z(d_1/b)$. Thus $Z(a/b) = Z(a/d_1b)$.

Now suppose that $d_1 \notin \mathrm{acl}(b)$. Then $d_1 \notin \mathrm{acl}(ab)$. So by 4.8(iii), $Z(d_1/ab) = Z(d_1/b) = 1$. So again we obtain $Z(a/b) = Z(a/d_1b)$. (*) is proved.

From (*) we obtain $Z(a/b) = Z(a/bd_1, \ldots, d_n)$. The same argument shows that $Z(a/bc) = Z(a/bcd_1 \ldots d_n)$. But $c \in \mathrm{dcl}(bd_1 \ldots d_n)$. It follows that $Z(a/bcd_1 \ldots d_n) = Z(a/bd_1 \ldots d_n)$, so $Z(a/b) = Z(a/bc)$ as required. ■

By virtue of Lemma 4.9, we can make the following definition without risk of ambiguity.

Definition 4.10. (i) *Let* $\mathrm{tp}(a/B)$ *be stationary (where* $\{a\} \cup B \subseteq D^{\mathrm{eq}}$*). Then* $Z(a/B) =_{\mathrm{def}} Z(a/b)$ *where b is some (any) element of B such that* $\mathrm{tp}(a/B)$ *does not fork over b and* $\mathrm{tp}(a/b)$ *is stationary. (Note that by Claim 2.2 and Remark 1.2.26 such b exists.)*

(ii) *For any* $a \in D^{\mathrm{eq}}$, $B \subseteq D^{\mathrm{eq}}$, $Z(a/B) =_{\mathrm{def}} \mathrm{mult}(\mathrm{tp}(a/B)).Z(a/\mathrm{acl}(B))$.

(iii) *For* $q = \mathrm{tp}(a/B)$, *define* $Z(q) = Z(a/B)$ *(this clearly is well-defined for* $Z(a/B)$ *depends only on* $\mathrm{tp}(a,B/\emptyset)$*).*

So note:

Remark 4.11. (i) *if* $\mathrm{tp}(a_1/B_1)$, $\mathrm{tp}(a_2/B_2)$ *are stationary and parallel (i.e. have a common non-forking extension), then* $Z(a_1/B_1) = Z(a_2/B_2)$.

(ii) *Suppose* $\mathrm{tp}(a_1/B_1)$ *and* $\mathrm{tp}(a_2/B_2)$ *have exactly the same non-forking extensions over* $\mathrm{acl}(B_1B_2)$; *then* $Z(a_1/B_1) = Z(a_2/B_2)$.

Finally we are able to define $Z(X)$ for ∞-definable subsets of D^{eq}, or equivalently $Z(\Phi)$ for the corresponding partial types in D^{eq}. Note that this is consistent with the case where Φ is a complete type.

Definition 4.12. *Let* Φ *be partial type in* D^{eq} *over the set* $B \subseteq D^{\mathrm{eq}}$. *Suppose* $U(X) = n$, *and let* $q_1, \ldots, q_m$ *be the distinct complete types over B extending* Φ *which have U-rank n. Then* $Z(\Phi) =_{\mathrm{def}} Z(q_1) + Z(q_2) + \cdots + Z(q_m)$.

LEMMA 4.13. (i) *Let $\Phi(x, y)$ be a partial type in D^{eq} over $B \subseteq D^{eq}$. Let $\Phi_2(y)$ be the partial type $(\exists x)(\Phi(x, y))$. Suppose that for all b realizing Φ_2, $U(\Phi(x, b)) = r$, $Z(\Phi(x, b)) = d$ (some fixed r, d). Then $Z(\Phi) = d.Z(\Phi_2)$.*

(ii) *Suppose that $\Phi(z)$, $\Psi(z)$ are partial types in D^{eq}. Suppose that $U(\Phi(z)) = U(\Psi(z)) = U(\Phi(z) \cup \Psi(z))$, and similarly $\text{mult}(\Phi(z)) = \text{mult}(\Psi(z)) = \text{mult}(\Phi(z) \cup \Psi(z))$ (i.e. $\Phi(z)$ and $\Psi(z)$ have the same stationary non-forking extensions of maximal rank). Then $Z(\Phi) = Z(\Psi)$.*

(iii) *If the partial type Φ is algebraic then $Z(\Phi)$ is exactly the number of solutions of Φ.*

Proof. (i) Suppose $U(\Phi(x, y)) = n$. Let $b_1, \ldots, b_k$ be realizations of the distinct complete types over B of maximal rank extending $\Phi_2(y)$. For each b_i, let $a^i_1, \ldots, a^i_{m(i)}$ be realizations of the distinct complete types over Bb_i which have U-rank r and extend $\Phi(x, b_i)$. It is then easy to check from the assumptions that the set of (a^i_t, b_i) is a set of realizations of the distinct complete types over B of maximal rank which extend $\Phi(x, y)$.

So $Z(\Phi) = \sum_i \{\Sigma\, \{Z(a^i_t, b_i/B)\colon t = 1, \ldots, m(i)\}\} = \sum_i \{\Sigma\{Z(a^i_t/b_iB).Z(b_i/B)\colon t = 1, \ldots, m(i)\}\} = \sum_i d.Z(b_i/B) = d(\sum_i Z(b_i/B)) = d.Z(\Phi_2)$, as required.

(ii) and (iii) follow from the definitions and 4.11. ■

Now we have all the machinery to start proving things.

PROPOSITION 4.14. (*Remember we assume D to be unimodular and we work in D^{eq}.) Let $A \subseteq D^{eq}$ be algebraically closed (so any complete type over A of an element in D^{eq} will be stationary.) Let a_1, a_2, a'_1, $a'_2 \in D$. Let $b = \text{Cb}(\text{stp}(a_1, a_2/Ab))$, $c = \text{Cb}(\text{stp}(a'_1, a'_2/Ac))$. Assume $U(b/A)$, $U(c/A)$ are both ≥ 2, and that b is independent from c over A, and $U(a_1, a_2/AB) = U(a'_1, a'_2/Ac) = 1$. Let $p(\mathbf{x}) = \text{tp}(a_1, a_2/Ab)$, $q(\mathbf{x}) = \text{tp}(a'_1, a'_2/Ac)$ (note these are both stationary types).*

Then the cardinality of the set of solutions of $p(\mathbf{x}) \cup q(\mathbf{x})$ is equal to $Z(p).Z(q)$.

Proof. Write $p(\mathbf{x})$ as $p(\mathbf{x}, b)$ and $q(\mathbf{x})$ as $q(\mathbf{x}, c)$ (where $p(\mathbf{x}, y_1)$, $q(\mathbf{x}, y_2)$ are then complete types over A). Let $r_1 = \text{tp}(b/A)$ and $r_2 = \text{tp}(c/A)$. Let $Q(\mathbf{x}, y_1, y_2)$ be the partial type '$p(\mathbf{x}, y_1) \cup q(\mathbf{x}, y_2) \cup$ "y_1 is independent from y_2 over A".' As $p(\mathbf{x}, b)$ and $q(\mathbf{x}, c)$ are U-rank 1 types which are not parallel, clearly $Q(\mathbf{x}, b, c)$ has at most finitely many solutions.

CLAIM 1. *Q is consistent.*

Proof. Note that $U(a_1, a_2/A) = U(a'_1, a'_2/A) = 2$ (for otherwise $U(b/A)$ or

$U(c/A)$ would be 0). Thus (as $\text{tp}(a_1, a_2/A) = \text{tp}(a'_1, a'_2/A)$) there is c' such that $q(a_1, a_2, c')$ and we may assume that c' is independent from b over Aa_1a_2. The same arguments as in Claims 1 and 2 of Proposition 3.2 shows that b is independent from c' over A. Thus $(\mathbf{a}, b, c')$ realizes $Q(\mathbf{x}, y_1, y_2)$, proving the claim. ■

By the claim and the fact that $\text{tp}(b, c/A) = \text{tp}(b', c'/A)$ whenever b' realizes r_1, c' realizes r_2, and b' is independent from c' over A, we may assume that $(\mathbf{a}, b, c)$ realizes Q. Note also:

CLAIM 2. *For any* c' such that $q(\mathbf{a}, c')$, *if* c'' *realizes the non-forking extension of* $\text{stp}(c'/\mathbf{a}A)$ *over* $\mathbf{a}bA$, *then also* $(\mathbf{a}, b, c'')$ *realizes* Q.

Thus we have:

CLAIM 3. $Z(Q(\mathbf{a}, b, y_2)) = Z(q(\mathbf{a}, y_2))$.

Proof. By Claim 2 and 4.13(ii).

Let $Q_2(y_1, y_2)$ be $(\exists \mathbf{x})(Q(\mathbf{x}, y_1, y^2))$. So Q_2 is stationary and is the type of (b, c) over A. Thus for any (b', c') realizing Q_2, $|Q(\mathbf{x}, b, c)| = |Q(\mathbf{x}, b', c')|$. So by Lemma 4.13(i) and (iii),

$$(4) \qquad Z(Q) = Z(Q(\mathbf{x}, b, c)).Z(Q_2).$$

On the other hand, as the projection of Q on the $\mathbf{x}$, y_1 coordinates is precisely the complete type $p(\mathbf{x}, y)$, we have, again by 4.13(i):

$$(5) \quad Z(Q) = Z(Q(\mathbf{a}, b, y_2)).Z(p(\mathbf{x}, y_1)) = Z(Q(\mathbf{a}, b, y_2)).Z(p(\mathbf{x}, b)).Z(r_1(y_1)).$$

Using Claim 3 we see from (5) that

$$(6) \qquad Z(Q) = Z(q(\mathbf{a}, y_2)).Z(p(\mathbf{x}, b)).Z(r_1(y_1)).$$

But

$$(7) \qquad Z(q(\mathbf{x}, y_2)) = Z(\mathbf{a}, c/A) = Z(\mathbf{a}/A).Z(q(\mathbf{a}, y_2)) = Z(q(\mathbf{a}, y_2))$$
$$(\text{as } Z(\mathbf{a}/A)) = 1 \text{ by } 4.8(\text{iii})).$$

Also

$$(8) \qquad Z(q(\mathbf{x}, y_2)) = Z(\mathbf{a}, c/A) = Z(c/A).Z(\mathbf{a}/cA).$$

From (7) and (8) we have

$$(9) \qquad Z(q(\mathbf{a}, y_2)) = Z(c/A).Z(q(\mathbf{x}, c))(=_{\text{def}} Z(r_2(y_2)).Z(q(\mathbf{x}, c))).$$

From (4) and (6) and (9) we have

(10) $\quad Z(Q(\mathbf{x}, b, c)).Z(Q_2) = Z(r_1(y_1)).Z(r_2(y_2)).Z(p(\mathbf{x}, b)).Z(q(\mathbf{x}, c)).$

But (from 4.11(i), 4.13(i), and as $Q_2 = r_1 \otimes r_2$) we see that $Z(Q_2) = Z(r_1).Z(r_2)$. Thus (10) yields $Z(Q(\mathbf{x}, b, c)) = Z(p(\mathbf{x}, b)).Z(q(\mathbf{x}, c))$. This completes the proof of the proposition. ■

The funny thing about Proposition 4.14 is that its hypotheses will turn out to be impossible, although not quite yet.

THEOREM 4.15. (*D unimodular*) *We work over* $\mathrm{acl}(\varnothing) \cap D^{\mathrm{eq}}$. *Let* $a_1, a_2 \in D$. *Suppose that* $U(a_1, a_2/B) = 1$, *and* $b = \mathrm{Cb}(\mathrm{stp}(a_1, a_2/B))$. *Then* $U(b/\varnothing) \leq 2$, (*Namely, D is '2-pseudolinear'.*)

Proof. Suppose by way of contradiction that $U(b/\varnothing) = n \geq 3$. *Let* $\mathbf{a} = (a_1, a_2)$. Let c realize $\mathrm{stp}(b/\mathbf{a})$ such that c is independent from b over $\mathbf{a}$. Then as in Claims 1 and 2 of the proof of 3.2, we have

($_*$) b is independent from c over $\varnothing$

and $\mathrm{tp}(b/\varnothing) = \mathrm{tp}(c/\varnothing)$, $U(b/\mathbf{a}) = U(c/\mathbf{a}) = n - 1$.

Let $p(x_1, x_2, b) = \mathrm{tp}(a_1, a_2/b)$. By Proposition 4.14, we have

(1) the number of solutions of $p(\mathbf{x}, b) \cup p(\mathbf{x}, c)$ is equal to

$Z(p(x, \mathbf{b})).Z(p(\mathbf{x}, c))(= (Z(p(\mathbf{x}, b)))^2$ as $\mathrm{tp}(b) = \mathrm{tp}(c))$.

Let $q(\mathbf{x})$ be the non-forking extension of $p(\mathbf{x}, b)$ over $\mathbf{a}b$. Then $\mathrm{Cb}(q)$ is still b. Moreover, we can write q as $q(\mathbf{x}, b)$ where q may contain additional parameters from $\mathbf{a}$. Clearly also $q(\mathbf{x}, c)$ is the non-forking extension of $p(\mathbf{x}, c)$ over $\mathbf{a}c$.

Then by 4.11 we have

(2) $\quad Z(q(\mathbf{x}, b)) = Z(p(\mathbf{x}, b)) = Z(p(\mathbf{x}, c)) = Z(q(\mathbf{x}, c)).$

As c is independent from b over $\mathrm{acl}(\mathbf{a})$, and (by ($_*$)) $U(b/\mathbf{a}) = U(c/\mathbf{a}) \geq 2$, we have again using 4.14

(3) the number of solutions of $q(\mathbf{x}, b) \cup q(\mathbf{x}, c)$ equals $Z(q(\mathbf{x}, b)).Z(q(\mathbf{x}, c))$.

So by (1) and (2), $p(\mathbf{x}, b) \cup p(\mathbf{x}, c)$ has the same (finite number of) solutions as $q(\mathbf{x}, b) \cup q(\mathbf{x}, c)$. This is a contradiction, as $q(\mathbf{x}, b) \cup q(\mathbf{x}, c) \models p(\mathbf{x}, b) \cup p(\mathbf{x}, c)$, and $\mathbf{a}$ is a solution of the right-hand but not of the left-hand side.

This contradiction proves the theorem. ■

REMARK 4.16. *We point out the interpretation of unimodularity and the Z-function in terms of Haar measure.*

Recall first that if G is a locally compact group, then a (right) Haar measure on G is a non-zero Borel measure μ on G such that all compact subsets of G have measure $< \infty$ and for any μ-measurable subset A of G, $\mu(A) = \mu(A.g)$. Such a (right) Haar measure always exists. Similarly for left Haar measures. If φ is a topological automorphism of G, and μ is a right Haar measure then so is $\varphi(\mu)$, where $\varphi(\mu)(A)$ is defined to be $\mu(\varphi(A))$. In fact, given φ, there is unique $c \in \mathbf{R}^+$ such that for every *μ, $\varphi(\mu) = c.\mu$. Now for $x \in G$, Inn_x, conjugation by x, is a topological automorphism of G, and the map taking x to the c corresponding to Inn_x is called the* modular *function. G is said to be* unimodular *if the modular function has constant value* 1.

Now let D be a minimal set defined over $\varnothing$. As above, for each finite-dimensional algebraically closed subset X of D, let G(X) be the group of elementary permutations of X. Then G(X) is a locally compact topological group: for each tuple **a** *from X define $G(X, \mathbf{a}) = \{\sigma \in G(X) \colon \sigma(\mathbf{a}) = \mathbf{a}\}$. Then the subgroups $G(X, \mathbf{a})$ are taken as a neighbourhood basis of the identity of the topology on G(X). Note that if $\mathbf{X} = \mathrm{acl}(\mathbf{d}) \cap D$, then $G(X, \mathbf{d})$ is profinite and thus compact, This shows the topology on G(X) to be locally compact. Moreover, any right Haar measure μ on G(X) is determined once we know $\mu(A)$ for some compact open subset of G(X).*

Then unimodularity of D, as defined in 4.2, *is equivalent to unimodularity of G(X) for all X.*

For suppose first that **d**, **e** *are both bases for finite-dimensional X. As remarked in the proof of* 4.5, $\mathrm{mlt}(\mathbf{d}/\mathbf{e})$ = *index of $G(X, \mathbf{de})$ in $G(X, \mathbf{e})$. Thus for any right Haar measure μ on G(X),*

$$(_*) \qquad \mu(G(X, \mathbf{e})) = (\mathrm{mlt}(\mathbf{e}/\mathbf{d})/\mathrm{mlt}(\mathbf{d}/\mathbf{e})).(\mu(G(X, \mathbf{d}))).$$

So, assuming D to be unimodular then $\mu(G(X, \mathbf{e})) = \mu(G(X, \mathbf{d}))$. But clearly $G(X, \mathbf{e}) = \sigma.G(X, \mathbf{d}).\sigma^{-1}$, where $\sigma \in G(X)$ takes **d** *to* **e** *(as they have the same type). Thus $\mu(G(X, \mathbf{d})) = \mu(G(X, \mathbf{d})^\sigma)$ for any $\sigma \in G(X)$ (by replacing* **e** *above by $\sigma(\mathbf{d})$), and we see that G(X) is unimodular. On the other hand, if G(X) is unimodular, then (for σ taking* **d** *to* **e**) $\mu(G(X, \mathbf{d})) = \mu(G(X, \mathbf{d})^\sigma) = \mu(G(X, \mathbf{e}))$, *so by* $(_*)$ $\mathrm{mlt}(\mathbf{d}/\mathbf{e}) = \mathrm{mlt}(\mathbf{e}/\mathbf{d})$. *As this is true for all* **d**, **e**, *D is unimodular.*

Assuming now unimodularity (in all equivalent senses), define $\mu(X)$ to be the unique Haar measure on G(X) such that $\mu(G(X, \mathbf{d})) = 1$ (for some, or equivalently all, bases **d** *of X). Then the Zilber function turns out to be the following. Let b be any element of D^{eq}. Let X be a finite-dimensional subset of D such that $c \in \mathrm{dcl}(X)$. Let* **d** *be a basis for X* over *c. Then $Z(c) = \mu(G(X, (c, \mathbf{d})))^{-1}$.*

We now aim towards proving:

THEOREM **4.17.** *Suppose D is a strongly minimal ω-categorical set with canonical parameter a, say. Then the pregeometry of D is locally modular.*

Proof. The proof will be a little long and will proceed through a few lemmas.

We assume for simplicity that D is defined over $\varnothing$.

Assume by way of contradiction that D is not locally modular. There will be two stages in getting the contradiction. The first will be to modify D so as to obtain a 2-dimensional family of definable (generic) functions from D to D. The second stage will be to show that the family of all such functions forms a definable group, and then to use results from the end of Chapter 1 on the nature of such groups. The constructions in both these stages will appear again in Chapter 5 in a more general context: the group configuration. In the present context, the ω-categoricity assumption trivializes some of the arguments.

DEFINITION **4.18.** *By a* generically defined invertible function from D to D, *we mean a strong type,* $\mathrm{stp}(a_1, a_2/B)$, *where* $a_1, a_2 \in D$, $a_1, a_2 \notin \mathrm{acl}(B)$, $a_1 \in \mathrm{dcl}(a_2, B)$, *and* $a_2 \in \mathrm{dcl}(a_1, B)$.

Such a generically defined invertible function will be determined by some $\mathrm{acl}(B)$-definable bijection f between cofinite subsets of D. For clearly there are $\mathrm{acl}(B)$-definable partial functions f, g such that $f(a_1) = a_2$ and $g(a_2) = a_1$. Then $\{(x, y) \in D^2 \colon f(x) = y \wedge g(y) = x\}$ is the graph of the required definable function. The point, however, of the definition above is that we wish to identify such functions when they agree 'almost everywhere' on D. This is done most conveniently by identifying the function with the strong type above (up to parallelism) so in fact with the canonical base of the strong type. In accordance with Zilber's original parlance, we shall call also a generically defined invertible function a *quasitranslation* of D.

LEMMA **4.19.** *There is a strongly minimal set D_1 in D^{eq}, definable over some $b \in D^{\mathrm{eq}}$, and there is a quasitranslation q of D_1 such that $U(\mathrm{Cb}(q)/b) = 2$.*

Proof. The first step is to replace D by a strongly minimal set whose pregeometry is actually a geometry. This is made possible by ω-categoricity: let $E(x, y)$ be the equivalence relation on $D - \mathrm{acl}(\varnothing)$: $x \in \mathrm{acl}(y)$. (This is an equivalence relation, by forking symmetry.) By ω-categoricity of D, both $D - \mathrm{acl}(\varnothing)$ and E are $\varnothing$-definable. Let D_0 be the set of E-classes. It is clear that D_0 is strongly minimal, ω-categorical, has a unique 1-type over $\varnothing$ (which we call p_0), and the pregeometry of D_0 is now a geometry (in fact it is the geometry associate to the pregeometry of D). Note D_0 lives in D^{eq}.

As D was assumed to be non-locally modular, so also D_0 is non-locally modular, By Proposition 2.6, D_0 is not linear. Thus there are $u, v \in D_0$, and some $b \in D^{eq}$, such that $b = \mathrm{Cb}(\mathrm{stp}(u, v/b))$, $U(u, v/b) = 1$, and $U(b/\varnothing) \geq 2$. By Theorem 4.15, $U(b/\varnothing) = 2$. Note that $u, v \notin \mathrm{acl}(b)$, $v \in \mathrm{acl}(u, b)$, and $u \in \mathrm{acl}(v, b)$. Let $p_1(x_1, x_2, y) = \mathrm{tp}(u, v, b/\varnothing)$. Thus $p(x_1, x_2, b) = \mathrm{tp}(u, v/b)$ is stationary. Let c realize $\mathrm{stp}(b)$ such that c is independent from $\{b, u, v\}$ over $\varnothing$. Note that $\mathrm{tp}(c, u/\varnothing) = \mathrm{tp}(b, u/\varnothing)$. So we can find v' such that (u, v') realizes $p_1(x_1, x_2, c)$. Let $d = \mathrm{Cb}(\mathrm{stp}(v, v'/bc))$. So $d \in \mathrm{acl}(b, c)$.

CLAIM 1. *$U(d/\varnothing) = 2$, $\{b, c, d\}$ is pairwise independent over $\varnothing$, but each of b, c, d is algebraic over the other two.*

Proof of Claim 1. Note v and v' are interalgebraic over d, and $U(v, v'/bc) = 1$. By 4.15, $U(d/\varnothing) \leq 2$. But clearly $\mathrm{stp}(uv'/bcd)$ is a non-forking extension of both $\mathrm{stp}(uv'/c)$ and $\mathrm{stp}(uv'/bd)$. As $c = \mathrm{Cb}(\mathrm{stp}(u, v'/c))$ it follows from 1.2.26 that $c \in \mathrm{acl}(b, d)$. But $U(c/b) = 2$. Thus $U(d/b) \geq 2$. So $U(d/b) = 2$ and d is independent from b over $\varnothing$. The rest of the claim follows in a similar fashion. ■

CLAIM 2. $v' \in \mathrm{dcl}(b, c, d, u, v)$.

Proof. If not, let $w \neq v'$, $\mathrm{tp}(w/b, c, d, u, v) = \mathrm{tp}(v'/b, c, d, u, v)$. By choice of D_0, w is independent from v' over $\varnothing$, that is $U(v', w/\varnothing) = 2$. Now clearly $U(c, u/\varnothing) = U(c, u, v', w/\varnothing) = U(c, v', w/\varnothing) = 3$. Thus $U(c/v', w) = 1$. Similarly $U(d, v/\varnothing) = U(d, v, v', w/\varnothing) = U(d, v', w/\varnothing) = 3$, so $U(d/v', w) = 1$. Thus $U(c, d/v', w) \leq 2$. But $b \in \mathrm{acl}(c, d)$ (by Claim 1); hence $U(b, c/v', w) \leq 2$. Altogether we obtain $U(b, c, v', w) \leq 4$. But $U(b, c/\varnothing) = 4$; thus $v' \in \mathrm{acl}(b, c)$. Then clearly also $u \in \mathrm{acl}(b, c)$, which is a contradiction. Claim 2 is proved. ■

Note that $v' \notin \mathrm{acl}(b)$; thus we can find $u' \in D_0$ such that (u', v') realizes $p(x_1, x_2, b)$. As in the proof of Claim 2 we obtain:

CLAIM 3. *$u' \in dcl(b, c, d, u, v)$.*

Altogether we have $(u', v') \in \mathrm{dcl}(u, v, b, c, d)$. A similar argument shows that $(u, v) \in \mathrm{dcl}(u', v', b, c, d)$ (although this could be also obtained from Remark 4.3 applied to $tp(u, v/b)$).

Let D_1 be the set of realizations of $\mathrm{tp}(u, v/b)$. D_1 is strongly minimal ω-categorical, and inside D^{eq}. Let $q = \mathrm{stp}((u, v), (u', v')/bcd)$. Note that $U(q) = 1$, $bcd = \mathrm{Cb}(q)$, and $U(bcd/b) = 2$. Also we have just seen that q is a quasi-translation of D_1. This proves the lemma. ■

Let D_1 and q be as given by Lemma 4.19. *We add a name for b to the language*, so assume D_1 to be defined over $\varnothing$.

LEMMA **4.20.** *For every quasitranslation r of* D_1, *there is some 4-tuple* **b** *from* D_1 *such that* $\mathrm{Cb}(r) \in \mathrm{acl}(\mathbf{b})$.

Proof. Looking back to the proof of (ii) → (iii) of Proposition 2.6, we see what was actually proved there was that if D is a linear minimal set defined over $\varnothing$, then for any strong type, $\mathrm{stp}(a_1, a_2/B)$ of U-rank 1, with $a_1, a_2 \in D$, there are two elements e, $d \in D$ such that $\mathrm{Cb}(\mathrm{stp}(a_1, a_2/B)) \in \mathrm{acl}(e, d)$ (although this was already known from 2.5). A similar proof (left as an exercise to the reader) shows that if D is 2-pseudolinear, then for every a_1, a_2, B as above, there are four elements e_1, e_2, e_3, e_4 in D such that $\mathrm{Cb}(\mathrm{stp}(a_1, a_2/B)) \in \mathrm{acl}(e_1, e_2, e_3, e_4)$. By 4.3 and 4.15, D_1 *is* 2-pseudolinear. So this holds for D_1. ■

Our aim now is to show that the group of all quasitranslations of D_1, together with the 'generic' action of this group on D_1, are definable, and that the group has Morley rank (or U-rank) 2.

Lemma 4.20 says that whenever $r = \mathrm{stp}(a_1, a_2/c)$ is a quasitranslation of D_1, and $c = \mathrm{Cb}(r)$, then there is some 4-tuple **b** from D_1 such that (a_1, a_2) is independent from **b** over c, and $c \in \mathrm{acl}(\mathbf{b})$. Thus $\mathrm{stp}(a_1, a_2/\mathbf{b})$ is the non-forking extension of r. Now as D_1 is ω-categorical, there are only finitely many types (over $\varnothing$) of the form $\mathrm{tp}(a_1, a_2, \mathbf{b})$ where $a_1, a_2 \in D_1$ and **b** is a 4-tuple from D_1. For each such $a_1, a_2, \mathbf{b}$, there are only finitely many extensions of $\mathrm{tp}(a_1, a_2/\mathbf{b})$ over $\mathrm{acl}(\mathbf{b})$ (see Lemma 2.1), and the canonical base of each such extension is in $\mathrm{acl}(\mathbf{b})$. It follows that there are finitely many complete types over $\varnothing$, $s_1(x_1, y_1, z_1), \ldots, s_n(x_1, y_1, z_n)$, such that for each quasitranslation r of D_1, there is $c \in D_1^{\mathrm{eq}}$ and i between 1 and n, such that r is parallel to $s_i(x_1, y_1, c)$, $c = \mathrm{Cb}(r)$, and, moreover, each s_i occurs in this way. Note that by ω-categoricity of D_1 each s_i is equivalent to a formula over $\varnothing$. Let $t_i(z_i)$ be $(\exists x_1, y_1)(s_i(x_1, y_1, z_i))$. We may assume that the types t_i are distinct, that for $i \neq j$, and c realizing t_i and any d realizing t_j, $s_i(x_1, y_1, c)$ is not parallel to $s_j(x_1, y_1, d)$, and for any i and c, d realizing t_i, $s_i(x_1, y_1, c)$ is parallel to $s_i(x_1, y_1, d)$ iff $c = d$ (by quotienting by a suitable $\varnothing$-invariant and thus $\varnothing$-definable equivalence relation). The disjoint union of the solution sets of the t_i then forms a $\varnothing$-definable set in D_1^{eq}, which we shall call G (for group). If $\sigma \in G$, σ realizing t_i say, and $a_1 \in D_1$ is independent from c over $\varnothing$, then there is unique $a_2 \in D$ such that (a_1, a_2, c) realizes $s_i(x_1, y_1, z_i)$. Similarly given a_2 there is unique such a_1. We write $a_2 = \sigma.a_1$, and $a_1 = \sigma^{-1}.a_2$. Note that the (partial) map $(\sigma, a) \to \sigma.a$ is then $\varnothing$-definable. Note also that if $\sigma, \tau \in G$ and $\sigma \neq \tau$, then for $a \in D_1$ such that

$a \notin \mathrm{acl}(\sigma, \tau)$, $\sigma.a \neq \tau.a$. (For otherwise $\mathrm{stp}(a, \sigma.a/\sigma)$ is parallel to $\mathrm{stp}(a, \sigma.a/\tau)$ whence $\sigma = \tau$ by our above assumptions on the t_i.) Now let $\sigma_1, \sigma_2 \in G$. Let $a \in D_1$, $a \notin \mathrm{acl}(\sigma_1, \sigma_2)$. Then $\sigma_1.a \notin \mathrm{acl}(\sigma_1, \sigma_2)$, and hence $\sigma_2.(\sigma_1.a)$ is defined (and independent from σ_1, σ_2). Then $\mathrm{stp}(a, \sigma_2.(\sigma_1.a)/\sigma_1, \sigma_2)$ is a quasitranslation of D_1 so is parallel to $s_i(x_1, y_1, \sigma)$ for some *unique* s_i and *unique* σ realizing t_i. The map which takes (σ_2, σ_1) to σ is invariant under automorphisms of the universe, and thus is, by ω-categoricity of D_1^{eq}, $\emptyset$-definable. We write $\sigma = \sigma_2 * \sigma_1$. We summarize:

Lemma 4.21. *$(G, *)$ is a $\emptyset$-definable group in D_1^{eq} of U-rank (or equivalently Morley rank) 2. There is a partial $\emptyset$-definable map $(\sigma, a) \to \sigma.a$ from $(G \times D_1)$ to D_1 such that*

(i) *for $a \notin \mathrm{acl}(\sigma)$, $\sigma.a$ is defined;*

(ii) *for $\sigma, \tau \in G$, $\sigma = \tau$ iff for some (any) $a \in D_1 - \mathrm{acl}(\sigma, \tau)$, $\sigma.a = \tau.a$;*

(iii) *for $\sigma_1, \sigma_2, \sigma \in G$, $\sigma_1 * \sigma_2 = \sigma$ iff for $a \in D_1 - \mathrm{acl}(\sigma_1, \sigma_2)$, $\sigma.a = \sigma_1.(\sigma_2.a)$.*

Proof. The only thing to be seen is that $U(G) = 2$. Well for any $\sigma \in G$, $U(\sigma/\emptyset) \leq 2$, by 2-pseudolinearity. On the other hand Lemma 4.19 provides us with some $\sigma \in G$ with $U(\sigma/\emptyset) = 2$. Thus $U(G)$ $(= \max(U(\mathrm{tp}(\sigma/\emptyset)): \sigma \in G) = 2$. ∎

We are now able to derive a contradiction. G is totally transcendental, so has a smallest definable subgroup of finite index, G^0, its connected component, which is also defined over $\emptyset$. $RM(G^0) = U(G^0) = 2$. G^0 being defined in D_1^{eq} is also ω-categorical. By Fact 1.6.24, and 4.21, G^0 is nilpotent, with $Z(G^0)$ = the centre of G^0, infinite.

Claim. *If $\sigma \in G^0$ and $a \in D_1 - \mathrm{acl}(\sigma)$, then $\sigma \in \mathrm{dcl}(a, \sigma.a)$.*

Proof. Let $\tau \in G^0$ be such that $\mathrm{tp}(\tau/a, \sigma.a) = \mathrm{tp}(\sigma/a, \sigma.a)$. In particular $\tau.a = \sigma.a$. Let $\sigma' \in Z(G^0)$ be such that $U(\sigma'/\emptyset) \geq 1$ (as $Z(G^0)$ is infinite) and σ' is independent from $\{\sigma, \tau, a\}$ over $\emptyset$. As $U(\sigma'/\emptyset) \geq 1$, it is not difficult to see that $\{a, \sigma'(a)\}$ forks with σ' over $\emptyset$. In particular,

(*) $\quad \sigma'.a \notin \mathrm{acl}(\sigma, \tau)$ and note also

(**) $\quad a \notin \mathrm{acl}(\sigma, \sigma')$, $a \notin \mathrm{acl}(\tau, \sigma')$.

Let $\sigma'.a = b$. Then

$$\begin{aligned}\sigma.b &= \sigma.(\sigma'(a))\\ &= (\sigma * \sigma').a \quad \text{(by (**) and 4.21(iii))}\\ &= (\sigma' * \sigma).a \quad \text{(as } \sigma' \in Z(G^0))\\ &= \sigma'.(\sigma.a) \quad \text{(again by (**) and 4.21(iii))}\\ &= \sigma'.(\tau.a)\\ &= (\sigma' * \tau).a \quad \text{(by (**) and 4.21(iii))}\\ &= (\tau * \sigma').a\\ &= \tau.(\sigma'.a)\\ &= \tau(b).\end{aligned}$$

By 4.21(ii), $\tau = \sigma$. The claim is proved. ■

In the above, let $\sigma \in G^0$ with $U(\sigma/\varnothing) = 2$. By the claim and the fact that σ is independent from a over $\varnothing$, we have $U(\sigma/\varnothing) = U(\sigma/a) \leq U(\sigma.a/a) \leq 1$, giving our much sought after contradiction.

This contradiction proves Theorem 4.17. ■

COROLLARY **4.22.** *Let* (S, cl) *be an infinite, locally finite, homogeneous, non-trivial geometry. Then* (S, cl) *is an infinite-dimensional affine or projective geometry over some finite field.*

Proof. Let L be the first-order language which has a relation symbol $R_n(x_1, \ldots, x_n, x)$ of arity $n + 1$ for each $n < \omega$. We make S into an L-structure by defining: $R_n(b_1, \ldots, b_n, a)$ iff $a \in \mathrm{cl}(b_1, \ldots, b_n)$. The reader is invited to prove, using the assumptions on (S, cl), that $(S, \{R_n : n < \omega\})$ is an ω-categorical, strongly minimal structure, whose theory has quantifier elimination, and whose pregeometry is precisely (S, cl). Now use Theorem 4.17 and Fact 1.11. ■

5 'Coordinatization' in theories of finite U-rank

In this section and the next we study how knowledge of the (geometric) structure of U-rank 1 types can be used to deduce global properties of T, if T has finite U-rank. In particular the important notion '1-based' will make its appearance in this section. The first general idea (see Lemma 5.1) is that U-rank 1 types form the 'building blocks' for theories of finite U-rank.

In this section T will be stable, unless stated otherwise.

Lemma 5.1. *Let $p(x) \in S(A)$ be a stationary non-algebraic type of finite U-rank. Then there is some stationary type q, such that $U(q) = 1$, and p is not orthogonal to q.*

Proof. (We use Lemma 1.3.18 throughout.) Suppose $U(p) = n > 0$. If $n = 1$ there is nothing to do (take $q = p$). Otherwise (by definition of U-rank) there is $B \supseteq A$ and a stationary extension $p'(x) \in S(B)$ of $p(x)$, such that $U(p') = n - 1$. The proof of Lemma 1.3.19 shows that there is some $c \in \mathbf{C}^{eq}$ such that c is interalgebraic with $\mathrm{Cb}(p')$ over A. (If the reader does not want to believe this then he or she may simply assume T to be superstable.)

As p' is a forking extension of p, $c \notin \mathrm{acl}(A)$. Thus $U(c/A) \geq 1$.

Let a realize $p'|Ac$. Let $E \supseteq A$ be a set such that $U(c/E) = 1$. We may choose E so that E is independent from a over Ac. Then $\mathrm{tp}(a/Ec)$ does not fork over Ac (forking symmetry), so

$(*)$ $\qquad$ c is interalgebraic with $\mathrm{Cb}(\mathrm{stp}(a/Ec))$ over A.

On the other hand $c \notin \mathrm{acl}(E)$. So by 1.2.26 and $(*)$, $\mathrm{tp}(a/Ec)$ forks over E. Thus $n = U(a/A) \geq U(a/E) > U(a/Ec) = n - 1$.

It follows that $U(a/E) = n$, so $\mathrm{stp}(a/E)$ is the non-forking extension of p. Also a forks with c over E, so p is non-orthogonal to $\mathrm{stp}(c/E)$, which has U-rank 1. The lemma is proved. ∎

We now try to understand more about non-orthogonality, under certain geometric assumptions.

Lemma 5.2. *Let D be a minimal set, defined over $\varnothing$, say. Assume the pregeometry $\mathbf{D}_{\varnothing}$ to be modular. Let $a \in \mathbf{C}^{eq}$ and $a \in \mathrm{acl}(D)$. Then $a \in \mathrm{acl}(\mathrm{acl}(a) \cap D)$.*

Proof. Let $\mathbf{b}$ be a tuple from D such that $a \in \mathrm{acl}(\mathbf{b})$. Let $\mathbf{c}$ be such that $\mathrm{stp}(\mathbf{c}/a) = \mathrm{stp}(\mathbf{b}/a)$ and $\mathrm{stp}(\mathbf{c}/\mathbf{b}a)$ does not fork over a.

Let $C = \mathrm{Cb}(\mathrm{tp}(\mathbf{c}/\mathrm{acl}(\mathbf{b})))$.

Claim 1. $\mathrm{acl}(a) = \mathrm{acl}(C)$.

First $C \subseteq \mathrm{acl}(a)$, as $\mathrm{stp}(\mathbf{c}/\mathrm{acl}(\mathbf{b}))$ does not fork over a. On the other hand $\mathbf{c}$ is independent with $\mathbf{b}$ over C so $\mathbf{c}$ is independent with a over C, but $a \in \mathrm{acl}(\mathbf{c})$ so $a \in \mathrm{acl}(C)$. The claim is proved.

Now let $A = \mathrm{acl}(\mathbf{b}) \cap \mathrm{acl}(\mathbf{c}) \cap D$. By modularity of D, $\mathbf{c}$ is independent from **b** over A, that is tp(**c**/acl(**b**)) does not fork over A. As in the proof of Claim 1 we have:

CLAIM 2. $\mathrm{acl}(A) = \mathrm{acl}(C)$.

Putting together Claims 1 and 2 we see that $\mathrm{acl}(a) = \mathrm{acl}(A)$, where $A \subseteq D$. ■

COROLLARY 5.3. *Let D be minimal, defined over $\varnothing$, say, and modular. Let $a \in \mathbf{C}$. Then a is independent from D over* $\mathrm{acl}(a) \cap D$.

Proof. It is not difficult to see that $\mathrm{Cb}(\mathrm{stp}(a/D)) \subseteq \mathrm{dcl}(a) \cap \mathrm{acl}(D)$. (Let p' be the non-forking extension of $\mathrm{stp}(a/D)$ over $\mathbf{C}$. Suppose f is an automorphism of $\mathbf{C}$ which fixes a; then, as D is ∞-definable over $\varnothing$, f fixes $\mathrm{tp}(a/\mathrm{acl}(D))$, so fixes the non-forking extension p' of this strong type. Now use 1.2.26.)

By 5.2, $\mathrm{dcl}(a) \cap \mathrm{acl}(D) \subseteq \mathrm{acl}(\mathrm{acl}(a) \cap D)$, whereby $\mathrm{tp}(a/D)$ does not fork over $\mathrm{acl}(a) \cap D$. ■

DEFINITION 5.4. *Let D_1, D_2 be minimal sets, defined over A, B respectively. Let $p(x) \in S(A)$, $q(y) \in S(B)$ be the minimal complete types extending D_1, D_2 respectively. We will say that D_1 is orthogonal to D_2 if $p(x)$ is orthogonal to $q(y)$. If $A = B$, we say D_1 is weakly orthogonal to D_2 if $p(x)$ is weakly orthogonal to $q(y)$.*

COROLLARY 5.5. *Let D_1, D_2 be modular minimal sets, defined over $\varnothing$. Assume they are non-orthogonal. Then they are non weakly-orthogonal, that is, there are $a \in D_1 - \mathrm{acl}(\varnothing)$, $b \in D_2 - \mathrm{acl}(\varnothing)$ such that $a \in \mathrm{acl}(b)$ (and so $b \in \mathrm{acl}(a)$).*

Proof. By Lemma 1.4.3.1(iii), there are tuples **a** from D_1, **b** from D_2, and some $b \in D_2$ such that $b \in \mathrm{acl}(\mathbf{ab})$, $b \notin \mathrm{acl}(\mathbf{b})$. Applying Lemma 5.2 to the modular pregeometry $(D_1)_{\mathbf{b}}$ (i.e. naming **b**), we find $a \in D_1$ such that $\mathrm{acl}(\mathbf{ab}) = \mathrm{acl}(b\mathbf{b})$. In particular $a \in \mathrm{acl}(b\mathbf{b}) - \mathrm{acl}(\varnothing)$. Applying 5.2 again, we find $c \in D_2$ such that $\mathrm{acl}(a) = \mathrm{acl}(c)$, as required. ■

DEFINITION 5.6. *The (stable) theory T is said to be* 1-based *if for any a, B,* $\mathrm{Cb}(\mathrm{stp}(a/B)) \subseteq \mathrm{acl}(a)$.

REMARK 5.7. *a, B above are supposed to be in $\mathbf{C}^{\mathrm{eq}}$. An obviously equivalent definition of* 1*-basedness is that for any sets A and B, A is independent from B over* $\mathrm{acl}(A) \cap \mathrm{acl}(B)$ *(acl(-) again in the sense of $\mathbf{C}^{\mathrm{eq}}$). In Chapter 4, a more thorough study of* 1*-based theories will be carried out. In this section we are mainly concerned with the case where T has finite U-rank.*

We say that T has finite U-rank if for *every* complete type p (in $\mathbf{C}^{eq}$), $U(p) < \omega$.

PROPOSITION **5.8.** *If T is 1-based then all minimal types are locally modular. The converse also holds if T has finite U-rank.*

Proof. Suppose first that T is 1-based. Let $p(x) \in S(A)$ be a minimal type (i.e. a stationary type of U-rank 1). We may assume $A = \emptyset$. We will show that p is linear. Let a, b be realizations of p, and suppose that $U(\mathrm{stp}(a, b/c)) = 1$ where $c = \mathrm{Cb}(\mathrm{stp}(a, b/c))$. By 1-basedness $c \in \mathrm{acl}(a, b)$, so $U(c/\emptyset) \leq 2$. We must show that $U(c/\emptyset) \leq 1$.

By 1.3.26,

$$U(abc/\emptyset) = U(c/ab) + U(ab/\emptyset) = 0 + U(ab/\emptyset) \leq 2.$$

Again by 1.3.26,

$$U(abc/\emptyset) = U(ab/c) + U(c/\emptyset) = 1 + U(c/\emptyset).$$

So $U(c/\emptyset) \leq 1$.

Thus p is linear, so by 2.6, locally modular.

Conversely assume that T has finite U-rank and that all minimal types are locally modular. Suppose by way of contradiction that a forks with B over $A = \mathrm{acl}(a) \cap \mathrm{acl}(B)$. Let $b \in B$ be such that a forks with b over A. Let M be an a-model containing A such that $\mathrm{tp}(ab/M)$ does not fork over A.

CLAIM **1.** $\mathrm{acl}(aM) \cap \mathrm{acl}(bM) = M$.

Proof. By 1.2.22, $\mathrm{tp}(ab/Ma)$ does not fork over Aa, so (by forking symmetry) $\mathrm{tp}(Ma/abA)$ does not fork over Aa. Similar $\mathrm{tp}(Mb/abA)$ does not fork over Ab. Thus if $c \in \mathrm{acl}(Ma) \cap \mathrm{acl}(Mb)$, then $\mathrm{Cb}(\mathrm{stp}(c/abA)) \subseteq \mathrm{acl}(Aa) \cap \mathrm{acl}(Ab) = A$. Let $C = \mathrm{acl}(Ma) \cap \mathrm{acl}(Mb)$. We have shown that C is independent from abA over A, that is $\mathrm{tp}(ab/C)$ does not fork over A. As $A \subseteq M \subseteq C$, it follows from 1.2.22 that $\mathrm{tp}(ab/C)$ does not fork over M, and thus $\mathrm{tp}(C/Mab)$ does not fork over M. But $C \subseteq \mathrm{acl}(Mab)$, so $C \subseteq \mathrm{acl}(M) = M$. Thus $C = M$. Claim 1 is proved. ■

Note also:

CLAIM **2.** *a forks with b over M.*

CLAIM **3.** *There is an M-independent sequence $(c_1, \ldots, c_n)$ such that $\mathrm{tp}(c_i/M)$ is minimal for each i, and a is domination equivalent to $(c_1, \ldots, c_n)$ over M.*

Similarly there is $(d_1, \ldots, d_m)$ *for* b. *Note* $c_i \in \mathrm{acl}(aM)$, $d_j \in \mathrm{acl}(bM)$ *for all* i, j.

Proof. By 1.4.5.7, $M[a] = M[c_1, \ldots, c_n]$ where $(c_1, \ldots, c_n)$ is some M-independent sequence of elements realizing regular types over M. By Lemma 5.1 (and the finite U-rank hypothesis), each $\mathrm{tp}(c_i/M)$ is non-orthogonal to some U-rank 1 (so regular) type r_i. The proof of 1.4.5.5 shows that r_i can be taken to be over M. By 1.4.3.4(iv), r_i is realized in $M[a]$ by some c_1' which forks with c_i over M; 1.4.5.7 shows that a dominates $(c_1', \ldots, c_n')$ over M and $(c_1', \ldots, c_n')$ dominates a over M.

As a forks with b over M, it follows that $(c_1, \ldots, c_n)$ forks with $(d_1, \ldots, d_m)$ over M. We may assume that for any types r, s from

$$\{\mathrm{tp}(c_i/M)\colon i \leq n\} \cup \mathrm{tp}(d_j/M)\colon j \leq m\},$$

$r = s$ or r is orthogonal to s. Thus clearly there is some minimal type q over M, and subsequences $\mathbf{c}'$, of $(c_1, \ldots, c_n)$ and $\mathbf{d}'$ of $(d_1, \ldots, d_n)$ such that each element of the tuple $(\mathbf{c}'\mathbf{d}')$ realizes q, and $\mathbf{c}'$ forks with $\mathbf{d}'$ over M. Now as M is an a-model, clearly q is modular. (Let $A_0 \subseteq M$ be finite such that q does not fork over M. Let q_0 be the restriction of q to $\mathrm{acl}(A_0)$. By hypothesis, q_0 is locally modular. But q_0 is realized in M. Thus q is an extension of a localization of q_0, so is modular.) Thus $\mathrm{acl}(\mathbf{c}'M) \cap \mathrm{acl}(\mathbf{d}'M) \cap q^{\mathrm{C}} \neq \emptyset$. Thus $\mathrm{acl}(aM) \cap \mathrm{acl}(bM)$ properly contains M. This contradicts Claim 1, and completes the proof of Proposition 5.8. ■

The next result shows how an element in a 1-based theory of finite U-rank can be 'understood' in terms of U-rank 1 elements.

Proposition 5.9. *Suppose T is 1-based of finite U-rank. Then for any a, A with $a \notin \mathrm{acl}(A)$, there is $c \in \mathrm{acl}(a)$ such that $U(c/A) = 1$.*

Proof. If $U(a/A) = 1$ there is nothing to do. So suppose $U(a/A) = n > 1$, and let $B \supseteq A$ be such that $U(a/B) = n - 1$. Let $C = \mathrm{Cb}(\mathrm{stp}(a/B))$, and let $c \in C$ be such that $\mathrm{tp}(a/C)$ does not fork over cA. Then $U(a/cA) = n - 1$, and $c \in \mathrm{acl}(a)$.

So $U(ac/A) = U(c/aA) + U(a/A) = U(a/cA) + U(c/A)$. Thus $U(c/A) = 1$. ■

Finally in this section we show that two reasonably important kinds of theories are 1-based of finite U-rank. Note first:

Remark 5.10. *Suppose $p \in S(A)$, $q \in S(B)$ are non-orthogonal minimal types. Then p is locally modular iff q is locally modular.*

Proof. Let $M \supseteq A \cup B$ be an a-model Let $p' \in S(M)$, $q' \in S(M)$ be the non-forking extensions of p, q respectively over M. By 1.4.3.1(ii), p' and q' are not weakly orthogonal. Thus for each a realizing p' there is b realizing q' such that $\mathrm{acl}(aM) = \mathrm{acl}(bM)$ (and vice versa). This shows that the geometries associated to p' and q' are isomorphic. In particular, by 2.6, p is locally modular iff p' is locally modular. Similarly for q. This proves the remark. ■

Proposition 5.11. *Suppose T is superstable, unidimensional, and not totally transcendental. Then T is 1-based of finite U-rank.*

Proof. By 1.5.14(i) T has finite U-rank. By 5.8 we must show that all (or by 5.10 at least one) minimal type is locally modular. Let $p(x) \in S(A)$ be minimal. If p is not locally modular, then by 3.3, $RM(p) = 1$. So $RM(\theta) = 1$ for some formula θ. By 1.5.14(iii), T is totally transcendental, a contradiction. ■

Theorem 5.12. *Let T be (countable) ω-categorical and ω-stable. Then T is 1-based of finite U-rank.*

Proof. It is worth remarking, to begin with, that by Lemma 1.3.21, $U(p) = RM(p) = R^{\infty}(p) < \infty$, for any complete type p. Note also that if $p \in S(A)$ is minimal (i.e. stationary with U-rank 1) then p is locally modular. We may assume A to be algebraically closed. By superstability, p is definable over some finite $A_0 \subseteq A$. Let $p_0 = p \restriction A_0$. So p_0 is the unique non-algebraic type extending some strongly minimal set D defined over A_0. As A_0 is finite, D is ω-categorical, so by Theorem 4.17, locally modular. Thus p_0, and so also p, are locally nodular.

So to prove the theorem, it suffices by 5.8 to show that every complete type has finite U-rank.

Suppose by way of contradiction that some type has U-rank $\geq \omega$. Then we can find (by superstability) a complete stationary type $p(x)$ over a finite set A, such that $U(p) = \omega$. By ω-categoricity, the set of realizations of p is an A-definable set X, say. Assume, for notational simplicity that $A = \emptyset$.

The U-rank inequality (1.3.26) shows that p is a regular type (1.3.26 shows that a type of U-rank ω is orthogonal to any type of finite U-rank, in particular to any forking extension of itself.) Thus the relation on X: x forks with y over $\emptyset$, is an equivalence relation. E is clearly invariant under automorphisms of $\mathbf{C}$ so by ω-categoricity is $\emptyset$-definable. Let $X' = X/E$. It should then be clear, again by the U-rank inequality, that X' is the set of realizations of a complete stationary type p' over $\emptyset$, and $U(p')\ (= U(X')) = \omega$. We now have the additional feature:

(∗) if $a, b \in X'$ with $a \neq b$ then a is independent from b over $\emptyset$.

Let $a \in X'$. Then there is B such that $U(a/B)$ is finite, but >1 (p', having U-rank ω, has extensions of arbitrarily large finite U-rank). Now as $\mathrm{Cb}(\mathrm{stp}(a/B))$ is in the definable closure of a Morley sequence in this type, there is (by superstability again) some finite tuple $\mathbf{c}$ from X' such that $U(a/\mathbf{c})$ is finite and >1. Let $(c_1, \ldots, c_n)$ be a maximal independent subtuple of $\mathbf{c}$. Then (by regularity of p') a forks with $(c_1, \ldots, c_n)$ over $\varnothing$, and clearly $U(a/c_1, \ldots, c_n)$ is still >1.

Let $Y = \{b \in X' \colon b \text{ forks with } (c_1, \ldots, c_n) \text{ over } \varnothing\}$. Then Y is (by ω-categoricity), a $(c_1, \ldots, c_n)$-defnable set containing a. So $U(Y) > 1$. Also clearly $U(Y) < \omega$ (as for all $b \in Y$, $U(b/c_1, \ldots, c_n) < \omega$, and $U(\text{-}) = RM(\text{-})$). Let $u \in \mathbf{C}^{\mathrm{eq}}$ be a canonical parameter for Y.

CLAIM. *Whenever a_1, $a_2 \in Y$, with $a_1 \neq a_2$, then a_1 is independent from a_2 over u.*

Proof. By $(*)$, a_1 is independent from a_2 over $\varnothing$. Extend a_1, a_2 to a maximal $\varnothing$-independent subset $\{a_1, a_2, \ldots, a_k\}$ of Y. As p' has weight 1, by Lemma 1.4.4.1(iii), $k = n$. (Note that $\{c_1, \ldots, c_n\}$ is a maximal independent subset of Y, and $Y \subseteq X'$.) Both $\mathbf{a}$ and $\mathbf{c}$ are then 'bases' for Y. As p' is stationary, $\mathrm{tp}(a_1, \ldots, a_n/\varnothing) = \mathrm{tp}(c_1, \ldots, c_n/\varnothing)$, so there is an automorphism f of $\mathbf{C}$ taking $\mathbf{a}$ to $\mathbf{c}$, and in particular taking (a_1, a_2) to (c_1, c_2). f must fix Y setwise, and hence fixes u. We have shown that $\mathrm{tp}(a_1, a_2/u) = \mathrm{tp}(c_1, c_2/u)$. However, we can find distinct $a_1, a_2 \in Y$ such that a_1 is independent from a_2 over u. What we have just shown implies this is true for all distinct $a_1, a_2 \in Y$.

Now Y has finite U-rank, is ω-categorical, and all minimal types in $(Y_u)^{\mathrm{eq}}$ are locally modular. We can apply 5.8 to Y to conclude that Y is 1-based, that is for any c, B in Y^{eq}, $\mathrm{Cb}(\mathrm{stp}(c/Bu) \subseteq \mathrm{acl}(cu)$.

In particular, let $a \in Y$ with $U(a/u) > 1$ (as $U(Y) > 1$).

Let $\mathrm{stp}(a/uB)$ be forking extension of $\mathrm{stp}(a/u)$, of U-rank > 0. Let a_1, a_2 be the beginning of a Morley sequence in $\mathrm{stp}(a/uB)$. Then $a_1 \neq a_2 \in Y$. But $c = \mathrm{Cb}(\mathrm{stp}(a/uB)) \in \mathrm{acl}(a_1, u) \cap \mathrm{acl}(a_2, u)$, $c \notin \mathrm{acl}(u)$. Thus a_1 forks with a_2 over u. This contradicts the claim, and proves the theorem. ■

It is worth making a final remark in this section which will be useful later. Suppose D is any ∞-definable set in a stable structure, defined over A, say. We will say that D *is* 1-*based*, if for every $a \in (D_A)^{\mathrm{eq}}$, and *any* set $B \supseteq A$, $\mathrm{Cb}(\mathrm{stp}(a/BA) \subseteq \mathrm{acl}(aA)$. Let us call a set X algebraically closed in $(D_A)^{\mathrm{eq}}$ if $A \subseteq X$ and $\mathrm{acl}(X) \cap (D_A)^{\mathrm{eq}} = X$. It is then not difficult to see that 1-basedness of D is equivalent to: for all X, Y algebraically closed in $(D_A)^{\mathrm{eq}}$, X is independent from Y over $X \cap Y$. Also 1-basedness of D will be independent of the parameter set A.

Remark 5.13. *Suppose D is minimal, locally modular, and defined over $\varnothing$, say. Then D is 1-based.*

Proof. This is a simple modification of the proof of 5.8. For example, suppose $a, b \in D^{\mathrm{eq}}$, $A = \mathrm{acl}(a) \cap \mathrm{acl}(b) \cap D^{\mathrm{eq}}$, and suppose for a contradiction that a forks with b over A. Let M be a very saturated model containing A such that $\mathrm{tp}(ab/M)$ does not fork over A.

As in the proof of 5.8, $\mathrm{tp}(ab/Ma)$ does not fork over Aa, and $\mathrm{tp}(ab/Mb)$ does not fork over Ab. Hence.

$$\mathrm{Cb}(\mathrm{stp}(ab/\mathrm{acl}(Ma) \cap \mathrm{acl}(Mb))) \subseteq \mathrm{acl}(Aa) \cap \mathrm{acl}(Ba) \cap D^{\mathrm{eq}} = A.$$

(Note the canonical base above *is* in D^{eq}, as it is in the definable closure of a Morley sequence in the type concerned, and this Morley sequence is in D^{eq}.) Thus

$$(*) \qquad \mathrm{acl}(Ma) \cap \mathrm{acl}(Mb) = M.$$

Now as M is saturated and $a \in D^{\mathrm{eq}}$, one can easily find some tuple $\mathbf{a}$ from D such that a is interalgebraic with $\mathbf{a}$ over D. Similarly find tuple $\mathbf{b}$ from D such that b is interalgebraic with $\mathbf{b}$ over D. Now a forks with b over M, so $\mathbf{a}$ forks with $\mathbf{b}$ over M. As the pregeometry $\mathbf{D}_M$ is modular, $\mathrm{acl}(\mathbf{a}M) \cap \mathrm{acl}(\mathbf{b}M) \cap D$ contains some $c \notin M$. This contradicts $(*)$. ■

6 Totally categorical structures

The countable theory T is said to be totally categorical if T is λ-categorical for all infinite cardinals λ. The aim of this section is to show that there is no finitely axiomatizable, totally categorical complete theory. In Chapter 3 we prove the same result for ω-stable, ω-categorical structures as part of a more general analysis. The proof here depends on showing that any totally categorical structure M can be 'smoothly approximated', that is can be written as an increasing union of finite 'homogeneous' substructures. So essentially M will be just a non-standard finite structure.

For the remainder of this section we assume that T is ω-categorical and stable.

We have seen in Theorem 4.17 (using Fact 1.11) that if D is a strongly minimal set defined over the finite set A, then the geometry associated to the pregeometry $\mathbf{D}_A$ is trivial or is an affine or projective geometry over a finite field. As pointed out in section 1, an affine geometry is not modular. We shall say that D is trivial over A, or affine over A, or projective over A, accordingly. It is important to note that if D_1, D_2 are strongly minimal sets defined over A, and for some $a \in D_1 - \mathrm{acl}(A)$, $b \in D_2 - \mathrm{acl}(A)$, $a \in \mathrm{acl}(b)$, then D_1 and D_2 have isomorphic geometries over A. If $\mathbf{D}_A$ is modular (i.e. trivial or projective) we say that D is modular over A. It will be rather important for us to understand the delicate matter of what happens to the (pre)geometry of an affine D after naming a finite set of parameters. Everything we say will also hold outside the ω-categorical framework (as we shall see in Chapter 5), but the ω-categorical assumption again makes the arguments simpler.

The following notion is convenient.

DEFINITION 6.1. *Let D be a strongly minimal set defined over a finite set A. We say that D is* strictly minimal over A, *if $D \cap \mathrm{acl}(A) = \varnothing$, and for any $a \in D$, $\mathrm{acl}(aA) \cap D = \{a\}$.*

So if D is strictly minimal over A, then the pregeometry $\mathbf{D}_A$ is a geometry. ω-categoricity allows us to associate to any strongly minimal set a certain minimal set as follows. Assuming D is defined over the set A, let $D_1 = D - \mathrm{acl}(A)$ (which is also A-definable, as $\mathrm{acl}(A) \cap D$ is finite). Let E be the A-definable equivalence relation on D_1 defined by $E(x, y)$ if $x \in \mathrm{acl}(yA)$. Then $D_2 = D_1/E$ is strictly minimal over A (and of course non-orthogonal to the original D).

Now suppose D to be strictly minimal, defined over $\varnothing$, and *affine*. We show how to associate naturally to D a strictly minimal, $\varnothing$-definable set which is projective (over the same field as D). Now D *is* an infinite-dimensional affine space over a finite field F, and algebraic closure in D is the same thing as F-affine closure. By a *line* L in D we will mean an algebraically closed subset of D of dimension 2, that is a set of the form $\mathrm{acl}(a, b) \cap D$, where $a, b \in D$ and a is independent from b over $\varnothing$. Such a line is therefore simply the translate of some 1-dimensional vector subspace of D (but note 0 is not named, so the F-vector-space structure is in our 'imagination'). It is clear that if L_1, L_2 are distinct such lines then $L_1 \cap L_2 = \varnothing$ iff L_1 and L_2 are translates of the *same* vector subspace. We shall say in this situation that L_1 and L_2 are *parallel*. Let E' be the equivalence relation on pairs of distinct elements of D defined as: (for $a \neq b$, $c \neq d$ in D) (a, b) E' (c, d) iff $\mathrm{acl}(a, b) \cap D$ and $\mathrm{acl}(c, d) \cap D$ are the same or parallel. By-ω-categoricity, E' is $\varnothing$-definable. Let D' be $D^{(2)}/E'$ (where $D^{(2)} = \{(a, b) \in D^2 \colon a \neq b\}$).

LEMMA 6.2. *D' is strictly minimal over $\varnothing$, and is projective geometry over F.*

Proof. Left to the reader.

Clearly D' is non-orthogonal to D; in fact given $a \in D$, for any $b \in D - \{a\}$, let L_b be $\operatorname{acl}(a, b) \cap D$. Then b is interalgebraic with L_b/E' over a.

Up to and including Lemma 6.5, let us fix D and D' as given above (D strictly minimal, affine, defined over $\varnothing$), and D' the associate strictly minimal projective set.

LEMMA 6.3. *Let A be a (finite) subset of D'. Then*

(i) $\operatorname{acl}(A) \cap D = \varnothing$.

(ii) $\mathbf{D}_A$ *is affine.*

(iii) *Let D_2, $(D')_2$ be the strictly minimal sets over A obtained from D, D' as in remarks following Definition 6.1 (i.e. localize at A, then quotient by algebraic closure over A). Then $(D')_2 = D'_2$.*

Proof. (All we use here is the modularity of D'.)

(i) Suppose $c \in \operatorname{acl}(A) \cap D$. By Lemma 5.2, $c \in \operatorname{acl}(\operatorname{acl}(c) \cap D')$ (as D' is modular over $\varnothing$). In particular $\operatorname{acl}(c) \cap D' \neq \varnothing$. But then D will be modular too (as there will be $d \in D'$ with $d \in \operatorname{acl}(c)$), a contradiction.

(ii) If $\mathbf{D}_A$ were not affine, then it would be modular, so by Corollary 5.5, non-weakly orthogonal to D' (over A). Thus there would be $c \in D'$ such that $\operatorname{acl}(Ac) \cap D \neq \varnothing$, contradicting (i).

(iii) Left to the reader. ■

LEMMA 6.4. *Let* $\mathbf{a}$, $\mathbf{b}$ *be a finite tuples from D. Suppose that* $\operatorname{acl}(\mathbf{a}) \cap \operatorname{acl}(\mathbf{b}) \cap D = \varnothing$. *Then* $\mathbf{a}$ *is independent from* $\mathbf{b}$ *over* $\operatorname{acl}(\mathbf{a}) \cap \operatorname{acl}(\mathbf{b}) \cap D'$.

Proof. It suffices to prove

(*) if $\mathbf{a}$ forks with $\mathbf{b}$ over $\varnothing$, then $\operatorname{acl}(\mathbf{a}) \cap \operatorname{acl}(\mathbf{b}) \cap D' \neq \varnothing$.

(That is, add names for $\operatorname{acl}(\mathbf{a}) \cap \operatorname{acl}(\mathbf{b}) \cap D'$, and use 6.3(iii).)

So, suppose $\mathbf{a}$ forks with $\mathbf{b}$ over $\varnothing$. Thus there is a subtuple $\mathbf{a}_1$ of $\mathbf{a}$, and an element a of $\mathbf{a}$ such that $\mathbf{a}_1$ is independent from $\mathbf{b}$ over $\varnothing$, but $a \in \operatorname{acl}(\mathbf{a}_1, \mathbf{b}) - \operatorname{acl}(\mathbf{a}_1)$. (Note that $\mathbf{a}_1 \neq \varnothing$, for otherwise $a \in \operatorname{acl}(\mathbf{b})$ contradicting the assumption that $\operatorname{acl}(\mathbf{a}) \cap \operatorname{acl}(\mathbf{b}) \cap D = \varnothing$.) It follows that $\mathbf{a}_1$ is not independent from $\mathbf{b}$ over a. But localizing at a makes D modular. Thus there is some $c \in \operatorname{acl}(a, \mathbf{a}_1) \cap \operatorname{acl}(a, \mathbf{b}) \cap D$, $c \notin \operatorname{acl}(a)$. So $\operatorname{acl}(a, c) \cap D$ is a line L_1 contained in $\operatorname{acl}(\mathbf{a})$

Repeating the argument by interchanging the roles of $\mathbf{a}$ and $\mathbf{b}$, we find $b \in \mathbf{b}$, and $d \neq b$ such that $\operatorname{acl}(b, d) \cap D$ is a line L_2 contained in $\operatorname{acl}(\mathbf{b})$.

Our assumption that $\operatorname{acl}(\mathbf{a}) \cap \operatorname{acl}(\mathbf{b}) \cap D = \varnothing$ implies that L_1 is parallel to L_2. Let $e = L_1/E' = L_2/E'$. So $e \in \operatorname{acl}(\mathbf{a}) \cap \operatorname{acl}(\mathbf{b}) \cap D'$, as required. ■

LEMMA 6.5. *Let B be any set (in* **C***) such that* $\operatorname{acl}(B) \cap D = \varnothing$.

(i) *Let* **a** *be a tuple from D. Then* $\operatorname{Cb}(\operatorname{stp}(\mathbf{a}/B)) \subseteq \operatorname{acl}(D' \cap \operatorname{acl}(B))$.

(ii) $\mathbf{D}_B$ *remains affine.*

Proof. (i) By for example 5.13, D is 1-based, and hence $E = \operatorname{Cb}(\operatorname{stp}(\mathbf{a}/B)) \subseteq \operatorname{acl}(\mathbf{a})$. Let **b** realize the non-forking extension of $\operatorname{stp}(\mathbf{a}/B)$ over $B\mathbf{a}$. So $E = \operatorname{Cb}(\operatorname{stp}(\mathbf{b}/\mathbf{a}))$. On the other hand, clearly $\operatorname{acl}(\mathbf{a}) \cap \operatorname{acl}(\mathbf{b}) \subseteq \operatorname{acl}(B)$; thus our hypothesis implies that $\operatorname{acl}(\mathbf{a}) \cap \operatorname{acl}(\mathbf{b}) \cap D = \varnothing$. By Lemma 6.4, **b** is independent from **a** over $\operatorname{acl}(\mathbf{a}) \cap \operatorname{acl}(\mathbf{b}) \cap D'$. Therefore

$$\operatorname{Cb}(\operatorname{stp}(a/B)) \subseteq \operatorname{acl}(D' \cap \operatorname{acl}(B)).$$

(ii) Let $A = \operatorname{acl}(B) \cap D'$. Part (i) shows that for any $a \in D$, and tuple **b** from D, $a \in \operatorname{acl}(\mathbf{b}, B)$ iff $a \in \operatorname{acl}(\mathbf{b}, A)$. Thus the pregeometries $\mathbf{D}_B$ and $\mathbf{D}_A$ are identical. By Lemma 6.3, $\mathbf{D}_B$ is affine. ■

We now aim towards:

THEOREM 6.6. *Let T be a totally categorical (countable) theory. Then T is not finite axiomatizable.*

Finite axiomatizability of T means that there is a single L-sentence σ such that the sentences in T are precisely the logical consequences of σ. This implies on the face of it that the language L is finite. Even if the language L is not finite, however, one could make sense of the finite axiomatizability question (and the subsequent quasifinite axiomatizability problem which has a *positive* answer), by choosing a different language. More will be said about this in Chapter 3. All we will use of the hypothesis: T is totally categorical, is the consequence: T is ω-categorical, ω-stable, and unidimensional. (See Theorem 1.5.18). By Theorem 5.12 T is also 1-based of finite U-rank.

For now, T remains ω-categorical. We work as usual in **C**, the big saturated model of T, but for our present purposes we could simply take **C** to be a (the) countable model of T. Up to now we have purposely confused T with T^{eq}, and **C** with $\mathbf{C}^{\mathrm{eq}}$. Here, however, we need to be clear about the distinction. So we assume that T is a 1-sorted theory. There is also no harm in assuming that the language L of T is relational. **C** is a model of T in the usual sense.

DEFINITION 6.7. *Let A be a subset of* $\mathbf{C}^{\mathrm{eq}}$. *We say that A is a* homogeneous substructure *of* $\mathbf{C}^{\mathrm{eq}}$ *if whenever* **a**, **b** *are finite tuples from A such that* $\operatorname{tp}(\mathbf{a}/\varnothing) = \operatorname{tp}(\mathbf{b}/\varnothing)$ *(in* $\mathbf{C}^{\mathrm{eq}}$*), and $c \in A$, then there is $d \in A$ such that* $\operatorname{tp}(\mathbf{a}, c/\varnothing) = \operatorname{tp}(\mathbf{b}, d/\varnothing)$ *(in* $\mathbf{C}^{\mathrm{eq}}$*).*

If A is a subset of **C**, *and the above holds, then we simply say A is a homogeneous substructure of* **C**.

REMARK 6.8. (i) *Suppose A is a homogeneous substructure of* $\mathbf{C}^{\text{eq}}$. *Then* $A \cap \mathbf{C}$ *is a homogeneous substructure of* **C**. (*Similarly for any sort S in* $\mathbf{C}^{\text{eq}}$, $A \cap S^{\mathbf{C}}$ *is a homogeneous substructure of* $\mathbf{C}^{\text{eq}}$.)

(ii) *Suppose D is a strongly minimal set in* $\mathbf{C}^{\text{eq}}$ *defined over* $\varnothing$. *Then any algebraically closed subset of D is a homogeneous substructure of* $\mathbf{C}^{\text{eq}}$.

Proof of (*ii*). Let A be an algebraically closed subset of D. Let $\mathbf{a}$, $\mathbf{b}$ be tuples from D with the same type over $\varnothing$. In particular $U(\mathbf{a}/\varnothing) = U(\mathbf{b}/\varnothing) = n$, say. Let $a \in A$. We seek $b \in A$ such that $\text{tp}(\mathbf{a}a) = \text{tp}(\mathbf{b}b)$. If $a \in \text{acl}(\mathbf{a})$ this is clear. Otherwise $a \notin \text{acl}(\mathbf{a})$. So $\dim(A)$ (=the unique cardinality of any maximal algebraically independent subset of A) is strictly greater than n. In particular there is $b \in A - \text{acl}(\mathbf{b})$. Clearly then $\text{tp}(\mathbf{a}a) = \text{tp}(\mathbf{b}b)$. ■

LEMMA 6.9. *Suppose that for any finite subset A of* **C**, *there is a finite homogeneous substructure B of* **C** *which contains A. Then T is not finitely axiomatizable.*

Proof. Let σ be an L-sentence in T. So $\mathbf{C} \models \sigma$. Assume the quantifier rank of σ is n. By ω-categoricity, there is a finite subset A of **C** such that all complete n-types of T (with respect to the original sort) are realized by tuples from A. By hypothesis we can find a finite homogeneous substructure B of **C** which includes A. Now (as L is relational), B is also an L-structure.

CLAIM. $B \models \sigma$.

Proof. It is clearly enough to show that if we have tuples $\mathbf{a}$, $\mathbf{b}$ of length $k < n$, with $\mathbf{a}$ from **C**, $\mathbf{b}$ from B such that $\text{tp}(\mathbf{a}/\varnothing) = \text{tp}(\mathbf{b}/\varnothing)$ in **C**, and $a \in \mathbf{C}$, then there is $b \in B$ such that $\text{tp}(\mathbf{a}a/\varnothing) = \text{tp}(\mathbf{b}b/\varnothing)$. As B realizes all n-types of T, there is some $k+1$-tuple $\mathbf{d}d$ from B such that $\text{tp}(\mathbf{a}a/\varnothing) = \text{tp}(\mathbf{d}d/\varnothing)$ in **C**. Then $\text{tp}(\mathbf{b}/\varnothing) = \text{tp}(\mathbf{d}/\varnothing)$ in **C**, so as both tuples are in B, the fact that B is a homogeneous substructure of **C** means that there is $b \in B$ with $\text{tp}(\mathbf{d}d/\varnothing) = \text{tp}(\mathbf{b}b/\varnothing)$ in **C**. Thus $\text{tp}(\mathbf{a}a/\varnothing) = \text{tp}(\mathbf{b}b/\varnothing)$ in **C**.

This proves the claim. ■

In particular we see that every sentence of T is true in some finite L-structure. As T is complete with no finite models, it follows that T cannot be finitely axiomatizable. ■

LEMMA **6.10.** *Suppose T to be (ω-categorical), ω-stable, and unidimensional. Then in $\mathbf{C}^{eq}$ there is some $\varnothing$-definable, modular, strictly minimal set.*

Proof. We know T is 1-based of finite U-rank, and all strongly minimal sets have modular or affine geometries. Let $b \in \mathbf{C} - \mathrm{acl}(\varnothing)$. By Proposition 5.9, there is $a \in \mathrm{acl}(b)$, $a \in \mathbf{C}^{eq}$, such that $U(a/\varnothing) = 1$. Let X be the set of realizations of $\mathrm{tp}(a/\varnothing)$. Then (by ω-categoricity, and as U-rank = Morley rank here), X is $\varnothing$-definable with $RM(X) = 1$. We may quotient X by the $\varnothing$-definable equivalence relation on X: $x \in \mathrm{acl}(y)$, to obtain X'. We will have $RM(X') = 1$, where X' is the set if realizations of a complete type p' over $\varnothing$, but now, for $a \neq b \in X'$, a is independent from b over $\varnothing$. Note that $X' \cap \mathrm{acl}(\varnothing) = \varnothing$.

By 1.3.24, p' has finitely many extensions $p_1, \ldots, p_n$ over $\mathrm{acl}(\varnothing)$. These are therefore distinguished from each other over a finite part, u say, of $\mathrm{acl}(\varnothing)$. By ω-categoricity each $p_i|u$ is isolated, so its set of realizations is a u-definable set D_i. It is then clear that X' is the disjoint union of $D_1, \ldots, D_n$. Also each D_i is *strictly minimal.* Moreover, for $i, j = 1, \ldots, n.$, there is an automorphism taking D_i to D_j. Thus all the D_i are affine, or they are all modular. If the D_i are affine, then replace each D_i by D_i' as in the discussion before Lemma 6.2. Then the disjoint union of the D_i' remains $\varnothing$-definable, and the D_i' remain pairwise automorphic.

So we may assume all the D_i are modular. By unidimensionality, the D_i are mutually non-orthogonal. By Corollary 5.5 and strict minimality, there is, for each $i, j = 1, \ldots, n$, a *unique* $\mathrm{acl}(\varnothing)$-definable bijection $f_{ij}\colon D_i \to D_j$. Thus the f_{ij} commute. We may thus quotient out $\{(x, y)$: for some $i \neq j$, $x \in D_i, y \in D_j\}$ by a suitable $\varnothing$-definable equivalence relation, to obtain a $\varnothing$-definable strictly minimal modular set D, as required. ■

We now fix T as in the hypotheses of 6.10 (ω-categorical, ω-stable, unidimensional), and fix some strictly minimal *modular* $\varnothing$-definable D which is given by 6.10. We use repeatedly the facts that T is 1-based of finite U-rank, and that all minimal typrs are locally modular.

DEFINITION **6.11.** (i) *Let $A \subseteq \mathbf{C}^{eq}$. By a D-*envelope *of A, we mean a maximal subset B of $\mathbf{C}^{eq}$ such that $B \supseteq A$ and* $\mathrm{acl}(B) \cap D = \mathrm{acl}(A) \cap D$.

(ii) *By a D-*envelope *we mean a D-envelope of some A.*

Note that if B is a D-envelope then B is algebraically closed (in $\mathbf{C}^{eq}$).

LEMMA **6.12.** *Suppose $A \subseteq \mathbf{C}^{eq}$ is* finite *and B is a D-envelope of A. Then for every sort S of $\mathbf{C}^{eq}$, $B \cap S$ is finite.*

Proof. Suppose $B \cap S$ is infinite. As finitely many types over $\emptyset$ are realized in S (by ω-categoricity), there is $c \in B \cap S$ such that infinitely many realizations of $\mathrm{tp}(c)$ are contained in B. In particular $U(c/\emptyset) \geq 1$. Now using Proposition 5.9 and the fact that $U(c/\emptyset)$ is finite, there are $c_0, \ldots, c_{n-1}$, $c_{n+1} = c$, such that for each i, $c_i \in \mathrm{acl}(c)$ and $U(c_{i+1}/c_0, \ldots, c_i) = 1$. As B is algebraically closed, there is, using ω-categoricity, some $i \leq n$ such that

$(*)$ infinitely many realizations of $\mathrm{tp}(c_{i+1}/c_0, \ldots, c_i)$ are contained in B.

Rename $c_{i+1} = c$, $(c_0, \ldots, c_i) = b$. We may assume that $\mathrm{tp}(c/b)$ is stationary. So $\mathrm{tp}(c/b)$ is minimal and locally modular. By $(*)$ and ω-categoricity, the non-forking extension p' of $\mathrm{tp}(c/b)$ over bc has infinitely many realizations in B. But p' is modular (we localized at c). Also $b \in B$. By Corollary 5.5 (and ω-categoricity and the fact that bc is finite), $\mathrm{acl}(B) \cap D$ is infinite. Thus $\mathrm{acl}(A) \cap D$ is infinite, which contradicts ω-categoricity and the finiteness of A. ■

Finally we need:

LEMMA **6.13.** *Any D-envelope is a homogeneous substructure of* $\mathbf{C}^{\mathrm{eq}}$.

Proof. Let B be a D-envelope. Let $A_0 = \mathrm{acl}(B) \cap D$ $(= B \cap D)$. Note that B will then be a D-envelope of A_0.

Let $\mathbf{b}_0$, $\mathbf{b}_1$ be tuples from B with the same type in $\mathbf{C}^{\mathrm{eq}}$. Let $e_0 \in B$. We must find $e_1 \in B$ such that $\mathbf{b}_0 e_0$ has the same type in $\mathbf{C}^{\mathrm{eq}}$ as $\mathbf{b}_1 e_1$. By 5.9 and induction on U-rank (of $\mathrm{tp}(e_0/\mathbf{b}_0)$) we may assume that $U(e_0/\mathbf{b}_0) = 1$. We may even assume that $\mathrm{tp}(e_0/\mathbf{b}_0)$ is strictly minimal. Also as B is algebraically closed (in $\mathbf{C}^{\mathrm{eq}}$!) and $\mathrm{mult}(\mathrm{tp}(e_0/\mathbf{b}_0))$ is finite, we may assume that $\mathrm{tp}(e_0/\mathbf{b}_0)$ is stationary, and hence a minimal type.

Case (i). $\mathrm{tp}(e_0/\mathbf{b}_0)$ is modular (over $\mathbf{b}_0$).

The non-forking extension of D over $\mathbf{b}_0$ is also modular, so by unidimensionally and 5.5, there is $d_0 \in D - \mathrm{acl}(\mathbf{b}_0)$ such that e_0 is interalgebraic with d_0 over $\mathbf{b}_0$. Let $\mathbf{c}_0 = \mathrm{acl}(\mathbf{b}_0) \cap D$. Then clearly $\mathbf{c}_0 d_0$ is contained in A_0. Let $\mathbf{c}_1$ be such that $\mathrm{tp}(\mathbf{c}_1 \mathbf{b}_1) = \mathrm{tp}(\mathbf{c}_0 \mathbf{b}_0)$. Note that then $\mathbf{c}_1 = \mathrm{acl}(\mathbf{b}_1) \cap D$, and so $\mathbf{c}_1 \subseteq A_0$. Now A_0 is an algebraically closed (in D) subset of the minimal set D, and $\mathbf{c}_0$, $\mathbf{c}_1$ are tuples from A_0 with the same type. By Remark 6.8(ii), there is $d_1 \in A_0$ such that $\mathrm{tp}(\mathbf{c}_0 d_0) = \mathrm{tp}(\mathbf{c}_1 d_1)$, that is $d_1 \notin \mathrm{acl}(\mathbf{c}_1)$. Thus $d_1 \in D - \mathrm{acl}(\mathbf{b}_1)$. So $\mathrm{tp}(\mathbf{b}_1 d_1) = \mathrm{tp}(\mathbf{b}_0 d_0)$, so there is e_1 such that $\mathrm{tp}(\mathbf{b}_0 d_0 e_0) = \mathrm{tp}(\mathbf{b}_1 d_1 e_1)$. But $e_0 \in \mathrm{acl}(\mathbf{b}_0, d_0)$. So $e_1 \in \mathrm{acl}(\mathbf{b}_1, d_1)$. As $\mathbf{b}_1$, $d_1 \subseteq \mathrm{acl}(B) = B$, also $e_1 \in B$.

Case (ii). $\mathrm{tp}(e_0/\mathbf{b}_0)$ is affine (over $\mathbf{b}_0$).

Let $q(x, \mathbf{y})$ be $\mathrm{tp}(e_0, \mathbf{b}_0)$. Then $q(x, \mathbf{b}_1)$ is also stationary, strictly minimal, and affine (so non-modular) over $\mathbf{b}_1$. Suppose by way of contradiction that $q(x, \mathbf{b}_1)$ is not realized in $B = \mathrm{acl}(B)$. Let c realize $q(x, \mathbf{b}_1)$. By 6.5(ii) (where D there corresponds to the set of realizations of $q(x, \mathbf{b}_1)$ here), $\mathrm{tp}(c/B)$ is affine (so non-modular) over B. In particular there is no $d \in D - \mathrm{acl}(B)$ which is interalgebraic with c over B (as $\mathbf{D}_B$ is modular). Thus $\mathrm{acl}(Bc) \cap D = \mathrm{acl}(B) \cap D = A_0$, contradicting the maximality of B in its definition as a D-envelope. Thus $q(x, \mathbf{b}_1)$ *is* realized by some $e_1 \in B$.

The lemma is proved. ■

Proof of Theorem 6.6 We claim that any finite subset A of $\mathbf{C}$ is contained in a finite homogeneous substructure of $\mathbf{C}$. For let B be a D-envelope of A. By 6.13, B is a homogeneous substructure of $\mathbf{C}^{\mathrm{eq}}$. In particular $B \cap \mathbf{C}$ is a homogeneous substructure of $\mathbf{C}$, containing A. By Lemma 6.12, $B \cap C$ is finite.

Now apply Lemma 6.9. ■

3
Quasifinite axiomatizability

We saw in Chapter 2 that totally categorical theories are not finitely axiomatizable. A slight extension of the proof yields non-finite axiomatizability of ω-categorical, ω-stable theories. At some point it was asked whether the only obstruction to the finite axiomatizability of, say, totally categorical theories was the non-finite expressibility of 'there are infinitely many elements', that is whether totally categorical theories could be finitely axiomatized modulo the 'axiom of infinity'. In this chapter we see that this is true. In fact we prove here that ω-categorical, ω-stable theories are 'quasi-finitely axiomatizable', that is finitely axiomatizable modulo a finite number of 'schemas of infinity'. The proof has very little to do with model theory, and depends on a certain combinatorial property of projective spaces over finite fields. Although the result implies that there are only countable many ω-categorical, ω-stable structures, not much additional information is obtained about the actual structures. In the first section of this chapter we take the opportunity to discuss automorphism groups of ω-categorical structures, and their connection with interpretability.

Throughout this chapter all models we study will be countable and ω-categorical (and in a countable language).

1 ω-categoricity and automorphism groups

Let M be a (ω-categorical) structure in a language L. Aut(M), the group of automorphisms of M, can be viewed in three ways, as an abstract group, as a topological group, and as a permutation group. The abstract group and permutation group structures are clear. Note that Aut(M) also acts on the set of n-tuples of M. The topological group structure is given by taking, as basic open sets, sets of the form $\{f \in \mathrm{Aut}(M)\colon f(\mathbf{a}) = \mathbf{b}\}$ for $\mathbf{a}$, $\mathbf{b}$ n-tubles from M, with n varying. Note that a base for the open neighbourhoods of the identity of Aut(M) is $\{\mathrm{Fix}(\mathbf{a})\colon \mathbf{a}$ some finite tuple from $M\}$ where $\mathrm{Fix}(\mathbf{a}) = \{g \in \mathrm{Aut}(M)\colon g(\mathbf{a}) = \mathbf{a}\}$. It is then clear that a homomorphism $f\colon \mathrm{Aut}(M) \to \mathrm{Aut}(N)$ is *continuous* just if for each finite tuple $\mathbf{a}$ from N, there is a finite tuple $\mathbf{b}$ from M such that $\mathrm{Fix}(\mathbf{b}) \subseteq f^{-1}(\mathrm{Fix}(\mathbf{a}))$ (for then the subgroup $f^{-1}(\mathrm{Fix}(\mathbf{a}))$ contains an open subgroup of Aut(M), so is itself open). One basic fact about an ω-categorical structure M, is that $X \subset M^n$ is $\emptyset$-definable in M iff X is Aut(M)-invariant. Also tp($\mathbf{a}$) = tp($\mathbf{b}$) in M iff there $\sigma \in \mathrm{Aut}(M)$ such that $\sigma(\mathbf{a}) = \mathbf{b}$. This is because there are only finitely many formulae (complete types) in n variables, for each n; and also because any countable model is ω-homogeneous.

Definition 1.1. (i) *Let M, N be L- and L'-structures with the same universe ω. Let $A \subseteq \omega$. We say that M and N are* A-bidefinable *if they have the same A-definable relations.*

(ii) *Let M, N be L-, L'-structures, and let $A \subseteq N$. An A-interpretation f of M in N consists of the following data: an A-definable set U in N^{eq}, and for each (n-ary) relation R of L a subset f_R of U^n which is A-definable in N^{eq}, and a bijection f of the universe of M with U, such that f yields an isomorphism $(M, R, \ldots) \cong (U, f_R, \ldots)$ of L-structures.*

Remark 1.2. (i) *In this section we will be basically concerned with the case where $A = \emptyset$. So (for the purposes of this chapter), we say M and N are bidefinable if they are $\emptyset$-bidefinable, and that f is an interpretation of M in N if it is a $\emptyset$-interpretation.*

(ii) *If M_1, M_2, M_3 are L_1, L_2, L_3-structures respectively, and f is an interpretation of M_1 in M_2 and g is an interpretation of M_2 in M_3, then it is not difficult to see that we can compose f and g to get an interpretation $g.f$ of M_1 in M_3.*

Definition 1.3. *Let M_1, M_2 be L_1-, L_2-structures respectively. We say that M_1 and M_2 are* bi-interpretable *if there are interpretations f of M_1 in M_2, and g of M_2 in M_1, such that the map $g.f$ is $\emptyset$-definable in M_1 and the map $f.g.$ is $\emptyset$-definable in M_2.*

A typical example is the bi-interpretability of M and M^{eq} (where we view M^{eq} as an ω-categorical many-sorted structure). We now use the ω-categorical hypothesis to get a nice characterization of bi-interpretability (as well as showing that it is a natural notion).

Proposition 1.4. *Let M_i be an L_i-structure for $i = 1, 2$.*

(i) *If M_1 and M_2 have the same universe, then they are bidefinable iff* Aut $(M_1) =$ Aut(M_2) *as permutation groups.*

(ii) *M_1 and M_2 are bi-interpretable iff* Aut(M_1) *and* Aut(M_2) *are homeomorphic as topological groups.*

Proof. (i) is clear by the previously mentioned characterization of $\varnothing$-definable sets in ω-categorical structures.

(ii) We first make a few remarks.

(I) If f is an interpretation of M_1 in M_2 then f induces a continuous homomorphism which we call $\mathrm{Aut}(f)$ from $\mathrm{Aut}(M_2)$ into $\mathrm{Aut}(M_1)$. We use the notation of Definition 1.1(ii). f is an isomorphism of M_1 with $(U, \{f_R : R \in L_1\})$ where the latter lives $\varnothing$-definably in M_2^{eq}. Any automorphism σ of M_2 preserves U and the f_R so induces an automorphism of $(U, \{f_R : R \in L_1\})$ and thus an automorphism, which we call $\mathrm{Aut}(f)(\sigma)$, of M_1. Formally, for $a \in M_1$ $(\mathrm{Aut}(f)(\sigma))(a) = f^{-1}(\sigma(f(a)))$. The map $\mathrm{Aut}(f)$ is clearly a homomorphism. For continuity, let $\mathbf{a}$ be a finite tuple from M_1, and let $\mathbf{b}$ be a finite tuple from M_2 such that $f(\mathbf{a}) \in \mathrm{dcl}(\mathbf{b})$ in M_2. Then $\mathrm{Aut}(f)^{-1}(\mathrm{Fix}(\mathbf{a}))$ contains $\mathrm{Fix}(\mathbf{b})$.

(II) $\mathrm{Aut}(g).\mathrm{Aut}(f) = \mathrm{Aut}(gf)$. (Clear.)

(III) Let f be an interpretation of M into M. Then f (as a map) is $\varnothing$-definable in M iff $\mathrm{Aut}(f) = \mathrm{id}_{\mathrm{Aut}(M)}$. If the map f is $\varnothing$-definable, then for $\sigma \in \mathrm{Aut}(M)$ and $a \in M$,

$$(\mathrm{Aut}(f)(\sigma))(a) = f^{-1}(\sigma(f(a))) = f^{-1}(f(\sigma(a))) = \sigma(a), \text{ so } \mathrm{Aut}(f)(\sigma) = \sigma.$$

Conversely, if $\mathrm{Aut}(f) = \mathrm{id}_{\mathrm{Aut}(M)}$, then the line above shows that for all $a \in M$, $\sigma(f(a)) = f(\sigma(a))$ for all $g \in \mathrm{Aut}(M)$, that is the graph of f is $\mathrm{Aut}(M)$-invariant, and thus $\varnothing$-definable.

(I), (II), and (III) clearly yields that bi-interpretability of M_1 and M_2 (via f_1, f_2) implies that $\mathrm{Aut}(M_1)$ is homeomorphic to $\mathrm{Aut}(M_2)$ (via $\mathrm{Aut}(f_1) = \mathrm{Aut}(f_2))^{-1}$).

We now want to show the reverse implication. So suppose $\varphi\colon \mathrm{Aut}(M_1) \to \mathrm{Aut}(M_2)$ to be a homeomorphism. We will construct an interpretation f of M_2 in M_1 such that $\varphi = \mathrm{Aut}(f)$. First assume M_2 to have a unique 1-type over $\varnothing$, that is $\mathrm{Aut}(M_2)$ acts transitively on M_2. Let $b \in M_2$ and let $\mathbf{a}$ be, say, an n-tuple in M_1 such that $\varphi(\mathrm{Fix}(\mathbf{a})) \subseteq \mathrm{Fix}(b)$. Let $X \subseteq M_1^n$ be the set of realizations of $\mathrm{tp}(\mathbf{a})$ in M_1, and define, for $\sigma \in \mathrm{Aut}(M_1)$, $h(\sigma(\mathbf{a})) = \varphi(\sigma)(b)$. Then h is a map from X onto M_2, and it is easy to check that the equivalence relation E on X: $E(\mathbf{a}_1, \mathbf{a}_2)$ if $h(\mathbf{a}_1) = h(\mathbf{a}_2)$, is $\mathrm{Aut}(M_1)$-invariant and thus $\varnothing$-definable in M_1. Let $U = X/E$ which is $\varnothing$-definable in M_1^{eq}. Then h induces a bijection between U and M_2, the inverse of which we call f. It is easy to check that the image under f of a $\varnothing$-definable set in M_2 is $\mathrm{Aut}(M_1)$-invariant, so $\varnothing$-definable in M_1. Thus f is an interpretation of M_2 in M_1. Looking back at (I) above, we also see that $\varphi = \mathrm{Aut}(f)$. In general M_2 has finitely many orbits under $\mathrm{Aut}(M_2)$, say $Y_1, \ldots, Y_m$. Define f_i as above on each Y_i. We may assume that $\mathrm{Im}(f_i)$ are disjoint, so we take $f = \cup f_i$, and everything works.

Similarly there is an interpretation g of M_2 in M_1 such that $\mathrm{Aut}(g) = \varphi^{-1}$. Thus (by (II)) $\mathrm{Aut}(g).\mathrm{Aut}(f) = \mathrm{Aut}(g.f) = \mathrm{id}$. Similarly $\mathrm{Aut}(f.g) = \mathrm{id}$. By (III) $g.f.$ and $f.g.$ are $\varnothing$-definable in M_2, M_1 respectively, so M_1 and M_2 are bi-interpretable. ∎

Before the next example, let us recall the notion of 'induced structure'. Suppose M to be an ω-categorical structure, and let U be a $\varnothing$-definable subset of M^n. By the *structure induced on U from M*, we mean the structure whose universe is U and whose $\varnothing$-definable relations are precisely the sets of the form $X \cap U^m$, as X ranges over $\varnothing$-definable subsets of M^{nm} (m varying). This structure is clearly unique up to bidefinability.

Example 1.5. Every (ω-categorical structure) M is bi-interpretable with an (ω-categorical) N-structure which has a unique 1-type over $\varnothing$.

Proof. Let $p_1, \ldots, p_n$ be the 1-types realized in M, and let a_i realize p_i. Let $U \subseteq M^n$ be the set of realizations of $\mathrm{tp}((a_1, \ldots, a_n))$, and let N be U with its induced structure. Then N is ω-categorical, every $\sigma \in \mathrm{Aut}(M)$ induces (by definition) an automorphism $\varphi(\sigma)$ of N, and the map φ $\mathrm{Aut}(M) \to \mathrm{Aut}(N)$ is easily seen to be a homeomorphism. ∎

For the purposes of this chapter we say that a language L is finite if it contains just finitely many relation symbols (and no other symbols).

Definition 1.6. (i) *We say that M* has a finite language *if there is a finite language L′ and an L′-structure N which is bidefinable with M.*

(ii) *We say that M is* quasifinitely axiomatizable *if there is a finite language L′ and an L′-structure N bidefinable with M, such that Th(N) can be axiomatized by some finite set of sentences, together with finitely many 'schemas of infinity', where*

(iii) *A schema of infinity in a language L is a schema of the form* $(\forall x)(\varphi(x) \to \exists^{\infty} y(\psi(x, y)))$, *where* $\varphi(x)$, $\psi(x, y)$ *are formulae of* L^{eq}.

So these notions, having a finite language and being quasifinitely axiomatizable, are by definition properties of the bidefinability type of M. We see now that these properties depend just on the bi-interpretability type of M.

Lemma 1.7. *Let M_1 and M_2 be bi-interpretable. Then*

(i) *M_1 has a finite language iff M_2 has a finite language.*

(ii) *M_1 is quasifinitely axiomatizable iff M_2 is.*

Proof. (i) Suppose M_2 has a finite language L_2. Let g be an interpretation of M_2 in M_1 and f an interpretation of M_1 in M_2 such that the map $g.f$ is $\emptyset$-definable in M_1. Let $g: M_2 \cong (U, \{f_R: R \in L_2\})$, where U and the f_R are $\emptyset$-definable in M_1^{eq}. Let L_1' be a sufficiently large finite sublanguage of L_1 such that U, the f_R for $R \in L_2$, and the map $g.f$ are $\emptyset$-definable in $(M_1 | L_1')^{\text{eq}}$. It is then easy to check that any permutation of M_1 which preserves the L_1'-relations is an automorphism of M_1. Thus M_1 is bidefinable with $M_1 | L_1'$, so M_1 has a finite language.

(ii) Suppose M_2 is quasifinitely axiomatizable (in finite language L_2). Let f, g, U, the f_R, and L_1' be as in the proof of (i). Each schema of infinity for M_2 translates into a schema of infinity for M_1 in L_1' (talking about $(U, \{f_R: R \in L_2\})$) and each of the remaining finitely many axioms for M_2 translates similarly into an axiom in L_1' for M_1. Add another axiom stating that $g.f$ is an isomorphism of L_1'-structures. Then any (countable) model of these axioms is easily seen to be isomorphic to $M_1 | L_1'$. ■

Finally in this section we try to pin down more precisely the 'structure' of an ω-categorical strictly minimal set D. This question is made precise by the earlier material of this section. For we are really interested in D up to interdefinability; that is, assuming D countable, we want to identify Aut(D) as a permutation group. Actually the strictly minimal sets D we are interested in typically arise as definable sets in a given stable structure M. So we are talking about the group of permutations of D which are elementary in the sense of M. We still call this Aut(D). In any case, ω-categoricity of D is sufficient for the earlier notions and results (1.3, 1.4) to make sense. Assume for now that D is *strictly minimal*, ω-categorical, defined over $\emptyset$ in the countable stable structure M.

The discussion will be a little long, but we summarize it in 1.9 and 1.10.

Theorem 2.4.17 (together with 2.1.11) tells us what the geometry of D is, namely trivial (disintegrated), or affine or projective over a finite field F. If the geometry is trivial, then clearly *any* permutation of D is elementary, so Aut(D) is the full symmetric group on the underlying set of D, so we have just one possibility.

Now suppose the geometry on D is projective over the finite field F. As D is ω-categorical, the relations on D: $R_n(x_1, \ldots, x_n, y) =$ '$y \in \text{acl}(x_1, \ldots, x_n)$' are $\emptyset$-definable for each n. Let D_0 be the reduct of D to these relations. Of course D_0 is an ω-categorical (and strictly minimal) structure in its own right, and the fundamental theorem of projective geometry says that: Aut(D_0) is exactly PΓL(V) (the projective general semi-linear group), where V is an (ω-dimensional) F-vector space, of which D is the projective space.

We should say what PΓL(V) is. First, by a *semi-linear* automorphism of

V we mean a *group* automorphism f of V such that there is an automorphism σ of F such that $f(\alpha.v) = \alpha^\sigma.f(v)$ for all $v \in V$ and $\alpha \in F$. $\Gamma\mathrm{L}(V)$ is by definition the group of all semi-linear automorphisms of V. D is (as a set) precisely the set of 1-dimensional F-linear subspaces of V. $\Gamma\mathrm{L}(V)$ also acts on D, and the resulting permutation group is called $\mathrm{P}\Gamma\mathrm{L}(V)$ (which we said above is $\mathrm{Aut}(D_0)$). Thus $\mathrm{Aut}(D)$ will be a *subgroup* of $\mathrm{P}\Gamma\mathrm{L}(V)$.

Now $\Gamma\mathrm{L}(V)$ is the semi-direct product of $\mathrm{GL}(V)$ (the group of F-linear automorphisms of V) with $\mathrm{Aut}(F)$, where $\mathrm{Aut}(F)$ acts on $\mathrm{GL}(V)$ in the natural way (after choosing a basis for V). So (as F is finite), $\mathrm{GL}(V)$ is a finite index subgroup of $\Gamma\mathrm{L}(V)$. Also $\mathrm{PGL}(V)$ denotes the group of permutations of projective space over V, induced by $\mathrm{GL}(V)$, and similarly $\mathrm{P}\Gamma\mathrm{L}(V)$ is the semi-direct product of $\mathrm{PGL}(V)$ with $\mathrm{Aut}(F)$. Consider the structure $\mathbf{V} = (V, +, \alpha)_{\alpha \in F}$, where α denotes scalar multiplication by α. This is ω-categorical, strongly minimal, with quantifier elimination, and its automorphism group is clearly $\mathrm{GL}(V)$. Let D_1 denote projective space over V, as a $\emptyset$-definable set in $\mathbf{V}^{\mathrm{eq}}$. D_1 is also strictly minimal. We may identify (canonically) the universe of D_1 with that of D (and also of D_0). Clearly $\mathrm{Aut}(D_1) = \mathrm{PGL}(V)$. In a moment we will get to the question of what is a 'natural language' for D_1. By looking at the automorphism groups, we see that D_0 is a reduct of D_1.

In fact we will show that D_1 is obtained from D_0 simply by naming some (algebraic) elements in $(D_0)^{\mathrm{eq}}$: let a_1, a_2, a_3, a_4 be distinct collinear points in D_0 (i.e. $a_3, a_4 \in \mathrm{acl}(a_1, a_2)$). Using the fact that F is commutative, one sees easily that there is *unique* $\alpha \in F$ such that for some representatives x_1, x_2, x_3, x_4 in V, *of* a_1, a_2, a_3, a_4 respectively, $x_3 = x_1 + x_2$ and $x_4 = \alpha.x_1 + x_2$. α is called the *cross-ratio* of (a_1, a_2, a_3, a_4), and each $\alpha \in F$ $(\alpha \neq 0, 1)$occurs this way. Let E be the equivalence relation on the set of distinct 4-tuples from D_0, given by having the same cross-ratio. Then E has finitely many classes. It can also be easily seen that E is $\mathrm{P}\Gamma\mathrm{L}(V)$ invariant, and thus $\emptyset$-definable in D_0. Also the group of automorphisms of D_0 which fix each E-class is precisely $\mathrm{PGL}(V)$. This shows that D_1 is obtained from D_0 by simply naming these E-classes. It also shows that $\mathrm{PGL}(V)$ is an *open* subgroup of $\mathrm{P}\Gamma\mathrm{L}(V)$, and thus also a closed subgroup.

The main point we wish to make is that D lies *in between* D_0 and D_1, or putting it another way, $\mathrm{Aut}(D)$ lies in between $\mathrm{P}\Gamma\mathrm{L}(V)$ and $\mathrm{PGL}(V)$. This could be seen by model-theoretic techniques developed in the next couple of chapters. But for now we will refer to a result of Cameron and Kantor. Here V denotes a countably infinite vector space over the finite field F, P the corresponding projective space. For V', a non-zero finite-dimensional F-subspace of V, P' will denote the corresponding finite-dimensional projective subspace of P. If G is a group of permutations of a set X, and $Y \subseteq X$, then $G|Y$ denotes the group of elements of G which fix setwise Y, acting on Y. $\mathrm{SL}(V')$ denotes the group of linear automorphisms of V' of determinant 1, and $\mathrm{PSL}(V')$ the induced action on P'.

Fact 1.8. [Ca–K] Let G be a subgroup of $\mathrm{P\Gamma L}(V)$. If V' is a finite subspace of V of dimension ≥ 3, and $G|P'$ acts 2-transitively on P', then $G|P'$ contains $\mathrm{PSL}(V')$.

With the previous notation we then have:

COROLLARY **1.9.** $\mathrm{PGL}(V) < \mathrm{Aut}(D) < \mathrm{P\Gamma L}(V)$.

Proof. Let $G = \mathrm{Aut}(D)$. Remember that the algebraically closed subsets of D are precisely the projective subspaces of D. Also, as D is strictly minimal, there is a unique 2-type (of distinct elements) realized in D. So the hypotheses of 1.8 hold. As G is *closed* in $\mathrm{P\Gamma L}(V)$, it suffices to prove that if $\sigma \in \mathrm{PGL}(V)$ $\mathbf{a}$, $\mathbf{b}$ are n-tuples from D, and $\sigma(\mathbf{a}) = \mathbf{b}$, then $\tau(\mathbf{a}) = \mathbf{b}$ for some $\tau \in G$.

So pick some $\mathbf{a}$, $\mathbf{b}$, and σ. We can find a finite-dimensional F-subspace V' of V, such that the corresponding projective subspace P' of D contains $\mathbf{a}$ and $\mathbf{b}$ and such that some $\sigma' \in \mathrm{PGL}(V')$ takes $\mathbf{a}$ to $\mathbf{b}$. By enlarging V' and P' we may assume that $\sigma' \in \mathrm{PSL}(V')$. By Fact 1.8, $G|P'$ contains $\mathrm{PSL}(V')$. Thus $\tau(\mathbf{a}) = \mathbf{b}$ for some $\tau \in G$. ■

It follows from 1.9 that, fixing F, there are only finitely many possibilities for $\mathrm{Aut}(D)$ (and thus for D). In fact the possibilities correspond to subfields F' of F. For each such subfield F', let $D_{F'}$ be obtained from D_0 by naming the E-classes corresponding to elements of F'. This can be seen by considering the structure of $\mathrm{Aut}(F)$.

It should be mentioned that an analogous result follows for strictly minimal sets whose geometries are affine over a finite field. If D is such, then we can identify the underlying set of D with the points of an F-vector space V. $\mathrm{GL}(V)$ and $\mathrm{\Gamma L}(V)$ are as before. $\mathrm{AGL}(V)$ denotes the group of permutations of V generated by $\mathrm{GL}(V)$ and the translations: $v \to v + a$ (for $a \in V$). This is also a semi-direct product $\mathrm{GL}(V)$ acting on V in the natural way. Similarly we obtain $\mathrm{A\Gamma L}(V)$. Using a result analogous to Fact 1.8 (replacing PSL by ASL), we have:

Fact 1.10. If D is a strictly minimal set whose geometry is affine over the finite field F, then $\mathrm{AGL}(V) < \mathrm{Aut}(D) < \mathrm{A\Gamma L}(V)$.

Again there are only finite many possibilities, and elements of F are coded up now by equivalence classes of triples from D. Details are left to the reader.

So for either the case of projective or affine strictly minimal sets D, having fixed the finite field F, there is a *weakest* structure D_0 (with largest automorphism group), and a *strongest* structure D_1 (with smallest automorphism group). D can always be expanded to D_0 by naming a finite set of elements in $\mathrm{acl}(\varnothing)$.

It is worth giving (natural) languages for the strongest structures: in the affine case, this is given by the ternary function $x + y - z$ and for each $\alpha \in F$, the binary function $\alpha.x + (1 - \alpha).y$. In the projective case, take the collinearity relation $(x, y, z) \in \mathrm{Coll}$, and for each $\alpha \in F$, the ternary function, which given collinear distinct x_1, x_2, x_3, yields the unique x_4 such that cross-ratio $(x_1, x_2, x_3, x_4) = \alpha$.

An important remark, which we will use in this chapter, is:

LEMMA **1.11.** *Suppose D_1 is a strictly minimal ω-categorical set (defined over $\varnothing$), whose geometry is projective over the finite field F, and which is the strongest such structure (i.e.* $\mathrm{Aut}(D_1) = \mathrm{PGL}(V)$). *Then for any $a_1, \ldots, a_n \in D$, $\mathrm{tp}(a_1, \ldots, a_n/\varnothing)$ is stationary.*

Proof. We can consider D_1 as $\varnothing$-definable in the structure $\mathbf{V} = (V, +, \alpha)_{\alpha \in F}$ (i.e. living in $\mathbf{V}^{\mathrm{eq}}$ as the collection of 1-dimensional subspaces). This is because $\mathrm{Aut}(\mathbf{A})$ is exactly $\mathrm{GL}(V)$, and thus the induced automorphism group on D_1 is $\mathrm{PGL}(V)$, which is what it should be.

Now $\mathbf{V}$ is easily seen to have quantifier elimination (in fact all we need of this is that for any subset A of V, $\mathrm{acl}(A) \cap V$ is the F-subspace generated by A, which can be seen by an automorphism argument). ■

CLAIM. *For each $a_1, \ldots, a_n$ in* $\mathbf{V}$, $\mathrm{tp}(a_1, \ldots, a_n/\varnothing)$ *is stationary.*

Proof. $\mathbf{V}$ is strongly minimal (so is connected as a group), so there is a *unique* generic type over $\varnothing$. Let $a_1, \ldots, a_n \in \mathbf{V}$. Without loss $\{a_1, \ldots, a_m\}$ is maximal algebraically independent over $\varnothing$. Thus $\mathrm{tp}(a_1, \ldots, a_m/\varnothing)$ is stationary. For $i > m$, $a_i \in \mathrm{acl}(a_1, \ldots, a_m)$. By quantifier elimination, a_i is in the F-linear closure of $(a_1, \ldots, a_m)$. As the elements of F are given by functions in the language, $a_i \in \mathrm{dcl}(a_1, \ldots, a_m)$. Thus clearly $\mathrm{tp}(a_1, \ldots, a_n/\varnothing)$ is stationary.

But if $\mathbf{b}$ is some tuple from D_1, then $\mathbf{b} \in \mathrm{dcl}(\mathbf{a})$ for some tuple $\mathbf{a}$ from $\mathbf{V}$. By the claim $\mathrm{tp}(\mathbf{a}/\varnothing)$ is stationary. So $\mathrm{tp}(\mathbf{b}/\varnothing)$ is also stationary. ■

We end up by remarking that if M is a (ω-categorical) structure, N is an 'expansion' of M, and $\mathrm{Aut}(N)$ is an open subgroup of $\mathrm{Aut}(M)$, then it is not difficult to prove (like in 1.7) that M is quasifinitely axiomatizable iff N is. Now the classical axioms for projective geometries show that the weakest structure D_0 is quasifinitely axiomatizable. It follows, using 1.9, that any strictly minimal projective ω-categorical structure is quasifinitely axiomatizable. Similarly for the affine case. We will not use this remark in our subsequent proofs.

2 ω-categorical, ω-stable structures

We now begin some preparatory work towards proving the quasifinite axiomatizability of ω-categorical, ω-stable structures.

The first step, which we carry out in this short section, is to replace our given structure M by a bi-interpretable structure M^* which is more friendly. M^* will have a 'tree structure' given by coordinatization.

M is from now on a fixed, countable, ω-categorical, ω-stable structure, in a language L, with theory T. By Theorem 2.5.12 T is 1-based of finite U-rank. Moreover, 1.3.21 says that U-rank = Morley rank. By virtue of 1.5 and 1.7 we assume that M has a unique 1-type, say p, over $\varnothing$.

LEMMA **2.1.** *Let $c \in M$ (so c realizes p). Then there are $a_1, \ldots, a_n \in M^{eq}$ such that*

(i) $c = a_n$.

(ii) *For each* $i = 1, \ldots, n-1$, $\mathrm{tp}(a_{i+1}/a_1 \ldots a_i)$ *is algebraic or strictly minimal.*

(iii) *If* $\mathrm{tp}(a_{i+1}/a_1 \ldots a_i)$ *is strictly minimal and affine, then it is non-orthogonal to* $\mathrm{tp}(a_{j+1}/a_1 \ldots a_j)$ *for some* $j < i$.

(iv) *if* $\mathrm{tp}(a_{i+1}/a_1 \ldots a_i)$ *is strictly minimal and projective, then the* $(a_1 \ldots a_i)$*-definable structure on the set of realizations of* $\mathrm{tp}(a_{i+1}/a_1 \ldots a_i)$ *is the strongest possible structure (see* 1.9 *and the discussion following* 1.10), *in particular (by* 1.11) *for any tuple* $\mathbf{b}$ *of realizations of this type,* $\mathrm{tp}(\mathbf{b}/a_1 \ldots a_i)$ *is stationary.*

Proof. We work in M^{eq}. Suppose we have already found $a_1, \ldots, a_i$. If $c \in \mathrm{acl}(a_1 \ldots a_i)$, put $a_{i+1} = c$ and stop.

Otherwise, by Proposition 2.5.9 (so-called coordinatization), find $d \in \mathrm{acl}(c, a_1, \ldots, a_i)$ such that $U(d/a_1, \ldots, a_i) = 1$. Thus $RM(d/a_1 \ldots a_i) = 1$. Note that $RM(c/a_1 \ldots a_i d) < RM(c/a_1 \ldots a_i)$ (by the U-rank equalities, for example). By the finite equivalence relation theorem (1.2.20(iv)) together with ω-stability, there is $a_{i+1} \in \mathrm{acl}(a_1 \ldots a_i)$ such that $\mathrm{tp}(d/a_1 \ldots a_{i+1})$ is strongly minimal (i.e. has Morley rank 1, and Morley degree 1), and after replacing d by something interalgebraic with itself over $\{a_1, \ldots, a_{i+1}\}$ we may assume $\mathrm{tp}(d/a_1, \ldots, a_{i+1})$ is strictly minimal. If this type is disintegrated, put $a_{i+2} = d$. If this type is projective, we can, by Lemma 1.9, add a finite part of $\mathrm{acl}(a_1, \ldots, a_{i+1})$ (i.e. the equivalence classes of the equivalence relation 'having the same cross-ratio') to a_{i+1} to satisfy (iv), and again put $a_{i+2} = d$. So suppose $\text{tp}(d/a_1 \ldots a_{i+1})$ is affine. If it is non-orthogonal to $\text{tp}(a_{j+1}/a_1 \ldots a_j)$ for some $j < i$, again put $a_{i+2} = d$. Otherwise , let a_{i+2} realize a strictly minimal projective type over $a_1 \ldots a_{i+1}$ which is non-orthogonal to $\text{tp}(d/a_1 \ldots a_{i+1})$ (e.g. let a_{i+2} be a parallelism class of lines in the affine geometry of realizations of $\text{tp}(d/a_1 \ldots a_{i+1})$). We may again (by fattening a_{i+1}) assume that (iv) is satisfied by a_{i+2}. Then $\text{tp}(d/a_1 \ldots a_{i+2})$ is still strongly minimal and affine (see 2.6.3(ii)), and can be assumed, as above, to be strictly minimal. Put $a_{i+3} = d$. This clearly proves the lemma. ■

DEFINITION 2.2. *Let $r_i = \text{tp}(a_1, \ldots, a_i)$, let L_i be the set of realizations of r_i in M^{eq} and let M^* be the structure whose universe is (the disjoint union) $L_1 \cup L_2 \cup \cdots \cup L_n$ with all structure induced from M^{eq}. (Note that then $\text{Aut}(M^*)$ is the set of permutations of $L_1 \cup \ldots \cup L_n$ induced by automorphisms of M^{eq}.) As in Example 1.5, M and M^* are bi-interpretable. It is convenient to adjoin to M^* a distinguished element a_0 (this changes nothing).*

We continue to denote elements of L_i as $(b_1, \ldots, b_i)$ (where $\text{tp}(b_1 \ldots b_i) = r_i$ in M^{eq}). Clearly the map $\pi_{i,j}$ from L_i to L_j ($j \le i$) taking $(b_1, \ldots, b_i)$ to $(b_1, \ldots, b_j)$ is $\varnothing$-definable in M^*. M^* thus has a $\varnothing$-definable tree structure: $(x_1, \ldots, x_j) \le (y_1, \ldots, y_i)$ iff $j \le i$ and $x_1 = y_1, \ldots, x_j = y_j$.

We consider the constant a_0 as a root of this tree (and the unique element in L_0).

For $(b_1, \ldots, b_{i-1}) \in L_{i-1}$, let $L(b_1, \ldots, b_{i-1})$ be the set of realizations $(x_1, \ldots, x_i)$ of r_i such that $\pi_{i,i-1}(x_1, \ldots, x_i) = (b_1, \ldots, b_{i-1})$. So clearly we can identify $L(a_1, \ldots, a_{i-1})$ with the set of realizations of $\text{tp}(a_i/a_1 \ldots a_{i-1})$ which is either a finite set or a strictly minimal set. Note that for any realization $\mathbf{b}$ of L_{i-1} there is an automorphism of M^* taking $L(a_1, \ldots, a_{i-1})$ to $L(\mathbf{b})$.

We now partition $I = \{1, \ldots, n\}$ into four sets.

DEFINITION 2.3. *We put $i \in I_{\text{new}}$ if $L(a_1, \ldots, a_{i-1})$ is modular (i.e. projective or trivial) and orthogonal to $L(a_1, \ldots, a_{j-1})$ for all $j < i$ (including $j = 1$).*

Put $i \in I_{\text{old}}$ if $L(a_1, \ldots, a_{i-1})$ is again modular but now non-orthogonal to some $L(a_1, \ldots, a_{j-1})$ for $j < i$.

Put $i \in I_{\text{aff}}$ if $L(a_1, \ldots, a_{i-1})$ is affine.

Finally put $i \in I_{\text{fin}}$ if $L(a_1, \ldots, a_{i-1})$ if finite.

By the above remarks, this partition is well defined (i.e. is a function of i not of $(a_1, \ldots, a_{i-1})$).

Finally for $i \in I_{\text{old}}$ or I_{aff} we will define a suitable $\varnothing$-definable function

f_i, witnessing the non-orthogonality of $L(a_1, \ldots, a_{j-1})$ with $L(a_1, \ldots, a_{i-1})$ *for some* $j < i$.

CONSTRUCTION **2.4.** (a) *Let* $i \in I_{\text{old}}$. *Let* $\alpha(i)$ *be the least* $j < i$ *such that* $L(a_1, \ldots, a_{i-1})$ *is non-orthogonal to* $L(a_1, \ldots, a_{j-1})$ *(so the latter is modular). For* $c \in L(a_1, \ldots, a_{j-1}) - \text{acl}(a_1, \ldots, a_{i-1})$ *there is clearly (by modularity of the relevant types and Lemma 2.5.5) unique* $d \in L(a_1, \ldots, a_{i-1})$ *such that* $d \in \text{acl}(c, a_1, \ldots, a_{i-1})$. *Clearly then* $d = f_i(a_1, \ldots, a_{i-1}, c)$ *for some* $\varnothing$-*definable function* f_i. *Note that* $f_i(a_1, \ldots, a_{i-1}, x)$: $(L(a_1, \ldots, a_{j-1}) - \text{acl}(a_1, \ldots, a_{i-1})) \to L(a_1, \ldots, a_{i-1})$ *is surjective.*

(b) *Now suppose* $i \in I_{\text{aff}}$. *Let* $\alpha(i)$ *be chosen as in* (a). *So* $L(a_1, \ldots, a_{\alpha(i)-1})$ *is projective. (For, by the above, whenever* $L(a_1, \ldots, a_{j-1})$ *is affine, then* $L(a_1, \ldots, a_{j-2})$ *is projective and non-orthogonal to* $L(a_1, \ldots, a_{j-1})$.) *Let* P *denote* $L(a_1, \ldots, a_{\alpha(i)-1})$, *and* A *denote* $L(a_1, \ldots, a_{i-1})$. *Let* $\mathbf{a}$ *denote* $(a_1, \ldots, a_{i-1})$. *Let* A^* *denote the* $\mathbf{a}$-*definable projective strictly minimal set consisting of the parallelism classes of lines (or directions) in* A *(see Lemma* 2.6.2). *So* P *is non-orthogonal to* A^*. *Let* g *denote some surjective* $\mathbf{a}$-*definable map from* $P - \text{acl}(\mathbf{a})$ *onto* A^*, *chosen as in part* (a). *Let* $\mathbf{b} = (b_0, \ldots, b_q)$ *be a tuple from* A *which enumerates some affine line. Then we obtain an* $\mathbf{ab}$-*definable map* h, *say, from* $(A^*)^3$ *onto* A *as follows. First we restrict our attention to elements* $d \in A^*$ *which are not the direction of the line* $\mathbf{b}$. *If* d_1, d_2, d_3 *are such, and* $d_1 \neq d_2$ *but* $d_1 = d_3$, *then let* $h(d_1, d_2, d_3)$ *be the unique element* $x \in A - \mathbf{b}$ *such that the line joining* b_0 *and* x *is in direction* d_1 *and the line joining* b_1 *and* x *is in direction* d_2. *All elements of* $A - \mathbf{b}$ *are obtained in this fashion. If* $d_1 \neq d_2$ *and* $d_1 \neq d_3$, *then let* $x \in A - \mathbf{b}$ *be as obtained from* d_1 *and* d_2 *in the previous sentence, and let* $h(d_1, d_2, d_3)$ *be the intersection point of the affine line* $\mathbf{b}$ *and the line passing through* x *in direction* d_3. *The image of* h *now also contains the elements* $b_1, \ldots, b_q$. *We now obtain also* b_0 *by defining* $h(d_1, d_2, d_3) = b_0$ *if some* d_i *is the direction of* $\mathbf{b}$, *or if* $d_1 = d_2$.

Clearly this gives an $\mathbf{ab}$-*definable map from* $(A^*)^3$ *onto* A. *Composing with* g *yields an* $\mathbf{ab}$-*definable map,* $f_i(\mathbf{a}, \mathbf{b}, -)$ *from* $(P - \text{acl}(\mathbf{a}))^3$ *onto* A. *Note this process was uniform in* $\mathbf{a}$ *and* $\mathbf{b}$.

So to sum up: for $i \in I_{\text{old}} \cup I_{\text{aff}}$ *there is a* $\varnothing$-*definable function* f_i *such that if* $i \in I_{\text{old}}$ *and* $(c_1, \ldots, c_{i-1}) \in L_{i-1}$, *then* $f_i(c_1, \ldots, c_{i-1}, -)$ *is a function from* $(L(c_1, \ldots, c_{\alpha(i)-1}) - \text{acl}(c_1, \ldots, c_{i-1}))$ *onto* $L(c_1, \ldots, c_{i-1})$. *And if* $i \in I_{\text{aff}}$, $(c_1, \ldots, c_{i-1}) \in L_{i-1}$ *and* $\mathbf{b}$ *is a tuple from* $L(c_1, \ldots, c_{i-1})$ *enumerating an affine line, then* $f_i(c_1, \ldots, c_{i-1}, \mathbf{d}, -)$ *is a function from* $(L(c_1, \ldots, c_{\alpha(i)-1}) - \text{acl}(c_1, \ldots, c_{i-1}))^3$ *onto* $L(c_1, \ldots, c_{i-1})$.

We will assume that the f_i *have been trivially 'extended' so that, in the case where* $i \in I_{\text{old}}$, *and* $(c_1, \ldots, c_{i-1}) \in L_{i-1}$, *then* $f_i(c_1, \ldots, c_{i-1}, -)$ *is defined on all of* $L(c_1, \ldots, c_{\alpha(i)-1})$ *and maps into* $L(c_1, \ldots, c_{i-1})$. *Similarly when* $i \in I_{\text{aff}}$.

3 Nice enumerations and a special case of the main theorem

In this section the combinatorial basis of the proof is given: the existence of 'nice' enumerations of projective spaces. The expression 'nice enumeration' comes from the paper of Ahlbrandt and Ziegler, where it has a specific meaning. For us it will refer to the general content of the next proposition, whose proof will comprise the first part of this section.

PROPOSITION **3.1.** *Let P be a (countable) strictly minimal ω-categorical structure. Then there is an ordering $\leq$ of P of order type ω with the following feature: for any n, if $(\mathbf{a}_i: i < \omega)$ is a sequence of n-tuples of P, then there are $i < j < \omega$ and some order-preserving elementary map $f: P \to P$ such that $f(\mathbf{a}_i) = \mathbf{a}_j$.*

If P has degenerate geometry then P is simply an infinite set with no structure, so any ordering of P of type ω works. If P has an affine geometry then P is a reduct of a vector space $(V, +, \lambda)_{\lambda \in F}$ (F a finite field). If the geometry of P is projective then P (as a structure) is a reduct of the projective space associated to the vector space $(V, +, \lambda)_{\lambda \in F}$. So it is enough to prove that a vector space over a finite field satisfies the conclusions of 3.1.

So we prove:

LEMMA **3.2.** *Let $(V, +, \lambda)_{\lambda \in F}$ be a (countably) infinite-dimensional vector space over a finite field. Then there is a total ordering of V, of order type ω, with the following property. For any n, and infinite sequence $\mathbf{x}_1, \mathbf{x}_2, \ldots$ of n-tuples from V, there are $i < j$ and an order-preserving linear injection $f: V \to V$ such that $f(\mathbf{x}_i) = \mathbf{x}_j$.*

The proof makes use of a combinatorial lemma due to Higman. Let us call a partial ordering $(\Sigma, \leq)$ a partial well ordering if for any subset S of Σ there is a finite subset S' of S such that for each $s \in S$ there is $s' \in S'$ such that $s' \leq s$.

REMARK **3.3.** *Let $(\Sigma, \leq)$ be a partial ordering. Then Σ is a partial well ordering iff for every sequence $(s_i: i < \omega)$ of members of Σ, there are $i < j < \omega$ such that $s_i \leq s_j$ iff $(\Sigma, \leq)$ is well founded with no infinite antichains.*

LEMMA **3.4.** (Higman) *Let Ω be a finite set of symbols, and let Σ be the set of words over Ω (i.e. finite sequences of elements of Ω). Define a partial ordering on Σ by: $w_1 \leq w_2$ if w_1 is a subword of w_2; that is, after deleting some members of w_2 we are left with w_1. Then $(\Sigma, \leq)$ is a partial well ordering.*

Proof. The lemma is proved by induction on $|\Omega|$. So suppose we know the lemma for all Ω' of cardinality strictly less than that of Ω.

SUBLEMMA. *Let w be a non-empty word in Σ, and let $(w_i\colon i < \omega)$ be a sequence of elements of Σ such that $\neg(w \leq w_i)$, for all $i < \omega$. Then there are $i < j$ such that $w_i \leq w_j$.*

Proof of sublemma. By induction on length of w. If length$(w) = 1$, then w is simply a symbol $\sigma \in \Omega$, whence $\sigma \notin w_i$ for all i, so each w_i is a word over the smaller alphabet $\Omega - \{\sigma\}$. We can then apply the global induction hypothesis to finish. Now suppose w is a word $\sigma w'$ where $\sigma \in \Omega$. We may assume again that each w_i contains the symbol σ. So we can write each w_i in the form $u_i v_i$ where $\sigma \notin u_i$ and $\neg(w' \leq v_i)$. (Simply let u_i be the greatest initial segment of w_i not containing σ.) By the global induction hypothesis, and Ramsey's theorem, we may assume (by thinning out the sequence) that $u_i \leq u_j$ whenever $i < j < \omega$. By the induction hypothesis (as length(w') < length(w)), there are $i < j$ such that $v_i \leq v_j$. Thus clearly $u_i v_i \leq u_j v_j$, that is $w_i \leq w_j$, as required. This proves the sublemma.

Now let $(w_i\colon i < \omega)$ be an arbitrary sequence of words. We want to show that $w_i \leq w_j$ for some $i < j$. If already $w_0 \leq w_j$ for some $j > 0$, we are fine. So we may assume that $\neg(w_0 \leq w_i)$ for all $i > 0$. But then we can apply the sublemma.

This completes the proof. ∎

Proof of Lemma 3.2. Fix an ordering $0 < 1 < \ldots$, of F. Let $\{v_i\colon i < \omega\}$ be a basis of V, and order V by the reverse lexicographic ordering: $\sum_i \alpha_i v_i < \sum_i \beta_i v_i$ iff for some j, $\alpha_j < \beta_j$ and $\alpha_i = \beta_i$ for all $i > j$. We will show this ordering works. First, to set notation, by the leading index, lead(v) of $v = \sum_i \alpha_i v_i$, we mean the greatest i such that $\alpha_i \neq 0$.

Now let $(\mathbf{x}_p\colon p < \omega)$ be a sequence of n-tuples from V. We may assume that the dimension of the subspace of V spanned by $\mathbf{x}_p$ is constant with p, with value $m \leq n$, say.

The first step is to replace each $\mathbf{x}_p$ by an m-tuple $\mathbf{u}_p = (u_{p,1}, \ldots, u_{p,m})$ which forms a basis of the subspace spanned by $\mathbf{x}_p$. We can easily choose $\mathbf{u}_p$ such that lead$(u_{p,1})$ < lead$(u_{p,2}) < \cdots <$ lead$(u_{p,m})$. We may also assume that $\mathbf{x}_p$ is uniformly linear in $\mathbf{u}_p$ as p varies (i.e. for some fixed matrix A over F, $\mathbf{x}_p = \mathbf{u}_p.A$ for all $p < \omega$).

So it suffices to find $p < q < \omega$ such that some order-preserving linear endomorphism of V takes $\mathbf{u}_p$ to $\mathbf{u}_q$.

Let us write $u_{p,i}$ as $\sum \{\alpha_{(p,j,i)} v_j\colon j \leq l_{p,i}\}$ where $l_{p,i}$ is lead$(u_{p,i})$.

In order to apply 3.4 we will code up the $\mathbf{u}_p$ by words in a suitable alphabet. Let F^+ denote the set of symbols $\{\alpha^+ : \alpha \in F\}$. The alphabet Ω will then consist of m-tuples from $F \cup F^+$. Σ will be the set of words on Ω. The word w_p corresponding to $\mathbf{u}_p$ will be of length $\text{lead}(u_{p,m}) + 1$. The $(j+1)$st symbol in w_p will be the m-tuple $(a_1, \ldots, a_m)$, say, where $a_i = \alpha^+_{(p,j,i)}$ if $j = l_{p,i}$, and $a_i = \alpha_{(p,j,i)}$ otherwise. Thus the $(j+1)$st symbol of w_p contains all the information about the coefficients of v_j in the various $u_{p,i}$, as well as telling us for which i, $j = \text{lead}(u_{p,i})$. By Lemma 3.4, there are $p < q < \omega$ such that w_p is a subword of w_q. So there is a non-decreasing embedding $\varepsilon\colon \omega \to \omega$ such that for $j < \text{length}(w_p)$, the $(j+1)$st symbol in w_p is the $(\varepsilon(j)+1)$st symbol in w_q. Rewrite $u_{p,i} = \sum \{\alpha_{j,i} v_j : j \le l(i)\}$, and $u_{q,i} = \sum \{\beta_{j,i} v_j : j \le m(i)\}$ (so $l(i) = \text{lead}(u_{p,i})$ and $m(i) = \text{lead}(u_{q,i})$).

Thus

(1) $$m(i) = \varepsilon(l(i)), \text{ for } i = 1, \ldots, m, \text{ and}$$

(2) $$\alpha_{j,i} = \beta_{\varepsilon(j),i}, \text{ for } i = 1, \ldots, m \text{ and } j \le l_i.$$

For each $i = 1, \ldots, m$, let us define

(3) $$y_i = \sum \{\beta_{j,i} v_j : j \le m(i), j \notin \text{range}(\varepsilon)\}.$$

From (3) and (2) we have

(4) $$u_{q,i} = \sum \{\alpha_{j,i} v_{\varepsilon(j)} : j \le l(i)\} + y_i, \text{ for each } i.$$

Now we can inductively find vectors $r_{l(1)}, \ldots, r_{l(m)}$ in V such that

$$(\alpha_{l(1),1}) r_{l(1)} = y_1$$

$$(\alpha_{l(1),2}) r_{l(1)} + (\alpha_{l(2),2}) r_{l(2)} = y_2$$

$$\vdots$$

$$(\alpha_{l(1),m}) r_{l(1)} + \cdots + (\alpha_{l(m),m}) r_{l(m)} = y_m.$$

Using (1) and (3) we then see that

(5) each $r_{l(i)}$ is in the linear span of $\{v_j : j < \varepsilon(l(i)), j \notin \text{range}(\varepsilon)\}$.

We now define the sought after order-preserving embedding of V into V as follows: $h(v_j) = v_{\varepsilon(j)} + r_j$, where r_j is defined to be 0 if $j \ne l(i)$ for all $i = 1, \ldots, m$. Extend h to V by linearity.

As $\varepsilon\colon \omega \to \omega$ is non-decreasing, our definition of the ordering $<$ on V together with (5) implies that h is order preserving.

On the other hand, for $i \leq m$.

$$\begin{aligned} h(u_{p,i}) &= h(\textstyle\sum \{\alpha_{j,i} v_j : j \leq l(i)\}) \\ &= \textstyle\sum \{\alpha_{j,i}(v_{\varepsilon(j)} + r_j) : j \leq l(i)\} \quad \text{(by the definition of } h) \\ &= \textstyle\sum \{\alpha_{j,i} v_{\varepsilon(j)} : j \leq l(i)\} + (\alpha_{l(1),i}) r_{l(1)} + \cdots + (\alpha_{l(i),i}) r_{l(i)} \\ &= \textstyle\sum \{\alpha_{j,i} v_{\varepsilon(j)} : j \leq l(i)\} + y_i \quad \text{(by above choice of the } r_{l(i)}) \\ &= u_{q,i} \quad \text{(by (4))}. \end{aligned}$$

Thus $h(\mathbf{u}_p) = \mathbf{u}_q$.

The proof of Lemma 3.2 is complete. ■

To conclude this section we will point out how Proposition 3.1 (actually a weak version of 3.1) is used to obtain quasifinite axiomatizability in a special case. This will, we hope, help to provide a better understanding of the rather complicated constructions carried out in the general case. Let us first give a little motivation for this and subsequent material. Suppose M is a ω-categorical strongly minimal structure and there is some number n, such that for any two tuples $\mathbf{a}$, $\mathbf{b}$ from M, tp($\mathbf{a}$) = tp($\mathbf{b}$) iff corresponding n-tuples from $\mathbf{a}$ and $\mathbf{b}$ have the same type. Add n-ary predicates $R_p(x_1, \ldots, x_n)$ for all complete n-types. (In these circumstances we say that M is 'homogeneous in a finite relational language'.) It is clear that M is bidefinable with a structure M^* which has only the relations R_p. With a little reflection one can see that $Th(M^*)$ is quasifinitely axiomatizable. One only needs to write down sufficient axioms to allow a back-and-forth argument. (If this is not clear now, it will become clear later.) A general, strongly minimal ω-categorical structure will *not* have the above feature (e.g. if M is a vector space over a finite field). The point of the combinatorics above, however, is that in the general case we *will* be able to find a certain *cofinal* collection W of (finite) subsets of M, and some number λ, such that for any set S in W, the type of S is determined by the collection of λ-tuples of subsets of S.

We go back to the set-up defined in section 2.

Assumptions. In the special case which we will consider now, M^* has two levels, L_1 and L_2 (so $I = \{1, 2\}$); $1 \in I_{\text{new}}$, and $2 \in I_{\text{fin}}$. That is, L_1 is a modular strictly minimal set, and for each $a \in L_1$, $L(a)$ is a finite set of fixed cardinality κ, say.

For S a subset of L_1, $L(S)$ will denote $\cup \{L(a): a \in S\}$. Think of $L(a)$ as a finite fibre above a. π will denote the projection map from L_2 onto L_1. Let us fix an ordering $<$ of L_1 given in Proposition 3.1.

Definition 3.5. *Let* $W = \{(S, a)$: *S is a finite subset of* L_1, $a \in L_1 - S$, *and there is an automorphism* σ *of* L_1 *such that* $\sigma(S \cup \{a\})$ *is an initial segment (under* $<$*) of* L_1 whose greatest element is $\sigma(a)\}$.

Lemma 3.6. *There is a some finite number* λ_0 *such that*

(i) *For all* $(S, a) \in W$, *and* $b \in L_1$, *if* $b \in \mathrm{acl}(S \cup \{a\})$ *then there is* $S_1 \subseteq S \cup \{a\}$ *such that* $|S_1| < \lambda_0$, $\mathrm{tp}(b/S_1) \vdash \mathrm{tp}(b/S \cup \{a\})$, *and* $\mathrm{tp}(b/L(S_1)) \vdash \mathrm{tp}(b/L(S \cup \{a\}))$.

(ii) *For all* $(S, a) \in W$, *there is* $S_1 \subseteq S$ *with* $|S_1| < \lambda_0$ *such that* $\mathrm{tp}(\mathbf{a}/L(S_1) \cup \{a\}) \vdash \mathrm{tp}(\mathbf{a}/L(S) \cup \{a\})$ *(where* **a** *is some enumeration of* $L(a)$), *and, moreover, if* $a \in \mathrm{acl}(S)$, *also* $\mathrm{tp}(a/L(S_1)) \vdash \mathrm{tp}(a/L(S))$.

Proof. (i) Suppose (by way of contradiction) that for each finite λ there is $(S_\lambda, a_\lambda) \in W$ and $b_\lambda \in L_1$ such that $b_\lambda \in \mathrm{acl}((S_\lambda \cup \{a_\lambda\})$, but

($*$) for every subset S_1 of $S_\lambda \cup \{a_\lambda\}$ of size λ, $\mathrm{tp}(b_\lambda/S_1)$ does not imply $\mathrm{tp}(b_\lambda/S_\lambda \cup \{a_\lambda\})$ or $\mathrm{tp}(b_\lambda/L(S_1))$ does not imply $\mathrm{tp}(b_\lambda/L(S_\lambda \cup \{a_\lambda\}))$.

By definition of W we may assume (after applying automorphisms) that for each λ, S_λ is the set of elements of L_1 strictly less than a_λ (in the chosen ordering $<$ of L_1). By Proposition 3.1 (for $n = 2$), and Ramsey's theorem, we may find an infinite subset J of ω such that for $\lambda < \lambda'$ in J, there is an order-preserving elementary map $f_{\lambda,\lambda'}$ of L_1 into L_1 such that $f_{\lambda,\lambda'}(a_\lambda, b_\lambda) = (a_{\lambda'}, b_{\lambda'})$. Thus $f_{\lambda,\lambda'}(S_\lambda) \subseteq S_{\lambda'}$. Note also that $f_{\lambda,\lambda'}$ extends to an elementary map (in M^*) taking $L(a_\lambda)$ onto $L(a_{\lambda'})$ and $L(S_\lambda)$ into $L(S_{\lambda'})$. (For L_1 and M^* both have the structure induced from M, so $f_{\lambda,\lambda'}$ is elementary in the sense of M^*, and thus extends to a partial elementary map in M^* defined on $\mathrm{acl}(L_1)$.)

It then follows that there are numbers $\lambda(0)$, m_1, and m_2, such that for all $\lambda \in J$ greater than $\lambda(0)$, $\mathrm{mult}(\mathrm{tp}(b_\lambda/S_\lambda \cup \{a_\lambda\})) = m_1$, and $\mathrm{mult}(\mathrm{tp}(b_\lambda/L(S_\lambda \cup \{a_\lambda\})) = m_2$.

Now choose λ in J with $\lambda > |S_{\lambda(0)}| + 1$. Then $\mathrm{tp}(b_\lambda/f_{\lambda(0),\lambda}(S_{\lambda(0)} \cup \{a_{\lambda(0)}\})) \vdash \mathrm{tp}(b_\lambda/S_\lambda \cup \{a_\lambda\})$ and similarly $\mathrm{tp}(b_\lambda/L(f_{\lambda(0),\lambda}(S_{\lambda(0)} \cup \{a_{\lambda(0)}\}))) \vdash \mathrm{tp}(b_\lambda/L(S_\lambda \cup \{a_\lambda\}))$. This contradicts ($*$), so proving (i).

(ii) is proved in a similar way (using 3.1 for $n = 1$). ■

Lemma 3.7. *For any* $\lambda < \omega$ *there is* $\mu(\lambda) < \omega$ *such that whenever* $(S, a) \in W$ *and* $T \subseteq S$ *with* $|T| < \lambda$ *then there is a subset* S_1 *of* S *containing* T *such that* $|S_1| < \mu$ *and* $(S_1, a) \in W$.

Proof. Fix λ. If there is no μ as required then we find, for each μ, some $(S_\mu, a_\mu) \in W$ and subset T_μ of S_μ of cardinality $< \lambda$ such that there is *no* subset

S_1 of S_μ which contains T_μ, has cardinality $< \mu$, and such that $(S_1, a_\mu) \in W$. We may assume that all T_μ have the same cardinality. Enumerate T_μ as a tuple $\mathbf{b}_\mu$, say. By definition of W, we may assume (after applying an automorphism of L_1) that each set $(S_\mu \cup \{a_\mu\})$ is an initial segment of L_1 under the chosen ordering, with maximal element a_μ. By Proposition 3.1 and Ramsey's theorem we may find an infinite subset J of ω such that for $\mu(1) < \mu(2) \in J$ there is an order-preserving elementary embedding $h_{\mu(1),\mu(2)}$: $L_1 \to L_2$ which takes $a_{\mu(1)}\mathbf{b}_{\mu(1)}$ to $a_{\mu(2)}\mathbf{b}_{\mu(2)}$. This mapping then takes $S_{\mu(1)}$ onto a subset of $S_{\mu(2)}$ which contains $T_{\mu(2)}$, and it is clear that $(h_{\mu(1),\mu(2)}(S_{\mu(1)}), a_{\mu(2)}) \in W$ also (as $h_{\mu(1),\mu(2)}|(S_{\mu(1)} \cup \{a_{\mu(1)}\})$ is elementary and so extends to an automorphism of L_1). But then choosing $\mu < \mu'$ in J with $\mu' > |S_\mu|$ gives a contradiction. This proves the lemma. ■

LEMMA **3.8.** *Let* $\lambda = \lambda_0 + 1$, *where* λ_0 *is as given by Lemma* 3.6. *Let* $(S, a) \in W$, *and let* $\mathbf{c}$ *enumerate* $S \cup \{a\}$. *Then*

(i) $\cup\{\mathrm{tp}(\mathbf{b})$: $\mathbf{b}$ is a subtuple of $\mathbf{c}$ of length at most $\lambda\} \vdash \mathrm{tp}(\mathbf{c})$; and

(ii) $\cup\{\mathrm{tp}(L(\mathbf{b}))$: $\mathbf{b}$ a subtuple of $\mathbf{c}$ of length at most $\lambda\} \vdash \mathrm{tp}(L(\mathbf{c}))$.

Proof. (As far as (ii) is concerned we equip $L(\mathbf{c})$ also with a fixed enumeration.)

Everything follows from Lemma 3.6. For (i) the proof proceeds by induction on $|S \cup \{a\}|$. Given $(S, a) \in W$, then $S = S' \cup \{a'\}$ for some S', a' such that $(S', a') \in W$, and the induction hypothesis applies to (S', a'). Now by 3.6(i), $a \in \mathrm{acl}(S)$ iff $a \in \mathrm{acl}(S_1)$ for some subset S_1 of S of cardinality at most λ_0, and, moreover, if $a \in \mathrm{acl}(S)$ then $\mathrm{tp}(a/S_1) \vdash \mathrm{tp}(a/S)$ for some such S_1. Moreover, this is true for *any* $b \in L_1$ in place of a. So (as L_1 is strongly minimal) $\cup\{\mathrm{tp}(a/S_1)$: $S_1 \subset S$ and $|S_1| \leq \lambda_0\} \vdash \mathrm{tp}(a/S)$. Together with the induction hypothesis, this yields (i).

(ii) is also proved in a similar fashion using Lemma 3.6(ii). ■

We now describe a finite language for M^* and a finite set of axioms in this language which together with the schema of infinity will axiomatize $Th(M^*)$. The language L^* will contain predicates for L_1 and L_2 and a $\kappa + 1$-ary relation symbol R_π representing the projection map π (so $R_\pi(\mathbf{y}, x)$ is interpreted as $\mathbf{y} = \pi^{-1}(x)$). Let λ be as chosen in Lemma 3.8. Let μ be $\mu(2\lambda)$ as given by 3.7. Add a relation symbol $R_p(x_1, \ldots, x_\lambda)$ for each complete λ-type p over $\varnothing$ realized in L_1, and a symbol $Q_q(\mathbf{y}_1, \ldots, \mathbf{y}_\lambda)$ for each complete $\kappa\lambda$-type q over $\varnothing$ realized in L_2 by some tuple $\mathbf{b}_1, \ldots, \mathbf{b}_\lambda$ where each $\mathbf{b}_i$ is an enumeration of some fibre. $(\exists^m x)(\ldots x \ldots)$ will mean 'there are exactly m x such that $\ldots x \ldots$'.

The axioms will be:

1. R_π is what it should be (i.e. there is a map π from L_2 onto L_1 with κ-sized fibres, and $R_\pi(\mathbf{y}, x)$ if $\mathbf{y}$ is the fibre above x).

2. $(\exists x_1, \ldots, x_\lambda)(R_p(x_1, \ldots, x_\lambda)) \wedge (\forall x_1, \ldots, x_\lambda)(R_p(x_1, \ldots, x_\lambda) \rightarrow \wedge L_1(x_i))$ for all p, $(\exists \mathbf{y}_1, \ldots, \mathbf{y}_\lambda)(Q_q(\mathbf{y}_1, \ldots, \mathbf{y}_\lambda)) \wedge (\forall \mathbf{y}_1, \ldots, \mathbf{y}_\lambda)(Q_q(\mathbf{y}_1, \ldots, \mathbf{y}_\lambda) \rightarrow \bigwedge_i (\exists x_i)(R_\pi(\mathbf{y}_1, \mathbf{x}_i))$, for all q. Also $(\forall x_1, \ldots, x_\lambda, \mathbf{y}_1, \ldots, \mathbf{y}_\lambda)(\bigwedge_i R_\pi(\mathbf{y}_i, x_i) \wedge Q_q(\mathbf{y}_1, \ldots, \mathbf{y}_\lambda) \rightarrow R_p(x_1, \ldots, x_\lambda))$, for all p and q for which this is true.

3. $(\forall x_1, \ldots, x_\mu, \mathbf{y}_1, \ldots, \mathbf{y}_\mu)(\varphi(x_1, \ldots, x_\mu, \mathbf{y}_1, \ldots, \mathbf{y}_\mu))$ whenever φ is a quantifier-free formula in L^*, and this sentence is true.

4. $(\forall x_1, \ldots, x_{\lambda-1}, \mathbf{y}_1, \ldots, \mathbf{y}_{\lambda-1})(\exists x_\lambda, \mathbf{y}_\lambda)(\varphi(x_1, \ldots, x_\lambda, \mathbf{y}_1, \ldots, \mathbf{y}_\lambda))$ whenever φ is a quantifier-free L^*-formula and this sentence is true in M^*. Also $(\forall x_1 \ldots x_{\lambda-1})(\varphi(x_1, \ldots, x_{\lambda-1}) \rightarrow (\exists^m x)(\psi(x_1, \ldots, x_{\lambda-1}, x))$, whenever φ, ψ are quantifier-free L^* formulae, and this sentence is true in M^*.

5. 'There are infinitely many elements in L_1'.

We first show that L^* is a finite language for M^*.

Lemma 3.9. *If h is a permutation of M^* which preserves the relations in L^* then h is an automorphism of M^*.*

Proof. By Lemma 3.8, and choice of λ, for any $(S, a) \in W$, $h \restriction (S \cup \{a\} \cup L(S \cup \{a\}))$ is a partial elementary map. So clearly h is an automorphism of M^*. ■

Thus M^* and $M^* \restriction L^*$ are bidefinable. So we may assume now that L^* is the language of M^*.

Lemma 3.10. *Let N^* be a model of the axioms 1–5 above. Then N^* is isomorphic to M^*.*

Proof. Let $\{a_1, a_2, \ldots\}$ be the chosen ordering of L_1. So for each n, $(\{a_i : i < n\}, a_n) \in W$. We will define a map $h: M^* \rightarrow N^*$ by defining it inductively on $\{a_1, \ldots, a_n\} \cup L(\{a_1, \ldots, a_n\})$ so as to preserve quantifier-free L^*-formulae, and in such a way that h ends up being surjective. Axiom set 2 allows us to get started.

For the inductive step, we suppose that S is a proper initial segment of L_1, and h is an L^*-partial isomorphism from $S \cup L(S)$ into N^*. a is the next element of L_1 and we want to extend h to h_1 on $S \cup \{a\} \cup L(S \cup \{a\})$.

Case (I). $a \in \mathrm{acl}(S)$. Let $\mathbf{a}$ enumerate $L(a)$. Let $S_1 \subseteq S$ be as given by Lemma 3.6(ii). Let $S_1 = \{b_1, \ldots, b_k\}$, say (with k $< \lambda$), and $S = \{b_1, \ldots, b_r\}$, say. For

$i = 1, \ldots, r$ *let* $\mathbf{b}_i$ be some enumeration of $\mathrm{L}(b_i)$. Let $Q(\mathbf{y}_1, \ldots, \mathbf{y}_k)$ be the predicate of L^* corresponding to $\mathrm{tp}(\mathbf{b}_1, \ldots, \mathbf{b}_k)$, and $Q'(\mathbf{y}_1, \ldots, \mathbf{y}_k, \mathbf{y})$ the predicate corresponding to $\mathrm{tp}(\mathbf{b}_1, \ldots, \mathbf{b}_k, \mathbf{a})$.

Then, by saturation of M*, the sentence $(\forall \mathbf{y}_1, \ldots, \mathbf{y}_k \exists \mathbf{y}(Q(\mathbf{y}_1, \ldots, \mathbf{y}_k) \rightarrow Q'(\mathbf{y}_1, \ldots, \mathbf{y}_k, \mathbf{y})))$ is true in M^* and is thus an axiom of type 4, so is also true in N^*. But $N^* \models Q(h(\mathbf{b}_1), \ldots, h(\mathbf{b}_k))$. Thus we can find $\mathbf{c}$ in N^* such that $N \models Q'(h(\mathbf{b}_1), \ldots, h(\mathbf{b}_k), \mathbf{c})$. Extend h to a map h_1 defined on $S \cup \{a\} \cup L(S \cup \{a\})$ by defining $h_1(\mathbf{a}) = \mathbf{c}$ and $h_1(a) = \pi(\mathbf{c})$.

CLAIM A. $h_1 \restriction (S_1 \cup \{a\} \cup L(S_1 \cup \{a\}))$ *is an* L^**-partial isomorphism.* This is because axioms 3 are true in N^*.

We need to check that h_1 is an L^*-partial isomorphism.

CLAIM B. *For* $i(1), \ldots, i(\lambda - 1) \leq r$, *and* Q *a basic predicate of* L^*, $M^* \models Q(\mathbf{b}_{i(1)}, \ldots, \mathbf{b}_{i(\lambda-1)}, \mathbf{a})$ *iff* $N^* \models Q(h_1(\mathbf{b}_{i(1)}), \ldots, h_1(\mathbf{b}_{i(\lambda-1)}), h_1(\mathbf{a}))$.

Proof. By Lemma 3.7 and the choice of μ, we can find a subset S_2 of S such that $|S_2| < \mu$, $\{b_1, \ldots, b_k, b_{i(1)}, \ldots, b_{i(\lambda-1)}\} \subseteq S_2$, and $(S_2, a) \in W$. Write S_2 as $\{b_1, \ldots, b_k, b_{i(1)}, \ldots, b_{i(\lambda-1)}, b_{j(1)}, \ldots, b_{j(t)}\}$, say.

By Lemma 3.8, $\mathrm{tp}(\mathbf{b}_1, \ldots, \mathbf{b}_k, \mathbf{b}_{i(1)}, \ldots, \mathbf{b}_{i(\lambda-1)}, \mathbf{b}_{j(1)}, \ldots, \mathbf{b}_{j(t)})$ in M^* is determined by its quantifier-free L^*-type $\varphi(\mathbf{y}_1, \ldots, \mathbf{y}_k, \mathbf{y}_{i(1)}, \ldots, \mathbf{y}_{i(\lambda-1)}, \mathbf{y}_{j(1)}, \ldots, \mathbf{y}_{j(t)})$.

By choice of S_1, $\mathrm{tp}(\mathbf{a}/\mathbf{b}_1 \ldots \mathbf{b}_k) \models \mathrm{tp}(\mathbf{a}/\mathbf{b}_1 \ldots \mathbf{b}_k \mathbf{b}_{i(1)} \ldots \mathbf{b}_{i(\lambda-1)} \mathbf{b}_{j(1)} \ldots \mathbf{b}_{j(t)})$. Let $\psi(\mathbf{y}_1, \ldots, \mathbf{y}_k, \mathbf{y})$ be the L^*-quantifier-free type of $(\mathbf{b}_1, \ldots, \mathbf{b}_k, \mathbf{a})$ in M^* and $\sigma(\mathbf{y}_{i(1)}, \ldots, \mathbf{y}_{i(\lambda-1)}, \mathbf{y})$ the quantifier-free L^*-type of $(\mathbf{b}_{i(1)}, \ldots, \mathbf{b}_{i(\lambda-1)}, \mathbf{a})$ in M^*. Then the sentence $(\forall \mathbf{y}_1, \ldots, \mathbf{k}_k, \mathbf{y}_{i(1)}, \ldots, \mathbf{y}_{i(\lambda-1)}, \mathbf{y}_{j(1)}, \ldots, \mathbf{y}_{j(t)}, \mathbf{y})(\varphi \wedge \psi \rightarrow \sigma)$ is true in M^* so is an axiom of type 3. Thus this sentence is true also in N^*. By Claim (A) ψ is true of $(h_1(\mathbf{b}_1), \ldots, h_1(\mathbf{b}_k), h(\mathbf{a}))$, and the fact that h is an L^*-partial isomorphism shows that φ is true of $(h_1(\mathbf{b}_1), \ldots, h_1(\mathbf{b}_k), h_1(\mathbf{b}_{i(1)}), \ldots, h_1(\mathbf{b}_{i(\lambda-1)}), h_1(\mathbf{b}_{j(1)}), \ldots, h_1(\mathbf{b}_{j(t)}))$. Thus σ is true of $(h_1(\mathbf{b}_{i(1)}), \ldots, h_1(\mathbf{b}_{i(\lambda-1)}), h_1(\mathbf{a}))$. Claim (B) is proved. ■

CLAIM C. h_1 *is an* L^**-partial isomorphism.*

Proof. Clearly h preserves π. By Claim (B) and axiom set 3, $h \restriction (\mathrm{S} \cup \{a\})$ preserves quantifier-free L^*-formulae. The same is true for $h \restriction L(S \cup \{a\})$, directly by Claim (B). This is enough. ■

So Case (I) is complete.

Case (II). $a \notin \mathrm{acl}(S)$.

For each tuple $\mathbf{c} = (a_{i(1)}, \ldots, a_{i(\lambda-1)})$ in S, let $R_{\mathbf{c}}(x_{i(1)}, \ldots, x_{i(\lambda-1)}, x)$ be the L^*-predicate corresponding to the complete type of $(a_{i(1)}, \ldots, a_{i(\lambda-1)}, a)$ in M^*, and let $\mathbf{R}_{\mathbf{c}}$ be the set of predicates $R(x_{i(1)}, \ldots, x_{i(\lambda-1)}, x)$ corresponding to tuples $(a_{i(1)}, \ldots, a_{i(\lambda-1)}, a')$ where $a' \in L_1$ and $a' \in \mathrm{acl}(a_{i(1)}, \ldots, a_{i(\lambda-1)})$. By axioms 1 and 4 and the infiniteness of $L_1^{N^*}$ there is $c \in L_1^{N^*}$ such that $N^* \models \neg R(h(a_{i(1)}), \ldots, h(a_{i(\lambda-1)}), c)$ for each tuple $\mathbf{c} = (a_{i(1)}, \ldots, a_{i(\lambda-1)})$ from S, and each $R \in \mathbf{R_c}$. Choose c to be the least possible such element in N^* under some fixed ordering of N^* of order type ω.

Axiom 3 then implies that $N^* \models R_{\mathbf{c}}(h(a_{i(1)}, \ldots, h(a_{i(\lambda-1)}), c)$ for all $\mathbf{c} = (a_{i(1)}, \ldots, a_{i(\lambda-1)})$.

Extend h to the element a by defining $h_1(a) = c$.

Claim D. *$h_1 \restriction (S \cup \{a\})$ is an L^*-partial isomorphism.*

Proof. Clear. ■

Let now $\mathbf{a}$ enumerate $L(a)$. By Lemma 3.6(ii) choose $S_1 \subseteq S$ such that $|S_1| < \lambda$ and $\mathrm{tp}(\mathbf{a}/L(S_1) \cup \{a\}) \models \mathrm{tp}(\mathbf{a}/L(S) \cup \{a\})$.

Let $S_1 = \{b_1, \ldots, b_k\}$, say, and $\mathbf{b}_i$ be the fibre above b_i. Let $\varphi(\mathbf{y}_1, \ldots, \mathbf{y}_k, \mathbf{y})$ be the quantifier-free L^*-type of $(\mathbf{b}_1, \ldots, \mathbf{b}_k, \mathbf{a})$, let $\psi(x_1, \ldots, x_k, x)$ be the quantifier-free type of $(b_1, \ldots, b_k, a)$, and let $\sigma(x_1, \ldots, x_k, \mathbf{y}_1, \ldots, \mathbf{y}_k)$ be the quantifier-free L^*-type of $(b_1, \ldots, b_k, \mathbf{b}_1, \ldots, \mathbf{b}_k)$.

Claim (E). *The sentence $(\forall x_1, \ldots, x_k, x, \mathbf{y}_1, \ldots, \mathbf{y}_k, \mathbf{y})((\bigwedge R_\pi(\mathbf{y}_i, x_i) \wedge \psi(x_1, \ldots, x_k, x) \wedge \sigma(x_1, \ldots, x_k, \mathbf{y}_1, \ldots, \mathbf{y}_k) \wedge R_\pi(\mathbf{y}, x)) \rightarrow \bigvee \{\varphi(\mathbf{y}_1, \ldots, \mathbf{y}_k, \tau(\mathbf{y})): \tau$ a permutation of $\mathbf{y}\})$ is an axiom of type 3.*

Proof. $a \notin \mathrm{acl}(S_1)$, so $\mathrm{tp}(a/S_1)$ is strongly minimal and thus has a unique extension over $S_1 \cup L(S_1)$. As the formulae ψ and σ determine complete types, it is clear that the sentence exhibited in the claim is true in M^*, and is thus an axiom. ■

By Claim (D) and the fact that h is an L^*-partial isomorphism we see that $N^* \models \bigwedge R_\pi(h_1(\mathbf{b}_i), h_1(b_i)) \wedge \psi(h_1(b_1), \ldots, h_1(b_k), h_1(a)) \wedge \sigma(h_1(b_1), \ldots, h_1(b_k), h_1(\mathbf{b}_1), \ldots, h_1(\mathbf{b}_k))$.

So by Claim (E) and axiom 1 we can find $\mathbf{d}$ in N^* such that $N^* \models R_\pi(\mathbf{d}, h_1(a)) \wedge \varphi(h_1(\mathbf{b}_1), \ldots, h_1(\mathbf{b}_k), \mathbf{d})$. Define $h_1(\mathbf{a}) = \mathbf{d}$. We see immediately that $h_1 \restriction (S_1 \cup \{a\} \cup L(S_1 \cup \{a\}))$ is an L^*-partial isomorphism.

Claim F. *h_1 is an L^*-partial isomorphism from $S \cup \{a\} \cup L(S \cup \{a\})$ into N^*.*

Proof. This is proved just like Claim (B) above using the fact that $\mathrm{tp}(\mathbf{a}/L(S_1) \cup \{a\}) \models \mathrm{tp}(\mathbf{a}/L(S) \cup \{a\})$. ■

We have completed the inductive step. Thus we obtain an L^*-embedding h of M^* into N^*. We claim this must be surjective. It is enough for $h \restriction L_1$ to be onto $L_1^{N^*}$. Suppose not. Then choose the least element d, say, in the aforementioned fixed ordering of $L_1^{N^*}$ which is not in range(h). We have two cases:

Case (*i*). There is a quantifier-free L^*-formula $\varphi(x_1, \ldots, x_{(\lambda-1)}, x)$ and $b_1, \ldots, b_{(\lambda-1)}$ in M^* such that $M^* \models (\exists_x^{=m})\varphi(b_1, \ldots, b_{(\lambda-1)}, x)$ for some $m < \omega$, and $N^* \models \varphi(h(b_1), \ldots, h(b_{(\lambda-1)}), d)$.

In this case let $\psi(x_1, \ldots, x_{(\lambda-1)})$ be the quantifier-free L^*-type of $(b_1, \ldots, b_{(\lambda-1)})$.

So $(\forall x_1, \ldots, x_{(\lambda-1)})(\psi(x_1, \ldots, x_{(\lambda-1)}) \rightarrow (\exists_x^{=m})(\varphi(x_1, \ldots, x_{(\lambda-1)}, x))$ is an axiom of sort 4, so is true in N^*. Thus $N^* \models (\exists_x^{=m})(\varphi(h(b_1), \ldots, h(b_{(\lambda-1)}), x))$. So clearly all the solutions in N^* of $\varphi(h(b_1), \ldots, h(b_{(\lambda-1)}), x)$ must be in the range of h; in particular d is, a contradiction.

Case (*ii*). Not case (i).

Now at some stage in the construction of h, we will have already defined h on $\{a_1, \ldots, a_n\}$, and $a_{n+1} \notin \mathrm{acl}(a_1, \ldots, a_n)$, and d will be the least element of N^* not in the range of $h \restriction \{a_1, \ldots, a_n\}$. By looking at the construction of h in this case (see Case (II) above), and using out current case assumption, we see that $h(a_n)$ will have to be d, again a contradiction.

This proves Lemma 3.10 and thus the quasifinite axiomatizability of $Th(M^*)$ in the special case we are considering. ■

Let us remark that all we used about L_1 was that it satisified Proposition 3.1, and that our proof recovers the quasifinite axiomatizability of infinite-dimensional projective space.

4 Pregood and good sets

The proof of quasifinite axiomatizability in the general (ω-categorical, ω-stable) case will have the same conceptual content as the proof of the special case in section 3. That is, we will find a certain cofinal collection of subsets of the model with nice uniform finiteness properties given by the combinatorial results of section 3, and with respect to which we can inductively define an isomorphism of the model with any other model of a suitable set of axioms. All that is required to transplant the machinery from the special case to the general case is some degree of organization which takes account of the tree structure of the model given in section 2.

We revert to the notation of section 2. In particular our structure is M^*, equipped with levels $L_1, \ldots, L_n$. Unless otherwise stated $\leq$ denotes the partial ordering on M^* corresponding to its tree structure. For the moment let $a_0, \ldots, a_n$ be as in section 2. For $i \in I_{\text{new}}$, fix some ordering $<(i)$ of $L(a_1, \ldots, a_{i-1})$ satisfying Proposition 3.1. Suppose $(b_1, \ldots, b_{i-1})$ realizes r_{i-1}: by a *standard ordering* of $L(b_1, \ldots, b_{i-1})$ we mean an ordering which is the image of $<(i)$ under some automorphism of M^* which takes $(a_1, \ldots, a_{i-1})$to $(b_1, \ldots, b_{i-1})$. Given an ordering $<$ of a structure P, of order-type ω, we obtain an ordering of P^3, also of order type ω as follows: $(a, b, c) < (a', b', c')$ iff either $\max(a, b, c) < \max(a', b', c')$, or $\max(a, b, c) = \max(a', b', c')$ and (a, b, c) is strictly less than (a', b', c') in the lexicographic ordering. Then any order-preserving embedding of P into P induces an order-preserving embedding from P^3 to P^3.

We slightly modify our earlier notation. $L_1, \ldots, L_n$ are as before but a_i will now typically denote an *element of* L_i (equivalently a realization of r_i). So for $i < n$, $L(a_i)$ denotes $\{x \in L_{i+1}\colon a_i \leq x\}$. Remember that a_0 is the unique element at the root of the tree.

We now define bases, pregood sets, and good sets.

Definition 4.1. (i) *A* base *is an ordered set of the form*

$$(a_0, S_1, a_1, \ldots, S_{k-1}, a_{k-1}, S_k)$$

where $k \leq n$, $S_i \cup \{a_i\}$ is a non-empty finite subset of $L(a_{i-1})$ (for $1 \leq i < k$), $a_i \notin S_i$, and S_k is a (possibly empty) finite proper *subset of $L(a_{k-1})$.*

$(a_0, \ldots, a_{k-1})$ is called the bounded *part of this base.*

By the underlying set *of such a base we mean the subset $\{a_0, \ldots, a_{k-1}\} \cup S_1 \cup \cdots \cup S_k$ of M^*. We may sometimes notationally confuse a base with its underlying set.*

(ii) *Let $(a_0, S_1, a_1, \ldots, S_k)$ be a base. The* set attached to *this base is the set $\{x \in M^*\colon x = a_i$ for some $i = 0, \ldots, k-1$,* or *for some $i = 1, \ldots, k$, and $y \in S_i$, $x \geq y\}$. By a* pregood set *we simply mean the set attached to some base $(a_0, S_1, a_1, \ldots, S_k)$.*

(iii) *We let $B_0 = B_0(M^*)$ denote the collection of bases, and $\mathbf{W}_0 = \mathbf{W}_0(M^*)$ the collection of pregood sets.*

Remark 4.2. (i) *It should be clear that if U is a pregood set, then there is a* unique *base B such that U is attached to B. (It is essential here that S_k be a* proper *subset of $L(a_{k-1})$.) We call B the base of U.*

(ii) *Also if N^* is another structure with a partial ordering $\leq'$ and levels L'_i for $i = 1, \ldots, n$, then $B_0(N^*)$ and $\mathbf{W}_0(N^*)$ make sense, and* (i) *also holds.*

(iii) *Note that $\{a_0\}$ is a pregood set with base $(a_0, \varnothing)$.*

(iv) *It is convenient to consider (the universe of) M^* as a pregood set with base $(\{a_0\})$. This notation is consistent with that above.*

DEFINITION 4.3. (I) Π *is the set of all objects of the form*

$$(a_0, S_1, a_1, \ldots, S_k, <_1, \ldots, <_k, \mathbf{d}_1, \ldots, \mathbf{d}_k, T_1, \ldots, T_k),$$

where $<_i$ is defined iff $i \in I_{\text{new}}$, $\mathbf{d}_i$ is defined iff $i \in I_{\text{aff}}$, T_i is defined iff $i \in I_{\text{old}} \cup I_{\text{aff}}$, and

(i) *$(a_0, S_1, a_1, \ldots, S_k)$ is a base.*

(ii) *Suppose $i \in I_{\text{new}}$. Then $<_i$ is a standard ordering of $L(a_{i-1})$. If, moreover, $i < k$, then $S_i \cup \{a_i\}$ is an $<_i$-initial segment of $L(a_{i-1})$ whose $<_i$-greatest element is a_i, and if $i = k$ then S_i is simply an $<_i$-initial segment of $L(a_{i-1})$.*

(iii) *Suppose $i \in I_{\text{old}}$. Then T_i is defined and is an $<_{\alpha(i)}$-initial segment of $L(a_{\alpha(i)-1})$. If $i < k$ then $S_i \cup \{a_i\}$ is the image of T_i under $f_i(a_{i-1}, -)$, and S_i is the image of some $<_{\alpha(i)}$-initial segment of T_i under $f_i(a_{i-1}, -)$. If $i = k$, we simply require that S_i is the image of T_i under $f_i(a_{i-1}, -)$.*

(iv) *If $i \in I_{\text{aff}}$, then T_i is defined and is an $<_{\alpha(i)}$-initial segment of $L(a_{\alpha(i)-1})^3$, and $\mathbf{d}_i$ is a tuple from $L(a_{i-1})$ enumerating an affine line. The same conditions as in* (iii) *hold except that the relevant function is now $f_i(a_{i-1}, \mathbf{d}_i, -)$.*

(II) *The set U* attached to *$p = (a_0, S_1, a_1, \ldots, S_k, <_1, \ldots, <_k, \mathbf{d}_1, \ldots, \mathbf{d}_k, T_1, \ldots, T_k)$ is simply the set attached to the base $(a_0, S_1, a_1, \ldots, S_k)$. We write $U = U_p$. By a* good set *we mean a set U_p for some p in Π. The collection of good sets is denoted* $\mathbf{W}$.

(III) *We define a partial ordering $\leq_\Pi$ on Π as follows. Let $p, q \in \Pi$, where $p = (a_0^p, S_1^p, \ldots, S_k^p, <_1^p, \ldots, <_k^p, \mathbf{d}_1^p, \mathbf{d}_k^p, T_1^p, \ldots, T_k^p)$ and similarly for q. Then $p \leq_\Pi q$ if $U_p \subseteq U_q$, and whenever $a_i^p = a_i^q$ then $<_{i+1}^p = <_{i+1}^q$* (if $i + 1 \in I_{\text{new}}$) *and $\mathbf{d}_{i+1}^p = \mathbf{d}_{i+1}^q$ (if $i + 1 \in I_{\text{aff}}$).*

REMARK. *Strictly speaking $\leq_\Pi$ is a prepartial ordering: we may have $p \leq_\Pi q$ and $q \leq_\Pi p$, but $p \neq q$ owing to some unimportant difference between the T_i^p and T_i^q.*

LEMMA 4.4. *Any increasing sequence (of length ω) $p(1) \leq_\Pi p(2) \leq_\Pi \cdots$ in Π has a least upper bound $p \in \Pi$, and U_p is the union of the $U_{p(i)}$.*

Proof. Write $p(i)$ as

$$(a_0^i, \ldots, S_{k(i)}^i, <_1^i, \ldots, <_{k(i)}^i, \mathbf{d}_1^i, \ldots, \mathbf{d}_{k(i)}^i, T_1^i, \ldots, T_{k(i)}^i).$$

Choose j largest such that a_j^i is eventually constant (with value a_j, say). It

easily follows that for each $r \leq j$, each of a_r^i, S_r^i (and thus also $<_{r+1}^i$, $\mathbf{d}_{r+1}^i$) is eventually constant, with values a_r, S_r ($<_{r+1}$ and d_{r+1}), say. We may assume then that for all i and $r \leq j$, $a_r^i = a_r$, $S_r^i = S_r$, $<_{r+1}^i = <_{r+1}$ and $\mathbf{d}_{r+1}^i = \mathbf{d}_{r+1}$.
Case (*i*). For arbitrary large i, $k(i) = j + 1$.
We may then clearly assume that for all i, $k(i) = j + 1$. So for all i, $S_{j+1}^i \subseteq S_{j+1}^{i+1}$. If the S_{j+1}^i stabilize, with eventual value S_{j+1}, then put

$$p = (a_0, \ldots, a_j, S_{j+1}, <_1, \ldots, <_{j+1}, \mathbf{d}_1, \ldots, \mathbf{d}_{j+1}, T_1, \ldots, T_{j+1})$$

(where $T_1, \ldots, T_{j+1}$ are suitable chosen). This choice of p clearly works.
If the S_{j+1}^i do not stablize then by conditions (ii), (iii), and (iv) in Definition 4.3, the union of the S_{j+1}^i must be all of $L(a_j)$. In this case put

$$p = (a_0, S_1, \ldots, a_{j-1}, S_j \cup \{a_j\}, <_1, \ldots, <_j, \mathbf{d}_1, \ldots, \mathbf{d}_j, T_1, \ldots, T_j)$$

(for a suitable choice of $T_1, \ldots, T_j$), which is easily seen to work.

Case (*ii*). For eventually all i, $k(i)$ is strictly greater than $j + 1$, and hence a_{j+1}^i is defined.
The maximal choice of j implies that

(∗) the sequence of a_{j+1}^i is not eventually constant.

Also

(∗∗) $S_{j+1}^i \subseteq S_{j+1}^{i+1}$ for all large enough i.

If $j + 1 \in I_{\text{fin}}$, then, using the fact that $U_{p(i)} \subseteq U_{p(i')}$ for $i < i'$, we easily contradict (∗). Thus $i \in I_{\text{new}} \cup I_{\text{old}} \cup I_{\text{aff}}$. If $i \in I_{\text{new}}$ then $S_{j+1}^i \cup \{a_{j+1}^i\}$ is a $<_{j+1}$-initial segment of $L(a_j)$ whose greatest element is a_{j+1}^i. By (∗) and (∗∗), the sequence of a_{j+1}^i must be cofinal in $L(a_j)$, so the union of the $S_{j+1}^i \cup \{a_{j+1}^i\}$ equals $L(a_j)$. In this case

$$p = (a_0, S_1, \ldots, S_j \cup \{a_j\}, <_1, \ldots, <_j, \mathbf{d}_1, \ldots, \mathbf{d}_j, T_1, \ldots, T_j)$$

works (for suitable $T_1, \ldots, T_j$). A similar argument (using (iii) and (iv) in Definition 4.3) shows that this choice of p also works if $i \in I_{\text{old}}$ or I_{aff}. ■

The following proposition gives a global consequence of Proposition 3.1. This will be the form in which 3.1 is applied.

Proposition 4.5. *Let* $\mathbf{x}$ *be a finite tuple from* M^*. *Let* $S_1, S_2, \ldots$ *be an* ω*-sequence of good sets (i.e. members of* $\mathbf{W}$*). Then there are* $i<j<\omega$ *and an automorphism* h *of* M^* *such that* $h(\mathbf{x}) = \mathbf{x}$ *and* $h(S_i) \subseteq S_j$.

We first prove a lemma, In the statement of the lemma and the subsequent notation,

$$p(j)=(a_0^{p(j)}, S_1^{p(j)}, \ldots, S_{k(j)}^{p(j)}, <_1^{p(j)}, \ldots, <_{k(j)}^{p(j)}, \mathbf{d}_1^{p(j)}, \ldots, \mathbf{d}_{k(j)}^{p(j)}, T_1^{p(j)}, \ldots, T_{k(j)}^{p(j)})$$

LEMMA **4.6.** *Let $k \leq n$. Let $p(1), p(2), \ldots$ be in Π, and $\mathbf{c}(1), \mathbf{c}(2), \ldots$ be tuples from M^* of equal length. Then there are $j < j' < \omega$, and an automorphism h of M^* such that*

(i) $h(\mathbf{c}(j)) = \mathbf{c}(j')$; *and*

(ii) for each $i \leq k$, $h(a_i^{p(j)}) = a_i^{p(j')}$, *and if $i \in I_{\text{aff}}$ then also* $h(\mathbf{d}_i^{p(j)}) = \mathbf{d}_i^{p(j')}$;

(iii) *for each* $i \leq k$, $h(S_i^{p(j)}) \subseteq S_i^{p(j')}$. *If $i \leq n$ and $\alpha(i) \leq k$, then also* $h(T_i^{p(j)}) \subseteq T_i^{p(j')}$. *If $i \leq k$ is in I_{fin}, then* $h(S_i^{p(j)}) = S_i^{p(j')}$.

(*In* (ii) *the notation '$h(a_i^{p(j)}) = a_i^{p(j')}$' for example, is supposed to mean that either both sides are undefined or both sides are defined and equal. Similarly with* (iii).)

Proof. The lemma is proved by induction on k. If $k = 0$ there is nothing to do, as only finitely many n-types are realized in M^* for any m. Assume now the lemma to hold for $k - 1$, and we prove it for k. So we are given $p(1)$, $p(2), \ldots$ in Π and tuples $\mathbf{c}(1), \mathbf{c}(2), \ldots$ of equal length. By our induction hypothesis and Ramsey's theorem, we may assume (after thinning out the sequence and applying various automorphisms of M^*) that

(i) $\mathbf{c}(j) = \mathbf{c}$ for all $h < \omega$;

(ii) *for each* $i < k$, $a_i^{p(j)} = a_i$ for all $j < \omega$, and if $i \in I_{\text{aff}}$, $\mathbf{d}_i^{p(j)} = \mathbf{d}_i$ for all $j < \omega$;

(iii) for each $i < k$ and $j < j' < \omega$, $S_i^{p(j)} \subseteq S_i^{p(j')}$ and if $i \in I_{\text{fin}}$, $S_i^{p(j)} = S_i^{p(j')}$. If $\alpha(i) < k$ and $j < j' < \omega$, then $T_i^{p(j)} \subseteq T_i^{p(j')}$.

Case I. $k \in I_{\text{fin}}$.
By thinning out the sequence we may assume that either all $a_k^{p(j)}$ are defined or none are defined, and that the $S_k^{p(j)}$ have equal cardinality. Let $\mathbf{e}(j)$ enumerate $S_k^{p(j)}$ together with $a_k^{p(j)}$ at the end if the latter is defined. Let $\mathbf{f}(j)$ be the sequence $\mathbf{ce}(j)$. Applying the induction hypothesis to $p(1), p(2), \ldots$ and $\mathbf{f}(1), \mathbf{f}(2), \ldots$ immediately yields the lemma for k.

Case II. $k \in I_{\text{old}} \cup I_{\text{aff}}$.
Suppose $k \in I_{\text{aff}}$ and we may suppose that for all j, $a_k^{p(j)}$ and $\mathbf{d}_k^{p(j)}$ are defined. Let $\mathbf{e}(j)$ be the sequence $\mathbf{cd}_k^{p(j)}a_k^{p(j)}$. Applying the induction hypothesis to $p(1)$, $p(2), \ldots$ and $\mathbf{e}(1), \mathbf{e}(2), \ldots$ easily yields the desired conclusion (using the function $f_k(-)$).

If $k \in I_{\text{old}}$, do the same thing but without the $\mathbf{d}_k^{p(j)}$.

Case III. $k \in I_{\text{new}}$.
Let $H = L(a_0) \cup L(a_1) \cup \cdots \cup L(a_{k-2})$. Let $P = L(a_{k-1})$ So P is a strictly minimal set (over a_{k-1}) whose geometry is either degenerate or projective with the strongest possible structure (by Lemma 2.1(iv)). As $k \in I_{\text{new}}$, P is orthogonal to $L(a_i)$ for each $i < k - 1$. It follows that for each $\mathbf{a}$ in P, $\text{tp}(\mathbf{a}/a_{k-1}) \vdash \text{tp}(\mathbf{a}/H)$. Thus any a_{k-1}-elementary permutation of P is H-elementary. In particular

$(*)$ Any a_{k-1}-elementary permutation of P extends to an automorphism of M^* which fixes H pointwise.

(This follows by ω-categoricity and ω-stability. In fact if M is any countable ω-stable, ω-categorical structure and X is some $\mathbf{a}$-definable set in M then M is prime (atomic) over $X \cup \mathbf{a}$. Assume without loss that $\mathbf{a} = \varnothing$. Let $\mathbf{c}$ be some tuple from M, and let the tuple $\mathbf{d}$ be chosen from X such that $\text{tp}(\mathbf{c}/X)$ is the unique non-forking extension of $\text{tp}(\mathbf{c}/\mathbf{d})$. Let $\varphi(\mathbf{x}, \mathbf{d})$ isolate $\text{tp}(\mathbf{c}/\mathbf{d})$. Let $\mathbf{c}'$ satisfy $\varphi(\mathbf{x}, \mathbf{d})$ in M. If $\text{tp}(\mathbf{c}'/X)$ forked over $\mathbf{d}$, then this is witnessed by a formula $\psi(\mathbf{c}', \mathbf{b}, \mathbf{d})$ with $\mathbf{d}$ in X. As X is $\varnothing$-definable, we can find $\mathbf{b}'$ in X such that $\text{tp}(\mathbf{b}'/\mathbf{d}) = \text{tp}(\mathbf{b}/\mathbf{d})$ and $\varphi(\mathbf{c}, \mathbf{d}', \mathbf{d})$, whereby $\text{tp}(\mathbf{c}/X)$ forks over $\mathbf{d}$, a contradiction. Thus $\text{tp}(\mathbf{c}'/X) = \text{tp}(\mathbf{c}/X)$, so $\varphi(\mathbf{x}, \mathbf{d})$ isolates $\text{tp}(\mathbf{c}/X)$.)

Now by the definition of Π, all the orderings $<_k^{p(j)}$ of P are conjugate under $\text{Aut}(M^*)$. By $(*)$ there are then H-automorphisms f_j of M^* such that the orderings $f_j(<_k^{p(j)})$ of P are all the same, say $<_k$. We may replace each $p(j)$ by $f_j(p(j))$. Then for the new $p(j)$, (ii) and (iii) above remain true, but now (i) is replaced by

(i)′ $\text{tp}(\mathbf{c}(j)/H) = \text{tp}(\mathbf{c}(j')/H)$ for all $j, j' < \omega$.

We can then find tuples $\mathbf{e}(j)$ in P, for $j < \omega$ such that

(iv) $\text{tp}(\mathbf{c}(j)\mathbf{e}(j)/H) = \text{tp}(\mathbf{c}(j')\mathbf{e}(j')/H)$ for all $j, j' < \omega$, and $\text{tp}(\mathbf{c}(j)/H\mathbf{e}(j)) \vdash \text{tp}(\mathbf{c}(j)/H \cup P)$ for all $j < \omega$.

(This is because, first, by ω-stability we can find a finite tuple $\mathbf{e}(1)$, say, from P such that $\text{tp}(\mathbf{c}(1)/H \cup P)$ is the unique non-forking extension over $H \cup P$ of $\text{tp}(\mathbf{c}(1)/H\mathbf{e}(1))$. As in the parenthetical remark above, $\text{tp}(\mathbf{c}(1)/H\mathbf{e}(1)) \vdash \text{tp}(\mathbf{c}(1)/H \cup P)$. The $\mathbf{e}(j)$ for $j > 1$ can be found using (i)′.)

Finally for each $t \leq n$ such that $\alpha(t) = k$, let $b_t^{p(j)} \in P(P^3)$ be such that $T_t^{p(j)} = \{x \in P(P^3)\colon x \leq_k b_t^{p(j)}\}$. (Remember that $T_t^{p(j)}$ is an $<_k$-initial segment of P or of P^3, depending on whether $t \in I_{\text{old}}$ or I_{aff}.)

We now apply Proposition 3.1 to find $j < j' < \omega$ and an a_{k-1}-elementary map $f\colon P \to P$ such that f preserves $<_k$ and

(v) $f(\mathbf{e}(j)) = \mathbf{e}(j')$;

(vi) $f(a_k^{p(j)}) = a_k^{p(j')}$; and

(vii) $f(b_t^{p(j)}) = b_t^{p(j')}$ for all relevant t.

As f preserves $<_k$ it follows from the definition of the $b_t^{p(j)}$ that

(viii) $f(T_t^{p(j)}) \subseteq T_t^{p(j')}$ for all relevant t.

Now we may find an a_{k-1}-elementary *permutation* h_1 of P satisfying also (v), (vi), and (viii) (although h need no longer preserve $>_k$). By $(*)$ we may assume that h_1 is induced by an automorphism h of M^* which fixes H pointwise. Then (by (ii), (iii), (vi), (viii)) h satisfies the requirements of the lemma, except possibly for the requirement that $h(\mathbf{c}(j)) = \mathbf{c}(j')$. However, $h(\mathbf{e}(j)) = \mathbf{e}(j')$, so using (iv) we see that $\mathrm{tp}(h(\mathbf{c}(j))/H \cup P) = \mathrm{tp}(\mathbf{c}(j')/H \cup P)$. Composing h with an automorphism of M^* which fixes $H \cup P$ pointwise and takes $h(\mathbf{c}(j))$ to $\mathbf{c}(j')$ preserves the old requirements of the lemma and now takes $\mathbf{c}(j)$ to $\mathbf{c}(j')$.

This completes the proof of the lemma. ■

Proof of Proposition 4.5. Let $S_i = U_{p(i)}$, where $p(i) \in \Pi$. Apply Lemma 4.6 to the sequence $p(1), p(2), \ldots$ and the constant sequence $\mathbf{x}, \mathbf{x}, \mathbf{x} \ldots$ to find an automorphism h of M^* and $j < j' < \omega$ such that $h(\mathbf{x}) = \mathbf{x}$ and (ii) and (iii) of the lemma are satisfied for $k = n$. It is then clear that $h(S_j) \subseteq S_{j'}$. ■

We now give analogues in the general case of Lemmas 3.6, 3.7, and 3.8.

LEMMA 4.7. *For each $k < \omega$ there is some $\lambda(k) < \omega$ such that for every k-tuple* $\mathbf{c}$ *from M^* and good set $U \in \mathbf{W}$, there is a subset F of U of cardinaility at most $\lambda(k)$ such that*

(i) $\mathrm{tp}(\mathbf{c}/U)$ *is the unique non-forking extension over U of* $\mathrm{tp}(\mathbf{c}/F)$; *and*

(ii) $\mathrm{tp}(\mathbf{c}/\mathrm{base}(U))$ *is the unique non-forking extension over* $\mathrm{base}(U)$ *of* $\mathrm{tp}(\mathbf{c}/F \cap \mathrm{base}(U))$.

Proof. Fix k. Suppose there is no such $\lambda(k)$. Then there are $\lambda(1) < \lambda(2) <, \ldots,$ k-tuples $\mathbf{c}_1, \mathbf{c}_2, \ldots$ and good sets $U_1, U_2, \ldots$ such that for each i, and subset F_i of U_i of cardinality at most $\lambda(i)$, (i) or (ii) fail (for $\mathbf{c}_i$ in place of $\mathbf{c}$, U_i in place of U, and F_i in place of F). For each $i < \omega$, let $\mathbf{a}_i$ be an enumeration of the bounded part of $\mathrm{base}(U_i)$. Then $\mathbf{a}_i$ is a tuple of length at most n. By ω-categoricity we may assume that $\mathrm{tp}(\mathbf{c}_i\mathbf{a}_i) = \mathrm{tp}(\mathbf{c}_j\mathbf{a}_j)$ for all i, $j < \omega$. So after applying some automorphisms of M^* we may assume that, for some $\mathbf{c}$ and $\mathbf{a}$, $\mathbf{c}_i = \mathbf{c}$ and $\mathbf{a}_i = \mathbf{a}$ for all i. By Proposition 4.5, we may, using Ramsey's theorem, and again applying automorphism of M^* (which fix $\mathbf{c}$ and $\mathbf{a}$) assume that $U_1 \subseteq U_2 \subseteq \ldots$. (By Ramsey's theorem and Proposition 4.5, we may find an infinite subset $J = \{j(1) < j(2) < \ldots\}$ of ω such that for each $j(i) < j(i')$ in J there is an automorphism of M^* which fixes $\mathbf{ac}$ and takes $U_{j(i)}$ into $U_{j(i')}$. There will then clearly be automorphisms $f_1, f_2, \ldots$ of M^*, each fixing $\mathbf{ac}$ and such that $f_1(U_{j(1)}) \subseteq f_2(U_{j(2)}) \subseteq \ldots$. Note that $\mathbf{a}$ is still the bounded part of base($f_i(U_{j(i)})$) for each i.)

As the bounded part of the bases of the U_i are all the same, it follows also that base(U_1) $\subseteq$ base(U_2) $\subseteq \ldots$. Let $X = \cup\{U_i: i < \omega\}$, and $Y = \cup$ {base(U_i): $i < \omega$}. By ω-stability there is some finite subset F of X such that tp($\mathbf{c}/X$) is the unique non-forking extension over X of tp($\mathbf{c}/F$) and tp($\mathbf{c}/Y$) is the unique non-forking extension over Y of tp($\mathbf{c}/F \cap Y$). Choosing i large enough so that $F \subseteq U_i$, $F \cap Y \subseteq$ base(U_i), and $\lambda(i) > |F|$ yields a contradiction to the original choice of U_i, $\mathbf{c}$, and $\lambda(i)$. The lemma is proved. ■

LEMMA **4.8.** *For any $\lambda < \omega$ there is $\mu < \omega$ such that whenever $U \in \mathbf{W}$ and S is a subset of U of cardinality at most λ, then there is $U' \in \mathbf{W}$ such that $S \subseteq U' \subseteq U$,* base($U'$) *is of cardinality at most μ and the bounded part of* base(U') *is the same as the bounded part of* base(U).

Proof. The proof is like that of 4.7 (and 3.7), so we are brief. Fix λ. Supposing there is no μ, we obtain $\mu(1) < \mu(2) < \ldots$, $U_1, U_2, \ldots$, in $\mathbf{W}$, and subsets S_i of U_i of cardinality at most λ, such that for each i there is *no* $U' \in \mathbf{W}$ with $S_i \subseteq U' \subseteq U_i$, base($U'$) of cardinality at most $\mu(i)$, and bounded part of base(U') = bounded part of base(U_i). Let $\mathbf{a}_i$ be an enumeration of the bounded part of base (U_i), and let $\mathbf{c}_i$ be an enumeration of S_i. As in the proof of 4.7, we may assume that $\mathbf{a}_i = \mathbf{a}_j = \mathbf{a}$ and $\mathbf{c}_i = \mathbf{c}_j = \mathbf{c}$ for all i, $j < \omega$. By 3.1 and Ramsey's theorem we may suppose in addition that $U_1 \subseteq U_2 \subseteq \ldots$. Now $\mathbf{c}$ is contained in U_1. Choose i large enough to that $\mu(i) >$ cardinality of base(U_1), and we have a contradiction. ■

LEMMA **4.9.** *There is a number $\lambda < \omega$, such that whenever $\mathbf{c}$ is an enumeration of the base of some good set, then* $\mathrm{tp}_\lambda(\mathbf{c}) \vdash \mathrm{tp}(\mathbf{c})$.

Proof. Let λ be $\lambda(1) + 1$ where $\lambda(1)$ is as in Lemma 4.7. We will prove that this choice of λ works, by induction on the length of the tuple $\mathbf{c}$. Fix $\mathbf{c}$, and suppose $\mathbf{c}$ is an enumeration of base(U_p), where $p \in \Pi$ is of the form $(a_0, S_1, \ldots, S_k, <_1, \ldots, <_k, \mathbf{d}_1, \ldots, \mathbf{d}_k, T_1, \ldots, T_k)$. So $\mathbf{c}$ is an enumeration of $\{a_0\} \cup S_1 \cup \{a_1\} \cup \cdots \cup S_k$.

Claim. $\mathbf{c}$ is (after reordering) of the form $\mathbf{de}$ where $\mathbf{d}$ is the base of some good set and e is an element of M^*.

Proof. This is easy by the definition of Π. For example, if $S_k \neq \varnothing$, and $k \in I_{\text{new}}$, then S_k is a $<_k$-initial segment of $L(a_{k-1})$, with last element e, say. Then $\{a_0\} \cup S_1 \cup \{a_1\} \cup \cdots \cup S_k - \{c\}$ is the base of some good set, and take $\mathbf{d}$ to be an enumeration of this base. The remaining cases are left to the reader. ■

Let now $\mathbf{c}' = \mathbf{d}'e'$ be a tuple (which we may take from M^*) such that $\text{tp}_\lambda(\mathbf{c}') = \text{tp}_\lambda(\mathbf{c})$, and we want to show that $\text{tp}(\mathbf{c}') = \text{tp}(\mathbf{c})$. Now also $\text{tp}_\lambda(\mathbf{d}') = \text{tp}_\lambda(\mathbf{d})$, so by the induction hypothesis, $\text{tp}(\mathbf{d}') = \text{tp}(\mathbf{d})$, whereby $\mathbf{d}'$ is also the base of some good set.

Case I. $RM(e'/\mathbf{d}') \leq RM(e/\mathbf{d})$.
By Lemma 4.7, the fact that $\mathbf{d}'$ is the base of a good set, and the choice λ, $\text{tp}(e'/\mathbf{d}')$ is the unique non-forking extension over $\mathbf{d}'$ of $\text{tp}(e'/\mathbf{d}'_1)$ where $\mathbf{d}'_1$ is some subtuple of $\mathbf{d}'$ of length strictly less than λ. Let $\mathbf{d}_1$ be the corresponding subtuple of $\mathbf{d}$. Then $\text{tp}(e\mathbf{d}_1) = \text{tp}(e'\mathbf{d}'_1)$. So $RM(e/\mathbf{d}) \leq RM(e/\mathbf{d}_1) = RM(e'/\mathbf{d}'_1) = RM(e'/\mathbf{d}')$. The case hypothesis then yields $RM(e/\mathbf{d}) = RM(e/\mathbf{d}_1)$, that is $\text{tp}(e/\mathbf{d})$ does not fork over $\mathbf{d}_1$. By automorphism, $\text{tp}(e/\mathbf{d}_1)$ has a unique non-forking extension over $\mathbf{d}$. Thus easily $\text{tp}(\mathbf{d}e) = \text{tp}(\mathbf{d}'e')$.

Case II. $RM(e/\mathbf{d}) \leq RM(e'/\mathbf{d}')$.
This is proved in the same way as Case I.

The lemma is proved. ■

We will point out that M^* has a finite language (and even more). Before this we record an easy observation about Π.

Fact 4.10. Let $p, q \in \Pi$, with $p <_\Pi q$. Then there is $p \in \Pi$ such that

(i) $p <_\Pi p' \leq_\Pi q$;

(ii) for some element $a \in M^*$, $U_{p'} = U_p \cup \{a\}$ and $\text{tp}(a/U_p)$ is algebraic or strongly minimal, and, moreover, there is a single element $b \in U_p$ such that $RM(a/b) \leq 1$.

Proof. Left to the reader. ■

Proposition 4.11. *There is $\lambda < \omega$ such that for any $U \in \mathbf{W}$, if h is a map from U into M^* which preserves all formulae over $\varnothing$ in at most λ variables, then h is a partial elementary map.*

Proof. We choose λ as in the proof of Lemma 4.9. Fix $q \in \Pi$, and let

$h: U_q \to M^*$ be a map preserving all formulae in at most λ variables. By Lemma 4.4, there is some $<_\Pi$-maximal $p \leq_\Pi q$ such that $h \restriction U_p$ is elementary. Suppose $p <_\Pi q$. Let $p' \in \Pi$ be as given by Fact 4.10. So $p <_\Pi p' \leq_\Pi q$, and $U_{p'} = U_p \cup \{a\}$ where a is a single element of M^*, with $RM(a/b) \leq 1$ for some $b \in U_p$. Let $V = h(U_p)$. Let $\mathbf{C}$ be an ω_1-saturated model of cardinality ω_1, containing M^*. Then $h \restriction U_p$ extends to an automorphism g of $\mathbf{C}$. Let M_1^* be $g(M^*)$. Then V is clearly a 'good set' in M_1^*.

We will show that $h \restriction U_{p'}$ is elementary, contradicting the maximal choice of p. To do this, it is enough to show:

CLAIM. $\mathrm{tp}(g(a)/V) = \mathrm{tp}(h(a)/V)$.

Proof of claim. As $\lambda \geq 2$, and $RM(a/b) \leq 1$ for some $b \in U_p$, also $RM(h(a)/V) \leq 1$. Suppose first that $RM(h(a)/V) = 0$, that is $h(a) \in \mathrm{acl}(V)$. Then in particular $h(a) \in M_1^*$. As V is a pregood set in M_1^* (or, if you wish, simply using the isomorphism g between M^* and M_1^*), there is by Lemma 4.7 some subset F of V of cardinality strictly less than λ, such that $\mathrm{tp}(h(a)/F) \vdash \mathrm{tp}(h(a)/V)$. But as $|h(a) \cup F| \leq \lambda$, $\mathrm{tp}(h(a)/F) = \mathrm{tp}(g(a)/F)$. Thus $\mathrm{tp}(h(a)/V) = \mathrm{tp}(g(a)/V)$, as required.

So we may suppose that $RM(h(a)/V) = 1$, and in particular $RM(g(a)/V) \leq RM(h(a)/V)$. Again we may find by Lemma 4.7 a subset F of V of cardinality strictly less than λ, such that $\mathrm{tp}(g(a)/V)$ is the unique non-forking extension over V of $\mathrm{tp}(g(a)/F)$. But $\mathrm{tp}(g(a)/F) = \mathrm{tp}(h(a)/F)$. Thus $h(a)$ must be independent from V over F, whereby $\mathrm{tp}(h(a)/V) = \mathrm{tp}(g(a)/V)$, as required. This proves the claim, and completes the proof of the proposition. ■

We will now determine a finite language for M^* in which we will (in the next section) give the axioms for M^*.

Let us now take λ to be $\lambda(1) + 1$ where $\lambda(1)$ is as given by 4.7. *Then λ also satisifies* 4.9 *and* 4.11.

Let L^* be the language consisting of a predicate for each complete k-type over $\varnothing$, for $k \leq \lambda$. Then L^* includes predicates for the levels $L_1, \ldots, L_n$, as well as for the tree ordering $\leq$ on M^*.

REMARK 4.12. *For each m there is a quantifier-free L^*-formula $\xi_m(x, y)$ such that whenever $U \in \mathbf{W}$ and* base(U) *has cardinality m, than for some enumeration* $\mathbf{d}$ *of* base(U), *U is defined in M^* by $\xi_m(x, \mathbf{d})$.*

Proof. Clear. (See Definition 4.1(ii).) ■

DEFINITION 4.13. *Let $U \in \mathbf{W}$. By a U-*restricted *L^*-formula, we mean a formula built up from the atomic L^*-formulae by Boolean connectives and the quantifiers* $(\exists x \in U)$ *and* $(\forall x \in U)$.

REMARK **4.14.** *Let $U \in \mathbf{W}$. Then any subset of U^m definable over* base(U) *in M^* is defined by a U-restricted L^*-formula with parameters from* base(U). *In particular any $\varnothing$-definable set in M^* is defined by an L-formula without parameters, whereby L^* is a finite language for M^*.*

Proof. Let U^* be the structure whose universe is U, equipped with constants for the elements of base(U), and predicates for all base(U)-definable (in M^*) relations on U. Then U^* is ω-categorical. By Proposition 4.11, any permutation of U which fixes base(U) pointwise and preserves the relations in L^* is an elementary map, from the point of view of M^*, and is thus an automorphism of the structure U^*. Thus any $\varnothing$-definable set in the structure U^* is defined in U^* by an L^*-formula with constants for elements of base(U), and is thus defined in M^* by a U-restricted L^*-formula with parameters from base(U). ■

Finally in this section we point out:

REMARK **4.15.** *Let $U \in \mathbf{W}$ and let* **d** *be an enumeration of* base(U). *Let a be an element of M^*. Suppose* **b** *is a tuple from U such that* tp(a/U) *is the unique non-forking extension over U of* tp$(a/\mathbf{b})$. *Then* tp$(a/\mathbf{bd}) \vdash$ tp(a/U).

Proof. We first show that for every $a' \in M^*$ with tp$(a'/\mathbf{bd})$ = tp$(a/\mathbf{bd})$, we have tp(a'/U) = tp(a/U). So suppose a' is such. There is an automorphism of M^* taking a to a' and fixing **bd**. This automorphism will fix U setwise, as U is **d**-definable. Thus tp(a'/U) is also the unique non-forking extension of tp(a/b) over U, whereby tp(a'/U) = tp(a/U).

As tp$(a/\mathbf{bd})$ is isolated, it is then easy to conclude that tp$(a/\mathbf{bd}) \vdash$ tp(a/U). ■

5 The proof of quasifinite axiomatizability

In this final section we give a 'quasifinite' set of axioms for M^* in the language L^*. Let λ be as chosen at the end of section 4 and let μ be as given in Lemma 4.8 for twice this value of λ.

We need to express certain facts of the form $\varphi(x, \mathbf{c}) \to q(x)$ (where q is a quantifier-free type over some suitable $U \in \mathbf{W}$) by 'nice' L^*-sentences. These L^*-sentences will play the role of the axioms 1–5 from section 3. As U is in general infinite, the required sentences are of a slightly more complicated form than in section 3. We work now in the language L^*.

We require a little more notation. See Remark 4.12 for the meaning of the formulae ξ_m.

Definition 5.1. *Let* $\mathbf{w}$ *be a sequence of variables of length* m. *By an* m–$\mathbf{w}$*-restricted formula, we mean a formula built up from the atomic formulae by means of Boolean connectives and the quantifiers*

$$(\exists z)(\xi_m(z, \mathbf{w}) \wedge \ldots, (\forall z)(\xi_m(\mathbf{z}, \mathbf{w}) \rightarrow \ldots),$$

where z *may vary.*

Let now $\mathbf{y}$, $\mathbf{z}$, $\mathbf{w}$ *be sequences of variables with* $l(\mathbf{y}) = l(\mathbf{z}) = \lambda - 1$, *and* $l(\mathbf{w}) = \mu$.

Let us fix a finite (by ω*-categoricity) collection* $\Phi(x, \mathbf{y}, \mathbf{z}, \mathbf{w})$ *of formulae such that whenever* $\mathbf{w}'$ *is a subtuple of* $\mathbf{w}$ *of length* m *for some* $m \leq \mu$, *and* φ *is an* m–$\mathbf{w}'$*-restricted formula whose variables are among* x, $\mathbf{y}$, $\mathbf{z}$, $\mathbf{w}'$, *then* φ *is equivalent (in* M^**) to a formula in* Φ. *Note here that 'x' represents a single variable.*

Our proposed set of axioms for $Th(M^*)$ *is the following*:

1. $\leq$ *is a partial ordering which induces a tree structure with levels* L_0, $L_1, \ldots, L_n$. L_0 *has a single element named by the constant* a_0. *For* $i \in I_{\mathrm{fin}}$, *and* $b \in L_{i-1}$, $L(b)(= \{x \in L_i : x \geq b\})$ *has* κ_i *elements (for suitable* κ_i*).*
2. *All sentences of the form* $(\forall \mathbf{y})(\psi(\mathbf{y}) \rightarrow (\exists_x^{=m})(\varphi(x, \mathbf{y})))$ *which are true in* M^*, *where* $l(\mathbf{y}) < \lambda$, *and both* $\psi(\mathbf{y})$, $\varphi(x, \mathbf{y})$, *are atomic formulae.*
3. *All sentences of the form* $(\forall x\mathbf{w})(\chi(x, \mathbf{w}) \rightarrow (\exists \mathbf{y})(\varphi(x, \mathbf{y}) \wedge \theta(\mathbf{y}, \mathbf{w})))$ *which are true in* M^*, *where* $l(\mathbf{w}) \leq \mu$, $l(\mathbf{y}) < \lambda$, $\chi(x, \mathbf{w})$ *and* $\varphi(x, \mathbf{y})$ *are quantifier free, and* $\theta(\mathbf{y}, \mathbf{w})$ *is in* Φ.
4. *All sentences of the form*

$$(\forall x\mathbf{yzw})((\eta(\mathbf{y}, \mathbf{w}) \wedge \varphi(x, \mathbf{y}, \mathbf{w}) \wedge \bigwedge\{\xi_m(z, \mathbf{w}) : z \in \mathbf{z}\}) \rightarrow (\psi(x, \mathbf{z}) \leftrightarrow \varepsilon(\mathbf{z}, \mathbf{y}, \mathbf{w})))$$

which are true in M^*, *where* $m \leq \mu$, $l(\mathbf{w}) = m$, $l(\mathbf{y}), l(\mathbf{z}) < \lambda$, $\varphi(x, \mathbf{y})$ *and* $\psi(x, \mathbf{z})$ *are quantifier free, and* $\eta(\mathbf{y}, \mathbf{w})$ *and* $\varepsilon(\mathbf{z}, \mathbf{y}, \mathbf{w})$ *are in* $\Phi(\mathbf{y}, \mathbf{z}, \mathbf{w})$.

5. *For each* $i \leq n$ *such that* $i \notin I_{\mathrm{fin}}$, *the schema*: $(\forall y)(L_{i-1}(y) \rightarrow$ '$L(y)$ *is infinite').*

Remark. *Clearly (fixing the variables* x, $\mathbf{y}$, $\mathbf{z}$, $\mathbf{w}$*) there are only finitely many axioms of sort* 1, 2, 3, *and* 4. *The axioms in* 5 *consist of finitely many 'schemas of infinity'.*

Remark 5.2. *Let us give a little explanation of the axioms in* 4. *We will refer back to this explanation in our proof of quasifinite axiomatizability.*

Let U be a good set in M^* such that base(U) has cardinality $m \leq \mu$. Let $\mathbf{d}$ be an enumeration of base(U) such that $\xi_m(z, \mathbf{d})$ defines U. We will consider two possible situations for an element $a \in M^*$, and show how these are expressed by an axiom of type 4.

First suppose $a \in M^*$ *and* $a \in \mathrm{acl}(U)$. By 4.7 there is some tuple $\mathbf{b}$ in U of cardinality $<\lambda$ such that *tp*(a/U) is the unique non-forking extension over U of $\mathrm{tp}(a/\mathbf{b})$ and thus $\mathrm{tp}(a/\mathbf{b}) \vdash \mathrm{tp}(a/U)$ in M^*.

Let $\varphi(x, \mathbf{y})$ be the atomic formula (of L^*) expressing $\mathrm{tp}(a, \mathbf{b})$. Then in M^* the formula $\varphi(x, \mathbf{b})$ isolates $\mathrm{tp}(a/U)$. We will point out that this (or rather the fact the the quantifier-free type of a over U is so isolated) is expressed by an axiom in 4.

Now clearly $\mathrm{tp}(a/U)$ is then also *definable* over $\mathbf{b}$ (this has nothing to do with stability); in particular for each atomic formula $\psi(x, \mathbf{z})$ there is some formula $\delta(\mathbf{z}, \mathbf{y})$ such that for $\mathbf{c}$ from U, $M^* \models \psi(a, \mathbf{c})$ iff $M^* \models \delta(\mathbf{c}, \mathbf{b})$. By Remark 4.14, the formula '$\delta(\mathbf{z}, \mathbf{b}) \wedge \mathbf{z} \subset U$' is equivalent in M^* to a formula $\varepsilon(\mathbf{z}, \mathbf{b}, \mathbf{d})$ where ε is in Φ.

So we have

(I) $M^* \models (\forall x)(\varphi(x, \mathbf{b}) \rightarrow (\forall \mathbf{z} \subset U)(\psi(x, \mathbf{z}) \leftrightarrow \varepsilon(\mathbf{z}, \mathbf{b}, \mathbf{d})))$.

On the other hand, by 4.14 again there is a formula $\rho(\mathbf{y}, \mathbf{w})$ in Φ, such that $\rho(\mathbf{y}, \mathbf{d})$ isolates $\mathrm{tp}(\mathbf{b}/\mathbf{d})$. Let $\tau(\mathbf{w})$ be the quantifier-free type of $\mathbf{d}$ in M^*. By 4.9, $\tau(\mathbf{w})$ isolates $\mathrm{tp}(\mathbf{d})$ in M^*. Let $\eta(\mathbf{y}, \mathbf{w})$ be the formula $\rho(\mathbf{y}, \mathbf{w}) \wedge \tau(\mathbf{w})$. Then $\eta(\mathbf{y}, \mathbf{w})$ is in Φ and isolates $\mathrm{tp}(\mathbf{bd})$ in M^*. So from (I) we have

(II) $M^* \models (\forall \mathbf{y}, \mathbf{w})(\eta(\mathbf{y}, \mathbf{w}) \rightarrow (\forall x)(\varphi(x, \mathbf{y}) \rightarrow (\forall \mathbf{z})(\bigwedge\{\xi_m(z, \mathbf{w}): z \in \mathbf{z}\} \rightarrow (\psi(x, \mathbf{z}) \leftrightarrow \varepsilon(\mathbf{z}, \mathbf{y}, \mathbf{w})))))$.

This latter sentence is then an axiom in 4, and as $\psi(x, \mathbf{z})$ ranges over atomic formulae, these sentences express the fact that $\varphi(x, \mathbf{b})$ isolates the quantifier-free type of a over U.

Second suppose $a \in M^*$, $\mathrm{tp}(a/\mathbf{d})$ *is strongly minimal, and* $\mathrm{tp}(a/U)$ *does not fork over* $\mathbf{d}$. By 4.7, $\mathrm{tp}(a/\mathbf{d}_0)$ is strongly minimal, for some subtuple $\mathbf{d}_0$ of $\mathbf{d}$ of cardinality $<\lambda$. (In fact in the applications $\mathbf{d}_0$ will be a single element from $\mathbf{d}$). Let $\varphi_0(x, \mathbf{w}_0)$ be an *atomic* formula isolating $\mathrm{tp}(a, \mathbf{d}_0)$. Then $\varphi(x, \mathbf{d}_0)$ is strongly minimal and $\mathrm{tp}(a/\mathbf{d})$ is isolated by $\{\varphi(x, \mathbf{d}_0)\} \cup$ '$x \notin \mathrm{acl}(\mathbf{d})$'. By 4.7(ii), the formula '$x \notin \mathrm{acl}(\mathbf{d})$' is equivalent in M^* to $\{x \notin \mathrm{acl}(\mathbf{d}_1)$: $\mathbf{d}_1$ a subtuple of $\mathbf{d}$ of cardinality at most $\lambda\}$, and thus $\mathrm{tp}(a/\mathbf{d})$ is isolated by a formula $\varphi(x, \mathbf{d})$ where $\varphi(x, \mathbf{w})$ is *quantifier free*. By Lemma 4.14 (or rather Remark 4.15), $\mathrm{tp}(a/\mathbf{d}) \vdash \mathrm{tp}(a/U)$ in M^*. Thus $\varphi(x, \mathbf{d})$ isolates $\mathrm{tp}(a/U)$ in M^*. Just as in the first part of this remark, there is an axiom of kind 4 which expresses this.

THEOREM **5.3.** *Let* N^* *be a countable* L^**-structure which is a model of the set of axioms* 1 *to* 5. *Then* N^* *is isomorphic to* M^*.

Proof. We will build inductively a family $h_p: U_p \rightarrow N^*$ of partial isomorphisms, where p ranges over a certain cofinal set in Π. It will be convenient first to fix such a cofinal subset of Π and record some properties.

LEMMA 5.4. *There is a subset Π_1 of Π which is well ordered by $<_\Pi$ and has the following features:*

(i) $(a_0, \varnothing, <_1, \mathbf{d}_1)$ *is in Π_1 for some $<_1, \mathbf{d}_1$, and is the first element of Π_1 (in fact $\mathbf{d}_1$ disappears here as L_1 is projective).*

(ii) $(\{a_0\})$ *is in Π_1 and is the greatest element of Π_1.*

(iii) If $p \in \Pi_1$ *is not the greatest element, then the successor p^+ of p in Π_1 is some immediate successor of p in Π.*

(iv) *If $p(1) <_\Pi p(2) <_\Pi \dots$ is a sequence of elements in Π_1 then the supremum of this sequence in Π_1 is the same as the supremum of the sequence in Π.*

(v) *Suppose $p \in \Pi_1$ is a limit element. Then p has the form*

$$(a_0, S_1, \dots, a_{k-1}, S_k, <_1, \dots, <_k, \mathbf{d}_1, \dots, \mathbf{d}_k, T_1, \dots, T_k),$$

where $S_k \neq \varnothing$, and $k + 1 \notin I_{\text{fin}}$. Moreover, there are $p(1), p(2), \dots$ in Π_1 such that p is the limit of the sequence of $p(j)$, and each $p(j)$ is of the form

$$(a_0, \dots, S'_k, a_k, X_j, <_1, \dots, <_{k+1}, \mathbf{d}_1, \dots, \mathbf{d}_k, \mathbf{d}_{k+1}, T_1, \dots, T_k, T_{k+1})$$

where $S_k = S'_k \cup \{a_k\}$, for each j, $X_{j+1} = X_j \cup \{b_j\}$ for some $b_j \notin X_j$, and $\cup\{X_j : j < \omega\} = L(a_k)$.

Proof. This is straightforward from the definitions of Π and $<_\Pi$, but the reader may wish to look at the proof of Lemma 4.4 for some clarification of (v)). ■

Let us fix in advance some ordering of N^* of order type ω. We will define inductively partial L^*-isomorphisms $h_p: U_p \to N^*$ for $p \in \Pi_1$ so as to satisfy:

(i) if $p <_\Pi p'$ then $h_{p'}$ extends h_p, and if p is the limit of $(p(j): j < \omega)$ then $h_p = \cup\{h_{p(j)} : j < \omega\}$.

(ii) $h_p(U_p)$ is in $\mathbf{W}_0(N^*)$, and $h_p(\text{base}(U_p)) = \text{base}(h_p(U_p))$.

The base stage

For $p(0)$ the minimal element of Π_1, $U_{p(0)} = \{a_0\}$, so define $h_{p(0)}(a_0) = a_0^{N^*}$.

The successor stage

Suppose h_p has been defined on $U = U_p$, and let p^+ be the successor of p in Π_1. We will define h_p^+. Suppose $\text{base}(U) = (a_0, S_1, \dots, a_{k-1}, S_k)$. *It is then clear that $U_p^+ = U \cup \{a_k\}$, where $a_k \in L(a_{k-1}) - S_k$.*

For x an object in domain(h_p), let $x^\sim$ denote the image of x under h_p.

Clearly $\mathrm{tp}(a_k/U)$ is algebraic or strongly minimal, and in the latter case already $\mathrm{tp}(a_k/a_{k-1})$ is strongly minimal.

Case I. $\mathrm{tp}(a_k/U)$ is algebraic.
By Lemma 4.7 there is a tuple $\mathbf{b}$ from U of length at most $\lambda - 1$ such that $\mathrm{tp}(a_k/\mathbf{b}) \vdash \mathrm{tp}(a_k/U)$. Let $\varphi(x, \mathbf{b})$ isolate $\mathrm{tp}(a_k/\mathbf{b})$ where $\varphi(x, \mathbf{y})$ may be chosen to be atomic. Let $\psi_0(\mathbf{y})$ be an atomic formula isolating $\mathrm{tp}(\mathbf{b})$ in M^*. Then for some $0 < m < \omega$, the sentence $(\forall \mathbf{y})(\psi_0(\mathbf{y}) \rightarrow (\exists_x^{=m})(\varphi(x, \mathbf{y})))$ is true in M^* so is an axiom of kind 2. Now as h_p is a partial isomorphism, $N^* \models \psi_0(\mathbf{b}^\sim)$. Thus we can find an element in N^* satisfying $\varphi(x, \mathbf{b}^\sim)$. We call this element $a_k^\sim$ and extend h_p to h_p^+ by defining $h_p^+(a_k) = a_k^\sim$.

We must show that h_{p^+} is a partial isomorphism. Let $\psi(x, \mathbf{z})$ be an atomic formula. (So $l(\mathbf{z}) < \lambda$.) Let $\mathbf{c}$ be a tuple in U of length that of $\mathbf{z}$. We have to show that $M^* \models \psi(a_k, \mathbf{c})$ iff $N^* \models \psi(a_k^\sim, \mathbf{c}^\sim)$.

By Lemma 4.8 again there is some good set U' contained in U and containing both $\mathbf{b}$ and $\mathbf{c}$, with base of cardinality at most μ. So $\varphi(x, \mathbf{b})$ isolates $\mathrm{tp}(a_k/U')$. We are now exactly in the first case considered in Remark 5.2, where U' and a_k here correspond to U and a there. Let $\mathbf{d}$ be a suitable enumeration of base(U'). Let the formulae $\eta(\mathbf{y}, \mathbf{w})$, $\varepsilon(\mathbf{z}, \mathbf{y}, \mathbf{w})$ be as in the sentence (II) in Remark 5.2 (which as remarked there is an axiom of kind 4). Now $M^* \models \eta(\mathbf{b}, \mathbf{d})$. As h_p is an isomorphism of U with its image $U^\sim$ in N^*, $U^\sim$ is a pregood set in N^* with base$(U^\sim) = h_p(\text{base}(U))$ and $\eta(\mathbf{y}, \mathbf{w})$ is an m–$\mathbf{w}$-restricted formula (for some $m < \mu$) it follows that $N^* \models \eta(\mathbf{b}^\sim, \mathbf{d}^\sim)$. As sentence (II) from 5.2 is true in N^*, and $N^* \models \varphi(a_k^\sim, \mathbf{b}^\sim)$ it follows that $N^* \models \psi(a_k^\sim, \mathbf{c}^\sim)$ iff $N^* \models \varepsilon(\mathbf{c}^\sim, \mathbf{b}^\sim, \mathbf{d}^\sim)$. Again by the fact that h_p is an isomorphism of U with $U^\sim$ and ε is an m–$\mathbf{w}$-restricted formula, $N^* \models \varepsilon(\mathbf{c}^\sim, \mathbf{b}^\sim, \mathbf{d}^\sim)$ iff $M^* \models \varepsilon(\mathbf{c}, \mathbf{b}, \mathbf{d})$. So clearly $M^* \models \psi(a_k, \mathbf{c})$ iff $N^* \models \psi(a_k^\sim, \mathbf{c}^\sim)$, as required.

Thus h_{p^+} is a partial isomorphism. The fact that $h_{p^+}(U_{p^+})$ is a pregood set in N^* with base the image of base (U_{p^+}) is immediate from our induction hypothesis.

Case II. $\mathrm{tp}(a_k/U)$ is not algebraic.
As already remarked $\mathrm{tp}(a_k/a_{k-1})$ is already strongly minimal. By Remark 4.15, $\mathrm{tp}(a_k/\text{base}(U)) \vdash \mathrm{tp}(a_k/U)$.

Moreover, $\mathrm{tp}(a_k/\text{base}(U))$ is determined by '$x \in L(a_{k-1})$'$\cup$'$x \in \mathrm{acl}(\text{base}(U)$'. By Lemma 4.7(ii), the formula '$x \notin \mathrm{acl}(\text{base}(U)$' is equivalent in M^* to $\Theta(x, \mathbf{e}^*) = \{\neg\theta(x, \mathbf{e})$: $\mathbf{e}$ a tuple from base(U) of cardinality $< \lambda$, and θ an atomic formula of L^* such that $\theta(x, \mathbf{e})$ is algebraic in $M^*\}$, where $\mathbf{e}^*$ enumerates base(U). By an axiom of kind 2, we have that $\theta(x, \mathbf{e}^\sim)$ has only finitely many solutions in N^*, for each of the formulae $\theta(x, \mathbf{e})$ above. But $L(a_{k-1}^\sim)$ is infinite (as clearly $k \notin I_{\mathrm{fin}}$); thus by axiom 5, there is an element in N^* satisfying '$x \in L(a_{k-1}^{\sim}) \wedge \bigwedge \Theta(x, (\mathbf{e}^*)^{\sim})$'. Choose $a_k^{\sim}$ (the value of $h_{p^+}(a_k)$) to be the *least* such element in the fixed ordering of N^*.

To show that h_{p^+} is a partial L^*-isomorphism, we must again show that for every *atomic* formula $\psi(x, \mathbf{z})$ and $\mathbf{c}$ in U, $M^* \models \psi(a_k, \mathbf{c})$ iff $N^* \models (a_k^{\sim}, \mathbf{c}^{\sim})$.

Fix such $\psi(x, \mathbf{z})$ and $\mathbf{c}$. So $\mathbf{c}$ has length at most $\lambda - 1$, whereby by Lemma 4.8 there is a good set $U' \subseteq U$ such that $\mathbf{c}$ is contained in U', base(U') has cardinality at most μ, and the bounded part of base (U') = the bounded part of base(U). In particular $a_{k-1} \in$ base(U'), so we conclude as above that '$x \in L(a_{k-1})$' $\cup \{\neg\theta(x, \mathbf{e})$: θ an atomic formula, $\mathbf{e}$ a tuple from base(U') of length $< \lambda$, $\theta(x, \mathbf{e})$ is algebraic$\} \vdash \mathrm{tp}(a_k/U')$ in M^*.

Let $\varphi(x, \mathbf{d})$ be the conjunction of the formulae in the left-hand side. So $\mathbf{d}$ is an enumeration of base(U'), and in M^*, $\varphi(x, \mathbf{d}) \vdash \mathrm{tp}(a_k/U')$.

Our assumptioms about U' imply that base (U') $\subseteq$ base(U). Thus

(*) $$N^* \models \varphi(a_k^{\sim}, \mathbf{d}^{\sim}).$$

We are now in the situation of the second case in Remark 5.2. As in Case I above, we conclude, using an axiom of kind 4, that $M^* \models \psi(a_k, \mathbf{c})$ iff $N^* \models \psi(a_k^{\sim}, \mathbf{c}^{\sim})$.

Thus h_{p^+} is a partial isomorphism, and again it is trivial to verify property (ii).

The limit stage

Let $p \in \Pi_1$ be a limit element. Define $h_p(U_p)$ to be $\cup \{h_{p(i)}(U_{p(i)})$: $i < \omega\}$ where $p(1) <_{\Pi} p(2) <_{\Pi} \cdots$ is any sequence in Π_1 with limit p. This is well defined by the induction hypothesis and Lemma 4.4. The main problem here is to show that (ii) holds. For this we make a more careful choice of the sequence p(i). In fact let us choose $p(1), p(2), \ldots$ as given by Lemma 5.4(v), and use the notation from there. Specifically base(p) = $(a_0, S_1, \ldots, S_k)$, base($p(j)$) = $(a_0, S_1, \ldots, S_k', a_k, X_j)$, where $S_k = S_k' \cup \{a_k\}$, $X_{j+1} = X_j \cup \{b_j\}$, and $\cup_j X_j = L(a_k)$. Note that $L(a_k)$ is infinite. Again we denote by $x^{\sim}$ the image in N^* under h_p of an object x in U_p.

Let $U = U_p$. We want to show that $h_p(U) = U^{\sim}$ is a pregood set in N^* with base $(a_0^{\sim}, S_1^{\sim}, \ldots, S_k^{\sim})$. By inductive hypothesis, for each $j < \omega$, $h_{p(j)}(U_{p(j)})$ is a pregood set in N^* with base $(a_0^{\sim}, S_1^{\sim}, \ldots, (S_k')^{\sim}, a_k^{\sim}, X_j^{\sim})$, so all that has to be shown is that

(**) $\quad U^{\sim} \supseteq L(a_k^{\sim})$ (and more specifically $\cup \{X_j^{\sim} : j < \omega\} = L(a_k^{\sim})$).

CLAIM **I.** *Let $d \in L(a_k^{\sim})$. Suppose there is an atomic formula $\varphi(x, \mathbf{z})$, $\mathbf{c}$ in U (of length $< \lambda$) such that in M^*, $\varphi(x, \mathbf{c})$ is algebraic with all solutions in $L(a_k)$, and $N^* \models \varphi(d, \mathbf{c}^{\sim})$. Then $d \in U^{\sim}$.*

Proof of Claim I. Let $\psi(\mathbf{z})$ be the quantifier-free type of $\mathbf{c}$ in M^*, and suppose $\varphi(x, \mathbf{c})$ has exactly m solutions in M^*. Thus the sentence $(\forall \mathbf{z})(\psi(\mathbf{z}) \rightarrow (\exists_x^{=m})(\varphi(x, \mathbf{z})))$ is an axiom of kind 2, so is true in N^*. As h_p is a partial isomorphism, $N^* \models \psi(\mathbf{c}^\sim)$, whereby $N^* \models (\exists_x^{=m})(\varphi(x, \mathbf{c}^\sim))$.

But $L(a_k) \subseteq U$, and h_p is a partial isomorphism, whereby clearly $d \in U^\sim$, as required. ■

Claim II. *Suppose that for some $j < \omega$, $L(a_k) \subseteq \mathrm{acl}(U_{p(j)})$. Then $L(a^\sim) \subseteq U_k^\sim$.*

Proof. Remember that base $(U_{p(j)}) = (a_0, S_1, \ldots, S'_k, a_k, X_j)$. Let $a \in L(a_k) - \mathrm{acl}(\text{base } (U_{p(j)}))$. By Lemma 4.7 there is some tuple $\mathbf{c}$ from $U_{p(j)}$ of cardinality $< \lambda$ such that $\mathrm{tp}(a/\mathbf{c})$ is algebraic and implies $\mathrm{tp}(a/U_{p(j)})$ in M^*. By Lemma 4.8, we can find a good set U' such that $\mathbf{c}$ is in U', $U' \subseteq U_{p(j)}$, $\mathrm{base}(U') \subseteq \mathrm{base}(U_{p(j)})$ and has cardinality $m \leq \mu$, and the bounded part of $\mathrm{base}(U')$ = the bounded part of $\mathrm{base}(U_{p(j)})$. Let $\mathbf{d}$ enumerate $\mathrm{base}(U')$. Let $\chi(x, \mathbf{w})$ be the conjunction of (a) the quantifier-free type of $\mathbf{d}$, (b) the formula $x \in L(a_{k-1})$, and (c) all formulae $\neg\delta(x, \mathbf{w}')$ where $\mathbf{w}'$ is a subtuple of $\mathbf{w}$ of length $< \lambda$ and $\delta(x, \mathbf{d}')$ is algebraic. By Lemma 4.9 together with Lemma 4.7(ii) we see that $\chi(x, \mathbf{w})$ isolates $\mathrm{tp}(a, \mathbf{d})$ in M^*. Let $\varphi(x, \mathbf{y})$ be the atomic type of $(a, \mathbf{c})$. Finally let $\theta(\mathbf{y}, \mathbf{w})$ be an m–$\mathbf{w}$-restricted formula in Φ which isolates $\mathrm{tp}(\mathbf{c}, \mathbf{d})$ in M^* (using 4.14 and 4.9). Then

$$M^* \models (\forall x\mathbf{w})(\chi(x, \mathbf{w}) \rightarrow (\exists \mathbf{y})(\varphi(x, \mathbf{y}) \wedge \theta(\mathbf{y}, \mathbf{w}))).$$

The above sentence is thus an axiom of type 3, so is true in N^*.

Suppose now that d is an element of $L(a_k^\sim)$ such that $N^* \models \chi(d, \mathbf{d}^\sim)$. Then $N^* \models \varphi(x, \mathbf{e}) \wedge \theta(\mathbf{e}, \mathbf{d}^\sim)$ for some $\mathbf{e}$ in N^*. We may assume that $\theta(\mathbf{y}, \mathbf{w})$ formally implies $\bigwedge\{\xi_m(y, \mathbf{w}) \colon y \in \mathbf{y}\}$. By the induction hypothesis $(U_{p(j)})^\sim$ is a pregood set in N^*, whose base is $\mathrm{base}(U_{p(j)})^\sim$, and, moreover, $h_{p(j)}$ is an isomorphism between $U_{p(j)}$ and $U_{p(j)}^\sim$. It is then easy to conclude that $\mathbf{e}$ is contained in $(U')^\sim$, and so is of the form $\mathbf{b}^\sim$ for some tuple $\mathbf{b}$ in U'. Again as $\theta(\mathbf{y}, \mathbf{w})$ is an m–$\mathbf{w}$-restricted formula, it follows that $M^* \models \theta(\mathbf{b}, \mathbf{d})$. By choice of $\theta(\mathbf{y}, \mathbf{w})$, $\mathrm{tp}(\mathbf{b}/\mathbf{d}) = \mathrm{tp}(\mathbf{c}/\mathbf{d})$. Thus $\varphi(x, \mathbf{b})$ is algebraic and all solutions of $\varphi(x, \mathbf{b})$ in M^* are in $L(a_k)$ (as the same is true for $\varphi(x, \mathbf{c})$). By Claim I, $d \in U^\sim$.

Suppose on the other hand that d is an $L(a_k^\sim)$, but does *not* satisfy $\chi(x, \mathbf{d}^\sim)$ in N^*. As $\mathbf{d}$ and $\mathbf{d}^\sim$ have the same quantifier-free type, and $d \in L(a_k^\sim)$, it must be that $N^* \models \delta(d, \mathbf{e}^\sim)$ for some atomic formula δ, and subtuple $\mathbf{e}$ of $\mathbf{d}$ such that $\delta(x, \mathbf{e})$ is algebraic. Again by Claim I, d must be in $U^\sim$.

We have shown that for all $d \in L(a_k^\sim)$, $d \in U^\sim$ holds, completing the proof of Claim II. ■

Thus (in order to prove $(**)$) we may assume: *for all $j < \omega$, $L(a_k)$ is not contained in* $\mathrm{acl}(U_{p(j)})$.

CLAIM **III.** *For all $j < \omega$, $L(a_k) \cap \mathrm{acl}(U_{p(j)}) = L(a_k) \cap \mathrm{acl}(\mathrm{base}(U_{p(j)})$.*

Proof. If not there is $a \in L(a_k) - \mathrm{acl}(\mathrm{base}(U_{p(j)}))$ and some tuple $\mathbf{c}$ from $U_{p(j)}$ such that $a \in \mathrm{acl}(\mathbf{c})$. But then (as $L(a_k)$ is strongly minimal and defined over $a_k \in \mathrm{base}(U_{p(j)})$, and $U_{p(j)}$ is definable over its base), for any $a' \in L(a_k) - \mathrm{acl}(\mathrm{base}(U_{p(j)}))$ there is some $\mathbf{c}'$ in $U_{p(j)}$ such that $a' \in \mathrm{acl}(\mathbf{c}')$. But then $L(a_k) \subseteq \mathrm{acl}(U_{p(j)})$, contradicting our assumption above. ■

Now suppose by way of contradiction that $L(a_k^\sim)$ is not contained in $U^\sim$, and let d be the least element (in the chosen ordering of N^*) which is in $L(a_k^\sim)$ but not in $U^\sim$. We can find $j < \omega$ such that all elements of $L(a_k^\sim)$ preceding d are in $X_j^\sim$. Now $\mathrm{base}(U_{p(j)}) = (a_0, S_1, \ldots, S_k', a_k, X_j)$, so we may also (using the infiniteness of $L(a_k)$) assume that $b_j \notin \mathrm{acl}(\mathrm{base}(U_{p(j)}))$. By Claim III, $b_j \notin \mathrm{acl}(U_{p(j)})$. Let $p(j)^+$ be the successor of $p(j)$ in Π_1. It is then clear that $U_{p(j)^+} = U_{p(j)} \cup \{b_j\}$. The definition of $h_{p(j)^+}$ in Case II of the successor stage dealt with above then calls for $h_{p(j)^+}(b_j)$ to be the least element in $L(a_k^\sim)$ satisfying $\neg\theta(x, \mathbf{e}^\sim)$, whenever $\theta(x, \mathbf{z})$ is a quantifier-free formula, $\mathbf{e}$ is from $\mathrm{base}(U_{p(j)})$, $\mathrm{length}(\mathbf{e}) < \lambda$, and $\theta(x, \mathbf{e})$ is algebraic. By Claim I and choice of d, d must be this element. So d is in $U^\sim$ after all. This contradiction proves that $L(a_k^\sim) \subseteq U^\sim$, and concludes the proof of the limit stage.

Thus the construction of the partial isomorphisms h_p can be carried out. When p is the maximum element $(\{a_0\})$ of Π_1, $h_p(U_p)$ must (by (ii)) be a pregood set in N^* with $\mathrm{base}(\{a_0^\sim\}) = (\{a_0^{N^*}\})$. Thus $h_p(U_p) = N^*$, and h_p yields the sought after isomorphism of M^* with N^*. This concludes the proof of Theorem 5.3. ■

COROLLARY **5.5.** *There are only countably many ω-categorical, ω-stable countable structures.*

Proof. Here we consider two countable structures M, N to be the same if $\mathrm{Aut}(M)$ and $\mathrm{Aut}(N)$ are isomorphic as permutation groups. The corollary follows then immediately from Theorem 5.3. ■

4

1-based theories and groups

The notion '1-based' turned up naturally in Chapter 2, as, for example, a way of describing the global behaviour of forking in a superstable theory of finite U-rank all of whose minimal types are locally modular. In this chapter we study arbitrary (stable) theories which are 1-based, and also (∞)-definable sets which are 1-based, without any assumption of finite U-rank, or even superstability. In section 1 we give various equivalences of the 1-based property. In sections 2 and 3 we show that under various additional assumptions (triviality, NDOP, few types) 1-basedness of a theory implies that the theory is 'superstable-like' in the sense that types have finite weight, and enough regular types exist. In section 4, we begin to study definable groups in 1-based theories, or even ∞-definable groups which are 1-based. These turn out to be essentially 'abelian structures', that is abelian groups equipped with predicates for subgroups. It will also be seen (in Chapter 5) that any 'non-triviality' of forking in a stable 1-based theory immediately gives rise to definable groups, so this analysis is quite pertinent. In section 5, we characterize the geometries attached to (locally) modular minimal groups.

Throughout this chapter T will be a complete stable theory, and $\mathbf{C}$ the monster model of T. As usual we work in $\mathbf{C}^{eq}$, unless otherwise stated.

1 Weakly normal formulae and pseudoplanes

DEFINITION **1.1.** *A definable set X is said to be* weakly normal *if for any $a \in X$, only finitely many distinct images of X under automorphisms of $\mathbf{C}$ contain a. We will call a formula $\varphi(x, b)$ weakly normal if the set it defines is weakly normal.*

REMARK **1.2.** *Let X be a definable set, with canonical parameter c. Then X is weakly normal iff* $c \in \operatorname{acl}(a)$, for any $a \in X$.

Example 1.3. Let G be a $\varnothing$-definable group, H an acl($\varnothing$)-definable subgroups of G, and X some left coset $a.H.$ of H in G. Then X is weakly normal.

In this first section we will be making use of 'local' stability theory, as expounded in Chapter 1.

The following observation seems to be (and is) a triviality, but is worth stating:

Remark 1.4. *The following are equivalent*:

(i) *T is* 1-*based*;

(ii) *for any L-formula $\delta(x, y)$, and any complete stationary δ-type $p(x)$, if a realizes p, then* $\mathrm{Cb}(p) \in \mathrm{acl}(a)$.

Proof. All one has to see is that for $q(x)$, a complete stationary type, $\mathrm{Cb}(q)$ is precisely $\{\mathrm{Cb}(q \restriction \delta)\colon \delta(x, y) \in L\}$. ∎

Proposition 1.5. *T is* 1-*based iff every definable set (in $\mathbf{C}^{\mathrm{eq}}$) is a (finite) Boolean combination of weakly normal definable sets.*

Proof. Assume the right-hand side.

First note

(1) if M is an ω-saturated model, then any M-definable set is a Boolean combination of weakly normal M-definable sets.

It follows that

(2) if $p(x), q(x) \in S(M)$ contain the same weakly normal formulae, then $p = q$.

Let now M be a reasonable saturated model, and $p(x) = \mathrm{tp}(a/M)$. Let $C = \mathrm{acl}(a) \cap M$ namely $\mathrm{acl}(a) \cap M^{\mathrm{eq}}$).

Claim 3. *Any weakly normal M-definable set in p is C-definable.*

Proof. Let X be such. Let c be the canonical parameter of X. As X is M-definable, $c \in M$. On the other hand, by 1.2, $c \in \mathrm{acl}(a)$, so $c \in C$.

Claim 4. *If f is an automorphism of M which fixes C pointwise, then* $f(p) = p$.

Proof. Suppose $f(p) = q \in S(M)$. As f fixes C pointwise, and f acts on the family of weakly normal formulae over M, it follows that q and p contain exactly the same weakly formulae. By (2), $p = q$.

As M is saturated, it follows that $\mathrm{Cb}(p) \subseteq C$. Thus $\mathrm{Cb}(p) \subseteq \mathrm{acl}(a)$. We have shown that T is 1-based.

Conversely, assume T is 1-based.

We are going to show that (2) above holds (for M saturated). We will make use of the local ranks $R_\Delta(\text{-})$ defined in section 3 of Chapter 1. If $\Delta = \{\delta(x, y)\}$ we write R_δ in place of R_Δ.

First:

LEMMA. *Let $p(x) \in S(A)$ be a stationary type, where $A = \mathrm{acl}(A)$. Let $\delta(x, y)$ be any L-formula. Then there is some weakly normal formula $\psi(x, c) \in p \restriction \delta$ such that $R_\delta(p \restriction \delta) = R_\delta(\psi(x, c))$ and* $\mathrm{mult}_\delta(\psi(x, c)) = 1$.

Proof. By 1.3.7, as $p \restriction \delta$ is stationary, $\mathrm{mult}_\delta(p \restriction \delta) = 1$.

Let $c = \mathrm{Cb}(p \restriction \delta)$ (= the code of a δ-definition for p). Then $p \restriction \delta$ does not fork over c, and the restriction $r(x, c)$ of $p \restriction \delta$ to c is a complete stationary δ-type over c, whose canonical base is c. (See Remark 1.2.10 for canonical bases of δ-types etc.) By 1.3.4, $R_\delta(p \restriction \delta) = R_\delta(r(x, c))$ and $\mathrm{mult}_\delta(r(x, c)) = 1$.

Now, by Remark 1.4, for any b realizing $r(x, c)$, $c \in \mathrm{acl}(b)$. A simple compactness argument, which is left to the reader, shows that there is a formula $\psi(x, c) \in r(x, c)$ such that for any $\psi'(x, c) \in r(x, c)$ which implies $\psi(x, c)$ we have

$$(*) \qquad \text{for any } b \text{ satisfying } \psi'(x, c),\ c \in \mathrm{acl}(b).$$

So we can choose such $\psi(x, c)$ with $\mathrm{mult}_\delta(\psi(x, c)) = 1$, and $R_\delta(\psi(x, c)) = R_\delta(r(x, c))$ $(= R_\Delta(p \restriction \delta))$.

By $(*)$, $\psi(x, c)$ is weakly normal, so is the required formula. ■

Now suppose $p(x)$, $q(x)$ are complete types over a saturated model M, which contain exactly the same weakly normal formulae. Fix $\delta(x, y) \in L$. Suppose $n = R_\delta(p)$. Let $\psi(x, c) \in p(x)$ be as given by the lemma. So $\psi(x, c) \in q$, whereby $R_\delta(q) \leq n$. By symmetry, $R_\delta(q) = R_\delta(p)$. We claim that $p \restriction \delta = q \restriction \delta$. If not there is $\chi(x) \in p \restriction \delta$, $\neg\chi(x) \in q \restriction \delta$. But then (as $\mathrm{mult}_\delta(\psi(x, c)) = 1$, and $R_\delta(\psi)) = n$, at least one of $\psi(x, c) \wedge \chi(x)$ (which is in $p \restriction \delta$) and $\psi(x, c) \wedge \neg\chi(x)$ (which is in $q \restriction \delta$) has R_δ-rank $< n$, a contradiction. So $p \restriction \delta = q \restriction \delta$. As δ was arbitrary, this shows that $p(x) = q(x)$. We have shown (2) above. Saturation of M, together with a simple compactness argument left to the reader, yields that *every* formula over M is equivalent to some finite Boolean combination of weakly normal formulae over M. This is enough to prove the right-hand side in the proposition (as M is ω-saturated).

Another way of viewing 1-basedness is to think of it as excluding the interpretation of certain reasonably complicated combinatorial objects. Such considerations played an important role in the work of Zilber and also the work of Lachlan. The notion of a quasidesign was introduced by Zilber, and that of a pseudoplane by Lachlan.

DEFINITION 1.6. *By a* complete-type-definable quasidesign (*over* $\emptyset$) *we mean the set of realizations of a complete type* $r(x, y) = \mathrm{tp}(b, c/\emptyset)$, *such that* (i)–(iii) *below hold.*

(i) $c \notin \mathrm{acl}(b)$.

(ii) $b \notin \mathrm{acl}(c)$.

(iii) *For any* $c \neq d$, $\{a\colon r(a, c) \wedge r(a, d)\}$ *is finite.*

If we also *have the dual of* (iii), *that is*

(iv) *for any* $b_1 \neq b_2$, $\{c\colon r(b_1, c) \wedge r(b_2, c)\}$ *is finite,*

then we call the quasidesign a complete-type-definable pseudoplane (over $\emptyset$).

PROPOSITION 1.7. *The following are equivalent*:

(i) *T is* 1-*based*;

(ii) *T has no complete-type-definable quasidesign*;

(iii) *T has no complete-type-definable pseudoplane.*

Proof. We first point out the equivalence of (ii) with (iii). (iii) trivially implies (ii). Suppose on the other hand that $r(x, y) = \mathrm{tp}(b, c/\emptyset)$ gives a quasidesign.

CLAIM. *Whenever* $B = \{b_1, b_2, \ldots\}$ *is an infinite set of elements such that* $\models r(b_i, c)$ *for all* $i < \omega$, *then* $c \in \mathrm{dcl}(B)$.

Proof of claim. Otherwise we have $c \neq c'$ such that $\models r(b_i, c')$ for all $i < \omega$. This contradicts (iii) in Definition 1.6. ∎

By the claim, there is, by compactness, some greatest n, such that there exist distinct $b_1, \ldots, b_n$ such that $\models r(b_i, c)$ for all $i = 1, \ldots, n$, and such that $c \notin \mathrm{acl}(b_1, \ldots, b_n)$. Let $b_1, \ldots, b_n$, be such elements. Let $d = \{b_1, \ldots, b_n\}$ (as a set, so an element of $\mathbf{C}^{\mathrm{eq}}$). Then it is easily checked that $r(x, w) = \mathrm{tp}(d, c/\emptyset)$ yields a complete-type-definable pseudoplane.

(i) → (ii). Suppose $r(x, y) = \mathrm{tp}(b, c/\emptyset)$ defines a quasidesign. We claim that $c \in \mathrm{Cb}(\mathrm{stp}(b/c))$. For note that $\mathrm{stp}(b/c)$ is not algebraic. But if $\mathrm{tp}(c') = \mathrm{tp}(c)$ and $c' \neq c$ then $r(x, c)$ and $r(x, c')$ have only finitely many realizations in common, so could not have a common non-forking extension. So $c \in \mathrm{Cb}(\mathrm{stp}(b/c))$ as claimed. But $c \notin \mathrm{acl}(b)$, contradicting 1-basedness.

(ii) → (i). Suppose T is not 1-based. Thus, by 1.4, there are a, M, and some L-formula $\delta(x, y)$, such that if c is the canonical base (i.e. δ-definition) of $\mathrm{tp}(a/M) \upharpoonright \delta$, then $c \notin \mathrm{acl}(a)$. Note that $R_\delta(\mathrm{tp}(a/M) \upharpoonright \delta)$ is then >0 (for otherwise this type is isolated by some δ-formula $\varphi(x)$, and as M is a model,

$\varphi(x)$ isolates a δ-type over $\mathbf{C}$; in particular $\varphi(x)$ is $\{a\}$-definable, whereby clearly $c \in \operatorname{dcl}(a)$). With $\delta(x, y)$ fixed, choose such a and M such that $n = R_\delta(\operatorname{tp}(a/M) \upharpoonright \delta)$ is least possible. There is clearly some $N \supseteq M$ such that $R_\delta(\operatorname{tp}(a/N) \upharpoonright \delta) < n$ (e.g. let N be a model containing $M \cup a$, and then as above $R_\delta(\operatorname{tp}(a/N) \upharpoonright \delta) = 0$). Choose $N \supseteq M$, and $m < n$ such that $R_\delta(\operatorname{tp}(a/N) \upharpoonright \delta) = m$, and m is greatest possible. We may suppose N is saturated. Let $p(x) = \operatorname{tp}(a/M) \upharpoonright \delta$, and $q(x) = \operatorname{tp}(a/N) \upharpoonright \delta$.

Let $b = \operatorname{Cb}(q)$. We will show that $r(w, z) = \operatorname{tp}(b, c/\emptyset)$ defines a quasi-design.

CLAIM **a.** $c \notin \operatorname{acl}(b)$.

Proof. $c \notin \operatorname{acl}(a)$, by choice of a and M. But $b \in \operatorname{acl}(a)$, as $m < n$ and n was minimized. ∎

CLAIM **b.** $b \notin \operatorname{acl}(c)$.

Proof. As $R_\delta(q) < R_\delta(p)$ q forks over M, so the δ-definition of q is not over M, in particular not in $\operatorname{acl}(c)$. ∎

CLAIM **c.** *If $c' \neq c$, and $r(b, c) \wedge r(b, c')$ then $b \in \operatorname{acl}(c, c')$.*

Proof. As N is saturated and contains b and c we may suppose $c' \in N$ too. Let $p_c(x)$ denote $p(x) \upharpoonright c$ (= the complete δ-type of a over c). Then as $c = \operatorname{Cb}(p_c(x))$ and $c' \neq c$, $p_{c'}(x)$ is not parallel to $p_c(x)$, that is they have no common non-forking extension. Properties of R_δ imply that $R_\delta(p_c(x) \cup p_{c'}(x)) < n$. As $b = \operatorname{Cb}(q(x))$, $q(x) = \operatorname{tp}(a/N) \upharpoonright \delta$, and $c, c' \in N$ have the same type over b it follows that a realizes $p_c(x) \cup p_{c'}(x)$. Thus, if $t(x) =$ the complete δ-type of a over cc' we have

$$(*) \qquad R_\delta(t(x)) < n.$$

If by way of contradiction there are distinct b_i for $i < \omega$, in N, such that $\operatorname{tp}(b_i/cc') = \operatorname{tp}(b/cc')$ for all i, then letting $q_i(x)$ be the conjugate of $q(x)$ under some cc' automorphism of N which takes b to b_i, we see that

$$(**) \qquad R_\delta(q_i) = m \text{ for all } i < \omega.$$

We also clearly have

$(***)$ for each $i < \omega$, q_i is a complete δ-type extending $t(x)$.

From $(**)$ and $(***)$ and the basic properties of the rank R_δ from section 3 of Chapter 1, we see that $R_\delta(t(x)) > m$. This together with $(*)$ contradicts the maximality in the choice of m. Thus Claim (c) is proved. ∎

By Claims (a), (b), and (c), $r(w, z)$ defines a quasidesign. This completes the proof of Proposition 1.7. ■

REMARK 1.8. (i) *Note that T is 1-based iff some (any) expansion of T by naming elements (adding constants) is 1-based. For suppose T is not 1-based, witnessed by:* $\mathrm{Cb}(\mathrm{stp}(a/B))$ *is not contained in* $\mathrm{acl}(a)$. *Let A be any small set of elements in $\mathbf{C}^{\mathrm{eq}}$. Let (a_1, B_1) realize a non-forking extension of* $\mathrm{tp}(a, B/\emptyset)$ *over A. Then (by automorphism), $C = \mathrm{Cb}(\mathrm{stp}(a_1/B_1))$ is not contained in* $\mathrm{acl}(a_1)$. *Now* $\mathrm{stp}(a_1/B_1A)$ *does not fork over B_1. Thus $C = \mathrm{Cb}(\mathrm{stp}(a_1/B_1A))$. $C \subseteq \mathrm{acl}(B_1)$. So (a_1, C) is independent from A over $\emptyset$. So $\mathrm{tp}(C/a_1A)$ does not fork over a_1. Thus C is not contained in* $\mathrm{acl}(a_1A)$ *(for then $C \subseteq \mathrm{acl}(a_1)$). This shows that naming elements of A preserves non-1-basedness.*

(ii) *At the end of section 5 of Chapter 2, we called an ∞-definable (over A) set X 1-based, if for $a \in (X_A)^{\mathrm{eq}}$, and any $B \supseteq A$,* $\mathrm{Cb}(\mathrm{stp}(a/B)) \subseteq \mathrm{acl}(aA)$.
As in (i), *this propery depends on X (not on A).*

(iii) *Proposition* 1.5, *is, in principle, an aid to determining if a given theory T is 1-based: if we understand reasonably the definable sets in models of T (via some sort of quantifier elimination), then we should be able to see if they are Boolean combinations of weakly normal sets. The problem, however, is that this must be done in $\mathbf{C}^{\mathrm{eq}}$. The next result (Lemma 1.9) shows that it is sufficient to look at definable subsets of $\mathbf{C}^n$ (as opposed to all of $\mathbf{C}^{\mathrm{eq}}$).*

(iv) *Note that Proposition 1.5 goes through 'sort by sort'. That is given sort S of T^{eq} the following are equivalent: (a) for any a of sort S, and any B,* $\mathrm{Cb}(\mathrm{stp}(a/B)) \subseteq \mathrm{acl}(a)$; *(b) any definable set of elements of sort S is a Boolean combination of weakly normal definable sets (of elements of sort S).*

LEMMA 1.9. *Suppose that T is a 1-sorted theory, and for any $\mathbf{a} \in \mathbf{C}^n$, and any B,* $\mathrm{Cb}(\mathrm{stp}(\mathbf{a}/B)) \subseteq \mathrm{acl}(a)$. *Then T is 1-based (i.e. this also holds for all $a \in \mathbf{C}^{\mathrm{eq}}$).*

Proof. Let S be some sort of T^{eq}. So $S = S_E$ where E is a $\emptyset$-definable equivalence relation on $\mathbf{C}^n$, say. Let b be an element of sort S. Let M be a model. We must show that $\mathrm{tp}(b/M)$ does not fork over $\mathrm{acl}(b) \cap M^{(\mathrm{eq})}$. Let $\mathbf{a} \in \mathbf{C}^n$ be such that $\mathbf{a}/E = b$ and $\mathrm{tp}(\mathbf{a}/Mb)$ does not fork over b. So $M \cap \mathrm{acl}(\mathbf{a}b) \subseteq \mathrm{acl}(b)$. Thus

$$(*) \qquad \mathrm{acl}(\mathbf{a}) \cap M^{\mathrm{eq}} = \mathrm{acl}(b) \cap M^{\mathrm{eq}} = A, \text{ say.}$$

(As $\mathrm{acl}(\mathbf{a}) \cap M \subseteq \mathrm{Cb}(\mathrm{tp}(\mathbf{a}/M)) \subseteq \mathrm{acl}(b)$.) By hypothesis, $\mathrm{tp}(\mathbf{a}/M)$ does not fork over A. As $b \in \mathrm{dcl}(\mathbf{a})$, also $\mathrm{tp}(\mathbf{b}/M)$ does not fork over A. We finish by $(*)$. ■

Example 1.10. Let M be the structure consisting of an abelian group $(A, +)$

together with predicates for a certain family Ω of subgroups of A^n (n varying). We may assume the family Ω is closed under projections, finite intersections, finite cartesian products, and also contains A^n as well as $\{(0, \ldots, 0)\}$ (n times) for all n. A generalization of the proof of '*pp*-elimination' in modules shows that the theory of M has quantifier elimination (in the language containing $+$ and predicates for the groups in Ω). We may assume M to be already saturated. Then every definable subset of M^n is a Boolean combination of cosets of $\emptyset$-definable subgroups. An n-type p over a model M is then determined by the information saying which (if any) coset of any $\emptyset$-definable subgroup of M^n is in p. Thus $Th(M)$ is stable (either by seeing that all types over models are definable, or by counting types). As the latter are weakly normal, using 1.8(iv), and 1.9, we conclude that $Th(M)$ is 1-based. $Th(M)$ is superstable iff there is no infinite descending chain B_i of ($\emptyset$)-definable subgroups of A such that B_{i+1} has infinite index in B_i. $Th(M)$ is totally transcendental if there is no infinite descending chain of ($\emptyset$)-definable subgroups of A. These last observations are left to the reader.

2 1-based trivial theories

1-based trivial theories are those which are in a sense as geometrically degenerate as possible. They include theories of finite U-rank, all minimal types of which are trivial. However, there *are* interesting stable, unsuperstable examples. We will see that assuming in addition NDOP (a fundamental 'non-structure' property, discovered by Shelah), such theories are 'superstable-like'.

Definition 2.1. (i) *T is said to be* trivial *if whenever a, b, c are elements (of $\mathbf{C}^{eq}$), A is a set, and $\{a, b, c\}$ is pairwise independent over A; then $\{a, b, c\}$ is A-independent.*

(ii) *The type $p(x) \in S(A)$ is* trivial *if whenever I is a set of realizations of p which is pairwise A-independent, then I is A-independent.*

Remark 2.2. *The definition of an arbitrary type being trivial is consistent with the case of minimal types (or their pregeometries) treated in Chapter 2. Also we leave it to the reader to check (by a padding argument) that T is trivial iff all types are trivial.*

Lemma 2.3. *Suppose T is superstable of finite U-rank. Then T is trivial if and only if* all *minimal types are trivial.*

Proof. The proof is similar to that of Proposition 2.5.8 so we are brief.

Suppose T is not trivial, and that this is witnessed by a, b, c, A. (So $\{a, b, c\}$ is pairwise A-independent, but A-dependent.) By taking non-forking extensions we may assume that A is an a-model, M say. Using finiteness of U-rank and Lemma 2.5.1, we can find tuples $\mathbf{a}$, $\mathbf{b}$, $\mathbf{c}$ such that each of $\mathbf{a}$, $\mathbf{b}$, $\mathbf{c}$ is a tuple of elements realizing minimal types over M, $\mathbf{a}$ is domination equivalent to a over M, and similarly for $\mathbf{b}$ and b, $\mathbf{c}$ and c. We may also assume that any two minimal types over M arising this way are the same or orthogonal. Then $\{\mathbf{a}, \mathbf{b}, \mathbf{c}\}$ is pairwise M-independent but M-dependent. By the above remarks, we may assume that each element of $\mathbf{a}$, $\mathbf{b}$, and $\mathbf{c}$ realises (over M) the same minimal type q, It is easy to see that we contradict triviality of q. ■

REMARK 2.4. *Similarly one can show that a superstable theory is trivial iff all regular types are trivial.*

Somewhat surprisingly Lemma 2.3 holds for 1-based theories without the finite rank hypothesis.

LEMMA 2.5. *Suppose T is* 1-*based. Then T is trivial of and only if all minimal types are trivial.*

Proof. We only have to prove the right to left direction.

Suppose T is not trivial. Let a, b, c, A witness this.

CLAIM. *We may suppose that $a \in \mathrm{acl}(bcA)$, $b \in \mathrm{acl}(acA)$ and $c \in \mathrm{acl}(abA)$ (and of course $\{a, b, c\}$ is still pairwise A-independent).*

Proof of claim. Let $a_1 \in \mathrm{Cb}(\mathrm{stp}(a/bcA)) - \mathrm{acl}(A)$ (such a_1 can be found as a forks with bc over A). Then $a_1 \in \mathrm{acl}(bcA) \cap \mathrm{acl}(a)$. Thus $\{a_1, b, c\}$ is still pairwise A-independent. Let $b_1 \in \mathrm{Cb}(\mathrm{stp}(b/a_1, c, A)) - \mathrm{acl}(A)$. Then $b_1 \in \mathrm{acl}(b) \cap \mathrm{acl}(a_1, c, A)$, and $\{a_1, b_1, c\}$ is pairwise A-independent but A-dependent. Finally let $c_1 \in \mathrm{Cb}(\mathrm{stp}(c/a_1 b_1 A)) - \mathrm{acl}(A)$, and a_1, b_1, c_1 are as required. ■

Now any non-algebraic type has an extension of U-rank 1 (take a 'maximal' non-algebraic extension). In particular, there is A_0 such that $U(\mathrm{tp}(a/AA_0)) = 1$. We may assume that $A_0 = \mathrm{Cb}(\mathrm{stp}(a/A_0 A))$, so $A_0 \subseteq \mathrm{acl}(a)$. Note that each of b, c is independent from aA_0 over A. So

(*) a is independent from each of b, c over AA_0 (and $b, c \notin \mathrm{acl}(AA_0)$).

Note also that

(**) $$b \notin \mathrm{acl}(AA_0 c),\ c \notin \mathrm{acl}(AA_0 b)$$

(for otherwise, we would have for example $a \in \mathrm{acl}(bAA_0)$, contradicting (*)).

Let $D_0 = \mathrm{acl}(AA_0b) \cap \mathrm{acl}(AA_0c)$. By (∗∗) and 1-basedness, b is independent from c over D_0, and each of $b, c \notin \mathrm{acl}(D_0)$.

By (∗) and choice of D_0, we also have that a is independent from each of bD_0, cD_0 over AA_0. So now we have

(∗∗∗) $\{a, b, c\}$ is pairwise D_0-independent, $a, b, c \notin \mathrm{acl}(D_0)$, each of a, b, c is algebraic over D_0 together with the other two, *and* $U(a/D_0) = 1$.

As $b \in \mathrm{acl}(acD_0)$, $U(b/cD_0) \leq 1$. But b is independent from c over D_0, so $U(b/D_0) = 1$. Similarly $U(c/D_0) = 1$. Now the types $\mathrm{stp}(a/D_0)$, $\mathrm{stp}(b/D_0)$, $\mathrm{stp}(c/D_0)$ are clearly non-orthogonal. Let M be a saturated model containing D_0 which is independent from $\{a, b, c\}$ over D_0. We can find b' interalgebraic with b over M, and c' interalgebraic with c over M such that $\mathrm{tp}(a/M) = \mathrm{tp}(b'/M) = \mathrm{tp}(c'/M)$. Thus this minimal type is non-trivial. We have found a non-trivial minimal type. ■

In 1-based theories, *weight* behaves in a nice way.

Lemma 2.6. *Suppose T is* 1-*based. Suppose* $\mathrm{wt}(\mathrm{tp}(a/A)) \geq n$. *Then there are* $c_1, \ldots, c_n \in \mathrm{acl}(aA) - \mathrm{acl}(A)$ *such that* $\{c_1, \ldots, c_n\}$ *is A-independent.*

Proof. As $\mathrm{wt}(\mathrm{tp}(a/A)) \geq n$, there is some $B \supseteq A$ and there are $d_1, \ldots, d_n$ such that $\mathrm{tp}(a/B)$ does not fork over A, $\{d_1, \ldots, d_n\}$ is B-independent and a forks with d_i over B for each i. Let $C_i = \mathrm{Cb}(\mathrm{stp}(a/Bd_i))$. As $\mathrm{tp}(a/Bd_i)$ forks over B, we can find $c_i \in C_i - \mathrm{acl}(B)$. Now $\{c_1, \ldots, c_n\}$ is clearly B-independent. But also $c_1, \ldots, c_n \in \mathrm{acl}(a)$ (by 1-basedness), and a is independent with B over A. Thus $\mathrm{tp}(c_1, \ldots, c_n/B)$ does not fork over A. It follows that $\{c_1, \ldots, c_n\}$ is also A-independent. This completes the proof. ■

In the 1-based context triviality has strong consequences:

Lemma 2.7. *Suppose T is* 1-*based and trivial. Let* a, $B_1 \subseteq B$, $C_1 \subseteq C$, *be such that* $\mathrm{tp}(a/B)$ *does not fork over* B_1 *and* $\mathrm{tp}(a/C)$ *does not fork over* C_1. *Then* $\mathrm{tp}(a/B \cup C)$ *does not fork over* $B_1 \cup C_1$.

Proof. We may suppose B, C to be algebraically closed, and $B_1 = \mathrm{Cb}(\mathrm{tp}(a/B))$, $C_1 = \mathrm{Cb}(\mathrm{tp}(a/C))$.

By 1-basedness, $B_1, C_1 \subseteq \mathrm{acl}(a)$. $\mathrm{tp}(B/\mathrm{acl}(a))$ does not fork over B_1, so does not fork over B_1C_1. Similarly $\mathrm{tp}(C/\mathrm{acl}(a))$ does not fork over B_1C_1. Thus

(∗) a is independent from each of B and C over B_1C_1.

Let $E = \mathrm{acl}(BB_1C_1) \cap \mathrm{acl}(CB_1C_1)$. Note that $E \supseteq B_1C_1$. By 1-basedness B is independent from C over E. By (∗) a is independent from each of B, C over E.

Thus $\{a, B, C\}$ is pairwise E-independent. By triviality a is independent from BC over E. As $E \subseteq \mathrm{acl}(BB_1C_1)$, by (*) a is independent from E over B_1C_1. Altogether a is independent from BC over B_1C_1 as required. ■

Note the following special case of 2.7:

COROLLARY **2.8.** *(T 1-based and trivial) Suppose $a \in \mathrm{acl}(b_1, \ldots, b_n)$. Then there are $c_1 \in \mathrm{acl}(a) \cap \mathrm{acl}(b_1), \ldots, c_n \in \mathrm{acl}(a) \cap \mathrm{acl}(b_n)$, such that $a \in \mathrm{acl}(c_1, \ldots, c_n)$.*

Proof. It is enough to prove it for $n = 2$. As $a \in \mathrm{acl}(b_1, b_2)$, $a = \mathrm{Cb}(\mathrm{stp}(a/b_1b_2))$. On the other hand $\mathrm{Cb}(\mathrm{stp}(a/b_i)) \in \mathrm{acl}(a) \cap \mathrm{acl}(b_i)$ (by 1-basedness). Now apply Lemma 2.7. ■

REMARK **2.9.** *If T satisfies the* conclusion *of 2.7, T is called (by Poizat)* perfectly trivial. *We will see an example in section 6 of a superstable perfectly trivial theory which is not 1-based. One can show that 2.8 also holds for perfectly trivial theories, if $\{b_1, \ldots, b_n\}$ is $\varnothing$-independent.*

Shelah introduced a property, the DOP (Dimensional Order Property), which is a kind of 'infinitary' instability property, and implies that T has the maximum number of models in cardinalities $> |T|$. The DOP was mainly used as a dividing line within superstable theories. However, the DOP is also meaningful for stable non-superstable theories, and its presence implies the maximum number of $|T|^+$-saturated models of suitable cardinalities. The DOP will also turn up in Chapter 8, where we give an equivalent definition. NDOP means not DOP.

DEFINITION **2.10.** (i) *By a V* of sets *we mean a triple $\mathbf{B} = (B_0, B_1, B_2)$, such that $B_0 \subseteq B_1$, $B_0 \subseteq B_2$, and B_1 is independent from B_2 over B_0.*

(ii) *T is said to have the DOP if there are some V of sets $\mathbf{B} = (B_0, B_1, B_2)$, there is a set $D \supseteq B_1 \cup B_2$ which is dominated by $B_1 \cup B_2$ over each of B_1, B_2, and there is some stationary (non-algebraic) type $p \in S(D)$ such that p is orthogonal to each of B_1, B_2.*

LEMMA **2.11.** *Suppose T is 1-based, trivial, and with NDOP. Let $A \subseteq B$ be algebraically closed sets. Let B_1 and B_2 be algebraically closed subsets of B such that (A, B_1, B_2) is a V of sets, and B_1, B_2 are maximal such (i.e. neither can be properly extended). Then $B = acl(B_1 \cup B_2)$.*

Proof. We begin with:

CLAIM. *$B_1 \cup B_2$ dominates B over each of B_1, B_2.*

Proof. Suppose X is independent from $B_1 \cup B_2$ over B_1. We must show that X is independent from B over B_1. If not, then by 1-basedness, there is $c \in \operatorname{acl}(X) \cap B$, $c \notin B_1$. So (by choice of X), $\operatorname{tp}(c/B_1 \cup B_2)$ does not fork over B_1. Thus $\operatorname{tp}(B_2/B_1 \cup c)$ does not fork over B_1, and thus does not fork over A. But then (A, B_1c, B_2) is a V of sets contained in B, contradicting maximality of (A, B_1, B_2). We have shown that $B_1 \cup B_2$ dominates B over B_1. Similarly $B_1 \cup B_2$ dominates B over B_2. The claim is proved. ■

Now suppose by way of contradiction that there is $b \in B - \operatorname{acl}(B_1 \cup B_2)$. By NDOP, $\operatorname{stp}(b/B_1 \cup B_2)$ is non-orthogonal to B_1 or to B_2. Without loss it is non-orthogonal to B_1. Let then $C \supseteq B_1 \cup B_2$ and c be such that b is independent from C over $B_1 \cup B_2$, c is independent from C over B_1, and b forks with c over C. Working over $B_1 \cup B_2$, and applying Lemma 2.7, we conclude that b forks with c over $B_1 \cup B_2$. As c is independent from $B_1 \cup B_2$ over B_1, this contradicts the claim. We have shown that $\operatorname{acl}(B_1 \cup B_2) = B$. ■

LEMMA **2.12.** (*T 1-based, trivial with NDOP*). *For any a, A, $\operatorname{acl}(aA)$ does not contain an infinite A-independent set.*

Proof. First, a downward Lowenheim–Skolem argument, which is left to the reader, allows us to assume that T and A are countable. We may now name A, that is assume that $A = \emptyset$. Assume by way of contradiction that $\operatorname{acl}(a)$ contains an infinite $\emptyset$-independent set $\{c_i : i < \omega\}$. For each $X \subseteq \omega$ we can choose algebraically closed subsets A_X, B_X of $\operatorname{acl}(a)$ such that: A_X contains $\{c_i : i \in X\}$, B_X contains $\{c_i : i \notin X\}$, A_X, $B_X \subseteq \operatorname{acl}(a)$, A_X is independent from B_X over $\emptyset$ and A_X, B_X are maximal such. By Lemma 2.11, for each $X \subseteq \omega$, $a \in \operatorname{acl}(A_X \cup B_X)$. Thus there are $a_X \in A_X$, and $b_X \in B_X$ such that $a \in \operatorname{acl}(a_X, b_X)$.

CLAIM. *For $X \neq Y$, $(a_X, b_X) \neq (a_Y, b_Y)$.*

Proof of claim. Without loss of generality, $Y - X \neq \emptyset$. *Let $n \in Y - X$.* If by way of contradiction $(a_X, b_X) = (a_Y, b_Y)$ then (as $c_n \in \operatorname{acl}(a)$), we have $c_n \in \operatorname{acl}(a_X, b_Y)$. However, $c_n \in B_X$ and is therefore independent from a_X over $\emptyset$. Also $c_n \in A_Y$ so is independent from b_Y over $\emptyset$. This contradicts Lemma 2.7, proving the claim. ■

As T is countable $\operatorname{acl}(a)$ must also be countable (even in $\mathbf{C}^{eq}$!). The claim gives us a contradiction. This proves the lemma. ■

PROPOSITION **2.13.** (*T1-based, trivial with NDOP*). *For any a, A with $a \notin \operatorname{acl}(A)$, there are $c_1, \ldots, c_n$ (some n) such that* $\operatorname{acl}(aA) = \operatorname{acl}(c_1, \ldots, c_n, A)$,

$\{c_1, \ldots, c_n\}$ is A-independent, and $\mathrm{stp}(c_i/A)$ *is regular for all i. In particular* $\mathrm{tp}(a/A)$ *has finite weight.*

Proof. The proof proceeds through a series of claims.

CLAIM 1. *There is* $c \in \mathrm{acl}(aA)$ *such that* $\mathrm{wt}(\mathrm{tp}(c/A)) = 1$.

Proof. Suppose not. In particular $\mathrm{wt}(\mathrm{tp}(a/A)) > 1$, so by Lemma 2.6, there are $c_1, d_1 \in \mathrm{acl}(aA) - \mathrm{acl}(A)$ such that c_1 is independent from d_1 over $\varnothing$. If $\mathrm{wt}(\mathrm{tp}(c_1/A)) = 1$ we are finished. Otherwise we can again find $c_2, d_2 \in \mathrm{acl}(c_1 A) - \mathrm{acl}(A)$ which are independent over $\varnothing$. Continuing this way, we either find the required c, or see that $\{d_i : i < \omega\}$ is A-independent and contained in $\mathrm{acl}(a)$, contradicting Lemma 2.12. ■

Now let C be a maximal A-independent set of elements in $\mathrm{acl}(aA)$, each realizing over A a type of weight 1. By 2.12, C is finite, $C = \{c_1, \ldots, c_n\}$, say. Choose $C_i \subseteq \mathrm{acl}(aA)$ for $i = 1, \ldots, n$ such that $c_i \in C_i$, $\{C_1, \ldots, C_n\}$ is A-independent, and the C_i are maximal such. So each C_i contains A and is algebraically closed.

CLAIM 2. $\mathrm{wt}(\mathrm{tp}(C_i/A)) = 1$ for all i.

Proof. Note that for each i, any $c \in C_i - \mathrm{acl}(A)$ must fork with c_i over A. For otherwise we could find, by Claim 1, some $d \in \mathrm{acl}(cA)$ with $\mathrm{wt}(\mathrm{tp}(d/A)) = 1$. But then clearly $\{c_1, \ldots, c_n, d\}$ contradicts the maximality of C. As $\mathrm{wt}(\mathrm{tp}(c_i/A)) = 1$, it follows that if $x, y \in C_i - \mathrm{acl}(A)$ then x forks with y over A. By Lemma 2.6, $\mathrm{wt}(\mathrm{tp}(C_i/A)) = 1$, as required. ■

CLAIM 3. $a \in \mathrm{acl}(C_1, \ldots, C_n)$.

Proof. We prove by induction on m that if b is a (possibly infinite) tuple and $B_1, \ldots, B_m \subseteq \mathrm{acl}(bA)$ are maximal subject to being algebraically closed, A-independent, and each of weight 1 over A, *then* $b \in \mathrm{acl}(B_1, \ldots, B_m)$. If $m = 1$, then clearly $B_1 = \mathrm{acl}(bA)$.

Suppose $m > 1$. Let B' be a maximal subset of $\mathrm{acl}(bA)$ which contains $B_1 \cup \cdots \cup B_{m-1}$ and is independent from B_m over A. By Lemma 2.11, $b \in \mathrm{acl}(B' \cup B_m)$. But from the induction hypothesis it follows that $B' = \mathrm{acl}(B_1 \cup \cdots \cup B_{m-1})$. Thus $b \in \mathrm{acl}(B_1 \cup \ldots \cup B_m)$, as required. ■

From Claim 3, we can find $d_i \in C_i$ for $i = 1, \ldots, n$ such that $\mathrm{acl}(aA) = \mathrm{acl}(d_1, \ldots, d_n, A)$. Clearly $\{d_1, \ldots, d_n\}$ is A-independent. Also by Claim 2,

$\text{wt}(\text{tp}(d_i/A)) = 1$ for each i. So all that is needed to complete the proof of the proposition is:

CLAIM **4.** *Suppose* $\text{wt}(\text{tp}(d/A)) = 1$. *Then* $\text{stp}(d/A)$ *is regular.*

Proof. Suppose that $B \supseteq A$, d is independent from B over A, and e realizes a forking extension of $\text{stp}(d/A)$ over B. As $\text{wt}(\text{tp}(e/A)) = 1$, e must be independent from d over A. So d is independent from each of B, Ae over A. By Lemma 2.7, d is independent from Be over A, and thus also over B. We have shown that $\text{stp}(d/A)$ is orthogonal to any forking extension of itself, that is it is regular. ■

REMARK **2.14.** *Proposition* 2.13 *also holds for perfectly trivial theories with NDOP.*

Example 2.15. (i) Let T be the theory of infinitely many equivalence relations $E_1, E_2, \ldots$ where each E_i-class is partitioned into infinitely many E_{i+1}-classes. Then T has quantifier elimination, is stable but not superstable, and is 1-based, trivial, with NDOP.

(ii) Let T be the theory of the 'free pairing function'. That is, T has just a 2-ary function symbol f which gives a bijection between $\mathbf{C}^2$ and $\mathbf{C}$, and for any term $t(x_1, \ldots, x_n)$, $T \models t(x_1, \ldots, x_n) \neq x_i$, for all i. This example is analysed in detail by Bouscaren and Poizat, who show that T is 1-based and trivial. Let $g(x)$, $h(x)$ be definable functions such that the function $x \to (g(x), h(x))$ is the inverse of f. For 'generic' $a \in \mathbf{C}$, the set $\{g.h^n(a): n < \omega\}$ will be an infinite $\varnothing$-independent set contained in $\text{acl}(a)$. By Lemma 2.12, T has the DOP.

3 Weight and theories with no dense forking chains

In the previous section we saw that in a trivial 1-based theory with NDOP all types have finite weight. We will prove in this short section the same result for 1-based countable theories with only countably many types over $\varnothing$. In fact this turns out to have very little to do with 'geometric' features of the theory. So we introduce and discuss some other properties: no dense forking chains, finite coding, and almost thinness.

DEFINITION **3.1.** *T is said to have* dense forking chains *if there are complete types $p_i(x) \in S(A_i)$ for $i \in \mathbf{Q}$ such that $i < j$* implies *$p_j(x)$ is a forking extension of $p_i(x)$, or equivalently if there are a, B, and subsets B_i of B for $i \in \mathbf{Q}$ such that $i < j$ implies $B_i \subseteq B_j$ and* $\text{stp}(a/B_j)$ *forks over B_i.*

We will say that T has NDFC if T has *no* dense forking chains. By Lemma 1.3.15 superstable theories have NDFC.

Remark 3.2. *It is easy to manufacture unsuperstable theories with NDFC. In fact, let T be a theory consisting of equivalence relations $E_i(x\ y)$ for $i \in I$, where I is some totally ordered set, and $i < j$ (in I) implies that E_j refines E_i and each E_i-class divides into infinitely many E_j-classes. The order types of forking chains in T correspond more or less to subsets of the 'completion' of I. In any case, if for example $I = \mathbf{Z}$ then T is unsuperstable but has NDFC. (Again these are all 1-based trivial theories.)*

Definition 3.3. *T is said to have* finite coding *if for any $|T|^+$-saturated model M (or a-model M) and $p(x) \in S(M)$, there is some* finite *$A \subseteq M$ such that p does not fork over A.*

Any superstable theory has finite coding.

Also:

Lemma 3.4. *Any 1-based theory has finite coding.*

Proof. Suppose T is 1-based and $p(x) = \mathrm{tp}(a/M)$, where M is $|T|^+$-saturated. Let $A \subseteq M$ be such that $|A| \leq |T|$ and p does not fork over A. Let $b \in M$ realize $\mathrm{stp}(a/A)$. Then $\mathrm{Cb}(\mathrm{stp}(a/M)) = \mathrm{Cb}(\mathrm{stp}(a/A)) = \mathrm{Cb}(\mathrm{stp}(b/A)) \subseteq \mathrm{acl}(b)$. That is, p does not fork over b. ■

Lemma 3.5. *Suppose T is countable, $S(\varnothing)$ is countable, and T has finite coding. Then T has NDFC.*

Proof. Suppose by way of contradiction that we have a and B_i for $i \in \mathbf{Q}$ such that for $i < j$, $B_i \subseteq B_j$ and $\mathrm{stp}(a/B_j)$ forks over B_i. Let M be an a-model containing $\bigcup_i B_i$. For $r \in \mathbf{R}$ (the real numbers) let $B_r = \cup\{B_i : i \leq r\}$. Note then that if $r < s$ are in $\mathbf{R}$ then $\mathrm{tp}(a/B_s)$ forks over B_r. For $r \in \mathbf{R}$, let $q_r(x) \in S(M)$ be the non-forking extension of $\mathrm{stp}(a/B_r)$ to M. Let (by finite coding) c_r be a finite tuple in M such that q_r does not fork over c_r. Let a_r realize the restriction of q_r to c_r. By Lemma 1.3.6, for each $r < s$ in $\mathbf{R}$ there is a finite set Δ of L-formulae (or even a single L-formula) such that $R_\Delta(\mathrm{tp}(a/B_s)) < R_\Delta(\mathrm{tp}(a/B_r))$.

On the other hand for any finite set Δ of L-formulae, and any $r \in \mathbf{R}$, $R_\Delta(\mathrm{tp}(a/B_r)) = R_\Delta(q_r(x)) = R_\Delta(\mathrm{tp}(a_r/c_r))$ (also by 1.3.6).

It follows that for $r < s$, $\mathrm{tp}(a_r c_r/\varnothing) \neq \mathrm{tp}(a_s c_s/\varnothing)$ and we have continuum many types over $\varnothing$. ■

DEFINITION **3.6.** (i) *We say that T is* thin *if every type has finite weight.*

(ii) *We say T is* almost thin *if for any a, A,* there does not exist *an infinite A-independent set $\{b_i : i < \omega\}$ such that a forks with b_i over A for each $i < \omega$.*

LEMMA **3.7.** *Suppose T has NDFC. Then T is almost thin.*

Proof. Suppose there exist such a, A, b_i, for $i < \omega$, violating almost thinness. Write $\{b_i : i \in \omega\}$ as $\{c_i : i \in \mathbf{Q}\}$. Then for $i < j$ in $\mathbf{Q}$, $\mathrm{tp}(a/\{c_k : k \leq j\} \cup A)$ forks over $\{c_k : k \leq i\} \cup A$, giving us a dense forking chain. ■

REMARK **3.8.** *If T is countable, almost thin, and not ω-categorical, then T has infinitely many countable models.*

Remark 3.8 is well known, and so shows (via 3.5 and 3.7) that a countable theory with finite coding (in particular a countable 1-based theory) which is not ω-categorical has infinitely many countable models.

Remarkably, almost thinness implies thinness, and the rest of this section is devoted to proving this.

LEMMA **3.9.** *Suppose T is almost thin. Then very (non-algebraic) type is non-orthogonal to some type of weight* 1.

Proof. Suppose by way of contradiction that $\mathrm{stp}(a/A)$ is orthogonal to all types of weight 1. We will construct inductively a-models $M_0 \subseteq M_1 \subseteq M_2 \ldots$, and elements $b_0, b_1, b_2, \ldots$ and $c_1, c_2, \ldots$ such that

(i) $A \subseteq M_0$;

(ii) $\mathrm{tp}(a/M_n)$ does not fork over A, for all n;

(iii) $\{b_0, \ldots, b_n, c_{n+1}\}$ is M_n-independent, for all n;

(iv) a forks with each element from $\{b_0, \ldots, b_n, c_{n+1}\}$ over M_n, for all n;

(v) $\mathrm{tp}(b_0, \ldots, b_{n-1}/M_n)$ does not fork over M_{n-1} for all $n > 0$.

Suppose we have already constructed $M_0, \ldots, M_n$, $b_0, \ldots, b_n$, c_{n+1}. We construct M_{n+1}, b_{n+1}, and c_{n+2}.

Choose $B \supseteq M_n$ such that

(a) $\mathrm{tp}(a/B)$ does not fork over M_n;

(b) $\mathrm{tp}(b_0, \ldots, b_n/Bc_{n+1})$ does not fork over M_n;

and whenever $B' \supseteq B$ also satifies (a) and (b) then $\mathrm{tp}(c_{n+1}/B')$ does not fork over B. (The existence of B is guaranteed, for if not, we find an increasing family of sets $\{B_\alpha: \alpha < |T|^+\}$ such that $\text{tp}(c_{n+1}/B_\beta)$ forks over B_α whenever $\alpha < \beta$, contradicting 1.4.2.1.)

We may assume that B is an a-model, which we call M_{n+1}. Now by (a) and the inductive assumptions, a forks with c_{n+1} over M_{n+1}. By (a) and the inductive assumptions, $\text{wt}(\text{tp}(c_{n+1}/M_{n+1}) \geq 2$. As M_{n+1} is an a-model, we can (by 1.4.4.1) find b_{n+1}, c_{n+2} such that $\{b_{n+1}, c_{n+2}\}$ is M_{n+1}-independent, and each forks with c_{n+1} over M_{n+1}. Without loss of generality

$(*)$ $\{b_{n+1}, c_{n+2}\}$ is independent from $M_{n+1} \cup \{b_0, \ldots, b_n\}$ over $M_{n+1} \cup \{c_{n+1}\}$.

By (b) and $(*)$ we have

$(**)$ $\text{tp}(b_0, \ldots, b_n/M_{n+1}, c_{n+1}, b_{n+1}, c_{n+2})$ does not fork over M_n.

By choice of $M_{n+1}(=B)$ and the fact that c_{n+1} forks with each of b_{n+1}, c_{n+2} over M_{n+1}, we obtain

(c) a forks with each of b_{n+1}, c_{n+2} over M_{n+1}.

On the other hand, from $(**)$ (and the M_{n+1}-independence of $\{b_{n+1}, c_{n+2}\}$ we obtain

(d) $\{b_0, \ldots, b_n, b_{n+1}, c_{n+2}\}$ is M_{n+1}-independent.

Clearly, by (b) and the inductive assumptions,

(e) a forks with each element from $\{b_0, \ldots, b_n\}$ over M_{n+1}.

So (a), (b), (c), (d), and (e) yield (i)–(v) for $n + 1$ in place of n.

Let $M = \cup\{M_i: i < \omega\}$. Then by (i) and (ii) $\text{tp}(a/M)$ does not fork over A. By (iii) and (v), $\{b_i: i < \omega\}$ is M-independent. By (ii), (iv), and (v), a forks with b_i over M for all i. We have contradicted the almost thinness of T (in fact with $\text{tp}(a/A)$). This proves the lemma. ■

PROPOSITION **3.10.** *Suppose T is almost thin. Then T is thin. Moreover, every (strong) type is domination equivalent to a finite product of types of weight* 1.

Proof. The proof is identical to that of Corollary 1.4.5.7, with weight 1 types replacing regular types. So we are brief. Assume T to be almost thin. Let M be an a-model, and $p = \text{tp}(a/M)$. Let $N = M[a]$. Let X be some maximal M-independent set of elements of N, each of whose type over M has weight 1. Each element of X forks with a over M, so X is finite. Let B be a maximal subset of N containing M such that X dominates B over M. Then B is also an a-model, say M_1. We claim that $M_1 = N$. If not, let $b \in N - M_1$. By Lemma 3.9, $\text{tp}(b/M_1)$ is non-orthogonal to some weight 1 type q. As M_1 is an a-model, we may assume q to be over M_1. It is easy to show (as q has weight 1), that $\text{tp}(b/N)$ dominates q whereby (Lemma 1.4.3.4(iv)), q is realized in N. So we may assume $\text{tp}(b/M_1)$ has weight 1. As in proof of 1.4.5.7, $\text{tp}(b/M_1)$ is nonorthogonal to M. As in the proof of Remark 1.4.5.5, q is non-orthogonal to some weight 1 type q' over M. By 1.4.4.2, q dominates q', and thus (by 1.4.3.4 again), the non-forking extension of q' over M_1 is realized in N. Thus we have some $c \in N - M_1$, such that $\text{tp}(c/M_1)$ does not fork over M, and $\text{tp}(c/M)$ has weight 1. This contradicts the maximality of X. We have shown that $M = N_1$. Thus a dominates X over M and X dominates a over M. In particular $\text{wt}(\text{tp}(a/M)) = \text{wt}(\text{tp}(X/M)) = |X| < \omega$. ■

4 1-based groups

Definition 4.1. *By a* 1-*based group we mean a group* $(G,.)$ *which is* ∞-*definable in a saturated model* $\mathbf{C}$ *of a stable theory, and such that the* ∞-*definable set* G *is* 1-*based. (See Remark* 1.8(ii).)

Recall that if G is defined over A, then 1-basedness of G means: for all $a \in (G_A)^{\text{eq}}$, and all $B \supseteq A$, $\text{Cb}(\text{stp}(a/B)) \subseteq \text{acl}(aA)$.

By Remark 1.8(ii), the 1-basedness of G does not depend on the particular set A over which G is defined. *We will usually assume that G (i.e. its underlying set) is defined over $\emptyset$ in $\mathbf{C}$ (i.e. is the solution set of some partial type $\Phi(x)$ over $\emptyset$).*

Of course we are interested in definable sets in G. By this we mean subsets of G (or G^{eq}) of the form $X \cap G^{\text{eq}}$ where X is definable in $\mathbf{C}$. Similarly for ∞-definable sets.

We will be working at the level of generality of Definition 4.1. The reader may, if he or she wishes, assume G to be just definable (rather than ∞-definable) or assume the global theory T to be 1-based (although in any case ∞-definable groups will enter the picture).

As usual we are working in $\mathbf{C}^{\text{eq}}$ where $\mathbf{C}$ is a big model of the stable theory T.

Before discussing 1-based groups proper, we make some additional remarks on stable group theory. The first lemma restates and strengthens Remark 1.6.20.

Lemma 4.2. *Suppose H is an ∞-definable group. Then there is some set A of parameters such that an automorphism f of $\mathbf{C}$ fixes H setwise (see Remark 4.3(i)) iff it fixes A pointwise.*

Proof. By Lemma 1.6.18, H is the intersection of a family $\{H_i : i \in I\}$ of *definable* groups, where $|I| \leq |T|$. This means that each H_i is a definable group, for $i, j \in I$, $H_i \cap H_j$ is a subgroup of each of H_i, H_j, and $H = \cap H_i$. Fix

$i \in I$. Then H_i is defined by some formula $\varphi_i(x, c)$, where $\varphi_i(x, y_i) \in L$, and the group operation on H_i is defined by some formula $\psi_i(x_1, x_2, x_3, c)$, with $\psi(x_1, x_2, x_3, z_i) \in L$. Let Λ be the family of groups K for which there is some d such that K is defined by $\varphi_i(x, d)$ and its group operation by $\psi_i(x_1, x_2, x_3, d)$, and such that $K \cap H_i$ is a subgroup of H_i which contains H. By Lemma 1.6.13, there is a finite subfamily $K_1, \ldots, K_n$, say, of Λ such that $H_i \cap (\cap \Lambda) = H_i \cap K_1 \cap \cdots \cap K_n = G_i$. Let c_i denote the pair consisting of the canonical parameter of the definable set G_i, and the canonical parameter of its group operation. It is evident that c_i is something canonical for H (i.e. an automorphism of **C** which fixes H as a set fixes c_i). On the other hand note that $H = \cap \{G_i : i \in I\}$. Thus an automorphism which fixes each c_i fixes H. Thus $A = \{c_i : i \in I\}$ is the required set. ■

REMARK **4.3.** (i) *So in* 4.2 *when we talk about an automorphism fixing a group we mean it fixes both the underlying set* and *the group operation setwise (so gives rise to a group automorphism of the group). For H an ∞-definable group we let $[H]$ denote some canonical set of parameters for H. Lemma* 4.2 *also holds for definable homogeneous spaces. In fact we only use it when X is a coset of ∞-definable H in some* acl($\varnothing$)*-definable group G. We again use $[X]$ to denote the canonical parameter set of X.*

(ii) *The proof of Lemma* 4.2 *shows*: *whenever H is an ∞-definable group, which is B-invariant, then H is the intersection of a family of* definable *groups, each defined over B. In particular, if H is an ∞-definable subgroup of a given ∞-definable group G, G is $\varnothing$-invariant, and H is B-invariant, then there is a family Λ of* relatively definable *subgroups of G, each defined over B such that $H = \cap \Lambda$.*

In section 6 of Chapter 1 we discussed homogeneous spaces and the theory of generic types for such objects. If G is a ∞-definable group, H an ∞-definable subgroup, and $X = a.H$ a left coset of H in G, then of course (H, X) is a homogeneous space (where the action is in the right), so the theory applies. Similarly if X is a right coset and the action is on the left.

LEMMA **4.4.** *Let G be an ∞-definable group, defined over $\varnothing$. Let $p(x) =$ stp(a/A) where $a \in G$. Let $H =$ Stab(p) be the left stabilizer of p. Suppose that the right coset $H.a = X$ is* acl(A)*-definable (i.e.* ∞-acl(A)*-definable). Then H is connected and $p(x)$ is the generic type of $H.a$.*

Proof. Let H^0 be the connected component of H (also defined over acl(A)). Let b be a generic point of H^0 over acl$(A \cup \{a\})$ (i.e. stp$(b/A \cup \{a\})$ is the generic type of H^0.) Then a is independent from b over A. So (as $b \in$ Stab(p)),

$\text{stp}(b.a/A) = p$. Let $Y = H^0.a$. Then Y is also acl(A)-definable. So by Lemma 1.6.11 (working over acl(A)), $\text{stp}(b.a/A)$ is the generic type of Y. Thus p is the generic type of Y. We claim that $Y = X$, that is $H^0 = H$: if H were not connected, then choosing $c \in H - H^0$ independent from a over A, we would have $c.a \notin Y$, so $\text{stp}(c.a/A) \neq p$, contradicting H being the left stabilizer of p. This proves the lemma. ■

We now begin the study of 1-based groups. As mentioned above, G will be an ∞-definable group, defined over $\varnothing$. (The more general case is dealt with by working over the canonical parameter set of G.)

Proposition 4.5. *Suppose G is 1-based. Let $a \in G$, and let A be any set of parameters. Then $p = \text{stp}(a/A)$ is the generic type of some* acl(A)*-definable right coset of an ∞-definable connected subgroup H of G, defined over* acl($\varnothing$). *(In fact H will be exactly the left stabilizer of p.)*

Proof. Let H be the left stabilizer of p. So H is acl(A)-definable. In order to prove the proposition we must prove two things:

(I) $X = H.a$ is acl(A)-definable (and then we can apply Lemma 4.4).

(II) H is acl($\varnothing$)-definable.

We first set about proving (I). Let b be generic in G over A, such that a is independent from b over A. Let M be a big model containing $A \cup \{b\}$ such that $\text{tp}(a/M)$ does not fork over A.

Claim (i). *Let f be an* acl(A)*-automorphism of M. Then f fixes* $\text{tp}(b.a/M)$ *iff f fixes $b.H$ (setwise).*

Proof. $\text{tp}(a/M)$ is definable over acl(A), and $H = \text{Stab}(\text{tp}(a/M))$ is defined over acl(A). Thus if f is an acl(A)-automorphism of M we have: $f(\text{tp}(b.a/M)) = \text{tp}(b.a/M)$ iff $\text{tp}(f(b).a/M) = \text{tp}(b.a/M)$ iff $\text{tp}(b^{-1}.f(b).a/M) = \text{tp}(a/M)$ iff $b^{-1}.f(b) \in H$ iff $f(b).H = b.H$ iff $f(b.H) = b.H$. This proves Claim (i). ■

From Claim (i) we obtain:

(ii) $\text{tp}(b.a/M)$ does not fork over $\text{acl}(A) \cup [b.H]$.

Now as in Remark 1.8(i), G remains 1-based after naming acl(A). Claim (i) says that $[b.H]$ is precisely $\text{Cb}(\text{stp}(b.a/M)$ after naming acl(A). Thus we have

(iii) $[b.H] \subseteq \text{acl}(A \cup \{b.a\})$.

As b is generic in G over $A \cup \{a\}$, we see by Lemma 1.6.9(iv) that a is independent from $b.a.$ over A. Thus (by (iii)), $\text{tp}(a/A \cup \{b.a\} \cup [b.H])$ does not fork over A, and in particular does not fork over $A \cup [b.H]$. By forking symmmetry

(iv) $\mathrm{stp}(b.a/A \cup \{a\} \cup [b.H])$ does not fork over $A \cup [b.H]$.

(ii) and (iv) say that $\mathrm{tp}(b.a/M)$ and $\mathrm{stp}(b.a/A \cup \{a\} \cup [b.H])$ are parallel, being common non-forking extensions of $\mathrm{stp}(b.a/A \cup [b.H])$. Let $\mathbf{q}(x)$ be their common non-forking extension over $\mathbf{C}$.

CLAIM (v). *$H.a.$ is M-definable.*

Proof. Let f be an M-automorphism of $\mathbf{C}$. Then (as $\mathbf{q}$ does not fork over M), $f(\mathbf{q}) = \mathbf{q}$. As $\mathrm{stp}(b.a/A \cup \{a\} \cup [b.H]) \subseteq q$, the partial type '$x \in b.H.a$' is contained in $\mathbf{q}$. As $b.H$ is M-definable, $x \in b.H.f(a)$ is in $f(\mathbf{q}) = \mathbf{q}$. Thus $b.H.a \cap b.H.f(a) \neq \varnothing$. Thus $H.a = H.f(a) = f(H.a)$. This shows that $H.a$ is M-definable. ∎

Finally note that $H.a$ is also $\mathrm{acl}(A) \cup \{a\}$-definable, and a is independent from M over $\mathrm{acl}(A)$. Thus $[H.a] \subseteq M \cap \mathrm{dcl}(\mathrm{acl}(A) \cup \{a\}) = \mathrm{acl}(A)$. So $[H.a]$ is $\mathrm{acl}(A)$-definable, proving (I).

By Lemma 4.4, p is the generic type of X and H is connected.

Now we prove (II). We continue with the previous notation. As $\mathrm{stp}(a/A \cup \{b\})$ is a non-forking extension of p, $H = \mathrm{Stab}(\mathrm{stp}(a/A \cup \{b\}))$. So clearly also $H = \mathrm{Stab}(\mathrm{stp}(a.b/A \cup \{b\}))$. By Lemma 1.6.16(i), H is defined over $\mathrm{Cb}(\mathrm{stp}(b.a/A \cup \{b\}))$. By 1-basedness of G, $[\mathrm{H}] \subseteq \mathrm{acl}(b.a)$. So $[H] \subseteq \mathrm{acl}(A) \cap \mathrm{acl}(b.a)$. But, as b was chosen generic in G over $A \cup \{a\}$, $b.a$ is independent from A over $\varnothing$. Thus $\mathrm{acl}(A) \cap \mathrm{acl}(b.a) = \mathrm{acl}(\varnothing)$. So $[H] \subseteq \mathrm{acl}(\varnothing)$, that is H is $\mathrm{acl}(\varnothing)$-definable.

COROLLARY **4.6.** *Suppose G is 1-based. Then any connected ∞-definable subgroup of G^n ($= G \times \cdots \times G$, n times) is $\mathrm{acl}(\varnothing)$-definable.*

Proof. Note that G^n is also 1-based, defined over $\varnothing$. Let H be some connected ∞-definable subgroup of G^n. Let p be the generic type of H. Clearly $\mathrm{Stab}(p)$ (in G^n) is precisely H, so by 4.5, H is $\mathrm{acl}(\varnothing)$-definable. ∎

COROLLARY **4.7.** *If G is 1-based then G is abelian by finite.*

Proof. Let G^0 be the connected component of G. G^0 is also ∞-definable over $\varnothing$ and 1-based. We will show that G^0 is abelian. For $g \in G$, let Inn_g be the automorphism of G^0 taking $h \in G^0$ to $g^{-1}.h.g$. The graph of Inn_g is a connected ∞-definable subgroup of $G^0 \times G^0$. By Corollary 4.6 Inn_g is $\text{acl}(\emptyset)$-definable. Thus

(*) if $\text{tp}(g/\text{acl}(\emptyset)) = \text{tp}(g'/\text{acl}(\emptyset))$ then $\text{Inn}_g = \text{Inn}_{g'}$.

Choose g, g' generic in G^0 over $\emptyset$, and independent over $\emptyset$. Then $\text{stp}(g/\emptyset) = \text{stp}(g'/\emptyset)$ so by (*), for all $h \in G^0$,

(**) $$g^{-1}.h.g = g'^{-1}.h.g'.$$

So for all h in G^0, $g'.g^{-1}.h = h.g'.g^{-1}$, that is $g'.g^{-1} \in Z(G^0)$. But $g'.g^{-1}$ is generic in G^0 over $\emptyset$, thus $g \in Z(G^0)$ for all generic $g \in G^0$ over $\emptyset$. As every element of G^0 is a product of generic elements, it follows that G^0 is abelian.

Now $H = C_G(G^0)$ is a relatively definable subgroup of G which contains G^0. Similarly $Z(H)$ is an abelian relatively definable subgroup of G containing G^0. By 6.14, $Z(H)$ has finite index in G. So G is abelian by finite. ■

COROLLARY **4.8.** *The following are equivalent*:

(i) *G is* 1-*based.*

(ii) *For every $n < \omega$, every (relatively) definable subset of G^n is a Boolean combination of cosets of (relatively)* $\text{acl}(\emptyset)$*-definable subgroups of G^n.*

Proof. (i) → (ii). Suppose G is 1-based. So also is G^n. Let M be a big model, and $p(\mathbf{x})$, $q(\mathbf{x})$ complete types over M extending the partial type '$\mathbf{x} \in G^n$'. We first show

(*) Suppose that p and q contain the same (right) cosets of relatively $\text{acl}(\emptyset)$-definable subgroups of G^n. Then $p = q$.

By 4.5, there are connected ∞-definable subgroups H_1, H_2 of G^n, both defined over $\text{acl}(\emptyset)$ (i.e. $\text{acl}(\emptyset)$-invariant), and $\mathbf{a}, \mathbf{b} \in (G^n)^M$, such that p is the general type of $H_1.\mathbf{a}$ over M and q is the generic type of $H_2.\mathbf{b}$ over M. By Remark 4.3(ii), H_1 is the intersection of a family Λ_1 of relatively definable subgroups of G^n each defined over $\text{acl}(\emptyset)$. Similarly we obtain Λ_2 for H_2. By our assumptions on p and q, p contains '$\mathbf{x} \in H_2.\mathbf{b}$'.

CLAIM. $H_1 = H_2$.

Proof. Let $\mathbf{c} \in H_1^M$. By 4.5 $\mathbf{c} \in \text{Stab}(p)$. Thus '$\mathbf{x} \in \mathbf{c}.H_2.\mathbf{b}$' $\in p$. As p is consistent, $\mathbf{c}.H_2 = H_2$, that is $\mathbf{c} \in H_2$. ■

It follows from the claim and our hypotheses (and the fact that the types p, q are consistent) that $H_1.\mathbf{a} = H_2.\mathbf{b}$. So $p = q$.

From ($*$) and a compactness argument we obtain (ii).

(ii) $\rightarrow$ (i). Let $p(x) = \mathrm{tp}(\mathbf{a}/M)$ be a type of an element $\mathbf{a} \in G^n$ over a big model M. p is, by hypothesis, determined by knowing, for each relatively $\mathrm{acl}(\varnothing)$-definable subgroup of G^n, which coset (if any) is contained in p. Any such coset is clearly $\mathrm{acl}(\mathbf{a})$-definable. Thus $\mathrm{Cb}(p) \subseteq \mathrm{acl}(\mathbf{a})$. As in 1.9, we conclude that G is 1-based. ■

Remark 4.9. *Corollary* 4.8 *can be strengthened to*: *G is* 1-*based iff for every n, every relatively* $\mathrm{acl}(\varnothing)$-*definable subset of* G^n *is a Boolean combination of cosets of relatively* $\mathrm{acl}(\varnothing)$-*definable subgroups of* G^n.

All that needs to be proved is the right to left direction which we sketch. We do it in two steps.

Step I. G is abelian by finite.
Our hypotheses imply that for $a \in G^0$, generic over $\varnothing$, $p(x, y) = \mathrm{stp}\,(a, a^{-1}/\varnothing)$ is the generic type of some coset (in $G \times G$). Thus if a, b, c are generic in G^0 over $\varnothing$ and $\{a, b, c\}$ is $\varnothing$-independent, we have: $(a.b.c, a^{-1}.b^{-1}.c^{-1})$ realizes p. Thus $a.b.c = c.b.a$. Let d be generic in G^0 over $\{a, b, c\}$. Then $a.b.c.d = c.b.a.d = c.d.a.b$. But $a.d$, $c.d$ are independent generics of G^0. We have shown that for x, y independent generic of G^0, $x.y = y.x$. It follows that G^0 is abelian, and as in Corollary 4.7, that G is abelian by finite.

Step II. G is 1-based.
Let G_1 be a relatively $\varnothing$-definable subgroup of G of finite index which is abelian. We will show that G_1 is 1-based. As G is contained in the definable closure of G_1 together with a finite set, it will follow that G is 1-based.

We show that any relatively definable subset of G_1^n is a Boolean combination of cosets of relatively $\mathrm{acl}(\varnothing)$-definable subgroups of G_1^n. Let X be a relatively definable subgroup of G^n. By stability X is of the form $\varphi(\mathbf{x}, \mathbf{c})^{\mathbf{C}} \cap G^n$, where $\varphi(\mathbf{x}, \mathbf{y})$ is a formula over $\varnothing$, and $\mathbf{c}$ is in G. Suppose $\psi(\mathbf{x}, \mathbf{y})$ defines (in G^{n+m}) one of the cosets of which $\varphi(\mathbf{x}, \mathbf{y})^{\mathbf{C}} \cap G^{n+m}$ is a Boolean combination, a coset of the subgroup $A \subseteq G^{n+m}$, say. As G_1 is abelian, $\psi(\mathbf{x}, \mathbf{c})$ defines (in G^n) a coset of $A_1 = \{\mathbf{x} \in G^n\colon (\mathbf{x}, 0) \in A\}$. If A is $\mathrm{acl}(\varnothing)$-definable so is A_1.

Now we can use 4.8. ■

Proposition 4.5 also yields considerable information on the structure of G^{eq} (and even more) if G is 1-based (although this can also be seen to follow directly from 4.9). We give two examples:

Lemma 4.10. *Suppose G is* 1-*based*, $a \in \mathbf{C}^{(\mathrm{eq})}$ *and* $a \in \mathrm{acl}(G)$. *Then there is some*

$n < \omega$, and some relatively acl($\emptyset$)*-definable subgroup H of G^n, such that a is interalgebraic with some $b \in G^n/H$.*

Proof. Let $\mathbf{c}$ be an n-tuple from G such that $a \in \text{acl}(\mathbf{c})$. Let $p(\mathbf{x}) = \text{stp}(\mathbf{c}/a)$. By 4.5, p is the generic type of a coset X of a connected ∞-acl($\emptyset$)-definable subgroup H of G^n. Clearly $[X] = \text{Cb}(p)$. But Cb(p) also interalgebraic with a. ($a \in \text{acl}(\text{Cb}(p))$, since $a \in \text{acl}(\mathbf{c})$ and a is independent from $\mathbf{c}$ over Cb(p) by forking symmetry.) So $[X] \subseteq \text{acl}(a)$ and $a \in \text{acl}[X]$. By 4.3 $[X] = \{c_i : i \in I\}$ where c_i is the canonical parameter of some coset X_i of a relatively acl($\emptyset$)-definable subgroup H_i of G. We may assume that $a \in \text{acl}(c_i)$ for some i. Thus a is interalgebraic with c_i over $\emptyset$. But c_i 'is' an element of G/H_i. ■

Lemma 4.11. *Suppose that G is 1-based, and H is a connected ∞-definable group, living in G^{eq}. Then there is $n < \omega$, some connected ∞-definable subgroup L of $(G^0)^n$ and a definable homomorphism f of L onto H. (So H is 'definably isomorphic' to L/K for some ∞-definable subgroup of L.)*

Proof. After naming some additional parameters we have that $G \subseteq \text{dcl}(G^0)$. Let us work over an algebraically closed set A of parameters which includes these as well as any parameters over which H is defined. So without loss of generality $G = G^0$. Note that (over A) H is a 1-based group, and so is $G^n \times H$, for any n. Let b realize the generic type of H over A. Let $\mathbf{c} \in G^n$ (some n) be such that $b \in \text{dcl}(\mathbf{c})$. Let $q = \text{stp}(\mathbf{c}, b/A)$. Then by 4.5, stp($\mathbf{c}, b/A$) is the generic type of a coset of some connected ∞-definable (over acl(A)) subgroup S of $G^n \times H$ (where $S = \text{Stab}(q)$). Now as b is generic in H over A, it is not difficult to see that the projection of S onto H must be all of H.

Claim. *S is the graph of a homomorphism (from some subgroup of G^n onto H).*

Proof of claim. Note that G, H, and so $G^n \times H$, are abelian (by 4.6 and connectedness). We write the group operations of both G^n, H additively. To prove the claim, we need only show that if $(\mathbf{0}, d) \in S$ then $d = 0$. So let $(\mathbf{0}, d) \in S$. We may assume d is independent from $(\mathbf{c}, b)$ over A. Thus (as $S = \text{Stab}(q)$), $(\mathbf{0}, d) + (\mathbf{c}, b)$ also realizes q. That is, $(\mathbf{c}, d + b)$ realizes q. As $b \in \text{dcl}(\mathbf{c}A)$, $d + b = b$. Thus $d = 0$. This proves the claim. ■

Let L be the projection of S onto G^n. Then L is connected and ∞-definable, and S gives a definable homomorphism from L onto H. ■

5 Minimal locally modular groups

For the purposes of this section we say that an ∞-definable group $(G, .)$ is minimal if its underlying set G is minimal , in the sense of 1.5.1, 1.5.2. This language is a little ambiguous as ∞-definable connected groups have also been called minimal if they have no infinite connected ∞-definable subgroups. Minimal in the first sense implies minimal in the second sense, but the converse fails (e.g. by considering simple abelian varieties of dimension >1, as definable groups in an algebraically closed field).

Until otherwise stated, G will denote a minimal ∞-definable group, defined over $\emptyset$.

Remark 5.1. *G is connected.*

Proof. Otherwise G has a proper relatively definable subgroup H of finite index. But then H and $G - H$ are both relatively definable infinite subsets of G, contradicting minimality. ■

We repeat 2.5.13.

Fact 5.2. If G is locally modular, then G is 1-based.

The main point of this section is to observe that if G is minimal and locally modular, then G is, from the point of view of all its definable structure, essentially just a vector space over some division ring. We should recall that, by 1.6.22, any minimal group G is abelian. So we use additive notation. We are going to define shortly the important notion of a (relatively) definable quasi-endomorphism of G. The notion is a little subtle so we make some preliminary comments. First, let us denote $\mathrm{acl}(\emptyset) \cap G$ by G_0. (So do not confuse G_0 with G^0, which here $= G$.) G_0 is a subgroup of G.

Consider the following three kinds of objects:

(i) Minimal subgroups S of $G \times G$, ∞-definable over $\mathrm{acl}(\emptyset)$, whose projection on the first coordinate is G.

(ii) Endomorphisms f of G/G_0, such that for some formula $\psi(x, y)$ over $\mathrm{acl}(\emptyset)$ we have: for all $a, b \in G$, $f(a + G_0) = b + G_0$ iff there are $c \in a + G_0$, $d \in b + G_0$ with $\models \psi(c, d)$.

(iii) Pairs (h, B) where B is a finite $\mathrm{acl}(\emptyset)$-definable subgroup of G, and h is an $\mathrm{acl}(\emptyset)$-definable homomorphism from G to G/B (note that G/B can naturally be considered as an ∞-definable group). (In saying that h is $\mathrm{acl}(\emptyset)$-definable, what we mean is that there is an $\mathrm{acl}(\emptyset)$-definable partial function h_1, whose domain contains G and whose restriction to G is precisely h.)

If B_1, B_2 are finite $\mathrm{acl}(\emptyset)$-definable subgroups of G then so is $B_1 + B_2$, and, moreover, we have canonical homomorphisms

$$\pi_1\colon G/B_1 \to G/(B_1 + B_2),\ \pi_2\colon G/B_2 \to G/(B_1 + B_2).$$

We define an equivalence relation $\approx$ on objects of kind (iii): $(h_1, B_1) \approx (h_2, B_2)$ if $\pi_1.h_1 = \pi_2.h_2$ (as homomorphisms from G to $G/(B_1 + B_2)$).

We show there is a natural identification between objects of kind (i), objects of kind (ii), and $\approx$-classes of objects of kind (iii).

LEMMA 5.3. *Let S be a minimal ∞-definable* $\mathrm{acl}(\varnothing)$*-invariant subgroup of $G \times G$ whose projection on the first coordinate is G. Then $\{(a + G_0, b + G_0): (a, b) \in S\}$ is the graph of an endomorphism f_S of G/G_0, and, moreover, there is some formula $\psi(x, y)$ over* $\mathrm{acl}(\varnothing)$ *such that for $a, b \in G$, $f_S(a + G_0) = b + G_0$ iff $\models \psi(c, d)$ for some $c \in a + G_0$, $d \in b + G_0$.*

Proof. We claim that for all $a \in G$, $\{b \in G: (a, b) \in S\}$ is finite.

This is clear for generic $a \in G$ (for otherwise we find, by compactness $b \in G\text{-}\mathrm{acl}(a)$, such that $(a, b) \in S$, whereby $U(a, b/\varnothing) = 2$, a contradiction). As S is a group, it follows that the 'cokernel' of $S = \{b \in G: (0, b) \in S\}$ is finite, so easily for *all* $a \in G$, $\{b \in G: (a, b) \in S\}$ is finite (with the same cardinality).

By compactness we easily find a formula $\psi(x, y)$ over $\mathrm{acl}(\varnothing)$ such that for $a \in G$, and *any* b, we have $(a, b) \in S$ iff $\models \psi(a, b)$.

To show that f_S is well defined we require: if $(a, b) \in S$ and $(a + c, d) \in S$, where $c \in G_0$ then $d - b \in G_0$. But as S is a group, $(c, d - b) \in S$. By the claim above, $d - b \in \mathrm{acl}(c)$. Thus $d - b \in G_0$, as required. This completes the proof. ■

LEMMA 5.4. *Let f be an endomorphism of G/G_0, such that for some formula $\psi(x, y)$ over* $\mathrm{acl}(\varnothing)$*, and for all $a, b \in G$, $f(a + G_0) = b + G_0$ iff there are $c \in a + G_0, d \in b + G_0$ with $\models \psi(c, d)$. Let $a_1 \in G$ be generic over $\varnothing$, and choose $b_1 \in G$ such that $\models \psi(a_1, b_1)$. Let $p =$ $\mathrm{stp}(a_1, b_1/\varnothing)$ and $S =$ $\mathrm{Stab}(p)$. Then S is minimal (and clearly ∞-definable over* $\mathrm{acl}(\varnothing)$*) and $f = f_S$.*

Proof. Note first that for any $x \in G$, there are only finitely many $y \in G$ such that $\models \psi(x, y)$. It follows that if $f(a + G_0) = b + G_0$ then $b \in \mathrm{acl}(a)$. Note also that if (a, b) realizes p, then $f(a + G_0) = b + G_0$.

In particular $p = \mathrm{stp}(a_1, b_1/\varnothing)$ has U-rank 1. Let (a_2, b_2) realize p such that (a_2, b_2) is independent from (a_1, b_1) over $\varnothing$. Thus $f(a_1 + G_0) = b_1 + G_0$, $f(a_2 + G_0) = b_2 + G_0$. As f is an endomorphism of G/G_0, we have $f((a_2 - a_1) + G_0) = (b_2 - b_1) + G_0$. Thus let $c = a_2 - a_1$, $d = b_2 - b_1$. So $d \in \mathrm{acl}(c)$. But c is independent with each of (a_1, b_1), (a_2, b_2) over $\varnothing$ (as for example a_1, a_2 are independent generics of G and $b_1 \in \mathrm{acl}(a_1)$, $b_2 \in \mathrm{acl}(b_2)$). Thus (c, d) is independent with each of (a_1, b_1), (a_2, b_2) over $\varnothing$. So $(c, d) \in \mathrm{Stab}(p) = S$. Thus $(a_1, b_1) + S = (a_2, b_2) + S$. So $(a_2, b_2) + S$ is $\mathrm{acl}(a_1, b_1)$-definable. But $\mathrm{stp}(a_2, b_2/a_1, b_1)$ is the non-forking extension of p. By Lemma

4.4, S is connected, and $\mathrm{stp}(a_2, b_2/a_1, b_1)(=p|\mathrm{acl}(a_1, b_1))$ is the generic type of a coset of S. The proof of 1.6.12(i) shows that $U(S) = 1$ and that the generic types of S are precisely the types (containing S) of U-rank 1. As S is connected, by 1.6.6, S has a unique generic (and so non-algebraic) type. Thus S is minimal. (Let us remark here that the last couple of lines have shown that an ∞-definable group is minimal iff it is connected and has a generic type of U-rank 1.)

Finally we claim that $f_S = f$: first suppose $f(a + G_0) = b + G_0$. Let (c, d) realize p, independently from (a, b). Then $f(c + G_0) = d + G_0$. As f is an endomorphism of G/G_0, $f((a + c) + G_0) = (b + d) + G_0$. As $a + c$ is generic in G over $\emptyset$, there is e, such that $\mathrm{stp}(a + c, e/\emptyset) = p$. As $\models \psi(a + c, e)$, we must have $(b + d) - e \in G_0$. Thus $b - (e - d) \in G_0$. So $f(a + G_0) = (e - d) + G_0$. As before $(a, e - d)$ is independent with each of (c, d), $(a + c, e)$ over $\emptyset$. Thus $(a, e - d) \in S = \mathrm{Stab}(p)$. We have $f_S(a + G_0) = (e - d) + G_0 = b + G_0 = f(a + G_0)$. The same argument shows that if $(a, b) \in S$ then $f(a + G_0) = b + G_0$. ■

Note that if S is an object of type (i), f_S is as obtained in Lemma 5.3, $a \in G$ is generic over $\emptyset$ and $f_S(a + G_0) = (b + G_0)$, *then* $S = \mathrm{Stab}(\mathrm{stp}(a, b/\emptyset))$. Thus, Lemmas 5.3 and 5.4 establish that $S \to f_S$ establishes a canonical bijection between objects of kind (i) and those of kind (ii). Also the proof of Lemma 5.3 shows that for any S of kind (i), there is a formula $\psi(x, y)$ over $\mathrm{acl}(\emptyset)$, defining a homomorphism $G \to G/B$ for some finite subgroup $B \subseteq G_0$ such that for generic $a \in G$ over $\emptyset$, there is b such that $\models \psi(a, b)$ and (a, b) realizes the generic type of S, in particular $S = \mathrm{Stab}(\mathrm{stp}(a, b/\emptyset))$. Now for any formula $\psi(x, y)$ which defines a homomorphism from G to G/B for some finite $\mathrm{acl}(\emptyset)$-definable subgroup B of G, by the above arguments $\psi(x, y)$ induces an endomorphism f_ψ of G/G_0: $f_\psi(a + G_0) = b + G_0$ iff for some $c \in a + G_0$, $d \in b + G_0$, $\models \psi(c, d)$. This is an object of kind (ii). Also it is not difficult to see that $f_\psi = f_{\psi'}$ iff $\psi \approx \psi'$. The reader can now fill in the additional details for our identification of objects of kind (i), objects of kind (ii), and $\approx$-classes of objects of kind (iii).

DEFINITION **5.5.** *By an* $\mathrm{acl}(\emptyset)$*-definable quasi-endomorphism of* G *we mean a minimal* ∞*-definable subgroup* S *of* $G \times G$*, defined over* $\mathrm{acl}(\emptyset)$*, whose projection on the first coordinate is* G*.* $D(G, \emptyset)$ *denotes the family of such objects.*

Finally we show, that $D(G, \emptyset)$ has the structure of a (division) ring. Note that if $S, S' \in D(G, \emptyset)$, then $f_S + f_{S'}$ is an object of kind (ii): if ψ corresponds to f_S and ψ' to $f_{S'}$ then the formula

$$\psi''(x, y) = (\exists z, w)(\psi(x, z) \wedge \psi'(x, w) \wedge (y = z + w))$$

corresponds to $f_S + f_{S'}$. Similarly the composition $f_S.f_{S'}$ is also of kind (ii). Thus $f_S + f_{S'} = f_{S''}$, and $f_{S'}.f_S = f_{S'''}$ for unique S'', S''' in $D(G, \emptyset)$. Note what S'', S''' are: if a is generic in G, and $(a, b) \in S$, $(a, c) \in S'$, then $S'' = \text{Stab}(\text{stp}(a, b + c/\emptyset))$, and if $(b, d) \in S'$ then $S''' = \text{Stab}(\text{stp}(a, d/\emptyset))$.

We write $S'' = S + S'$ and $S''' = S'.S$. So we see:

Remark 5.6. *G/G_0 is a $D(G, \emptyset)$ module, the action of $S \in D(G, \emptyset)$ on G/G_0 being given by f_S.*

Lemma 5.7. *$D(G, \emptyset)$ is a division ring.*

Proof. Note that the 0-element of $D(G, \emptyset)$ is $G \times \{0\}$.

Suppose that $S \in D(G, \emptyset)$ is not 0. Then the projection of S onto the second coordinate is all of G (as it is connected and non-zero). So f_S is surjective. Also if $b \in G$ and $(a, b) \in S$ then $a \in \text{acl}(b)$ (as in the proof of 5.3, do it for generic b first and then deduce for $b = 0$, so for all b). If $f_S(a + G_0) = f_S(b + G_0)$, then $f_S((a - b) + G_0) = G_0$. Thus there are c, $d \in G_0$ such that $(a - b + c, d) \in S$. By what we have just said, $a - b + c \in \text{acl}(G_0)$, so $(a - b) \in G_0$. So $a + G_0 = b + G_0$. So f_S is one to one and therefore bijective, so clearly has an inverse, which is also an object of type (ii). ■

The next proposition is crucial, and actually shows that, if G is locally modular, then G/G_0 equipped with all structure induced from $\mathbf{C}$ is simply a $D(G, \emptyset)$-vector space.

Proposition 5.8. *Suppose G is locally modular. Let $a_1, \ldots, a_n$, $b \in G$. Then $b \in \text{acl}(a_1, \ldots, a_n)$ iff there are $\alpha_1, \ldots, \alpha_n \in D(G, \emptyset)$ such that $(b + G_0) = \alpha_1(a_1 + G_0) + \cdots + \alpha_n(a_n + G_0)$.*

Proof. Remember to begin with that if $\alpha \in D(G, \emptyset)$ and $\alpha(a + G_0) = b + G_0$, then $b \in \text{acl}(a)$. (This was remarked at the beginning of the proof of 5.4). This yields the right to left direction of the proposition. Conversely, suppose $b \in \text{acl}(a_1, \ldots, a_n)$. By induction (on n) we may suppose that $a_i \notin G_0$ for all i, and $\{a_1, \ldots, a_n\}$ is $\emptyset$-independent. By Fact 5.2, G is 1-based. So by Proposition 4.5, $\text{stp}(a_1, \ldots, a_n, b/\emptyset)$ is the generic type of a coset X of some connected ∞-definable subgroup S of G^{n+1}, defined over $\text{acl}(\emptyset)$. So $U(X) = U(S) = n$. Note that X must also be $\text{acl}(\emptyset)$-definable (as it is something invariant for $\text{stp}(a_1, \ldots, a_n, b/\emptyset)$). As $a_1, \ldots, a_n$ are generic independent in G, the projection of X onto the first n coordinates must be G^n.

Thus the projection of S onto the first n coordinates is G^n. (Let Y be this projection. Y is an acl($\emptyset$)-definable coset, so is a translate of some acl($\emptyset$)-definable subgroup H of G^n. Clearly there are independent generic elements $b_1, \ldots, b_n$ of G such that $(b_1, \ldots, b_n) \in H$. Now G^n is connected with generic type stp$(b_1, \ldots, b_n/\emptyset)$. By 1.6.16, G^n is the stabilizer of this type. Thus $G^n \subseteq H$. So also $Y = G^n$.) Thus $X = S + (0, \ldots, 0, d)$ for some $d \in G$.

CLAIM 1. $d \in G_0$.

Proof. Both X and S are defined over acl($\emptyset$). So if d were generic in G, then for *any* generic d' in G, $X = S + (0, \ldots, 0, d')$. Then clearly $U(X) = n + 1$, which is impossible. Thus $d \in G_0$. ■

For $i = 1, \ldots, n$, let

$$S_i = \{(x_i, y)\colon (0, \ldots, 0, x_i, 0, \ldots, 0, y) \in S\},$$

an ∞-definable subgroup of $G \times G$. Let S_i^0 denote the connected component of S_i.

CLAIM 2. $S_i^0 \in D(G, \emptyset)$.

Proof. As S_i is defined over acl($\emptyset$), so is S_i^0. All we need to show is that $U(S_i) = 1$ and the projection of S_i onto the first coordinate is G. Note that if $(a_1', \ldots, a_n', b') \in S$ then $b' \in \text{acl}(a_1', \ldots, a_n')$ (this is true for generic $a_1', \ldots, a_n'$ in G, so 'cokernel of S' $= \{y \in G\colon (0, \ldots, 0, y) \in S\}$ is finite, so it is true for all $a_1', \ldots, a_n'$). Thus

$$(*) \qquad \text{if } (x_i, y) \in S_i \text{ then } y \in \text{acl}(x_i).$$

On the other hand the projection of S onto the first n coordinates is all of G^n, so there is $(0, \ldots, 0, x_i, 0, \ldots, 0, y)$ in S with x_i generic in G. That is, there is (x_i, y) in S_i with $U(\text{stp}(x_i, y)/\emptyset) \geq 1$. By $(*)$ $U(S_i) = 1$. We have also shown that the projection of S_i (and so S_i^0) on the first coordinate is G. ■

Write S_i^0 as α_i. Now, for each i, let $c_i \in G$ be such that $(a_i, c_i) \in S_i^0$. So $\alpha_i(a_i + G_0) = c_i + G_0$. Moreover, $(a_1, \ldots, a_n, c_1 + \cdots + c_n) \in S$. As $X = S + (0, \ldots, 0, d)$ we also have $(a_1, \ldots, a_n, b - d) \in S$. Thus substracting we have $(0, \ldots, 0, c_1 + \cdots + c_n - b + d) \in S$. But, as remarked earlier the cokernel of S is finite, so contained in G_0. Thus $c_1 + \cdots + c_n - b + d \in G_0$. By Claim 1, $d \in G_0$; thus $c_1 + \cdots + c_n - b \in G_0$. Thus $b + G_0 = \alpha_1(a_1) + G_0) + \cdots + \alpha_n(a_n + G_0)$, completing the proof of the proposition. ■

By vitrue of the correspondence established between objects of type (i) and those of type (iii) we can restate Proposition 5.8, in the following more concrete form:

COROLLARY **5.9.** *Suppose G is locally modular, $a_1, \ldots, a_n$, $b \in G$ and $b \in$ $\mathrm{acl}(a_1, \ldots, a_n)$. Then there is some finite $\mathrm{acl}(\varnothing)$-definable subgroup B of G, there are $\mathrm{acl}(\varnothing)$-definable homomorphisms $\alpha_1, \ldots, \alpha_n$ from G to G/B, and there is $d \in G_0$ such that $\alpha_1(a_1) + \cdots + \alpha_n(a_n) = (b + d)/B$ (or in previous notation $=$ $(b + B) + (d + B)$).*

Also:

COROLLARY **5.10.** *Suppose G is locally modular; then G is modular.*

Proof. Let p be the generic type of G. Proposition 5.8 shows that the geometry attached to the pregeometry $\mathbf{G}_{\varnothing}$ is precisely (infinite-dimensional) projective geometry over $D(G, \varnothing)$. ■

We should remark that if G is (locally) modular then the above definition of the division ring associated to G does *not* depend on the base set of parameters (which was $\varnothing$ in the above discussion). That is to say, if G is minimal, locally modular, and defined over $\varnothing$, then for any set of parameters A, $D(G, A) = D(G, \varnothing)$. The reason is simply Corollary 4.6 which says that any connected ∞-definable subgroup of $G \times G$ is $\mathrm{acl}(\varnothing)$-definable.

Of course without the locally modular assumption, this division ring can increase radically when we adjoin parameters.

Finally we add a remark on weakly minimal groups. We drop our previous assumptions on G. G is said to be a *weakly minimal* group if G is a *definable* group whose underlying set is *weakly minimal*, that is $R^{\infty}(G) = 1$. Then G^0, the connected component, is a minimal ∞-definable group, defined over $\varnothing$, and everything we have said above applies to G^0.

REMARK **5.11.** *Suppose G is a weakly minimal group defined over $\varnothing$. Then G is* 1-*based iff G^0 is* 1-*based.*

Proof. This can be seen in a couple of ways. On the one hand note that *all* generic types of G (i.e. non-algebraic stationary types containing '$x \in G$') are non-orthogonal (see 1.6.6); thus one is locally modular if all are locally modular. But the generic type of G^0 is one of these generic types. Now use 5.2 and 2.5.8 (working in the structure G which has finite U-rank).

On the other hand we can remark that after naming a small set of parameters (with cardinality the index of G^0 in G), $G \subseteq \mathrm{dcl}(G^0)$. So if G^0 is

1-based then so is G after naming some parameters, but then we can apply Remark 1.8 to see that G is 1-based over $\varnothing$. ■

In Chapter 6, we will be studying in detail weakly minimal 1-based groups. An important observation will be that any member S of $D(G^0, \varnothing)$ is the restriction to G^0 of a weakly minimal acl($\varnothing$)-definable subgroup of $G \times G$.

6 Reducts of 1-based theories

In this brief and final section we discuss the problem of when a reduct of a 1-based theory is 1-based. We interpret reducts in a general sense: we assume that any 1-sorted theory T we deal with is 'Morleyized', that is for every L-formula $\varphi(\mathbf{x})$ (where $\mathbf{x}$ is a tuple of variables from the basic sort), there is a relation symbol R_φ in L such that $T \vDash (\forall \mathbf{x})(R_\varphi(\mathbf{x}) \leftrightarrow \varphi(\mathbf{x}))$. A reduct T^* of T is then determined by the choice of a sublanguage L^* of L: T^* is the set of L^*-sentences in T. Note that if M is a κ-saturated model of T, then M^*, the L^*-reduct of M, is a κ-saturated model, of T^*. It is not difficult to see that stability of T implies stability of T^* (similarly for λ-stability, superstablity, total transcendentality, etc.). This can be seen by counting types, or looking at the various ranks. However, it is less clear how the geometry of forking behaves under reducts.

We start with an elementary example of a non-1-based (trivial) theory.

Example 6.1. The language L consists just of a binary relation R. The axioms of T say: R is symmetric; R is irreflexive; for any x there are infinitely many y such that $R(x, y)$; for each $n > 2$ there do *not* exist distinct $x_1, \ldots, x_n$ such that $R(x_1, x_2)$, $R(x_2, x_3), \ldots, R(x_{n-1}, x_n)$, and $R(x_n, x_1)$.

A back-and-forth argument shows that T is complete and has quantifier elimination after adding a predicate $R_n(x, y)$ for each formula:

$$(\exists z_1, \ldots, z_n)(R(x, z_1) \wedge R(z_1, z_2) \wedge \cdots \wedge R(z_n, y)).$$

T is ω-stable, for if $M < N$ are models and $a \in N - M$, then *either* for no $n < \omega$ is there $b \in M$ with $R_n(a, b)$, in which case tp(a/M) is uniquely determined, *or* there is such n, and assuming n to be least such, then tp(a/M) is definable over $\{b\}$.

One can show that the formula '$x = x$' has Morley rank ω. However, T is *not* 1-based: the relation R is in fact a complete 2-type over $\varnothing$ for T, and by definition it defines a pseudoplane. By 1.7, T is not 1-based.

We now give an example of a 1-based theory T' such that the theory T from 6.1 is a reduct of T'.

Example 6.2. Let L' consist of again of a single binary relation S. The axioms of T' are: S is irreflexive; $S(x, y) \to \neg S(y, x)$; for each y, there is *exactly one* x such that $S(x, y)$; for each x there are *infinitely many* y such that $S(x, y)$; for each $n > 1$, there do not exist $x_1, \ldots, x_n$ such that $S(x_1, x_2), \ldots, S(x_{n-1}, x_n), \ldots, S(x_n, x_1)$.

Think of T' as the theory of a tree where each element has infinitely many 'immediate successors', and S is the 'immediate successor' relation.

After adding a function symbol f for the function taking y to the unique x such that $R(x, y)$, one sees (again by back-and-forth) that T' is complete with quantifier elimination. As in Example 6.1, T' is ω-stable, with Morley rank (and also U-rank) ω. However, T' is 1-based. For let $p(x) = \text{tp}(a/M)$. If $f^n(a) \notin M$ for all n then p is definable over $\emptyset$. Otherwise p is definable over $f^n(a)$, where n is least such that $f^n(a) \in M$. Thus $\text{Cb}(p) \in \text{dcl}(a)$. The same can be seen for n-types in place of 1-types. By Lemma 1.9, T' is 1-based.

Let R be the relation $S(x, y) \vee S(y, x)$. It is clear that R satisfies the axioms on Example 6.1. Thus T is a reduct of T'.

However, in the following special cases, 1-basedness is preserved under taking reducts.

Proposition **6.3.** *Suppose T has finite U-rank (which remember means that all types have finite U-rank) and is 1-based. Then any reduct of T is 1-based.*

Proof. Let T^* be a reduct of T. Let $\mathbf{C}$ be the big saturated model of T, and let $\mathbf{C}^*$ be its reduct to a model of T. Suppose that T is 1-based but T^* is not 1-based. We seek a contradiction.

By Proposition 1.7, there is a complete type definable quasidesign in $\mathbf{C}^*$, given by the complete type $r(x, y)$ of T^*. Now $r(x, y)$ is a *partial* type over $\emptyset$, as far as T is concerned. We will work with T (equivalently in $\mathbf{C}$).

Claim. *If* $\mathbf{C} \models r(a, b)$, *then* $a \in \text{acl}(b)$ *or* $b \in \text{acl}(a)$.

Proof. Otherwise $\text{tp}(a, b/\emptyset)$ yields a complete type definable quasidesign for T, contradicting 1-basedness of T. ■

By the claim, and compactness (of T), there are formulae $\varphi_1(x, y)$ and $\varphi_2(x, y)$ of $L(T)$, such that

(i) $T \models (\forall x)(\exists^{<\omega} y)(\varphi_1(x, y)) \wedge (\forall y)(\exists^{<\omega} x)(\varphi_2(x, y))$ (where $\exists^{<\omega}$ means $\exists^k$ for some definite but unspecified $k < \omega$);

and

(ii) $T \cup r(x, y) \cup$ '$x \notin \operatorname{acl}(y)$' $\models \varphi_1(x, y)$;

(iii) $T \cup r(x, y) \cup$ '$y \notin \operatorname{acl}(x)$' $\models \varphi_2(x, y)$.

Now we construct $a_0, b_0, a_1, b_1, \ldots,$ in $\mathbf{C}$ such that (a_i, b_i) and (a_{i+1}, b_i) satisfy $r(x, y)$ for each $i < \omega$, and such that $a_{i+1} \notin \operatorname{acl}(a_0, b_0, \ldots, a_i, b_i)$, $b_i \notin \operatorname{acl}(a_0, b_0, \ldots, a_{i-1}, b_{i-1}, a_i)$. This can be done because whenever (a, b) satisfies $r(x, y)$ in $\mathbf{C}$, there are infinitely many y such that $C \models r(a, y)$, and infinitely many x such that $\mathbf{C} \models r(x, b)$ (as this is true in the reduct $\mathbf{C}^*$ of $\mathbf{C}$).

By (ii) and (iii), it follows that for each i.

(iv) $\mathbf{C} \models \varphi_1(a_{i+1}, b_i) \wedge \varphi_2(a_i, b_i)$.

By compactness we can find elements c_k for $k \leq 0$ such that for each $k \leq 0$, $c_k \notin \operatorname{acl}(c_i\colon i < k)$, and for *even* $k \leq 0$, $\models \varphi_1(c_k, c_{k-1})$ and for *odd* $k \leq 0$, $\models \varphi_2(c_{k-1}, c_k)$. Thus (by (i) above) $c_k \in \operatorname{acl}(c_0)$ for all $k < 0$. But then clearly, for each $k < 0$, c_k forks with c_0 over $\{c_i\colon i < k\}$, so $\operatorname{tp}(c_0/\{c_i\colon i \leq k\})$ forks over $\{c_i\colon i < k\}$. Now $U(c_0/\emptyset) \geq U(c_0/\{c_i\colon i \leq k\})$ for all $k < 0$, but the latter has arbitrarily large finite value (as k varies). So $U(c_0/\emptyset) \geq \omega$, contradicting our hypothesis on T. ■

Proposition 6.4. *Suppose T is a 1-based theory, let T^* be a reduct of T, and suppose that some $L(T^*)$-formula defines a group operation on the universe. Then T^* is also 1-based.*

Proof. Let $\mathbf{C}$ be the monster model of T, and $\mathbf{C}^*$ its reduct to L^* (the language of T^*). Let M^* be a saturated model of T^*. We may assume M^* is the L^*-reduct of a saturated model M of T. Let $\mathbf{a}$ be some n-tuple from $\mathbf{C}$. Let $p = \operatorname{tp}(\mathbf{a}/M)$ (in the sense of $\mathbf{C}$), and $p^* = \operatorname{tp}(\mathbf{a}/M^*)$ in the sense of $\mathbf{C}^*$. We will show that $\mathbf{p}^*$ is the generic type of a right coset of some ∞-definable over M^* (in $\mathbf{C}^*$) subgroup of $(\mathbf{C}^*)^n$ (with respect to the specified group operation). Now, by 4.5, in $\mathbf{C}$, p *is* the generic type of a right coset X of some ∞-definable over M subgroup S of C^n, where actually S is connected and equals the left stabilizer of p. Let S^* be the left stabilizer of p^* (in the sense of $\mathbf{C}^*$). Clearly S is a subgroup of S^*. Let X^* be the unique coset of S^* which contains X. We claim that X^* is M^*-definable. Well S^* is M^*-definable (being the stabilizer of $p^* \in S(M^*)$); also as X is M-definable and M is saturated M contains a point $\mathbf{b}$ of X. Thus $X^* = S^*.\mathbf{b}$ is M^*-definable. But $\mathbf{a} \in X$, so $\mathbf{a} \in X^*$. By Lemma 4.4, p^* is the generic type of X^*.

Now if p^* happened to be the non-forking extension of some strong type over $\varnothing$ to M^*, then both X^* and S^* would be defined over acl($\varnothing$) in $\mathbf{C}^*$ (as they are invariants attached to p^*). We easily obtain (as in 4.8) that every acl($\varnothing$)-definable subset of $(\mathbf{C}^*)^n$ is a Boolean combination of acl($\varnothing$)-definable cosets of acl($\varnothing$)-definable subgroups of $(\mathbf{C}^*)^n$. By 4.9, T^* is 1-based. ■

5

Groups and geometries

If D is a ω-categorical, non-trivial strongly minimal set, we know from Chapter 2 that D is locally modular and its associated geometry is affine or projective geometry over a finite field F. Either way, as in section 6 of Chapter 2, we obtain a strictly minimal set D_1 whose pregeometry is projective geometry over F. The well-known 'coordinatization theorem' (see Artin's *Geometric Algebra* [A]) produces a definable infinite-dimensional vector space over F.

Now suppose that we have a minimal set D, whose geometry is locally modular and non-trivial (D may be ∞-definable and there is no ω-categoricity assumption). The question arises of what the associated geometry is. In fact we will obtain the same answer: affine or projective geometry over a division ring. The proof will go in the reverse direction from the ω-categorical case. We first obtain a definable group, and then use the results of Chapter 4 to characterize the original geometry. In this construction the issue of what parameters have to be added to obtain the group is rather important.

In section 3 we define pseudolinearity and prove, using a group construction, that a pseudolinear minimal set is locally modular. This proves (in conjunction with the results of Chapter 2) that a unimodular minimal set is locally modular.

In fact it turns out that one can recognize the presence of an ∞-definable group (and also an ∞-definable homogeneous space) in a stable theory from certain configuration of points *vis-à-vis* algebraic closure. This is called the 'group configuration theorem' and will be proved in section 4.

So in fact, in this chapter we carry out a group construction three times. This will involve a certain repetition of material, which we hope does not irritate the reader too much.

T is a stable theory in this chapter.

1 Locally modular non-trivial minimal types: construction of a group

We will prove:

Theorem 1.1. *Let D be minimal and defined over $\varnothing$. Suppose D is locally modular and non-trivial. Then in D^{eq} there is a (∞)-definable connected group*

G with $U(G) = 1$. Moreover, there is some $e \in D^{eq}$ with $U(e/\varnothing) = 1$, such that G is e-definable.

The proof will be rather long, and we introduce a few notions as we go along. The basic idea has already been seen in the construction of the group of quasitranslations in the proof of Zilber's theorem (although there we were aiming for a contradiction). Non-ω-categoricity presents the main technical problem.

Let D be as in the assumptions of Theorem 1.1. *We will work entirely in D^{eq}.*

Recall from Remark 5.13 in Chapter 2 that D is 1-based, and in particular, for any algebraically closed sets A, B, in D^{eq}, A is independent from B over $A \cap B$. This will be used repeatedly.

LEMMA **1.2.** *There exist a, b, e, c, with $U(e/\varnothing) = U(c/e) = U(a/e) = U(b/e) = 1$, a independent from c over e, b independent from c over e, with a and b interdefinable over ec (i.e. $a \in \mathrm{dcl}(bec)$, $b \in \mathrm{dcl}(aec)$).*

Proof. First as D is non-trivial there are $A \subset D$ and $a_1, a_2 \in D$ with $\{A, a_1, a_2\}$ pairwise independent but with $a_2 \in \mathrm{acl}(A, a_1)$. By 1-basedness of D, $\{a_1, a_2\}$ and A are independent over $\mathrm{acl}(a_1, a_2) \cap \mathrm{acl}(A)$. Thus there is an $e \in D^{eq} \cap \mathrm{acl}(A)$ with $U(e/\varnothing) = 1$, and $\{e, a_1, a_2\}$ pairwise independent but dependent. Let $b_1, b_2 \in D$ be such that $\mathrm{stp}(b_1 b_2/e) = \mathrm{stp}(a_1 a_2/e)$, and $a_1 a_2$ and $b_1 b_2$ are independent over e. Thus $U(e, a_1, a_2, b_1, b_2/\varnothing) = 3$.

CLAIM **1.** *$\{a_1, b_2, e\}$ is $\varnothing$-independent. Similarly for $\{a_2, b_1, e\}$, $\{a_2, b_2, e\}$ and $\{a_1, b_1, e\}$.*

Proof. b_2 is independent from $a_1 e$ over e. So a_1 is independent from $b_2 e$ over e. But a_1 is independent from e over $\varnothing$. Thus a_1 is independent from $b_2 e$ over $\varnothing$. Together with the fact that b_2 is independent from e over $\varnothing$ this yields the $\varnothing$-independence of $\{a_1, b_2, e\}$. The rest is similar. ■

On the other hand $U(a_1, b_2, a_2, b_1/\varnothing) = 3$, so (a_1, b_2) forks with (a_2, b_1) over $\varnothing$. By 1-basedness of D, there is $d \in \mathrm{acl}(a_1, b_2) \cap \mathrm{acl}(a_2, b_1)$, $d \notin \mathrm{acl}(\varnothing)$. It is easily checked that $U(d/\varnothing) = 1$, and each of $\{d, a_1, b_2\}$, $\{d, a_2, b_1\}$ is pairwise independent but not independent over $\varnothing$. By Claim 1, d is independent from e over $\varnothing$, and a_1 is independent from b_1 over $\varnothing$. Again by 1-basedness of D, there is $d_1 \in \mathrm{acl}(d, e) \cap \mathrm{acl}(a_1, b_1)$, $d_1 \notin \mathrm{acl}(\varnothing)$. Again $U(d_1/\varnothing) = 1$.

In fact we have the following diagram:

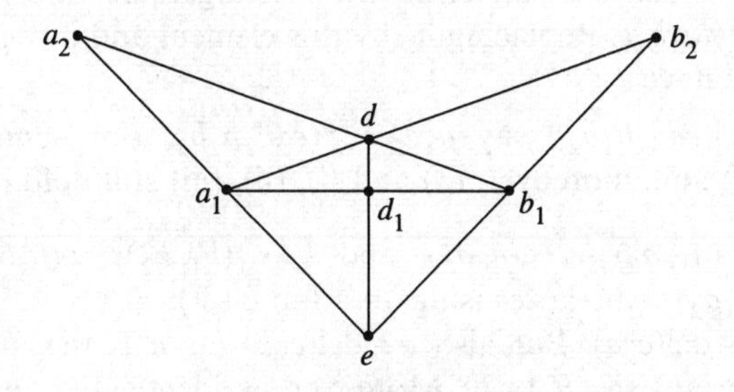

The reader should check that

($*$) Each point (or rather its type) has U-rank 1 over $\varnothing$. Each line has U-rank 2 over $\varnothing$. Also whenever x, y, z are distinct points from the diagram such that x and y are collinear but x, y, z are not collinear, then $\{x, y, z\}$ is $\varnothing$-independent (so has U-rank 3 over $\varnothing$). Also the U-rank of the whole set of points over $\varnothing$ is 3.

We will use, among other things, the following consequences of ($*$):

(i) $U(a_1a_2/e) = U(b_1b_2/e) = U(dd_1/e) = 1$.

(ii) a_1a_2 and dd_1 are independent over e; b_1b_2 and dd_1 are independent over e.

(iii) a_1a_2 and b_1b_2 are interalgebraic over edd_1.

So we could put $a = a_1a_2$, $b = b_1b_2$, and $c = dd_1$ and all the requirements of the lemma would be satisified, except we have only interalgebraicity of a and b over edd_1, not interdefinability. We will now remedy this.

Claim 2. *Let* $\text{tp}(a_1'/dd_1b_1b_2) = \text{tp}(a_1/dd_1b_1b_2)$. *Then* a_1' *and* a_1 *are interalgebraic over* $\varnothing$.

Proof of claim. Suppose for a contradiction that $U(a_1, a_1'/\varnothing) = 2$. Now (by ($*$)), $a_1, a_1' \in \text{acl}(d, b_2)$. So $U(a_1, a_1', d, b_2) = 2$, so $d \in \text{acl}(a_1, a_1')$. Similarly $d_1 \in \text{acl}(a_1, a_1')$. As $U(d, d_1/\varnothing) = 2$, this implies that $a_1 \in \text{acl}(d, d_1)$, contradicting ($*$) and proving the claim. ∎

Now the (finite) set $\{x: \text{tp}(x/dd_1b_1b_2) = \text{tp}(a_1/dd_1b_1b_2)\}$ may be regarded as an element of D^{eq}, and is, by the claim, interalgebraic with a_1. Thus replacing a_1 by this element preserves ($*$).

Now (the new) $a_1 \in \text{dcl}(dd_1b_1b_2)$.

An identical proof shows that $\{x: \mathrm{tp}(x/da_1b_1e) = \mathrm{tp}(a_2/da_1b_1e)\}$ (which we again view as an element of D^{eq}) is interalgebraic with a_2, and clearly definable over da_1b_1e. Replacing a_2 by this element and using transitivity of dcl(-), we now have

(I) a_1, $a_2 \in \mathrm{dcl}(edd_1b_1b_2)$, say $a_1a_2 = f(edd_1b_1b_2)$ (for some $\emptyset$-definable function f) and, moreover, $(*)$ and (i), (ii), (iii) still hold.

Now let $a = (a_1, a_2)$, $c = (d, d_1)$, and $b = \{(b'_1, b'_2)\colon \mathrm{tp}(b'_1b'_2/edd_1a_1a_2) = \mathrm{tp}(b_1b_2/edd_1a_1a_2)\}$ (the latter is again a finite set).

Clearly $b \in \mathrm{dcl}(eca)$. But also $a \in \mathrm{dcl}(ecb)$ (as a is the unique value of $f(edd_1b'_1b'_2)$ for any $(b'_1, b'_2) \in b$). Moreover, an identical argument to that of the claim shows that if $(b'_1, b'_2) \in b$ then b_1 and b'_1 are interalgebraic. It follows that b and b_1 are interalgebraic over e, whereby from (ii) above we conclude that b and c are independent over e. This proves the lemma. ■

We now work with the elements given by Lemma 1.2. Note that $c \in \mathrm{acl}(a, b, e)$. Thus c is interalgebraic *over* e with $\mathrm{Cb}(\mathrm{stp}(a, b/ce))$. Without loss of generality $c = \mathrm{Cb}(\mathrm{stp}(ab/ce))$ (up to interdefinability over $\emptyset$). (Note this preserves $a \in \mathrm{dcl}(bce)$, $b \in \mathrm{dcl}(ace)$, etc.)

We will suppress the parameter e (i.e. add it temporarily to the language).

Let $p = \mathrm{stp}(a/\emptyset)$, $q = \mathrm{stp}(b/\emptyset)$, and $r = \mathrm{stp}(ab/c)$. Using the language of Chapter 2 (section 6). we call the parallelism class of r, *a* generically defined invertible function from p to q, or also the *germ* of a definable invertible function from p to q.

As germs of functions will play a major role in this chapter, we take the opportunity again to give the general definition. Recall first that if r is a strong type, and B is a set of parameters containing $\mathrm{Cb}(r)$, then $r|B$ denotes the unique strong type over B which is parallel to r.

Definition 1.3. *Let p_1, q_1 be stationary types over $\emptyset$, say. By a germ of a definable function from p_1 to q_1 we mean (the parallelism class of) a strong type $r(x, y)$ which has the following features:*

(i) $r(x, y) \to p_1(x) \wedge q_1(y)$.

(ii) *For any (or equivalently some) B over which r is defined, if (a, b) realizes $r|B$ then a is independent from B, b is independent from B, and $b \in \mathrm{dcl}(a, B)$. If also $a \in \mathrm{dcl}(b, B)$ we say the germ is invertible.*

We will denote germs of functions by σ, τ, etc. The main point is that we can represent germs by elements, and the 'generic' action of the germ becomes definable. Simply represent the germ determined by some r as in the above definition, by $\mathrm{Cb}(r)$ (which is easily seen to be the definable closure of a single element). This can be made more 'explicit' in the following way. Suppose $r = \text{tp}(a, b/B)$, and suppose $f_1(a, d) = b$ where $d \in B$ and $f_1(\text{-},\text{-})$ is a partial function. Define $d_1 \, E \, d_2$ if for a' realizing $p_1 | \{d, d'\}$, $f_1(a', d_1) = f_1(a', d_2)$. Then d/E is a canonical base for r (and is in D^{eq}). We will usually consider germs of definable functions to be presented in this way. In any case, if σ is this germ and a realizes $p_1 | \sigma$ then $\sigma.a$ is the unique b such (a, b) realizes $r | \sigma$. Note that there is a $\varnothing$-definable function $f(\text{-},\text{-})$ such that for a realizing $p_1 | \sigma$, $\sigma.a = f(\sigma, a)$. Note also that if B contains σ and a realizes $p_1 | B$ then $\sigma.a$ realizes $q_1 | B$. (For $\sigma.a$ is independent with B over σ.)

Remark 1.4. *Suppose $r_1(x, y)$, $r_2(x, y)$ both satisfy* Definition 1.3 *(for p_1, q_1). Let $\sigma = \text{Cb}(r_1)$, $\tau = \text{Cb}(r_2)$. Then r_1 is parallel to r_2 (i.e. define the same germ from p_1 to q_1) if and only if for some a realizing $p_1 | \{\sigma, \tau\}$, $\sigma.a = \tau.a$ (i.e. there is b such that $\models r_1(a, b)$ and $\models r_2(a, b)$).*

Proof. Left to right is clear. Assume now the right-hand side. Then as $b \in \text{dcl}(\sigma, a)$, (a, b) is independent from τ over σ, so (a, b) realizes $r_1 | \{\sigma, \tau\}$. Similarly (a, b) realizes $r_2 | \{\sigma, \tau\}$. Thus r_1 is parallel to r_2. ■

We will usually be concerned with the family of conjugates of a strong type $r(x, y)$ as in Definition 1.3. If $r_1(x, y)$ is such a conjugate, σ is the germ corresponding to τ as defined above, and σ_1 is the corresponding conjugate of σ (under an automorphism taking r to r_1), then clearly σ_1 is the germ corresponding to r_1, and, moreover, $\sigma = \sigma_1$ if r is parallel to r_1.

It is important to note that germs can be composed. That is, suppose q_2 is another stationary type over $\varnothing$, σ is a germ from p_1 to q_1, τ is a germ from q_1 to q_2; then $\tau.\sigma$ denotes the germ from p_1 to q_2 determined by: for a realizing $p_1 | (\sigma, \tau\}$, $\tau.\sigma$ is the parallelism class of $\text{stp}((a, \tau(\sigma(a)))/\{\sigma, \tau\})$, or alternatively a suitable canonical base of this strong type. Note that $\text{tp}((a, \tau(\sigma(a)))/\{\sigma, \tau\})$ is stationary; thus its canonical base is in $\text{dcl}(\sigma, \tau)$. Thus with the above notation.

Remark 1.5. $\tau.\sigma \in \text{dcl}(\sigma, \tau)$.

There may be several choices for the element to represent $\tau.\sigma$, but if we are dealing with a finite family of germs, this can be done uniformly (i.e. in such a way that identical germs are represented by identical elements).

Let us now return to the situation at hand. We have p, q stationary types of U-rank 1, and $r(x, y) = \text{stp}(a, b/c)$ defining the germ of a definable invertible function from p to q. Let σ be the germ corresponding to r (so we could take σ to be c). Let s be $\text{stp}(\sigma)$. We denote by σ^{-1} the inverse of σ, that is the germ corresponding to $\text{stp}(b, a/c)$. By the above remarks, for any τ realizing s, $\tau^{-1}.\sigma$ is the germ of an invertible function from p to p (the inverse being $\sigma^{-1}.\tau$), and note that if σ and τ have the same type then also $\tau^{-1}.\sigma$ and its inverse have the same type.

Lemma 1.6. *Let σ, τ be independent realizations of s. Let s′ be* $\mathrm{stp}(\tau^{-1}.\sigma/\emptyset)$. *Then* $U(s')$ $(=U(\mathrm{stp}(\tau^{-1}.\sigma/\emptyset))) = 1$. *Moreover, if σ_1, σ_2 are independent realizations of s′ then $\sigma_1.\sigma_2$ also realizes $s'|\sigma_i$ for $i = 1, 2$.*

Proof. $U(s') = 1$, as s' is the strong type of the canonical base of a plane curve over p, and p is locally modular (see 2.2.6).

For the rest, first note that if $\sigma_1 = \tau^{-1}.\sigma$ (where σ, τ are independent realizations of s), then $\{\sigma, \tau, \sigma_1\}$ is pairwise independent. (For $U(\sigma/\emptyset) = U(\tau/\emptyset) = U(\sigma_1/\emptyset) = 1$, and by Remark 1.5 each of σ, τ, σ_1 is algebraic over the other two.)

Let σ_1, σ_2 be our independent realizations of s'. Let σ realize s, with σ independent of $\{\sigma_1, \sigma_2\}$. By what we have just said, there are τ, τ' realizing s with each of $\{\sigma, \tau\}$, $\{\sigma, \tau'\}$ independent, such that $\sigma_1 = \tau^{-1}.\sigma$ and $\sigma^2 = \sigma^{-1}.\tau'$. Clearly τ and τ' are independent, and $\sigma_1.\sigma_2 = \tau^{-1}.\tau'$, so realizes s'.

U-rank considerations (together with 1.5) show that $\sigma_1.\sigma_2$ is independent from each of σ_1, σ_2 over $\emptyset$. ■

Lemma 1.6 tells us that the set of realizations of s' is closed under 'generic composition'. As we shall see now this is enough to imply that the group of germs of functions from p into q generated by the realizations of $'$ is a connected (∞)-definable group (over $\mathrm{acl}(\emptyset)$) with generic type s'.

Lemma 1.7. *Let G be the semi-group of germs of functions from p to p generated by $\{\sigma\colon \sigma$ realizes $s'\}$. Then*

(i) *G is a group; and*

(ii) *for any $\tau \in G$ there are σ_1, σ_2 realizations of s′ such that $\tau = \sigma_1.\sigma_2$.*

Proof. (i) is because $\{\sigma\colon \sigma$ realizes $s'\}$ is closed under inverses. (ii) It is enough to see that if τ_i realize s' for $i = 1, 2, 3$ then there are σ_1, σ_2 realizing s' such that $\tau_1.\tau_2.\tau_3 = \sigma_1.\sigma_2$. So let σ realize s' with σ independent of $\{\tau_1, \tau_2, \tau_3\}$. Then clearly $\sigma^{-1}.\tau_2$ realizes s' and is independent from τ_3. Thus $\sigma^{-1}.\tau_2.\tau_3$ realizes s'. Also $\tau_1.\sigma$ realizes s'. Thus $\tau_1.\tau_2.\tau_3 = (\tau_1.\sigma).(\sigma^{-1}.\tau_2.\tau_3)$. ■

Corollary 1.8. *Let G be as in Lemma* 1.7. *Then G is (∞)-definable over $\emptyset$, connected, and minimal, with generic type s′. Moreover, the generic action of G on p is also definable.*

Proof. First for the definability. Identify G with $\{(\sigma_1, \sigma_2)/E\colon \sigma_1, \sigma_2$ realize $s'\}$, where E is the $\emptyset$-definable equivalence relation: $\sigma_1.\sigma_2 = \tau_1.\tau_2$ (E says: for a realizing $p|\{\sigma_1, \sigma_2, \tau_1, \tau_2\}$, $\sigma_1.(\sigma_2.a\} = \tau_1.(\tau_2.a))$. Also the generic action of G on p is $\varnothing$-definable. This easily follows from what we have just said and the fact that $\{(\sigma, a, b)$: $\sigma \in G$, a realizes $p|\sigma$, and $\sigma.a = b\}$ is invariant over $\varnothing$. Thus the multiplication on G is also $\varnothing$-definable: $\sigma_1.\sigma_2 = \sigma_3$ iff for a realizing $p|\{\sigma_1, \sigma_2, \sigma_3\}$, $\sigma_1.(\sigma_2.a) = \sigma_3.a$. Finally we need to know that s' is the unique generic type of G. It is easy to check that for any $\tau \in G$ and σ realizing $s'|\tau$, $\sigma.\tau$ realizes $s'|\tau$. This is enough, using Lemmas 1.6.6(ii) and 1.6.9(iv). ■

The proof of Theorem 1.1 is complete. We make some additional remarks.

REMARK **1.9.** *G is abelian (as it is minimal). As D is 1-based and G lives in D^{eq}, G is also 1-based. Thus, forking (algebraic closure) on the generic type s′ of G is explained by Proposition 4.5.8. It is linear dependence with respect to the division ring F of* acl(*e*)*-definable quasi-endomorphisms of G. The geometry associated to the type s′ is thus projective geometry over the division ring F. It follows that the geometry of any modular minimal type non-orthogonal to the original minimal set D is projective geometry over F.*

REMARK **1.10.** *The generic action of G on p is generically regular. In fact for any a, b realizing p there is unique $\sigma \in G$ such that σ is independent with a and $\sigma.a = b$. This is proved using the fact that G is abelian, as in the final claim of the proof of Theorem 2.4.17. Moreover, there is an* (∞-)*definable (over e) set P which contains the set of realizations of p, such that G acts definably and regularly on P, extending the generic action of G on p. For (working over e) let P be the set of pairs* (σ, a) *where $\sigma \in G$ and a realizes p, where we definably identify* (σ_1, a) *and* (σ_2, a) *if for τ realizing* $s'|\{\sigma_1, \sigma_2, a\}$, $(\tau\sigma_1).a = (\tau\sigma_2).a$. *Identifying a with the class of* $(1, a)$, *and defining the natural action of G on P, everything works.*

By Lemma 1.6.20, *one can also find a definable group G_1 containing G, a definable set P_1 containing P, and a definable regular action of G_1 on P_1 extending the action of G on P.*

Finally we can apply the group existence theorem above to the case of general 1-based theories. In section 4 we obtain a stronger result.

COROLLARY **1.11.** *Suppose T is* 1*-based. and non-trivial. Then some infinite group is definable in T.*

Proof. By Lemma 4.2.5, some minimal set D is non-trivial. By Theorem 1.1 (as D is locally modular), an ∞-definable infinite group is defined. By Lemma 1.6.18, there is a *definable* infinite group. ■

2 The geometry of a locally modular type

The aim of this section is to decribe the geometry of a locally modular minimal strong type p over $\emptyset$. Note that this is more than the question of describing the *eventual* geometry of p, that is the geometry of a non-forking extension (and clearly such considerations are crucial, as adjoining a single point makes the geometry modular).

If p is trivial, there is nothing to do. If p is non-trivial and modular then by Fact 2.1.12, the geometry is projective geometry over a division ring, although even in this case we have obtained a model-theoretic proof (see Remark 1.9). If p is non-trivial and non-modular, some more work is required. The proof here that the geometry is affine geometry over a division ring is rather indirect, and consists of (a) finding a modular type q over $\emptyset$, non-orthogonal to p, (b) finding a minimal type p' non-weakly orthogonal to p, such that the group $\mathrm{Aut}(p'^N/q^N)$ is $\emptyset$-definable and infinite. This group will be minimal, modular, and so satisfies the properties in Remark 1.9. Its action on p' gives p' the structure of affine space over the associated division ring.

So our first aim is to prove:

Proposition 2.1. *Let $p \in S(\emptyset)$ be a stationary, locally modular minimal type. Then there is some stationary modular minimal type q over* $\mathrm{acl}(\emptyset)$ *which is non-orthogonal to p. (In fact q can be found in p^{eq}.)*

Proof. Again we work completely in p^{eq}. We use dim and U-rank interchangeably. We may clearly assume p to be non-trivial. By Theorem 1.1 there is a minimal group G defined over some e with $U(e) = 1$, where, moreover, if a realizes the generic of G, then $U(e, a) = 2$. Write G as $G(e)$, and let $r = \mathrm{stp}(e)$. Let e' realize r with e' and e independent. By Remark 2.5.13, $G(e) \times G(e')$ is 1-based (over $\{e, e'\}$). As the generic types of $G(e)$ and $G(e')$ are both modular and as each of these is non-orthogonal to p, we conclude, by Lemma 2.5.5, that the generic types of $G(e)$ and $G(e')$ are non-weakly orthogonal over $\{e, e'\}$. By the results of Chapter 4 there is some minimal (i.e. connected and of U-rank 1) subgroup H of $G(e) \times G(e')$, such that H is neither $G(e) \times \{0\}$ nor $\{0\} \times G(e')$. (Explanation: $G(e) \times G(e')$ is 1-based over ee'. Let a_1 be generic of $G(e)$ over ee' and b_1 generic of $G(e')$ over ee' such that a_1 and b_1 are interalgebraic over ee'. Then by Proposition 4.4.5, $\mathrm{stp}(a_1, b_1/e, e')$ is a translate of the generic type of some $H \subseteq G(e) \times G(e')$. H is connected by 4.4.5. By choice of a_1 and b_1 H has generic type of U-rank 1. Thus H is minimal, and is clearly of the required form.) By 4.4.6, H is $\mathrm{acl}(e, e')$-definable. Let $(a, b) \in H$ be generic over $\{e, e'\}$. Then $U(a, e) = 2$, $U(a, e/b, e') = 1$. 1-basedness of $G(e) \times G(e')$ implies that $\mathrm{Cb}(\mathrm{stp}(a, e/b, e'))$ has U-rank 1. Let d be in this canonical base, $d \notin \mathrm{acl}(\varnothing)$. So $U(d) = 1$. We will show that $\mathrm{stp}(d)$ is modular. (As $d \in p^{\mathrm{eq}}$, $\mathrm{stp}(d/\varnothing)$ is non-orthogonal to p, so this will prove the proposition.) Let $q = \mathrm{stp}(d)$ and $q_1 = \mathrm{stp}(e, a, d)$. We write $H = H(e, e')$ (although it depends on $\mathrm{acl}(e, e')$).

CLAIM I. *$\{e, e', d\}$ is independent.*

Proof. Otherwise, $d \in \mathrm{acl}(e, e')$. But $a \in \mathrm{acl}(e, d)$, thus $a \in \mathrm{acl}(e, e')$, contradicting choice of H and (a, b). ■

Notation. $G_0(e) = \mathrm{acl}(e) \cap G(e)$. $D(G(e))$ is the division ring of (necessarily $\mathrm{acl}(e)$-)definable quasi-endomorphisms of $G(e)$ (called $D(G(e), e)$ in section 5 of Chapter 4). Elements of $D(G(e))$ 'are' connected 1-dimensional subgroups of $G(e) \times G(e)$. If α is such we write $\alpha.a = b \pmod{G_0(e)}$, if $\alpha(a + G_0(e)) = b + G_0(e)$.

So if $\alpha = H_1 \in D(G(e))$ is non-zero, then $H(e, e').H_1.H(e, e')^{-1}$ (composition of groups) is easily seen to be a U-rank 1 subgroup of $G(e') \times G(e')$, so its connected component is minimal (and so $\mathrm{acl}(e')$-definable). We call this $f_{e,e'}(\alpha) \in D(G(e'))$. ($f_{e,e'}$ is actually an isomorphism of rings.) What we are doing here is simply regarding $H(e, e')$ as an isomorphism between $G(e)/G_0(e)$ and $G(e')/G_0(e')$, which thus induces an isomorphism $f_{e,e'}$ between $D(G(e))$ and $D(G(e'))$.

We now do a rather funny 'construction'. Let M be a saturated model containing d say, such that e is independent from M over $\varnothing$. Let $Q = q^M$ (= the set of realizations of $\mathrm{stp}(d)$ which are in M). Let $X(e) = \{a' \in G(e) - G_0(e)$: there is $d' \in Q$, with $d' \in \mathrm{acl}(e, a')\}$. We will show

$(*)$ $X(e)$ is algebraically closed (inside $G(e) - G_0(e)$)

from which it will follow that $\{e\}$-algebraic closure on Q is modular. As Q is independent from e, it will follow that Q is modular over $\varnothing$. Let $Y(e) = \{a' \in G(e) - G_0(e)$: there is $d' \in Q$, $\mathrm{stp}(e, a', d') = q_1\}$. Note that $Y(e) \subseteq X(e)$, and that for any $a' \in X(e)$ there is $a'' \in Y(e)$ such that a' and a'' are interalgebraic over e. (Let $a' \in X(e)$, and $d' \in Q \cap \mathrm{acl}(e, a')$. As $\mathrm{stp}(d', e) = \mathrm{stp}(d, e)$ there is $a'' \in G(e)$ with $\mathrm{stp}(e, a'', d') = q_1.a'' \in \mathrm{acl}(d', e)$, so $a'' \in \mathrm{acl}(d, a')$.) So to establish $(*)$ it suffices to prove

$(**)$ $\mathrm{acl}(Y(e)) \subseteq X(e)$ (inside $G(e) - G_0(e)$).

CLAIM II. *If $a_1, a_2 \in Y(e)$, then $a_1 + a_2 \in X(e)$ (if $a_1 + a_2$ is generic over e), and if a_1 and a_2 are independent over e then $a_1 + a_2 \in Y(e)$.*

Proof. If $a_2 \in \mathrm{acl}(a_1 e)$, then also $a_1 + a_2 \in \mathrm{acl}(a_1 e)$, so is clearly in $X(e)$. We may therefore suppose a_1 and a_2 to be independent over e. Let d_i for $i = 1$, 2 be in Q such that $\mathrm{stp}(e, a_i, d_i) = q_1$, for $i = 1, 2$. Let e'' be a realization of r in M such that e'' is independent from $\{d_1, d_2\}$. In particular (by choice of e), e'' is independent from $\{e, d_1, d_2, a_1, a_2\}$. Thus we can clearly find $b_i \in G(e'')$ for $i = 1$, 2 such that $\mathrm{stp}(e, a_i, e'', b_i, d_i) = \mathrm{stp}(e, a, e', b, d)$ for $i = 1, 2$. In particular each of (a_i, b_i) is a generic of $H(e, e'')$. (Actually one should exercise a little care here, as $H(e, e')$ was $\mathrm{acl}(e, e')$-definable. However, one can fix an elementary map f taking $\mathrm{acl}(e, e')$ to $\mathrm{acl}(e, e'')$ and, fixing $\mathrm{acl}(e)$, choose b_i such that $\mathrm{tp}(f(\mathrm{acl}(e, e'), a_i, b_i, d_i)) = \mathrm{tp}(\mathrm{acl}(e, e'), a, b, d)$, and let $H(e, e'')$ be the image of $H(e, e')$ under f.)

It is clear that (a_1, b_1) and (a_2, b_2) are independent over ee''. Thus $(a_1 + a_2, b_1 + b_2) = (a', b')$ is also a generic of $H(e, e'')$ over ee''. Thus $\mathrm{tp}(f(\mathrm{acl}(e, e')), a', b') = \mathrm{tp}(\mathrm{acl}(e, e'), a, b)$ and we find d' realizing q such that $\mathrm{stp}(e, a', e'', b', d') = \mathrm{stp}(e, a, e', b, d)$. In particular $\mathrm{stp}(e, a', d') = q_1$, so $d' \in Q$ as long as we can show $d' \in M$. However, note that $b_i \in \mathrm{acl}(e'', d_i)$, so is in M. Thus $b' \in M$. As $d' \in \mathrm{acl}(e'', b')$, also $d' \in M$. Thus $d' \in Q$ and so in fact $a_1 + a_2 \in Y(e)$. ■

Claim III. *Let $\alpha \in D(G(e))$ be non-zero. Let $a_1 \in Y(e)$. Then there is $a_2 \in Y(e)$ such that $\alpha.a_1 = a_2 (\mathrm{mod}\ G_0(e))$.*

Proof. Let $d_1 \in Q$ witness that $a_1 \in Y(e)$ (i.e. $\mathrm{stp}(e, a_1, d_1) = q_1$). Again choose $e'' \in M$ independent with d_1 and thus also independent with $\{e, a_1\}$. Fix as before a suitable copy $H(e, e'')$ of $H(e, e')$. Let $b_1 \in G(e'')$ be such that (a_1, b_1) is a generic of $H(e, e'')$ over $\{e, e''\}$ and, moreover, such that $\mathrm{stp}(e, a, e', b, d) = \mathrm{stp}(e, a_1, e'', b_1, d_1)$. So note that $b_1 \in \mathrm{acl}(d_1, e'')$, so is in M. It is now not difficult to find $a_2 \in G(e)$ and $b_2 \in G(e'')$ such that $a_2 = \alpha.a_1 \bmod G_0(e)$, $b_2 = f_{e,e''}(\alpha).b_1 \bmod G_0(e'')$ and (a_2, b_2) is a generic of $H(e, e'')$ (over ee''). As $b_2 \in \mathrm{acl}(e'', b_1)$ (via $f_{e,e''}(\alpha)$), also $b_2 \in M$. Let d_2 be such that $\mathrm{stp}(e, a, e', b, d) = \mathrm{stp}(e, a_2, e'', b_2, d_2)$. Then $d_2 \in \mathrm{acl}(e'', b_2)$, so $d_2 \in Q$. Moreover, $\mathrm{stp}(e, a_2, d_2) = q_1$, whereby $a_2 \in Y(e)$. ■

We can now prove (**). Let $a' \in \mathrm{acl}(Y(e))$, a' generic of $G(e)$. By 4.5.8, there are $x_1, \ldots, x_n \in Y(e)$, $y_1, \ldots, y_n \in G(e)$, and non-zero $\alpha_1, \ldots, \alpha_n \in D(G(e))$ such that $y_i = \alpha_i.x_i \bmod G_0(e)$ and $a' = (y_1\sigma + \cdots + y_n) \bmod G_0(e)$, with $\{y_1, \ldots, y_n\}$ e-independent. By Claim III for each i there is $z_i \in Y(e)$ such that $z_i = \alpha_i.x_i$ mod $G_0(e)$, and thus $z_i - y_i \in G_0(e)$. Thus $a' - (z_1 + \cdots + z_n) \in G_0(e)$. So it is enough to show that $z_1 + \cdots + z_n \in X(e)$, which follows from Claim II by induction on n.

Thus $X(e)$ (with algebraic closure over e) is a subpregeometry of $G(e)$, and so is modular. Thus the geometry on Q is modular over e. As e is independent with Q, the geometry on $Q = q^M$ is modular. So q is modular. This completes the proof of Proposition 2.1. ■

Before continuing towards the main goal of this section, we point out an interesting fact which falls out of the proof of the above proposition.

PROPOSITION 2.2. *There is no minimal type $p \in S(\emptyset)$ whose pregeometry is projective geometry over an infinite division ring.*

Proof. We can restate the proposition as: there is no minimal modular $p \in S(\emptyset)$, whose associated geometry has infinite division ring F and such that for any a realizing p. $\mathrm{acl}(a) = \{a\}$ (inside p).

So suppose otherwise. Let $G(e)$, q, q_1, M, Q, $X(e)$, and $Y(e)$ be as given by the proof of Proposition 2.1. Let $P = p^M$. So p and q are both modular and thus by 2.5.5, non-weakly orthogonal. As, moreover, for a realizing p, $\mathrm{acl}(a) = \{a\}$ inside p, it follows that there is a $\emptyset$-definable function f such that for b realizing q, $f(b)$ realizes p. Fix $a \in Y(e)$. Thus there is $d \in Q$ with $\mathrm{stp}(e, a, d) = q_1$. Let $f(d) = c$. So $c \in P$. Let $Z = \{x: x \in G(e) - G_0(e)$, and for some y, $\mathrm{stp}(e, x, y) = q_1$ and $f(y) = c\}$. Clearly Z is an ∞-definable subset of $G(e)$. On the other hand also for any $x \in Z$, $x \in \mathrm{acl}(e, d)$ and thus Z is finite (by compactness, as $\mathrm{acl}(e, d)$ has a fixed size). We will also show that Z is infinite, yielding a contradiction. Note that $a \in Z$. The division ring F associated to the geometry on p is the same as $D(G(e))$, so the latter is infinite. Now for $\alpha \in D(G(e))$, there is by Claim III in the proof of Proposition 2.1 some $a_\alpha \in Y(e)$ such that $a_\alpha = \alpha.a \bmod G_0(e)$. In particular as $a_\alpha \in Y(e)$ there is some $d' \in Q$ with $\mathrm{stp}(e, a_\alpha, d') = q_1$. Let $f(d') = c' \in P$. Thus easily $c' \in \mathrm{acl}(e, c)$, As e is independent with M, $c' \in \mathrm{acl}(c)$ and thus $c' = c$. We have shown that $a_\alpha \in Z$. Clearly $\alpha \neq \beta$ implies $a_\alpha \neq a_\beta$ (in fact a_α and a_β are different mod $G_0(e)$). This shows Z to be infinite, giving the required contradiction. ■

We now aim towards.

PROPOSITION 2.3. *Let $p = \mathrm{stp}(a/\emptyset)$ be a locally modular non-modular minimal type. Then there is $a' \in \mathrm{acl}(a) - \mathrm{acl}(\emptyset)$, an $\mathrm{acl}(\emptyset)$-definable minimal group G, and an $\mathrm{acl}(\emptyset)$-definable regular action of G on $P = \{a'': \mathrm{stp}(a'') = \mathrm{stp}(a')\}$. G acts as a group of automorphisms ($\mathrm{acl}(\emptyset)$-elementary maps) of P.*

Proof. Let us work over $\mathrm{acl}(\emptyset)$. Again the proof goes through a number of stages, some of which are already done in some form in texts such as Poizat's *Groupes stables* [Po2]. In any case, the main point is definability of the so-called 'binding group'. As here the question of parameters (namely, their elimination) is crucial, we go through the whole proof. In the definability of this binding group, local modularity and minimality play no role.

First by Proposition 2.1 there is a modular minimal type $q \in S(\emptyset)$, non-orthogonal to p. We recall from Corollary 2.5.3:

($*$) if a realizes p, and B is any set of realizations of q, then a and B are independent (i.e. $a \notin \mathrm{acl}(B)$).

STEP I. *There are $a' \in \mathrm{acl}(a) - \mathrm{acl}(\emptyset)$, c independent of a', and a tuple* $\mathbf{b}$ *of realizations of q, such that $a' \in \mathrm{dcl}(c, \mathbf{b})$.*

Proof. Let a_1 realize p with a and a_1 independent. So $\mathrm{stp}(a/a_1)$ is modular and thus non-weakly orthogonal to $q|a_1$. There is b_1 realizing $q|a_1$ such that a and b_1 are interalgebraic over a_1. (Note that it is not necessarily the case that $a \in \mathrm{dcl}(a_1, b_1)$.) Let $a' \in \mathrm{Cb}(a_1, b_1/a) - \mathrm{acl}(\emptyset)$. Then a' is in the definable closure of a Morley sequence of $\mathrm{stp}(a_1, b_1/a)$. Let $((a_1, b_1), (a_2, b_2), \ldots)$ be such a Morley sequence, so for some n. $a' \in \mathrm{dcl}(a_1, b_1, a_2, b_2, \ldots, a_n, b_n)$. Then $\mathbf{b} = (b_1, \ldots, b_n)$ is a tuple of realizations of q, and clearly a' is independent from $(a_1, \ldots, a_n)$ over $\emptyset$. ■

Let $p' = \mathrm{tp}(a')$. So p' is locally modular, non-modular (and non-weakly orthogonal to p).

STEP II. *There is a sequence* $\mathbf{c}$ *of realizations of p' and a $\emptyset$-definable function $f(\mathbf{x}, \mathbf{y})$ such that for any a'' realizing p' there is a sequence* $\mathbf{b}$ *of realizations of q such that $a'' = f(\mathbf{c}, \mathbf{b})$.*

Proof. We first find $\mathbf{c}$ such that the above holds for a'' independent of $\mathbf{c}$. Let a', c, $\mathbf{b}$ be as given by Step I. Again considering a Morley sequence of $\mathrm{stp}(a', \mathbf{b}/c)$, we find that $a' \in \mathrm{dcl}(a_1, a_2, \ldots, a_n, \mathbf{b}_1, \mathbf{b}_2, \ldots, \mathbf{b}_n)$, where a' is independent from $\{a_1, a_2, \ldots, a_n\}$ and the a_i realize p'. So we can write $a' = g(\mathbf{a}, \mathbf{b})$ $(\mathbf{a} = (a_1, a_2, \ldots, a_n),\ \mathbf{b} = (\mathbf{b}_1, \mathbf{b}_2, \ldots, \mathbf{b}_n))$. Now let a'' be an arbitrary realization of p'. Let $\mathbf{a}'$ realize $\mathrm{tp}(\mathbf{a})$ such that $\mathbf{a}'$ is independent from $\{\mathbf{a}, a''\}$. In particular $\mathrm{tp}(a'', \mathbf{a}') = \mathrm{tp}(a', \mathbf{a})$ so there is $\mathbf{b}'$ such that $a'' = g(\mathbf{a}', \mathbf{b}')$. On the other hand each coordinate $\mathbf{a}_i'$ of $\mathbf{a}'$ is independent of $\mathbf{a}$ so again is of the form $g(\mathbf{a}, \mathbf{b}^i)$ for some $\mathbf{b}^i$. Composing these functions we see that $a'' = f(\mathbf{a}, \mathbf{b}', \mathbf{b}^1, \mathbf{b}^2)$, where $\mathbf{b}', \mathbf{b}^1, \mathbf{b}^2$ are sequences of realizations of q. Clearly the $\emptyset$-definable function f does not depend on a''. This finishes Step II. ■

Let $r = \mathrm{tp}(\mathbf{c})$ from Step II. Let $Q = \mathrm{acl}\{b\colon b \text{ realizes } q\}$. (We could take Q to be just the realizations of q, but never mind.) Let P be the set of realizations of p'. $\mathrm{Aut}(P/Q)$ denotes the (permutation) group of automorphisms of the universe which fix Q pointwise, acting on P. By ($*$), $\mathrm{Aut}(P/Q)$ is transitive.

As p' is stationary, for a', a'' realizing p', $\mathrm{tp}(a'/Q) = \mathrm{tp}(a''/Q)$. A back-and-forth argument using definability of types then shows that there is an automorphism of the universe fixing Q pointwise and taking a' to a''. In fact this shows that $\mathrm{Aut}(P/Q)$ is the same as the group of permutations of P which extend to Q-elementary maps.)

STEP **III.** $\mathrm{Aut}(P/Q)$ *is* $\varnothing$*-definable. That is, there is a* (∞)*-*$\varnothing$*-definable group* G*, and* $\varnothing$*-definable action of* G *on* P*, isomorphic to the action of* $\mathrm{Aut}(P/Q)$.

Proof. Let $\sigma \in \mathrm{Aut}(P/Q)$. We first claim that σ is determined by its effect on $\mathbf{c}$. For if $a'' \in P$, then by Step II there is $\mathbf{b} \in Q$ ($\mathbf{b}$ a sequence of realizations of q) such that $a'' = f(\mathbf{c}, \mathbf{b})$. So $\sigma(a'') = f(\sigma(\mathbf{c}), \mathbf{b})$. Now some $\mathbf{c}'$ realizing r is of the form $\sigma(\mathbf{c})$ for some σ iff $\mathrm{tp}(\mathbf{c}'/Q) = \mathrm{tp}(\mathbf{c}/Q)$. Let d be $\mathrm{Cb}(\mathrm{stp}(\mathbf{c}/Q))$. It is easy to see (by automorphisms) that $d \in \mathrm{dcl}(\mathbf{c})$, say $d = h(\mathbf{c})$. (This has nothing to do with 1-basedness. d is a single element because of our current assumptions (see Claim 2.2.2). In general d would be a set.) Let $r_0 = \mathrm{tp}(\mathbf{c}/d)$. So clearly $\mathrm{tp}(\mathbf{c}'/\mathbf{d}) = r_0$ is equivalent to $\mathrm{tp}(\mathbf{c}'/Q) = \mathrm{tp}(\mathbf{c}/Q)$. Moreover, note that $\mathrm{tp}(\mathbf{c}'/d) = r_0$ iff $\mathrm{tp}(\mathbf{c}') = r$ and $h(\mathbf{c}') = d$.

Thus, having fixed $\mathbf{c}$, $\mathrm{Aut}(P/Q)$ is in one-to-one correspondence with $\{\mathbf{c}' : \mathbf{c}' \text{ realizes } r_0\}$ (by $\sigma \to \sigma(\mathbf{c})$). Moreover, the action of σ on P is uniformly $\{\mathbf{c}, \sigma(\mathbf{c})\}$-definable: applying compactness to the implication, '$\mathbf{y}_1$, $\mathbf{y}_2$ are sequences of realizations of q, $\mathbf{x}$ realizes r_0, and $f(\mathbf{c}, \mathbf{y}_1)$, $f(\mathbf{c}, \mathbf{y}_2)$ are both defined and equal' $\to$ '$f(\mathbf{x}, \mathbf{y}_1) = f(\mathbf{x}, \mathbf{y}_2)$', gives us a formula $\phi(\mathbf{c}, \mathbf{x}, y, z)$ such that for any $a'' \in P$ and $\sigma \in \mathrm{Aut}(P/Q)$, there is unique z with $\phi(\mathbf{c}, \sigma(\mathbf{c}), a'', z)$, and this z is $\sigma(a'')$.

It is clear at this point that we have $\mathrm{Aut}(P/Q)$ and its action $\mathbf{c}$-definable. To get rid of the parameter $\mathbf{c}$, note that also for any $\mathbf{c}_1$ realizing r, any σ is determined by $\sigma(\mathbf{c}_1)$ and again $\phi(\mathbf{c}_1, \sigma(\mathbf{c}_1), y, z)$ defines the action of σ. Clearly the set $X = \{(\mathbf{c}_1, \sigma(\mathbf{c}_1)) : \mathbf{c}_1 \text{ realizes } r, \sigma \in \mathrm{Aut}(P/Q)\}$ equals $\{(\mathbf{c}_1, \mathbf{c}_2) : \mathbf{c}_1, \mathbf{c}_2 \text{ realizes } r \text{ and } h(\mathbf{c}_1) = h(\mathbf{c}_2)\}$, and so is ∞-definable.

The main remaining point is:

CLAIM. *Let* $(\mathbf{c}_1, \sigma(\mathbf{c}_1)), (\mathbf{c}_2, \tau(\mathbf{c}_2)) \in X$. *Then* $\sigma = \tau$ *iff for* $a'' \in P$ *independent of* $\{\mathbf{c}_1, \sigma(\mathbf{c}_1), \mathbf{c}_2, \sigma(\mathbf{c}_2)\}$, $\sigma(a'') = \tau(a'')$ (*i.e.*

$$(\forall z)(\phi(\mathbf{c}_1, \sigma(\mathbf{c}_1), a'', z) \leftrightarrow \phi(\mathbf{c}_2, \sigma(\mathbf{c}_2), a'', z))).$$

Proof. This is like the last part of Step II. Let $\mathbf{c}_3$ realize r independently of everything. By assumption $\sigma(\mathbf{c}_3) = \tau(\mathbf{c}_3)$. Let $a'' \in P$ be arbitrary. There is $\mathbf{b} \in Q$ with $f(\mathbf{c}_3, \mathbf{b}) = a''$. Thus $\sigma(a'') = \tau(a'')$. ■

By the claim and definability of types (specifically of p') there is a $\emptyset$-definable equivalence relation E on X such that $(\mathbf{c}_1, \mathbf{c}_2)$ and $(\mathbf{c}_3, \mathbf{c}_4)$ in X correspond to the same σ iff they are E-equivalent. It is then clear from the above that both the action of X/E on P and the group operation on X/E are $\emptyset$-definable. We call X/E with the group operation, G. This completes Step III. ■

Finally:

STEP IV. *G is minimal (i.e. connected and of U-rank 1).*

Proof. Let $\mathbf{c}$ be as above (given by Step II). We have seen that any $\sigma \in G$ is determined by its effect on $\mathbf{c}$. Note that for any $c \in P$, $P \subset \text{acl}(c, Q)$ (as $p'|c$ and $q|c$ are not weakly orthogonal). Thus $U(\mathbf{c}/Q) = 1$. Recall that $\text{tp}(\mathbf{c}/Q)$ was definable over d. Let σ realize a generic type of G over $\mathbf{c}, d$. Thus σ and $\sigma(\mathbf{c})$ are interdefinable over $\mathbf{c}d$. As $U(\sigma(\mathbf{c})/\mathbf{c}d)$ is at most 1, it follows that $U(\sigma(\mathbf{c})/\mathbf{c}, d) = 1$. Thus $U(\sigma/\mathbf{c}, d) = 1$. Moreover, as $\mathbf{c}$ and $\sigma(\mathbf{c})$ are independent realizations of the same (stationary) type over d (as $U(\sigma(\mathbf{c})/\mathbf{c}, d) = U(\sigma(\mathbf{c})/d) = 1$), $\text{stp}(\mathbf{c}, \sigma(\mathbf{c})/d)$ is unique. So (as $\sigma \in \text{dcl}(\mathbf{c}, \sigma(\mathbf{c}))$ in a fixed way), $\text{stp}(\sigma)$ is unique. G has a unique generic type of U-rank 1. So G is connected of U-rank 1. ■

G is abelian by minimality (again this does not depend on 1-basedness), and thus acts strictly transitively (regularly) on P (any transitive action of an abelian group is striclty transitive). This completes the proof of Proposition 2.3. ■

We can now conclude:

PROPOSITION 2.4. *Let $p \in S(\emptyset)$ be a stationary, minimal locally modular type. Then p is trivial, or p is non-trivial modular (in which case the geometry on p is projective over a division ring), or p is non-modular in which case the geometry associated to p (over $\emptyset$) is affine geometry over a division ring.*

Proof. We only need consider the non-modular case. Again we may work over $\text{acl}(\emptyset)$. Let a realize p. Let a' be as in Proposition 2.3, and $p' = \text{tp}(a')$. It is enough to show that the geometry on p' is affine. Let G be the $\emptyset$-definable minimal group which acts $\emptyset$-definably and regularly on the set P of realizations of p'. As the generic of G is non-orthogonal to p, G is locally modular and thus algebraic closure on G is as described by 4.5.8, that is linear dependence with respect to the division ring $F = D(G)$, mod G_0. In particular G is modular, and thus for any $A \subset G$ and $a'' \in P$,

(∗) a'' and A are independent.

For $x, y \in P$ let $y - x$ denote the unique $\sigma \in G$ such that $\sigma(x) = y$. (So σ and y are interalgebraic over x, and σ is independent with x). Thus x and y (in P) fork with each other iff $y - x \in G_0$. Moreover, for $x_0, x_1, \ldots, x_n$ in P we have

(∗∗) $\{x_0, \ldots, x_n\}$ is independent over $\emptyset$ iff $\{x_1 - x_0, \ldots, x_n - x_0\}$ is an independent (over $\emptyset$) set of generic points of G.

(Proof. $\{x_0, \ldots, x_n\}$ is independent over $\emptyset$ iff $U(x_0, \ldots, x_n) = n + 1$ iff $U(x_1, \ldots, x_n/x_0) = n$ iff $U(x_1 - x_0, \ldots, x_n - x_0/x_0) = n$ iff (by (∗))

$$U(x_1 - x_0, \ldots, x_n - x_0) = n \text{ iff } \{x_1 - x_0, \ldots, x_n - x_0\}$$

is an independent set of generic types of G over $\emptyset$.)

Let $\approx$ denote the equivalence relation of interalgebraity on P. Let x' denote $x/\approx$ for $x \in P$, and $\sigma' = \sigma/G_0$ for $\sigma \in G$. Then from the above G/G_0 acts regularly on $P/\approx$, and, moreover, $y' - x' = (y - x)'$. Thus by (∗∗) and our understanding of algebraic closure on G we have that $\{x_0, \ldots, x_n\}$ (in P) is independent iff $\{x'_1 - x'_0, \ldots, x'_n - x'_0\}$ is linearly dependent with respect to F in G/G_0. Thus the geometry on P is precisely that of an affine space over F. This completes the proof. ■

The following will be an important tool in the next chapter (classification of models of unidimensional superstable theories).

COROLLARY **2.5.** *Let $p \in S(\emptyset)$ be stationary, minimal, locally modular, and non-modular. Let A be any set. Then either there is a realizing p such that $a \in \operatorname{acl}(A)$ (in which case $p|A$ is modular), or $p|A$ is non-modular.*

Proof. This is like the proof of Lemma 2.6.5. We nevertheless give the proof. We may assume A is algebraically closed. The first case is clear, for $p|A$ is the non-forking extension of $p|a$ which is modular. Assume now that A contains no realization of p. Let $p' = \operatorname{stp}(a')$ where a' is as in Proposition 2.3. Let also G be as there. So p and p' are not weakly orthogonal. It is clearly enough to show that $p'|A$ is non-modular. Note that we still have that A contains no realization of p'. Let P = the set of realizations of p'. We will show

(∗) if B is a finite subset of P, then B is dependent over A iff B is dependent over $A \cap G$.

Let $B \subset P$ be such that B is dependent over A. We assume B is independent over $\varnothing$. Let B_1 be such that $\text{stp}(B_1/A) = \text{stp}(B/A)$ and B and B_1 are independent over A. Then (by 1-basedness) B is dependent over B_1. Also $\text{acl}(B) \cap \text{acl}(B_1) \subseteq A$. The same argument as that proving Lemma 2.6.4(ii) shows that there are independent b, c in $\text{acl}(B) \cap P$ and independent b', c' in $\text{acl}(B_1) \cap P$ such that $\dim(b, c, b', c') = 3$, and $\text{acl}(b, c) \cap \text{acl}(b', c') \cap P = \varnothing$. It easily follows from $(**)$ of Proposition 2.4 (with notation there) (i.e. two lines in a plane either meet or are parallel) that $c' - b' = \alpha(c - b)$ (mod G_0) for some non-zero $\alpha \in D(G)$. Thus $c - b \in \text{acl}(B) \cap \text{acl}(B_1) \subseteq A$. We have shown that B is dependent over $A \cap G$, proving $(*)$.

But $p'|A \cap G$ is clearly non-modular, for otherwise there is a'' realizing $p'|A \cap G$ and b generic of G over $A \cap G$ such that $a'' \in \text{acl}((A \cap G), b)$ contradicting $(*)$ in the proof of 2.4. Thus by $(**)$ $p'|A$ is non-modular. ■

3 A variant: pseudolinearity

Let D be a minimal set, defined over $\varnothing$, say. By a *plane curve* over D we mean some minimal type $q = \text{stp}(a_1, a_2/A)$, where $a_1, a_2 \in D$. We will say that a minimal set D is pseudolinear if there is k such that every plane curve over D has canonical base with U-rank (over $\varnothing$) at most k. An important step in the proof of Zilber's theorem in Chapter 2 was to show that D (there strongly minimal and ω-categorical) was pseudolinear with $k = 2$, from which linearity was deduced. A similar result holds in general: pseudolinearity implies linearity. Again the proof goes through finding a group, in this case a group of dimension k acting on a 1-dimensional set. We will then refer to Fact 1.6.25, in which the classification of such situations is given, and deduce that $k = 1$.

Definition 3.1. *Let D be a minimal set (∞)-defined over $\varnothing$. We say that D is k-pseudolinear if there is some plane curve q over D with* $\dim(\text{Cb}(q)) = k$, *and for every plane curve q over D,* $\dim(\text{Cb}(q)) \leq k$.

We leave it to the reader to show that k-pseudolinearity is preserved under non-orthogonality, and, moreover, that k-pseudolinearity is equivalent to 'eventual' k-modularity: that is there is a finite set A of parameters such that for a, $b \in D$ and $B \subseteq D$, if $a \in \text{acl}(b, B, A)$ then there is a subset B_1 of $\text{acl}(B, A) \cap D$ of cardinality at most k, such that $a \in \text{acl}(b, B_1, A)$.

We will prove:

Proposition 3.2. *If D is k-pseudolinear for some k then D is linear (i.e. locally modular).*

We first find a group (of dimension k acting on a 1-dimensional set). This involves a certain amount of repetition of arguments from section 1. We again work entirely in D^{eq}, and *over* $\mathrm{acl}(\varnothing)$.

Lemma 3.3. *There exist a, b, c, e, with $U(e) = k$, $U(a/e) = U(b/e) = 1$, a and b interdefinable over ce, $U(c/e) = k$, and $c = \mathrm{Cb}(\mathrm{stp}(a, b/c, e))$.*

Proof. First let $q = \mathrm{stp}(a_1, a_2/e)$ be a plane curve over D, with canonical base e and $U(e/\varnothing) = k$. Write q as q_e. Note that each of a_1, a_2 is independent from e over $\varnothing$. Let e' be independent from ea_1a_2, with the same type as e. Let b_2 be such that $\mathrm{stp}(a_1, b_2/ee') = q_{e'}|ee'$ (where $q_{e'}$ is the copy of q over e'). In particular b_2 is independent from ee'; thus we can find b_1 such that $\mathrm{stp}(b_1, b_2/ee') = q|ee'$. Let $E = \mathrm{acl}(e, e')$.

It is then clear that

Claim I. $\dim(a_1, a_2, b_1, b_2, E) = 2k + 1$.

Note that $\mathrm{stp}((a_1, a_2), (b_1, b_2)/E)$ is a plane curve over q_e. Let c be the canonical base of this plane curve.

Claim II. $U(c/e) = k$.
For by k-pseudolinearity, $U(c/e) \le k$. On the other hand $e' = \mathrm{Cb}(\mathrm{stp}(a_1, b_2/E))$, and $\mathrm{stp}(a_1, b_2/E)$ does not fork over c, so $e' \subseteq \mathrm{acl}(c)$. As $U(e'/e) = k$, $U(c/e) \ge k$.

We now attempt to replace (a_1, a_2) by a and (b_1, b_2) by b, in such a way that still $\dim(a/E) = \dim(b/E) = 1$, $c = \mathrm{Cb}(a, b/E)$, but such that now a and b are interdefinable over E.

Claim III. *Let b_2' be such that $\mathrm{tp}(b_2'/a_1, a_2, E) = \mathrm{tp}(b_2/a_1, a_2, E)$. Then b_2 and b_2' are interalgebraic (over $\varnothing$).*

Proof. Let $d = \mathrm{Cb}(\mathrm{stp}(a_2, b_2/E))$. Noting that $a_1 \in \mathrm{acl}(d, e', a_2)$ (so $\mathrm{tp}(a_1, a_2/E)$ does not fork over $\{d, e'\}$) we see that $e = \mathrm{Cb}(\mathrm{stp}(a_1, a_2/E)) \subseteq \mathrm{acl}(d, e')$. Thus $U(d/e') \ge k$. On the other hand k-pseudomodularity implies that $\dim(d) \le k$. Thus

$$(*) \qquad U(d) = k \text{ (and } d \text{ and } e' \text{ are independent).}$$

Now suppose by way of contradiction that b_2 is independent from b_2' over $\varnothing$. Now $b_2, b_2' \in \mathrm{acl}(e', a_1)$, $U(e') = k$, and $U(e', a_1) = k + 1$. Thus $U(e'/b_2b_2') = k - 1$. Similarly, using $(*)$ we see that $U(d/b_2b_2') = k - 1$. It follows that $U(e', d, b_2, b_2') \le 2(k - 1) + 2 = 2k$. As $\mathrm{acl}(e', d)$ contains e, this is a contradiction to I. ■

Replace b_2 by its (finite, by compactness) set of a_1a_2E-conjugates.

Claim IV. *Let b_1' be such that $\mathrm{tp}(b_1'/a_1b_2E) = \mathrm{tp}(b_1/a_1b_2E)$. Then b_1' is interalgebraic with b_1.*

Proof. As in Claim III. The role of d there is now played by $\mathrm{Cb}(\mathrm{stp}(b_1a_1/E))$. ■

So replacing b_1 by its finite set of a_1b_2E-conjugates, we have that I still holds and now $(b_1, b_2) \in \mathrm{dcl}(a_1a_2E)$. A similar argument shows that if a is the set of b_1b_2E-conjugates of (a_1, a_2), then $a \in \mathrm{dcl}(Eb_1b_2)$, $(b_1, b_2) \in \mathrm{dcl}(aE)$, and $a \in \mathrm{acl}(e, a_1)$. Put $b = (b_1, b_2)$. We now clearly have $U(a/e) = U(b/e) = 1$. $c' = \mathrm{Cb}(\mathrm{stp}(a, b/E))$ is interalgebraic with c over e and so $\dim(c'/e) = k$, a and b are interdefinable over $c'e$ (in fact over c'). The lemma is proved. ■

We add e to the language. Let $q_1 = \mathrm{stp}(a)$, $q_2 = \mathrm{stp}(b)$. Then $r =$ (the parallelism class of) $\mathrm{stp}(a, b/c)$ is clearly the germ of a definable invertible function from q_1 to q_2. We write this germ as σ, and put $s = \mathrm{stp}(\sigma)$. Note that $\dim(\sigma) = k$. If σ_1 and σ_2 are independent realizations of s, then $\sigma_1^{-1}.\sigma_2 = \tau$ is the germ of a definable invertible function from q_1 to itself. As q_1 is minimal and also k-pseudomodular, $\dim(\tau) \le k$ (as τ is a plane curve over q_1). As σ_i and τ are interdefinable over σ_{1-i}, it follows that $\dim(\tau) = k$, and $\{\sigma_1, \sigma_2, \tau\}$ is pairwise independent. Let $s' = \mathrm{stp}(\tau)$. As in Lemma 1.6, for τ_1, τ_2 independent realizations of s', $\tau_1.\tau_2$ realizes $s'|\tau_i$ $(i = 1, 2)$. As in Lemma 1.7, the group G of germs of definable functions from q_1 to itself, generated by the realizations of s', is ∞-definable (over $\varnothing$), connected, and has generic type s'. Moreover, the generic action of G on q_1 is $\varnothing$-definable and transitive. As in Remark 1.10, q_1 is contained in an ∞-definable set Q over $\varnothing$ such that the generic action of G on q_1 extends to a transitive effective action of G on Q, and q_1 is the generic type of Q with respect to this action. That is, (G, Q) is an ∞-definable homogeneous space with $U(Q) = 1$. Note that $U(G) = k$. Fact 1.6.25 says that $k = 1, 2$, or 3, and that if $k \neq 1$ then a minimal (connected of U-rank 1) algebraically closed field K is ∞-definable in D^{eq}. As D is assumed to be pseudolinear, so also K will be pseudolinear (pseudolinearity is preserved under non-orthogonality). We now point out that this is impossible (so k must be 1 and D is linear, so locally modular).

Remark 3.4. *Let K be a minimal set with a definable field structure (which must be that of an algebraically closed field). Then K is not pseudolinear.*

Proof. For every k we will find a plane curve p_k over K whose canonical base has dimension $\geq k$. Assume everything is defined over $\emptyset$. Let $\mathbf{a} = (a_0, a_1, \ldots, a_k)$ be independent generics of K. Let b be generic and independent of $\mathbf{a}$. Let $c = a_0 + a_1 b + \cdots + a_k b^k$. Then $\mathrm{stp}(b, c/\mathbf{a}) = r_{\mathbf{a}}$ is minimal. We claim that $\mathbf{a} \in \mathrm{Cb}(r_{\mathbf{a}})$. Let $\mathbf{a}'$ be a conjugate of $\mathbf{a}$. Suppose that $r_{\mathbf{a}}$ and $r'_{\mathbf{a}}$ are parallel (have a common non-forking extension). Then there is b, generic of K and independent from $\{\mathbf{a}, \mathbf{a}'\}$, such that $a_0 + a_1 b + \cdots + a_k b^k = a'_0 + a'_1 + \cdots + a'_k b^k$. This forces $\mathbf{a} = \mathbf{a}'$, so $\mathbf{a} \in \mathrm{Cb}(r_{\mathbf{a}})$. So $U(\mathrm{Cb}(r_{\mathbf{a}})) = k$. ■

The proof of Proposition 3.2 is complete.

Corollary 3.5. *Suppose that $p \in S(\emptyset)$ is minimal and not locally modular. Then for every k, there is a finite set B of realizations of p, and there is a realizing p such that $a \in \mathrm{acl}(B)$ but* $\mathrm{mult}(a/B) \geq k$. ($\mathrm{mult}(a/B)$ = *number of conjugates of a over B.)*

Proof. Work over $\mathrm{acl}(\emptyset)$. Fix $m > 1$. By Proposition 3.2, let q be a plane curve over p such that $\mathrm{Cb}(q)$ has dimension k. We may choose $q = \mathrm{tp}(a_1, a_2/d)$ where d is a finite set of realizations of p and, moreover, such that $c = \mathrm{Cb}(q) \in \mathrm{dcl}(d)$. (Note that a_1 and a_2 are independent over $\emptyset$). Let $\mathbf{a}^1, \ldots, \mathbf{a}^{k-1}, \mathbf{a}^k$ be the beginning of a Morley sequence in $\mathrm{tp}(a_1, a_2/d)$ (which we have chosen to be stationary). Note that by subadditivity, $U(c, \mathbf{a}^1, \ldots, \mathbf{a}^k\} = 2k$.

Claim I. *$\{\mathbf{a}^1, \ldots, \mathbf{a}^k\}$ is independent over $\emptyset$.*

Proof. Suppose not. Then $U(\mathbf{a}^k/\mathbf{a}^1, \ldots, \mathbf{a}^{k-1}\} = 1$. Thus $c \in \mathrm{acl}\{\mathbf{a}^1, \ldots, \mathbf{a}^{k-1}\}$. But then $U(c, \mathbf{a}^1, \ldots, \mathbf{a}^k) = U(\mathbf{a}^1, \ldots, \mathbf{a}^k) \leq 2(k-1) + 1 = 2k - 1$, contradicting the above. ■

Claim II. $U(c/\mathbf{a}^1, \ldots, \mathbf{a}^{k-1}) = 1$.

Proof. On the one hand $U(c, \mathbf{a}^1, \ldots, \mathbf{a}^{k-1}) = k + (k-1) = 2k - 1$ (subadditivity). By Claim I, $U(\mathbf{a}^1, \ldots, \mathbf{a}^{k-1}) = 2k - 2$. Now use subadditivity. ■

Now let c', d' be such that $\mathrm{stp}(c'd'/\mathbf{a}^1, \ldots, \mathbf{a}^{k-1}) = \mathrm{stp}(cd/\mathbf{a}^1, \ldots, \mathbf{a}^{k-1})$, and $c'd'$ and cd are independent over $\{\mathbf{a}^1, \ldots, \mathbf{a}^{k-1}\}$. By Claim II, $c \neq c'$, in fact $c' \notin \mathrm{acl}(c)$. As $c = \mathrm{Cb}(\mathrm{stp}(\mathbf{a}^1/c))$ and $c' = \mathrm{Cb}(\mathrm{stp}(\mathbf{a}^1/c'))$ it follows that $\mathbf{a}^1$ forks with c over c', and so $\mathbf{a}^1 \in \mathrm{acl}(cc')$. In particular $\mathbf{a}^1 \in \mathrm{acl}(dd')$. Let α be a permutation of $\{1, \ldots, k-1\}$. Then (as the $\mathbf{a}^i$ form a Morley sequence over cd), $\mathrm{tp}(\mathbf{a}^{\alpha(1)}, \ldots, \mathbf{a}^{\alpha(k-1)}, c, d) = \mathrm{tp}(\mathbf{a}^1, \ldots, \mathbf{a}^{k-1}, c, d)$. Similarly with c, d replaced by c', d'. As cd and $c'd'$ have the same strong type over, and are independent over, $\{\mathbf{a}^{\alpha(1)}, \ldots, \mathbf{a}^{\alpha(k-1)}\}$ it easily follows that $\mathrm{tp}(\mathbf{a}^1, \ldots, \mathbf{a}^{k-1}/cdc'd') = \mathrm{tp}(\mathbf{a}^{\alpha(1)}, \ldots, \mathbf{a}^{\alpha(k-1)}/cdc'd')$. As α was arbitrary and $c, c' \in \mathrm{dcl}(d, d')$ it follows that $\mathbf{a}^1$ has at least $k-1$ conjugates over $\{d, d'\}$, that is $\mathbf{a}^1, \ldots, \mathbf{a}^{k-1}$. This completes the proof (take B to be $\{d, d'\}$ and a to be the first coordinate of $\mathbf{a}^1$). ■

4 The group configuration for stable theories

So far we have produced groups under certain hypotheses (local modularity, pseudolinearity) by, in the sense, *ad hoc* methods (although the methods originate in the classical reconstruction of a vector space from a projective geometry). In these cases the group was found defined over a parameter of small rank (e.g. of rank 1 in the locally modular case), this being crucial for the classification of the geometries attached to locally modular types. It turns out that the existence of an (infinite) ∞-definable homogeneous space in a model $\mathbf{C}$ of a stable theory is equivalent to a certain combinatorial/model-theoretic property; the existence of a particular finite collection of points with specific algebraicity relations between them. In this section we prove this. A systematic use of canonical bases replaces the earlier U-rank calculations. However, in contradistinction with the locally modular case, given such a 'group configuration' over $\emptyset$ say, we will have less control over the additional parameters over which the homogeneous space is defined.

Again the fundamental work is to find a (infinite) family of germs of invertible definable functions from some strong type to itself, which is closed under generic composition. There will be a little repetition of arguments from section 1 and 3 of this chapter.

We begin with some remarks on homogeneous spaces. Let (G, S) be an ∞-definable homogeneous space, defined over $\emptyset$, say. Suppose G is connected and G acts effectively on S in the sense that if $g \in G$ fixes every $a \in S$ then $g = 1$. Let p be the generic type of S over $\emptyset$ (which is stationary as G is connected). Then G acts generically on p, in the sense that, for a realizing p, such that a is independent with g over $\emptyset$, then $g.a$ is also independent from g over $\emptyset$ (by 1.6.9(iv)). Thus any $g \in G$ defines the germ of an invertible function from p to p. Call this germ $g^\wedge$.

Remark 4.1. *Any $g \in G$ is determined by $g^\wedge$ (and consequently, in these circumstances g and $g^\wedge$ are* interdefinable).

Proof. It suffices to show that $g^\wedge$ is trivial (i.e. for a realizing $p|g$, $g.a = a$) if and only if $g = 1$. So suppose $g^\wedge$ is trivial. Let $a \in S$ be arbitrary. Let $b \in S$ be generic, and $h.b = a$. Let g_1 be generic in G over $\{g, h, a, b\}$. Then $g_1.g.h$ is generic in G over b. Note that the germ of $g_1.g.h$ at p equals the germ of

$g_1.h$ at p. Thus b is independent from (the 1-element set!) $\{(g_1.g.h)^\wedge, (g_1.h)^\wedge\}$ over $\varnothing$ (and b realizes p). Thus $(g_1.g.h).b = (g_1.h).b$. So $g.(h.b) = (h.b)$, that is $g.a = a$. As a was arbitrary, $g = 1$. ■

Let g_1, g_2 be generic A-independent elements of G. Let a_1 be generic in S over $A \cup \{g_1, g_2\}$. Let $a_2 = g_1.a_1$, $a_3 = g_2.a_2$, $g_3 = g_2.g_1$. Note that also $a_3 = g_3.a_1$. We have the following digram:

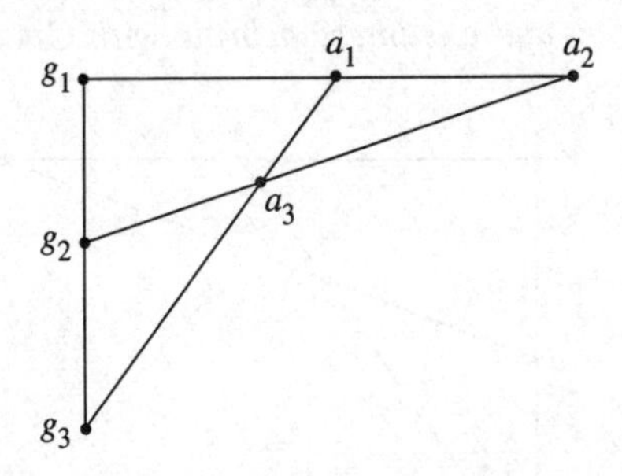

Properties of generic types in groups and homogeneous spaces (see section 6 of Chapter1) as well as Remark 4.1 imply

(i) for any *non-collinear* triple (x, y, z) of points in the diagram, $\{x, y, z\}$ is A-independent;

(ii) $\mathrm{acl}(A, g_1, g_2) = \mathrm{acl}(A, g_2, g_3) = \mathrm{acl}(A, g_1, g_3)$;

(iii) $\mathrm{acl}(A, g_1, a_1) = \mathrm{acl}(A, g_1, g_2)$, $\mathrm{acl}(A, g_2, a_3) = \mathrm{acl}(A, g_2, a_2)$, and

$$\mathrm{acl}(A, g_3, a_3) = \mathrm{acl}(A, g_3, a_1);$$

(iv) g_1 is interalgebraic over A with $\mathrm{Cb}(\mathrm{stp}(a_1, a_2/g_1, A))$, g_2 is interalgebraic over A with $\mathrm{Cb}(\mathrm{stp}(a_3, a_2/A, g_2))$, and g_3 is interalgebraic over A with $\mathrm{Cb}(\mathrm{stp}(a_3, a_1/A, g_3))$.

What we do is show that any diagram above satisfying (i), (ii), (iii), and (iv) comes from an ∞-definable homogeneous space.

We repeatedly make use of the following, which has been used before:

Fact 4.2. Suppose $c \in \mathrm{acl}(a) \cap \mathrm{acl}(b)$. Then $c \in \mathrm{acl}(\mathrm{Cb}(\mathrm{stp}(a/b)))$.

Proof. Let $B = \mathrm{Cb}(\mathrm{stp}(a/b))$. Thus (as $c \in \mathrm{acl}(a)$), c is independent from b over B. But $c \in \mathrm{acl}(b)$, and thus $c \in \mathrm{acl}(B)$. ■

Definition 4.3. *By a strict algebraic partial quadrangle over A we mean* *a 6-tuple of points (in* $\mathbf{C}^{\mathrm{eq}}$*),* (a, b, c, x, y, z)*, such that*

(i) *any triple of non-collinear points is A-independent (see the diagram below);*

(ii) $\mathrm{acl}(A, a, b) = \mathrm{acl}(A, a, c) = \mathrm{acl}(A, b, c)$;

(iii) *x and y are interalgebraic over Aa, y and z are interalgebraic over Ab, and z and x are interalgebraic over Ac.*

(iv) *a is interalgebraic with* $\mathrm{Cb}(\mathrm{stp}(xy/Aa))$ *over A, b is interalgebraic with* $\mathrm{Cb}(\mathrm{stp}(yz/Ab))$ *over A, and c is interalgebraic with* $\mathrm{Cb}(\mathrm{stp}((zx/Ac))$ *over A.*

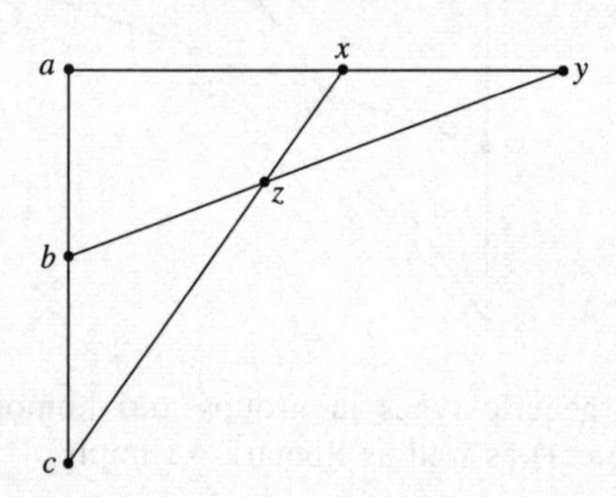

Remark 4.4. *Note that if each of a, b, c, x, y, z is replaced by an element interalgebraic with it over A, then the new 6-tuple* $(a_1, b_1, c_1, x_1, y_1, z_1)$ *is also a strict algebraic partial quadrangle over A. We say that this new partial quadrangle is* algebraically equivalent *to the first one over A.*

Theorem 4.5. *Suppose M to be a* $|T|^+$*-saturated model, and suppose* (a, b, c, x, y, z) *is a strict algebraic partial quadrangle over M. Then there is an* ∞*-definable homogeneous space (G, S) defined over M, with G connected, and there are elements* a', b', c' *of G, each generic over M, and* x', y', z' *of S, each generic over M, such that each primed element is interalgebraic with the corresponding non-primed element over M, and* $a'.x' = y'$, $b'.y' = z'$, $c'.x' = z'$*, and* $c' = b'.a'$.

Proof. As in the previous group constructions the first job is to replace the elements a, b, c, x, y, z by elements interalgebraic over M, such that z and y become *interdefinable* over $M \cup \{b\}$. Then we begin composing germs, and we construct a homogeneous space. The properties of the configuration are crucial in both steps.

So we begin with

Lemma 4.6. *There is a strict algebraic partial quadrangle* $(a_1, b_1, c_1, x_1, y_1, z_1)$, *algebraically equivalent to* (a, b, c, x, y, z) over M, such that $y_1 \in \mathrm{dcl}(b_1, z_1, M)$ and $z_1 \in \mathrm{dcl}(b_1, y_1, M)$.

Proof. Let $E \subseteq M$ be of cardinality at most $|T|$ such that $\mathrm{tp}(a, b, c, x, y, z/M)$ is definable over E. Then clearly (a, b, c, x, y, z) is a strict algebraic partial quadrangle over E. Let $a' \in M$ be such that $\mathrm{stp}(a'/E) = \mathrm{stp}(a/E)$. Then a' is independent from $\{b, z, y\}$ over E and thus $\mathrm{stp}(a', b, z, y/E) = \mathrm{stp}(a, b, z, y/E)$. We can then find c', x' such that $\mathrm{stp}(a', b, c', x', y, z/E) = \mathrm{stp}(a, b, c, x, y, z/E)$. So (i)–(iv) in Definition 4.3 also hold for (a', b, c', x', y, z) over E.

CLAIM I. *If* $\mathrm{tp}(z'/b, y, c', x', E) = \mathrm{tp}(z/b, y, c', x', E)$ *then* $\mathrm{acl}(E, z) = \mathrm{acl}(E, z')$.

Proof. Note that $\{b, c', z\}$ is E-independent. In particular b is independent from c' over zE. But $y \in \mathrm{acl}(b, z, E)$, and $x' \in \mathrm{acl}(c', z, E)$. Thus (b, y) is independent from (c', x') over zE. But $z \in \mathrm{acl}(b, y, E) \cap \mathrm{acl}(c', x', E)$. By Fact 4.2, z is interalgebraic (over E) with $\mathrm{Cb}(\mathrm{stp}(b, y/c', x', E))$. The same is true of z' (by automorphism). Thus z and z' are interalgebraic over E, proving the claim. ■

Note that (as $a' \in M$), $c' \in \mathrm{acl}(bM)$, $x' \in \mathrm{acl}(yM)$. Thus (a, bc', c, x, yx', z) is algebraically equivalent to (a, b, c, x, y, z) over M. Let z' denote the finite set (in $\mathbf{C}^{\mathrm{eq}}$) of conjugates of z over $M \cup \{bc', yx'\}$. By Claim I, z' is interalgebraic with z over M, and clearly $z' \in \mathrm{dcl}(bc'yx'M)$.

Rename $bc' = b$, $yx' = y$, $z' = z$, and we have that the new (a, b, c, x, y, z) is algebraically equivalent to the first one but now

II. $z \in \mathrm{dcl}(b, y, M)$.

By exactly the same procedure, we can enlarge a, y, and replace x by an appropriate finite set of conjugates, to obtain a new algebraically equivalent (over M) (a, b, c, x, y, z) such that

III. $x \in \mathrm{dcl}(a, y, M)$.

As z was not touched in this last step and a, y were only enlarged, II still holds.

Finally, let again $E \subseteq M$ be small such that $\mathrm{tp}(a, b, c, x, y, z/M)$ is definable over E. So (a, b, c, x, y, z) is a strict algebraic partial quadrangle over E, and III holds with M replaced by E. Let $c' \in M$ realize $\mathrm{tp}(c/E)$. As above, find a', x' such that $\mathrm{tp}(a', b, c', x', y, z/E) = \mathrm{tp}(a, b, c, x, y, z/E)$. So

$$(*) \qquad x' \in \mathrm{dcl}(a', y, E).$$

As in Claim I, if $\mathrm{tp}(y'/a', x', b, z, E) = \mathrm{tp}(y/a', x', b, z, E)$ then y' is interalgebraic with y over E. Now (a, ba', c, x, y, zx') is algebraically equivalent to

(a, b, c, x, y, z) over M (as $c' \in M$). Let y' be the finite set of conjugates of y over $M \cup \{a', x', b, z\}$. So y' is interalgebraic with y over M, and (a, ba', c, x, y', zx') is algebraically equivalent to (a, b, c, x, y, z) over M.

CLAIM IV. $y' \in \mathrm{dcl}(ba', zx', M)$.
(By choice of y'.)

CLAIM V. $zx' \in \mathrm{dcl}(ba', y', M)$.

Proof. By II and $(*)$, $zx' \in \mathrm{dcl}(b, a', y, M)$. As y' is the (finite) set of all conjugates of y over $M \cup \{z, x', b, a'\}$, we get Claim V. ■

Now put $a_1 = a$, $b_1 = ba'$, $c_1 = c$, $x_1 = x$, $y_1 = y'$, and $z_1 = zx'$. By Claims IV and V, the lemma is proved. ■

As far as proving the theorem is concerned we may as well use a, b, c, x, y, z to denote the new partial quadrangle given by 4.6.

We continue with the proof of Theorem 4.5. Let $q_1 = \mathrm{tp}(y/M)$, $q_2 = \mathrm{tp}(z/M)$. We will consider b as the germ of an invertible definable function from q_1 to q_2, and then show that for generic independent b_1, b_2 realizing $\mathrm{tp}(b/M)$, $b_1^{-1}.b_2$ is a germ of a definable invertible function from q_1 to q_1, and that the family of such germs is closed under generic composition.

First remember that by (iv) in Definition 4.3, b is interalgebraic (over M) with $\mathrm{Cb}(\mathrm{stp}(yz/bM))$. Also clearly $\mathrm{Cb}(\mathrm{stp}(yz/b, M))$ is interdefinable (over M) with a single element. Thus we may assume that b *is* this canonical base. So in terms of Definition 1.3 (the parallelism class of) $\mathrm{stp}(yz/bM)$ is the germ of an invertible definable function from q_1 to q_2, whose canonical base is b. Let $r = \mathrm{tp}(b/M)$. If b_1, b_2 realize r, then by $b_1^{-1}.b_2$ we mean (as in section 1) the germ of the invertible function from q_1 to q_1 obtained by applying b_2, then b_1^{-1}. Otherwise said, let y_1 realize $q_1|(M, b_1, b_2)$, and let $z_1 = b_2.y_1$. So z_1 realizes $q_2|(M, b_1, b_2)$. Let $y_2 = b_1^{-1}.z_1$ (i.e. $z_1 = b_1.y_1$); then $b_1^{-1}.b_2 = \mathrm{Cb}(\mathrm{stp}((y_1, y_2)/M, b_1, b_2))$ (or is rather interdefinable with it, over M).

LEMMA 4.7. *Let b_1, b_2 realize r $(= \mathrm{tp}(b/M))$ such that b_1 is independent from b_2 over M. Then $b_1^{-1}.b_2$ is independent from each of b_1, b_2 over M (and clearly each of b_1, b_2, $b_1^{-1}.b_2$ is M-definable over the other two).*

Proof. Without loss of generality, $b_2 = b$, and b_1 is independent from (a, b, c, x, y, z) over M. Thus $\mathrm{tp}(b, c, z, x/M) = \mathrm{tp}(b_1, c, z, x/M)$, so there are

a_1, y_1 such that $\text{tp}(a_1, b_1, c, x, y_1, z/M) = \text{tp}(a, b, c, x, y, z/M)$. The following diagram may help the reader:

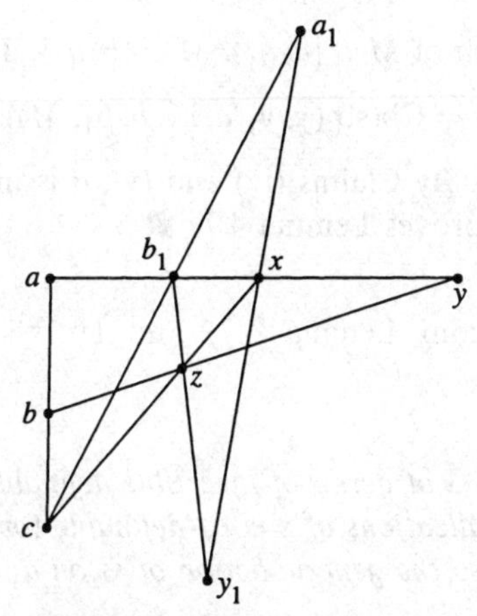

Claim i. *(a, a_1, b, b_1) is independent from y over M.*

Proof. y is independent from (a, b, c) over M. Also (a, b, c, y) is independent from b_1 over M. By forking calculus (a, b, c, b_1) is independent from y over M. But $a_1 \in \text{acl}(b_1, c, M)$. Thus y is independent from (a, a_1, b, b_1) over M. ■

Claim ii. $y_1 \in \text{acl}(a, a_1, y, M)$.

Proof. $x \in \text{acl}(a, y, M)$, and $y_1 \in \text{acl}(a_1, x, M)$. ■

Claim iii. $y_1 = (b_1^{-1}.b).y$.

Proof. Clear as (from (i)) y is independent from $\{b, b_1\}$ over M. ■

Claim iv. *(a, a_1) is independent from b over M.*

Proof. (a, b, c) is independent from b_1 over M (by choice of b_1), so (a, b) is independent from b_1 over $M \cup \{c\}$. As $a_1 \in \text{acl}(M, c, b_1)$, (a, b) is independent from a_1 over c. But a_1 is independent from c over M, so (a, b) is independent from a_1 over M. As a is independent from b over M, we conclude that (a, a_1) is independent from b over M. ■

Claim v. *(a, a_1) is independent from b_1 over M.*

Proof. Like Claim (iv). ■

Let us denote $b_1^{-1}.b$ by σ. Now, by Claims (i), (ii), and (iii),

$$\text{stp}(y, y_1/a, a_1, b, b_1, M)$$

does not fork over each of $M \cup \{a, a_1\}$, $M \cup \{b, b_1\}$. Thus

$$\sigma = \text{Cb}(\text{stp}(y, y_1/a, a_1, b, b_1, M)),$$

so $\sigma \in \text{acl}(M \cup \{a, a_1\})$. By Claims (iv) and (v), σ is independent from each of b, b_1 over M. This proves Lemma 4.7. ■

Let $\sigma = b_1^{-1}.b_2$ (from Lemma 4.7) and let $s = \text{tp}(\sigma/M)$. Note that $\text{tp}(\sigma^{-1}/M) = s$ also.

LEMMA **4.8.** *The group G of germs of invertible definable functions from q_1 to q_1 generated by the realizations of s is ∞-definable (over M), connected, and with generic type s. Also the generic action of G on q_1 is M-definable.*

Proof. We must check that the proofs of Lemmas 1.6 and 1.7 go through without using U-rank arguments.

1. We need (as in 1.6) that if σ_1, σ_2 are M-independent realizations of s, then $\sigma_1.\sigma_2$ realizes $s|M\sigma_i$ for $i = 1, 2$.

As in proof of 1.6 we choose σ realizing $r|(M \cup \{\sigma_1, \sigma_2\})$, and find (using 4.7) M-independent τ, τ' realizing r such that $\sigma_1 = \tau^{-1}.\sigma$ and $\sigma_2 = \sigma^{-1}.\tau'$. So $\sigma_1.\sigma_2 = \tau^{-1}.\tau'$ which realizes s. As σ_1 is independent from σ_2 over $M\sigma$, also τ is independent from τ' over $M\sigma$. By 4.7, τ is independent from σ over M, and thus $\{\tau, \tau', \sigma\}$ is M-independent. By 4.7, $\tau^{-1}.\tau'$ is independent from τ, and thus also from $\{\tau, \sigma\}$ over M. Thus $\tau^{-1}.\tau'$ is independent from σ_1 over M. That is, $\sigma_1.\sigma_2$ is independent from σ_1 over M. Similarly $\sigma_1.\sigma_2$ is independent from σ_2 over M.

2. We need (as in 1.7) that if τ_1, τ_2, τ_3 are realizations of s, then there are σ_1, σ_2 realizing s such that $\tau_1.\tau_2\tau_3 = \sigma_1.\sigma_2$.

Let again σ realize $s|(M \cup \{\tau_1, \tau_2, \tau_3\})$. By 1, $\sigma^{-1}.\tau_2$ realizes $s|(M \cup \{\tau_2\})$. Now $\{\sigma, \tau_2\}$ is independent from $\{\tau_1, \tau_2, \tau_3\}$ over $M\tau_2$. Thus $\sigma^{-1}.\tau_2$ is independent from $\{\tau_1, \tau_2, \tau_3\}$ over M. By 1, $\sigma^{-1}.\tau_2.\tau_3$ realizes s. Thus (as $\tau_1.\sigma$ also realizes s). we can take $\sigma_1 = \tau_1.\sigma$, and $\sigma_2 = \sigma^{-1}.\tau_2.\tau_3$.

From here on the proof of 1.8 goes through to yield Lemma 4.8. ■

Finally, as in Remark 1.10 we find an ∞-definable homogeneous space S for

G, whose generic type is q_1, and such that the action of G on S extends the generic action of G on q_1.

The only thing left to prove in Theorem 4.5 is the algebraic correspondence between the original partial quadrangle (a, b, c, x, y, z) and a natural one obtained from (G, S).

We may assume that (a, b, c, x, y, z) are as given in the italicized statement following the proof of Lemma 4.6, and also that $b = \mathrm{Cb}(\mathrm{stp}((y, z)/bM))$ (up to definability over M). So b realizes r, y realizes q_1, and z realizes q_2.

Let $E \subseteq M$ be small such that $\mathrm{tp}(a, b, c, x, y, z/M)$ is definable over E. Then, as above, the homogeneous space (G, S) is defined over E. Let $b_1 \in M$ (not to be confused with b_1 from Lemma 4.7) realize $\mathrm{tp}(b/E)$ $(=r \restriction E)$. Then $b_1^{-1}.b = \sigma$, say, realizes $s \restriction E$, the generic type of G over E. It easily follows (as σ is also independent from b_1 over E) that σ realizes s, the generic type of G over M, and σ is interdefinable with b over M.

As b_1 is independent from z over E we can find y_1 such that $\mathrm{tp}(b, z, y/E) = \mathrm{tp}(b_1, z, y_1/E)$. Then σ is independent from y over M and $\sigma.y = y_1$. Also y_1 is interdefinable with z over M. Thus, replacing b by σ and z by y_1, we have

new (a, b, c, x, y, z) *such that* b *realizes* s, z *realizes* q_1, *and* $b.y = z$.

Similarly we can choose a, x such that

a *realizes* s, x *realizes* q_1, *and* $a.x = y$.

This is done as follows. Choose $E \subseteq M$ as before, and let c_1 realize $\mathrm{tp}(c/E)$. Let b_1, z_1 be such that $\mathrm{tp}(a, b_1, c_1, x, y, z_1/E) = \mathrm{tp}(a, b, c, x, y, z/E)$. Note that a is interalgebraic with b_1 over M, and x is interalgebraic with z_1 over M. Clearly b_1 realizes s, z_1 realizes q_1 and $b_1.y = z_1$. Thus $b_1^{-1}.z_1 = y$. Replace a by b_1^{-1}, and x by z_1.

Finally, we still have that c is interalgebraic (over M) with

$$\mathrm{Cb}(\mathrm{stp}(z, x/c, M)).$$

But clearly $b.a.$ is this canonical base (up to M-algebraicity). So replace c by $b.a$. The new partial quadrangle satisfies the conditions on the primed elements stated in Theorem 4.5. The proof of Theorem 4.5 is complete. ■

Remark **4.9.** *Suppose that* (a, b, c, x, y, z) *and* A *satisfy Definition* 4.3, *except for* (iv), *but assume that we also have* $a \in \mathrm{acl}(x, y, A)$, $b \in \mathrm{acl}(z, y, A)$, *and* $c \in \mathrm{acl}(z, x, A)$. *Then by Fact* 4.2, *in fact* (iv) *is satisfied too. So assuming that* A *is a saturated model* M, *Theorem* 4.5 *applies, yielding* (G, S). *However, the additional hypotheses on the quadrangle (i.e.* $a \in \mathrm{acl}(x, y, A)$ *etc.) allow us to choose the algebraically equivalent (over* M*) quadrangle* (a', b', c', x', z') *such that all elements realize the generic type* s *of* G, *and* $a.x = y$, $b.y = z$, $c.x = z$.

(This is accomplished by modifying the final part of the proof of Theorem 4.5, *and is left to the reader, the main point being that if σ realizes s, y realizes q_1, and $z = \sigma.y$, then $\sigma \in \mathrm{acl}(y, z, M)$.)*

REMARK 4.10. *Suppose D is a minimal set defined over the saturated model M, and (a, b, c, x, y, z) is a strict algebraic partial quadrangle over M, all of whose points are in D^{eq} (or in D^{eq}_M). Of course Theorem* 4.5 *applies. But suppose in addition that $U(y/M) = 1$. Then $U(x/M)$, $U(z/M)$ are also* 1, *and the homogeneous space (G, S) obtained satisfies $U(S) = 1$. The possibilities are described in Fact* 1.6.25.

If D is a strongly minimal set, M need only be a model (not a saturated model). (This is also left to the reader.)

COROLLARY 4.11. *Suppose D is a minimal set, defined over the saturated model M (or D is strongly minimal, defined over a model M). Suppose (a, b, c, x, y, z) is a strict algebraic partial quadrangle over M, such that each element is in D^{eq} (so, for example, a tuple from D). Suppose the U-rank over M of each of these points is* 1. Then *there is $d \in \mathrm{acl}(a, z, M) \cap \mathrm{acl}(b, x, M)$, $d \notin M$ (so geometrically speaking the lines az and by intersect).*

Proof. As in Remark 4.9, we may assume that the points a, b, c, x, y, z are generic points (over M) of some connected ∞-definable group G, and $a.x = y$, $b.y = z$. But as $U(G) = 1$ (and G is connected) G is abelian. Hence $z = (b.a).x = (a.b).x$, and so $a^{-1}.z = b.x$ (a generic point of G over M). ■

REMARK 4.12. Corollary 4.11 *apparently gives new information even in the special case where D is an algebraically closed field of positive characteristic.*

COROLLARY 4.13. *Suppose that T is a 1-based theory. Let $p(x) \in S(A)$ be a non-trivial type. Then there are $B \supseteq A$, some non-forking extension $p' = \mathrm{tp}(a/B)$ of p, and some connected non-trivial ∞-definable group G defined over B, such that some generic element (over B) of G is algebraic over $B \cup \{a\}$.*

Proof. Assume for notational simplicity that $A = \varnothing$. As p is non-trivial, there is some c, and realizations a_1, a_2 of $p|c$ such that a_1 is independent from a_2 over $\varnothing$ but a_1 forks with a_2 over c. Choose $c' \in \mathrm{Cb}(\mathrm{stp}(a_1, a_2/c))$ such that a_1 forks with a_2 over c'. By 1-basedness $c' \in \mathrm{acl}(a_1, a_2)$. Choose $a_1' \in \mathrm{Cb}(\mathrm{stp}(a_2, c'/a_1))$ such that $c' \in \mathrm{acl}(a_2, a_1')$. Then $a_1' \in \mathrm{acl}(a_2, c')$. Finally choose $a_2' \in \mathrm{Cb}(\mathrm{stp}(a_1', c'/a_2))$ such that a_1' and c' are interalgebraic over a_2'. Again $a_2' \in \mathrm{acl}(a_1', c')$. Now $a_1' \in \mathrm{acl}(a_1) - \mathrm{acl}(\varnothing)$, $a_2' \in \mathrm{acl}(a_2) - \mathrm{acl}(\varnothing)$, and

$c' \in \mathrm{acl}(c) - \mathrm{acl}(\varnothing)$. Thus $\{a_1', a_2', c'\}$ is pairwise independent over $\varnothing$, but each is in the algebraic closure of the other two.

Let a_2'', c'' be chosen independent from $\{a_1', a_2', c'\}$ over a_1' such that $\mathrm{stp}(a_2'', c''/a_1') = \mathrm{stp}(a_2', c'/a_1')$. Then $\{a_2', c'', c', a_2''\}$ is 3-wise independent, but each element is in the algebraic closure of the other three. By 1-basedness (a_2', c'') is independent from (c', a_2'') over $\mathrm{acl}(a_2', c'') \cap \mathrm{acl}(c, a_2'')$. Clearly there is an element d which is interalgebraic with this last set over $\varnothing$. Then it is easy to see that $(a_1', a_2', c', a_2'', d, c'')$ is a strict algebraic partial quadrangle over $\varnothing$, such that, moreover, each line is algebraic over any two points on the line. Let M be a saturated model independent from $(a_1', a_2', c', a_2'', d, c'')$ over $\varnothing$. By Theorem 4.5 there is a connected ∞-definable group over M and some generic element g of G over M such that g is interalgebraic with a_1'. Thus $g \in \mathrm{acl}(a_1, M)$, and $\mathrm{tp}(a_1/M)$ is the required non-forking extension p. ■

Remark 4.14. *In fact, with notation from the proof of* 4.13, *and using* 1-*basedness, one can choose the* (*principal*) *homogeneous space* (G, S) *to be defined over* $\mathrm{acl}(A \cup \{a_1\})$, *using methods as in the proof of Theorem* 1.1.

6
Unidimensional theories

Zilber's theorem (proved in section 4 of Chapter 2) played a fundamental role in proving 'global' results about totally categorical theories and/or ω-categorical, ω-stable theories. Similarly the information gained in Chapters 4 and 5 about the structure of locally modular minimal types (together with the strongly minimal/locally modular dichotomy theorem from section 3 of Chapter 2) can be used to help prove 'global' results about theories of finite U-rank. The theorems we prove in this chapter belong to 'classical' classification theory, that is concern the number of models in certain cardinalities of certain theories. However, the geometric information is crucial in their proof. Basically in this chapter we are concerned with unidimensional superstable theories. There are two main problems here: (i) the spectrum function for uncountable unidimensional theories, (ii) the number of countable models of countable unidimensional theories. We shall prove partial results in both cases. Regarding (i) note that special cases of unidimensional theories are (by Theorem 1.5.18) those T which are λ-categorical for some $\lambda > |T|$. We classify all the models of such T. In regard to (ii) we prove that a countable unidimensional theory *of U-rank* 1 has countable many or continuum many countable models. The first result follows fairly directly from the local theory developed in Chapter 5, whereas the second requires more work.

T is simply a stable theory, unless other hypotheses are explicitly stated.

1 Groups and envelopes

The main aim of this section is to prove an omitting types theorem for superstable unidimensional theories, which in some contexts gives prime models. If T is unidimensional, not totally transcendental, and nontrivial, then for any set A of parameters there is a model M containing A which omits any non-modular rank 1 type $p \in S(\mathrm{acl}(A))$. This will require yet another group existence theorem (at least in the case of uncountable T).

The typical example of a superstable, unidimensional, non-trivial, non-totally transcendental theory is the following. Let V be an infinite-dimensional vector space over $\mathbf{F}_2$ equipped with an infinite descending chain of subgroups $V = V_0 > V_1 > \ldots$, where V_{i+1} has index 2 in V_i for all i.

The theory of the resulting structure has quantifier elimination (in the language with symbols $+$, 0, and predicates for the V_i).

Apart from the geometric theory we will make constant reference to the material in section 5 of Chapter 1 on unidimensional theories, specifically Proposition 5.14 and Theorem 5.18. We also recall Proposition 2.5.11: if T is superstable, unidimensional, and not totally transcendental, then T is 1-based of finite U-rank.

In section 2 of Chapter 4 we defined the notion of a *trivial* (stable) theory. Note that (by Lemma 4.2.3) if T is superstable and unidimensional, then T is trivial iff for some (any) stationary type of U-rank 1 is trivial.

Proposition 1.1. *Let T be countable, superstable, unidimensional, and not ω-stable. Then T is nontrivial, and, moreover, there is weakly minimal group G defined over* $\operatorname{acl}(\emptyset)$.

Proof. We know T is 1-based. Let $\phi(x)$ be a formula over $\emptyset$ with $R^\infty(\phi) = 1$. So ϕ does not have Morley rank (by 1.5.14(iii)) whereby easily there are continuum many non-algebraic types $p \in S(\operatorname{acl}(\emptyset))$ which contain ϕ. If these types (all of U-rank 1) are all modular, then by Corollary 2.5.5, they are weakly non-orthogonal; that is to say, if a realizes one of them, then $\operatorname{acl}(a)$ contains realizations of all of them. But then $\operatorname{acl}(a)$ is uncountable, which contradicts the countability of T. Thus we have some type $p \in S(\operatorname{acl}(\emptyset))$ of U-rank 1, which is non-modular. In particular p is non-trivial, and thus T is non-trivial. Moreover, by Proposition 5.2.3, there is an $\operatorname{acl}(\emptyset)$-($\infty$-)definable minimal group G. By Remark 1.6.21(ii) G is a subgroup of a group H which is definable (by a single formula) over $\operatorname{acl}(\emptyset)$, and with $U(H) = 1$. It easily follows that H is weakly minimal. (The reader should check that any formula with U-rank 1 must be weakly minimal, whether the ambient theory is unidimensional or not.) ■

On the other hand an uncountable, unidimensional, non-totally transcendental theory may very well be trivial, as the following example shows. Let $G = ({}^{\omega}2, +)$. Let M be the structure whose universe is ${}^{\omega}2$, and which has unary predicates $\{P_\sigma : \sigma \in {}^{\omega>}2\}$ and unary functions $\{f_g : g \in G)$. $x \in M$ satisfies P_σ if σ is an initial segment of x, and for $g \in G$, $x \in M$, $f_g(x) = x + g$. By the dimension of a model N of $Th(N)$ we mean $\dim(p, N)$ for any rank 1 type p. This is well defined. A model is determined by its dimension, and any dimension ≥ 1 can occur. As we see now this is typical.

Proposition 1.2. *Let T be unidimensional, non-totally transcendental, and trivial. Let $p(y) \in S(\operatorname{acl}(\emptyset))$ with $U(p) = 1$. Then for any model M of T which* *realizes* p, $M = \mathrm{acl}(p^M)$. *Moreover, if* A *is any non-empty set of realizations of* p *then* $\mathrm{acl}((A)$ *is a model (i.e. an elementary substructure of* $\mathbf{C}$).

Proof. Note T is 1-based of finite U-rank. We prove the moreover clause first. Since the hypotheses are preserved under naming elements, and we can proceed by transfinite induction, we may assume that A is a single realization a of p. Let $B = \mathrm{acl}(a)$. Let $\varphi(x, b)$ ($b \in B$) be consistent. We want to show that $\varphi(x, b)$ is realized in B.

CLAIM. *We may assume that* $R^\infty(\varphi(x, b)) = 1$.

Proof of claim. Suppose c realizes $\varphi(x, b)$ and $R^\infty(c/B) = n$. By 2.5.9 there is $d \in \mathrm{acl}(cB)$ with $U(d/B)$ $(=R^\infty(d/B)) = 1$. Thus $R^\infty(c/dB) = n - 1$, and (by 1.5.14(ii)) there is a formula $\chi(x, d, b')$ ($b' \in B$) of ∞-rank $n - 1$ true of c, and a formula $\psi(w, b')$ true of d, such that if d' realizes $\psi(w, b')$ then $R^\infty(\chi(x, d', b')) = n - 1$. We may assume that $\models \chi(x, w, b') \rightarrow \varphi(x, b)$ and that $R^\infty(\psi(w, b')) = 1$. So if we could satisfy $\psi(w, b')$ in B, then we could also satisfy $\varphi(x, b)$ in B, by induction on n. ∎

Now if c is any realization of $\varphi(x, b)$ with $c \notin B$, then $\mathrm{tp}(c/B)$ (which has U-rank 1) is, by 2.5.5 and triviality, non-weakly orthogonal to $p|B$, that is c is interalgebraic over B with some realization d of p, witnessed by some formula $\chi(c, d)$ over B. (So $\chi(x, y)$ implies $x \in \mathrm{acl}(y, B) \wedge y \in \mathrm{acl}(x, B)$.) Let $\sigma(y)$ be $(\exists x)(\varphi(x, b) \wedge \chi(x, y))$. Then $\sigma(y) \in p|B$. *So it is enough to show that* $\sigma(y)$ *is realized in* B.

Without loss of generality we write $\sigma(y)$ as $\sigma(y, b)$ (where the 'new' b is in B, extends a, and $\sigma(y, z)$ is over $\varnothing$). We may assume (as $R^\infty(p) = 1$) that $\models \sigma(y, b) \rightarrow \sigma_1(y)$, for some σ_1 over $\varnothing$ with $R^\infty(\sigma_1) = 1$. Let $q(z) = \mathrm{stp}(b/\varnothing)$. Let $\delta(y)$ be the $\sigma(y, z)$-definition of $q(z)$. Then $\delta(y)$ is in $p(y)$, and also $\models \delta(y) \rightarrow \sigma_1(y)$; hence $R^\infty(\delta(y)) = 1$. As T is non-totally transcendental, $\delta(y)$ extends to infinitely many rank 1 strong types. All of these are non-weakly orthogonal to p, so are realized in $B = \mathrm{acl}(a)$. On the other hand, clearly as $R^\infty(\delta(y)) = 1$, $\models \delta(y) \wedge y \notin \mathrm{acl}(b) \rightarrow \sigma(y, b)$. By compactness, all but finitely many realizations of $\delta(y)$ satisfy $\sigma(y, b)$. Thus $\sigma(y, b)$ is realized in B. So we have proved the moreover clause.

For the rest: let M be a model of T. By what we have just done $\mathrm{acl}(p^M)$ is an elementary substructure M_1 of M. If $M_1 \neq M$, then by 2.5.9, there is some $c \in M - M_1$ such that $U(c/M_1) = 1$. By triviality c is interalgebraic over M_1 with a realization of p, which contradicts $p^M \subseteq M_1$. ∎

We see that if T satisfies the hypotheses of the above proposition, then any model M of T is determined by the cardinality κ of a maximal independent set of realizations of p in N, where p is any rank 1 strong type over $\varnothing$ (and by 2.5.9, there does exist such a p). We call $\kappa = \dim(M)$. It follows that T is κ-categorical for $\kappa > |T|$. As any non-zero dimension can occur, we also know the number of models of T of cardinality at most $|T|$. (Of course there are examples where $\operatorname{acl}(\emptyset)$ is also a model: for example, name some realization of p.)

We now prove some lemmas which will be of use in the non-trivial situation.

Proposition 1.3. *Let $p \in S(\operatorname{acl}(\emptyset))$ be weakly minimal, non-trivial, and without Morley rank. Then there is a weakly minimal group G defined over* $\operatorname{acl}(\emptyset)$, *whose generic types are non-orthogonal to p.*

Proof. We will work inside θ^{eq} for some weakly minimal formula $\theta \in p$. So the results 1.5.8–1.5.13 (following Assumption 1.5.7) (as adapted in 1.5.16) can be (and will be) used.

Note that if T is countable, then this already follows from (the proof of) Proposition 1.1.

The point is to find an invertible germ from a rank 1 type to a rank 1 type, defined over $\operatorname{acl}(\emptyset)$.

First, by Proposition 2.3.2, p is locally modular.

Claim I. *There is a formula $\varphi(x) \in p$ with $R^\infty(\varphi) = 1$, such that every non-algebraic strong type containing φ is non-orthogonal to p.*

Proof of claim. By non-triviality of p, we can find a finite set A of realizations of p, and realizations a, b of p, such that $\{a, b, A\}$ is pairwise independent, but dependent. By linearity and the (finite) U-rank equalities $\operatorname{Cb}(\operatorname{stp}(a, b/A))$ is interalgebraic over $\emptyset$ with some c such that $U(\operatorname{tp}(c/\emptyset)) = 1$. Thus $\{a, b, c\}$ is pairwise $\emptyset$-independent, but dependent (in fact $\operatorname{acl}(a, b) = \operatorname{acl}(a, c) = \operatorname{acl}(b, c)$). Let $\chi(x, y, z)$ (over $\operatorname{acl}(\emptyset)$) be a formula true of (a, b, c) which 'says' that x and y are interalgebraic over z. Let $\psi(z)$ be a formula over $\operatorname{acl}(\emptyset)$ true of c, with $R^\infty(\psi) = 1$. Recall that $\theta(x)$ was chosen at the beginning of the proof to be a weakly minimal formula in p. Let $\varphi(x)$ be a formula over $\operatorname{acl}(\emptyset)$ expressing: '$\theta(x)$ and, for y realizing $p|x$, there is z satisfying ψ such that $\chi(x, y, z)$'. This formula exists by definability of the type p. We show that φ satisfies the claim. For let a' satisfy φ, $a' \notin \operatorname{acl}(\emptyset)$. Let b' realize $p|a'$, and c' realize ψ such that $\chi(a', b', c')$. So a' and b' are independent, and a' and b' are interalgebraic over c'. Thus $c' \notin \operatorname{acl}(a')$. So (as both $U(a'/\emptyset), U(c'/\emptyset) = 1$), we have $a' \notin \operatorname{acl}(c')$, that is a' is indepenent from c' over $\emptyset$. Similarly b' is independent with c' over $\emptyset$. We see that $\operatorname{stp}(a')$ is non-orthogonal to p, and the claim is proved. ■

We now work inside φ^{eq} and we also work over $\operatorname{acl}(\emptyset)$ (i.e. we name elements of acl($\emptyset$)). As in the beginning of the proof of the claim, find a, b, c (in φ^{eq}) each of ∞-rank 1, pairwise independent, but dependent. Choose such a, b, c with mult(tp(c/ab)) minimized.

CLAIM II. *If* tp(c'/ab) = tp(c/ab) then c and c' are interalgebraic over $\emptyset$.

Proof. Let mult(c/ab) = k, witnessed by a formula $\chi(x, y, z)$ (over $\emptyset$) true of (a, b, c) (i.e for all a', b' there are at most k elements z such that $\models \chi(a', b', z)$). Let σ_1 be a weakly minimal formula true of a, and σ_2 a weakly minimal formula true of b (both over $\emptyset$). Note that if tp(c'/ab) = tp(c/ab) then c and c' are independent over $\emptyset$ if and only $a, b \in \mathrm{acl}(c, c')$ (by computing the U-rank of tp(a, b, c, c') two ways). It follows that if Claim II fails then there are

(i) a formula $\psi(x, y, w, z)$ over $\emptyset$ which implies '$x, y \in \mathrm{acl}(w, z)$'; and

(ii) a formula $\sigma(x)$ over $\emptyset$ which is true of a and says: '$\sigma_1(x)$ and, for z_1 realizing stp($c/\emptyset$)$|x$, there is y satisfying $\sigma_2(x)$ with $\chi(x, y, z_1)$, and, moreover, if $z_2, \ldots, z_k$ are distinct and also different from z_1 and if $\chi(x, y, z_i)$ for all i, then for some $j = 2, \ldots, k$, $\psi(x, y, z_1, z_j)$'.

SUBCLAIM. *The formula* $\sigma(x)$ *has Morley rank* 1.

For let a' satisfy σ with $a' \notin \mathrm{acl}(\emptyset)$. Let c_1 realize stp($c/\emptyset$)$|a'$. Let b' be such that $\models \sigma_2(b')$ and $\models \chi(a', b', c_1)$. It easily follows that $\{a', b', c_1\}$ are pairwise independent, but each is algebraic over the other two. Moreover, each element has ∞-rank 1. By our minimal choice of k, mult($c_1/a', b'$) $\geq k$. So we can find distinct $c_2, \ldots, c_k$ such that tp($c_i/a'b'$) = tp($c_2/a'b'$) for $i = 2, \ldots, k$. As a' satisfies σ, there is i such that $\models \psi(a', b', c_1, c_i)$. So $a', b' \in \mathrm{acl}\{c_1, c_i\}$ whereby c_1 and c_i are independent over $\emptyset$. Thus c_1 and c_i are independent realizations of the stationary type p, stp(c_1, c_i) is fixed, and, moreover, $a' \in \mathrm{acl}(c_1, c_i)$ via a fixed formula. Thus there are only finitely many choices for tp(a') (and thus for stp(a') as we are working over acl($\emptyset$)). Thus σ has Morley rank 1. But any non-algebraic type containing σ is non-orthogonal to stp($c/\emptyset$) and hence non-orthogonal to p. Thus p has Morley rank 1, a contradiction. Claim II is proved. ■

Thus if we let c_1 denote the (finite) set of conjugates of tp(c/ab), it follows that $R^\infty(\mathrm{tp}(c_1)) = 1$, $\{a, b, c_1\}$ are pairwise independent and dependent, but now $c_1 \in \mathrm{dcl}(a, b)$. Similarly we can replace b by b_1 such that $b_1 \in \mathrm{dcl}(a, c_1)$. Replace a by Cb(stp($b_1, c_1/a$)) = a_1.

a_1 is the germ of an invertible definable function from stp($b_1/\emptyset$) to stp($c_1/\emptyset$), and $U(a_1/\emptyset) = 1$. Now inside φ^{eq} (where we are working) all U-rank 1 types are non-orthogonal (by Claim I) and thus locally modular. Lemmas 5.1.6, 5.1.7, and 5.1.8 in the proof of Theorem 5.1.1 then apply to our situation, and yield a connected minimal group G (in φ^{eq}) defined over $\emptyset$.

By 1.6.21(ii), G embeds in a $\emptyset$-definable weakly minimal group H (also in φ^{eq}). Thus the generic types of H are non-orthogonal to p. ■

The next lemma will permit us to build models omitting modular types (in the right context).

LEMMA **1.4.** *Let G be a weakly minimal abelian group defined over $\emptyset$, such that some (every) generic type of G is locally modular and such that G does not have Morley rank 1. Let $\varphi(x)$ be a non-algebraic formula over $\emptyset$ such that $\models \varphi(x) \rightarrow$ '$x \in G$' and such that $\models \neg\varphi(a)$ for all $a \in \operatorname{acl}(\emptyset) \cap G$. Then there is some non-algebraic formula $\psi(x)$ over $\operatorname{acl}(\emptyset)$ such that $\models \psi(x) \rightarrow \varphi(x)$ and for all a satisfying $\psi(x)$, $\operatorname{stp}(a/\emptyset)$ is non-modular.*

Proof. We emphasize here that G is assumed to be defined by a single formula. Also recall from Chapter 4, section 5, that the generic type $p \in S(\emptyset)$ of G^0 is *modular*, and G is 1-based. (G^0 is 1-based as its unique non-algebraic type is modular, so by 4.5.11 so is G.) Note also that as $RM(G) \neq 1$, G is not connected by finite, so in particular, every definable subgroup of finite index of G has a proper subgroup which is $\operatorname{acl}(\emptyset)$-definable and of finite index.

We will first prove the lemma in the following special case:

Case I. $\varphi(x) \in p$ (i.e. φ is true of every $a \in G^0 - \operatorname{acl}(\emptyset)$). By Corollary 4.4.8, $\varphi(x)$ is equivalent to a disjunction of conjunctions of formulae, each of which is over $\operatorname{acl}(\emptyset)$ and defines a coset or the complement of a coset of an $\operatorname{acl}(\emptyset)$-definable subgroup of G. Clearly any definable subgroup of G is finite or has finite index. Thus if we choose a disjunct satisfied by a generic of G^0, this disjunct must have the form '$x \in G_1 \wedge x \notin C$', where G_1 is an $\operatorname{acl}(\emptyset)$-definable subgroup of finite index, and C is a finite $\operatorname{acl}(\emptyset)$-definable subset of G. As G is not connected by finite (G does not have Morley rank 1), there is an $\operatorname{acl}(\emptyset)$-definable subgroup G_2 of finite index in G which is contained in G_1 and such that

$$(*) \qquad \models x \in G_2 \wedge x \notin G^0 \rightarrow x \notin C.$$

As $x \in G_1 \wedge x \notin C$ implies $\varphi(x)$ and $\varphi(x)$ is not satisfied by any algebraic elements, it follows that

$$(**) \qquad \models x \in G_2 \wedge x \notin G^0 \text{ implies } x \notin \operatorname{acl}(\emptyset).$$

Let now G_3 be a proper $\operatorname{acl}(\emptyset)$-definable subgroup of G_2, of finite index. Thus by $(*)$ the formula $\psi(x) =$ '$x \in G_2 \wedge x \notin G_3$' is over $\operatorname{acl}(\emptyset)$, non-algebraic, and implies $\varphi(x)$. We will show that ψ works. We work in the 1-based weakly minimal group G_2. Suppose by way of contradiction that there is b satisfying $\psi(x)$ such that $\operatorname{stp}(b/\emptyset)$ is modular. Thus $\operatorname{stp}(b)$ and p

are not weakly orthogonal, so there is a realization a of p such that a and b are interalgebraic over $\emptyset$.

By Proposition 4.4.5, $\operatorname{stp}(a, b/\emptyset)$ is the generic type of a ($\operatorname{acl}(\emptyset)$-definable) coset X of a connected $\operatorname{acl}(\emptyset)$-$\infty$-definable subgroup H of $G_2 \times G_2$. By connectedness of H, clearly

(i) $H \subset G^0 \times G^0$.

As $a, b \notin \operatorname{acl}(\emptyset)$, both the first and second projections of H on G^0 are equal to G^0, and as H is minimal

(ii) $H(0, y)$ is a finite subgroup of G^0.

As $a \in G^0$, clearly X is of the form $H + (0, c)$. As X is $\operatorname{acl}(\emptyset)$-definable, by (ii) $c \in \operatorname{acl}(\emptyset) \cap G_2$. By $(**)$, $c \in G^0$. But then, $b - c \notin G^0$, and, moreover, $(a, b - c) \in H$. This contradicts (i). We have proven the lemma in Case I.

Now let $\varphi(x)$ be an arbitrary formula satisfying the hypotheses of the lemma. If already for any a satisfying φ, $\operatorname{stp}(a/\emptyset)$ is non-modular, then there is nothing to do. Otherwise suppose a satisfies φ, and $\operatorname{stp}(a/\emptyset)$ is modular. So a is interalgebraic over $\emptyset$ with a realization b of p, via a formula $\chi(x, y)$. Let $\sigma(y) = \exists x(\varphi(x) \wedge \chi(x, y) \wedge y \in G)$. Then $\sigma(y)$ satisfies the hypotheses of the lemma and also of Case I. So by what we have just done there is non-algebraic $\sigma_1(y)$ implying σ such that for every b realizing σ_1, $\operatorname{stp}(b/\emptyset)$ is non-modular. Let $\psi(x)$ be the formula $\varphi(x) \wedge \exists y(\sigma_1(y) \wedge \chi(x, y))$. This clearly works (as if $\operatorname{stp}(b)$ is non-modular and $a \in \operatorname{acl}(b) - \operatorname{acl}(\emptyset)$ then also $\operatorname{stp}(a)$ is non-modular). ■

COROLLARY **1.5.** *Let T be unidimensional, non-totally transcendental, and non-trivial. Let A be any set. Let $\varphi(x)$ be a weakly minimal formula over A. Then either φ is realized in* $\operatorname{acl}(A)$ *or φ is not realized in* $\operatorname{acl}(A)$ *and there is a weakly minimal formula $\psi(x)$ over* $\operatorname{acl}(A)$, *such that $\models \psi(x) \rightarrow \varphi(x)$, and for all realizations a of ψ,* $\operatorname{stp}(a/A)$ *is non-modular.*

Proof. By our assumptions and Proposition 1.3, there is a weakly minimal group G defined over $\operatorname{acl}(\emptyset)$. Let $p \in S(\operatorname{acl}(\emptyset))$ be the generic type of G^0 and $p_1 = p|\operatorname{acl}(A)$. We may assume that φ is not realized in $\operatorname{acl}(A)$ and that for some a realizing φ, $\operatorname{stp}(a/A)$ is modular. But then $\operatorname{stp}(a/A)$ is now weakly orthogonal to the modular type p_1. The corollary now follows from Lemma 1.4 (working over A) by a similar argument to that in the last part of the proof of 1.4. ■

We now return to the notion of envelopes, which appeared in section 6 of Chapter 2 in the context of totally categorical theories.

Definition 1.6. *Let T be superstable, unidimensional, non-totally transcendental, and non-trivial. Let p be a fixed strong type over $\varnothing$, such that $U(p) = 1$ and p is modular. (Note that there always is such a type. For as T is 1-based of finite rank, Proposition 2.5.9 gives a U-rank 1 type over $\varnothing$. Now we can apply Proposition 1.3).*

(i) *Let M be a model and $A \subseteq M$. A p-envelope of A in M is a maximal set $B \subseteq M$ such that B contains A and $p|A$ is not realized in* $\mathrm{acl}(B)$.

(ii) *For any A, a p-envelope of A is a p-envelope of A in the universe* $\mathbf{C}$.

Proposition 1.7. (T *and* p *as in Definition* 1.6) (i) *If $A \subseteq M$, then any p-envelope of A in M is a model (which by convention means an elementary submodel of* $\mathbf{C}$).

(ii) *If M_1 is a p-envelope of A, then M_1 is ∞-homogeneous over A. Moreover, if M_2 is another p-envelope of A then M_1 and M_2 are isomorphic over A (by an elementary map in the sense of the universe).*

Proof. (i) Let B be a p-envelope of A in M. Clearly B is algebraically closed. To show B is a model, we show that for every consistent formula $\varphi(x)$ over B, φ is satisfied in B, by induction on $R^{\infty}(\varphi) = n$. If $n = 0$, this is clear. Suppose $n = 1$. If φ is not satisfied in B, then let $\psi(x)$ over B be given by Corollary 1.5. Let a realize ψ in M. So $\mathrm{stp}(a/B)$ has U-rank 1 and is non-modular. So clearly the modular type $p|B$ is not realized in $\mathrm{acl}(Ba) \subseteq M$. This contradicts maximality of B. Definability of ∞-rank and coordinatization enable us to continue the induction, as in the proof of Proposition 1.2. Details are left to the reader.

(ii) Let M_1 be a p-envelope of A (which is a model by (i)). Let $B \subseteq M_1$ and let f be an A-elementary map with domain B and with $f(B) \subseteq M_1$. Let $a \in M_1$. We want to extend f to $B \cup \{a\}$. We may clearly assume B (and so $f(B)$) to be algebraically closed. We may also assume, by induction on U-rank etc., that $U(a/B) = 1$. Let $r = \mathrm{tp}(a/B)$. Then r is not modular (as otherwise it is not weakly orthogonal to $p|B$ which would also be realized in M_1). Let $q = f(r)$. So, by automorphism $q \in S(f(B))$ is a stationary, U-rank 1, non-modular, locally modular type. By 5.2.5, if q is not realized in M_1, then $p|M_1$ is non-modular. But then, if b realizes $q|M_1$, $p|M_1$ is not realized in $\mathrm{acl}(M_1 b)$. This contradicts maximality of M_1. Thus q is realized in M_1 by some b, and f can be extended by sending a to b.

The proof of the rest of (ii) is similar, using a back-and-forth argument, and is left to the reader. ■

2 Theories categorical in a higher power

Here we give a Baldwin–Lachlan type structure theorem for the models of a theory T which is categorical in some $\lambda > |T|$, meaning that we generalize

Theorem 1.5.19(ii) to uncountable theories. In fact we also prove Theorem 1.5.19.

We redefine the cardinality of a complete theory T (in a language L) to be the number of L-formulae modulo T-equivalence. We may also clearly assume that $|L| = |T|$ (so the number of sorts in T^{eq} has also cardinality at most $|T|$.)

We again refer to section 5 of Chapter 1 (1.5.14 and 1.5.18) for the basic facts about unidimensional theories and theories categorical in a higher power.

We will prove:

THEOREM 2.1. *Suppose T is λ-categorical for some $\lambda > |T|$. Then either T is totally categorical (and countable) or there are models M_κ of T for $\kappa \geq 0$ such that*

(i) *M_0 is prime and minimal (i.e. has no proper elementary substructure);*

(ii) *M_κ is elementarily embeddable in M_λ iff $\kappa \leq \lambda$;*

(iii) *every model of T is isomorphic to M_κ for some (unique) κ; and*

(iv) *M_κ has cardinality $|T| + \kappa$ for all infinite κ (and in fact if κ is infinite then M_κ is ω-saturated).*

Also, if T is non-totally transcendental and non-trivial, then for some modular weakly minimal $p \in S(\mathrm{acl}(\varnothing))$, M_κ is the p-envelope of a Morley sequence $\{a_i : i < \kappa\}$ in p (also for $\kappa = 0$).

The proof separates into cases.

Case I. T is totally transcendental.
This is basically just a rewrite of the Baldwin–Lachlan proof in the uncountable setting, so we are brief and refer to the exposition in our earlier book [P1] for more details.

First recall that a *Vaughtian pair* for a theory T is a pair of models $M < N$ together with formula $\varphi(x)$ over M such that φ^N is infinite, but contained in M. It is well known that a stable unidimensional theory has no Vaughtian pairs. (Sketch proof: Suppose that M, N, φ are a Vaughtian pair for T. Let $a \in N - M$, and let $p(x) = \mathrm{tp}(a/M)$. Let M' be some a-model containing M, and $p'(x) \in S(M')$ the non-forking extension of p. It is then easy, using definability of types, to show that no realization of p' can fork over M' with any element satisfying φ. Now as M' is an a-model, we can easily find a type $q(x) \in S(M')$ which contains $\varphi(x)$ and has U-rank 1 (i.e. q is non-algebraic but any forking extension of q is algebraic). Let a' realize p' and let N' be a-prime over $M' \cup a'$. As for all $c \in N' - M'$, a' forks with c over M', it follows that $\varphi^{N'} \subseteq M'$, and in particular q is not realized in M'. By 1.4.3.4(iv), p' does not dominate q. As q is regular, it follows from 1.4.4.2 that p' is orthogonal to q, contradicting unidimensionality.)

In any case, returning to our case hypothesis, we know that T has no Vaughtian pairs. T also (being totally transcendental) has a prime constructible (so atomic) model over any set A, which is, moreover, unique up to A-isomorphism. Let M_0 be the prime model of T. Just as in the countable case there is a strongly minimal formula $\theta(x, a)$ over M_0, with associated 'generic' type $p \in S(a)$. So we can make the following assertions:

(a) If $M < N$ are models of T and $a \in M$, then $\theta(x, a)^N$ properly contains $\theta(x, a)^M$. (Moreover, if $b \in N - M$ satisfies $\theta(x, a)$ then $\mathrm{tp}(b/M) = p|M$.)

(b) So for every model M containing a, M is prime and minimal over $\theta(x, a)^M$.

As $\theta(x, a)^M$ is contained in the algebraic closure of $\{a\}$ together with a maximal $\{a\}$-algebraically independent subset of $\theta(x, a)^M$, it follows that

(c) For any M containing a, if I is a p-basis for M (i.e. a maximal a-independent set of realizations of p), then M is prime over $\{a\} \cup I$.

As p is a stationary type, $\mathrm{tp}(I/a)$ is determined by $|I|$, that is by $\dim(p, M)$.

We must get rid of the dependence on the parameter a. Note that $\mathrm{tp}(a)$, being isolated, is realized in any model. If b realizes $\mathrm{tp}(a)$, p_b denotes the copy of p over b. (So $p_a = p$.) As in the countable case one proves

(d) For any model M, and realizations b, c in M of $\mathrm{tp}(a)$, $\dim(p_b, M) = \dim(p_c, M)$.

We now proceed with the proof.

Subcase Ia. $\dim(p_a, M_0)$ is infinite.
Then actually $\dim(p_a, M_0)$ is countable. For let I be a countable Morley sequence in p_a, and let M be prime (and so atomic) over $\{a\} \cup I$. It is easy to see that $p_a|I \cup \{a\}$ is not realized in M (as I being infinite, this type could *not* be isolated), whereby $\dim(p_a, M)$ is countable. As M_0 is embedded in M, by (d) we see that $\dim(p_a, M_0)$ is countable.

Moreover, also note that for every finite $\{a\}$-independent set A of realizations of p_a, $p_a|A \cup \{a\}$ is isolated (as A is without loss contained in M_0, M_0 is atomic over $\{a\} \cup A$, and M_0 realizes $p_a|A$). Thus for every such A, $\mathrm{acl}(Aa) \cap \theta(x, a)^{\mathbf{C}}$ is finite. Thus if I is the countably infinite basis of p_a in M_0, $\theta(x, a)^{M_0} = \mathrm{acl}(aI) \cap \theta(x, a)^{\mathbf{C}}$ is countable. By 'no Vaughtian pairs', M_0 is countable.

CLAIM. *M_0 is ω-saturated.*

Let $B \subset M_0$ be finite, and $r \in S(B)$. We will show r is realized in M_0, by induction on $U(r)$. Suppose $U(r) = n$. Let r' be the non-forking extension of r to M_0. By (a) above, if c is a realization of r' then there is d realizing $p_a|M_0$, such that $d \in \operatorname{acl}(cM_0)$. We may assume that B contains a and that, moreover, $d \in \operatorname{acl}(cB)$. As $\dim(p_a, M_0)$ is infinite and B is finite we can (by superstability) find $d' \in M_0$ realizing p_a such that d' is independent from B over a, whereby $\operatorname{tp}(Bd) = \operatorname{tp}(Bd')$. Let c' be such that $\operatorname{tp}(Bcd) = \operatorname{tp}(Bc'd')$. Then $\mathrm{U}(\operatorname{tp}(c'/Bd')) = n - 1$, and by induction $\operatorname{tp}(c'/Bd')$ and thus also r is realized in M_0. This proves the claim.

By the claim (and using $|T| \leq |S(\emptyset)|$), we see that $|M_0| \geq |T|$. Thus T is countable. Clearly by (c), (d), and our subcase assumption T is totally categorical.

Subcase Ib. $\dim(p_a, M_0) = k$, for some $k < \omega$.
Then M_0 is minimal. For if M is a proper elementary extension of M_0, then by (a) above $p_a|M_0$ is realized in M, whereby $\dim(p_a|M) > k$, and so by (d) M is not isomorphic to M_0.

Let I be a p_a-basis of M_0 (necessarily of cardinality k). Let $q = p_a|M_0$. Let $\kappa \geq 0$ and let $\{a_i: i < \kappa\}$ be a Morley sequence in q of length κ. Finally let M_κ be prime over $M_0 \cup \{a_i: i < \kappa\}$. It is easy to check that $I \cup \{a_i: i < \kappa\}$ is a p_a-basis for M_κ. By (c) and (d) above we obtain (i), (ii), (iii) of the theorem. Just as in the proof of the above claim, one shows that for infinite κ, M_κ is ω-saturated, and thus of cardinality $\geq |T|$, giving (iv).

Case II. T is not totally transcendental, but is trivial.
Almost everything is given to us by Proposition 1.2. We know T to be 1-based, so by 2.5.9 there is a type $p \in S(\operatorname{acl}(\emptyset))$, *with* $R^\infty(p) = 1$.

Claim. *If M is a model which omits p, then $M = \operatorname{acl}(\emptyset)$ (so is the prime model M_0 of T).*

Proof. Suppose $c \in M$, $c \notin \operatorname{acl}(\emptyset)$. By 2.5.9 there is $d \in \operatorname{acl}(c)$, such that $U(d/\emptyset) = 1$. By triviality and unidimensionality, $\operatorname{stp}(d/\emptyset)$ is non-weakly orthogonal to p, and thus there is e realizing p such that $e \in \operatorname{acl}(d)$, so $e \in M$. ■

Thus, if p is omitted in some model, we can define, for any cardinal κ, $M_\kappa = \operatorname{acl}\{a_i: i < \kappa\}$ where $\{a_i: i < \kappa\}$ is an independent set of realizations of p. By 1.2 we see that (i), (ii), and (iii) of the theorem are satisfied. Otherwise p is realized in every model of T. By 1.2, the prime model M_0 is $\operatorname{acl}(a)$ for some realization a of p, and we define $M_\kappa = \operatorname{acl}(\{a\} \cup \{a_i: i < \kappa\})$, where the $\{a_i: i < \kappa\}$ is an independent set of realizations of $p|a$. It is easy to see that M_0 is minimal. By 1.2, (i), (ii), and (iii) of the theorem again hold.

The fact that $|M_\kappa| \geq |T|$ for infinite κ follows as before. In fact in this trivial non-totally transcendental case we can sharpen the cardinality condition. That is, we claim that if M is any model properly containing $\mathrm{acl}(\emptyset)$ then $|M| \geq |T|$. For let $\theta(x)$ be a weakly minimal formula over $\emptyset$ which is contained in p. We know (by 1-cardinality) that $|\theta^M| = |M|$ for any model M of T. So, if $M = M_\omega = \mathrm{acl}(a_i: i < \omega)$ (with notation as above) then $|\theta^M| = |T|$, and, moreover, by triviality of θ, $\theta^M \subset \cup\{\mathrm{acl}(a_i): i < \omega\}$. Thus $|M| \leq |\mathrm{acl}(a_0)|.\omega \leq |M_1|.\omega = |M_1|$. ■

Note that in the above two cases, we did not assume categoricity of T in a higher power (only superstability and unidimensionality).

Case III. T is not totally transcendental, and non-trivial.
This is the interesting case, in which categoricity is not a necessary consequence of unidimensionality (see the example at the beginning of section 1). We are assuming categoricity of T in $\lambda > |T|$. We use the results of section 1, specifically 1.3 onwards. Recall that T is 1-based, so all U-rank types are locally modular. Moreover, there exists a type $p(x)$ over $\mathrm{acl}(\emptyset)$ such that $U(p) = 1$ and p is *modular*. (By 2.5.9 and 1.3 or 5.2.1).

The main point is:

CLAIM **a.** *For any model M of T and* any $q \in S(M)$ *with* $U(q) = 1$, q *is modular.*

Proof of claim. Suppose not. We will contradict λ-categoricity. So we have M and minimal $q \in S(M)$, such that q is non-modular. Let $A \subset M$ be finite such that q does not fork over A. Let M_1 be an elementary substructure of M containing A and of cardinality at most $|T|$. Then $q_1 = q \restriction M_1$ is rank 1, and non-modular (but locally modular). Let I be a Morley sequence in $p|M_1$ of length λ.

SUBCLAIM. *There is a model* M_2 *of cardinality* λ, *which contains* $M_1 \cup I$ *and is dominated by I over M.*

Proof of subclaim. Choose $A_1 \supseteq M \cup I$ dominated by I over M and maximal such. If A_1 is not a model, then there is a consistent formula $\varphi(x, a)$ over A_1, of least ∞-rank n say, which is not satisfied in M. Let c satisfy φ. By maximality of A_1 (and transitivity of domination) there is d independent with A_1 over M such that $\mathrm{tp}(c/A_1d)$ forks over A_1. So there is a formula $\psi(x, y, z) \in L$ and a' in A_1 such that $\models \psi(c, a', d)$ and $R^\infty(\psi(x, a', d)) < n$. We may assume $\models \psi(x, a', z) \rightarrow \varphi(x, a)$. By definability of ∞-rank (1.5.14(ii)) and the independence of d with A_1 over M, we can find $d' \in M$, such that $R^\infty(\psi(x, a', d')) < n$ and $\psi(x, a', d')$ is consistent. By minimal choice of n, $\psi(x, a', d')$ and thus $\varphi(x, a)$ is realized in A_1, a contradiction. So A_1 is a model, and if it has cardinality $> \lambda$, we can use the Lowenheim–Skolem theorem to obtain M_2.

Let M_2 be as given in the subclaim. So for all $c \in M_2 - M_1$, c forks with I over M_1. Now q_1, being non-modular, is weakly orthogonal to $p|M_1$. Applying Corollary 5.2.5 repeatedly we see that q_1 is weakly orthogonal to $\text{tp}(I/M_1)$. Thus q_1 is not realized in M_2. Remember that $A \subseteq M_1$ was a finite set such that q_1 does not fork over A. In the last part of the proof of 1.5.18 we saw that T has model M_3 of cardinality λ such that any strong type over a finite subset A_0 of M_3 has a Morley sequence over A_0 of length λ, contained in M_3. By λ-categoricity this is true of M_2. Thus there is a Morley sequence J of $q_1|\text{acl}(A)$ such that $|J| = \lambda$ and J is contained in M_2. As $|M_1| \leq |T| < \lambda$, J is not contained in M, so there is $c \in J$, $c \notin M$. Thus c realizes q_1, contradicting what we said above. This contradiction proves Claim (a). ■

CLAIM **b.** *For any model M of T, if I is a maximal independent set of realizations of p in M, then M is a p-envelope of I. Moreover, M is prime and minimal over I.*

Proof. By definition $p|I$ is not realized in M. Let A be a p-envelope of I containing M. If M is properly contained in A, then by coordinatization, there is $c \in \text{acl}(A)$ such that $U(c/M) = 1$. By Claim (a), $\text{tp}(c/M)$ is modular, so not weakly orthogonal to $p|M$. Thus $p|M$ is realized in $\text{acl}(A)$, so $p|I$ is realized in $\text{acl}(A)$, a contradiction. Thus M *is* the p-envelope of I. It follows immediately that M is minimal over I. To show that M is prime over I, let M' be another model containing I. Let (by Proposition 1.7(i)) M'' be a p-envelope of I in M'. So $p|I$ is not realized in M''. By the first part of this claim M'' is a p-envelope of I, and thus, by 1.7(ii), isomorphic to M over I. Thus M is embeddable in M' over I. So M is prime over I. ■

The theorem (in Case III) now follows from Claim (b) and Proposition 1.7, except for (iv) which is proved as earlier. ■

COROLLARY **2.2.** *Let $|T| = \aleph_\alpha$, and suppose T is λ-categorical for some $\lambda > |T|$ but that T is not both countable and totally categorical. Then*

(i) *T is κ-categorical for all $\kappa > |T|$;*
(ii) *T has $\aleph_0 + |\alpha|$ many models of cardinality $|T|$.*

Proof. (i) follows directly from the theorem.

(ii) is also immediate if α is infinite. So suppose α to be finite. By (iv) of the theorem $|M_\omega| = \aleph_\alpha$. Moreover, clearly (by (ii) and (iii) of the theorem) $M_\omega = \cup\{M_i : i < \omega\}$. Thus for some $i < \omega$, $|M_i| = \aleph_\alpha$. (ii) follows. ■

We have seen that if T is unidimensional and either totally transcendental or trivial then T is κ-categorical for all $\kappa > |T|$. In the non-totally transcendental, non-trivial case, T may have 2^κ models for each $\kappa > |T|$. However, we now point out a structure theorem in the case where T is weakly minimal. Note that we have been working throughout in T^{eq}. We will say that T is weakly minimal if there is a weakly minimal formula $\varphi(x)$ over $\emptyset$, such that every model M of T is in the definable closure of φ^M. Let $\lambda(T)$ denote the first cardinal in which T is stable (so $|T| \le \lambda(T) \le 2^{|T|}$). Note also that $|T| \le |S(\emptyset)|$ always holds.)

Proposition 2.3. *Let T be unidimensional, non-totally transcendental, and non-trivial. Let p be a modular U-rank 1 strong type over $\emptyset$. Let M_0 denote the p-envelope of $\emptyset$. Then*

(i) $|M_0| \le \lambda(T)$.

(ii) *If T is weakly minimal, then for every model M of T, if M_1 is a p-envelope of $\emptyset$ in M, and I is a $p|M_1$-basis of M (or equivalently a p-basis of M) then* $M = \mathrm{acl}(M_1 \cup I)$.

Proof. (i) Let φ be an arbitrary weakly minimal formula over $\emptyset$. If r is an arbitrary non-algebraic strong type in φ, then M_0 does not contain two independent realizations of r (for if a, b are such then $\mathrm{stp}(b/a)$ is modular, so non-weakly orthogonal to p, whereby p would be realized in M_0). Thus $|\varphi^{M_0}| \le$ number of strong types in $\varphi \le \lambda(T)$. Now if by way of contradiction $|M_0| > \lambda(T)$, then (as $|T| \le \lambda(T)$) we could find a proper elementary substructure M' of M_0 which contains φ^{M_0}, yielding a Vaughtian pair, which we know to be impossible as T is unidimensional.

(ii) Let φ be a weakly minimal formula over $\emptyset$ witnessing the weak minimality of T. Let M, M_1, and I be as given. We claim that for all $a \in \varphi^M - M_1$, a forks with I over M_1. For $\mathrm{tp}(a/M_1)$ is modular (by maximality of M_1 in M) so there is $b \in \mathrm{acl}(aM_1)$ realizing $p|M_1$. So $b \in \mathrm{acl}(M_1 \cup I)$ and therefore $a \in \mathrm{acl}(M_1 \cup I)$. Thus $\varphi^M \subseteq \mathrm{acl}(M_1 \cup I)$, whereby $M = \mathrm{acl}(M_1 \cup I)$. ■

Corollary 2.4. *Let T be weakly minimal and unidimensional. Then for any κ, T has at most $2^{\lambda(T)}$ models of cardinality κ.*

Proof. If $\kappa \le \lambda(T)$ this is clear. For $\kappa > \lambda(T)$, this follows from Proposition 2.3. Let M be a model of power κ, M_1 a p-envelope of $\emptyset$ inside M, and I a p-basis of $|M|$. By 2.3(i), $|M| \le \lambda(T)$. By 2.3(ii), $|I| = \kappa$. As $\mathrm{tp}(I/\emptyset)$ is stationary (and I independent from M_1 over $\emptyset$), by 2.3(ii), M is determined by (κ) and the isomorphism type of M_1. There are at most $2^{\lambda(T)}$ possibilities. ■

3 Countable small unidimensional theories

We now begin an analysis of countable unidimensional theories, working towards classifying the countable models of a weakly minimal such theory, and also shedding some light on the general case. Many things we shall say are true with some local assumption in place of unidimensionality, and we leave the appropriate restatement to the reader.

Let T be a countable unidimensional theory. By Proposition 1.1, exactly one of the following holds: T is $\aleph_1$-categorical, or T is non-totally transcendental and non-trivial. (Again the example at the beginning of section 1 satisfies the second alternative.) As the Baldwin–Lachlan theorem (which is included in the proof of Case I of Theorem 2.1) says that in the first case T has either 1 or countably many countable models, we shall from now on assume that T is non-totally transcendental and non-trivial. If $S(T)$, the set of complete types over $\emptyset$, is uncountable, then it has cardinality 2^{ω}, so T has continuum many countable models. So we will assume $S(T)$ is countable (or in current parlance, T is small, or has few types). So from now on, until stated otherwise, we assume:

ASSUMPTION **3.1.** *T is countable, unidimensional, non-totally transcendental non-trivial, and small.*

We will use freely the geometric consequences of our assumptions (T-1-based, coordinatization, all U-rank types are locally modular, etc.). As T is small, also $S(A)$ is countable for any finite set A. In particular, for any sort S and finite set A, Cantor–Bendixson rank is well defined on the space of complete types over A in sort S. We note the Cantor–Bendixson rank of some such $p \in S(A)$, by $\mathrm{CB}(p)$, and if p is realized by a, by $\mathrm{CB}(a/A)$.

Fact 3.2. If $a \in \mathrm{acl}(bA)$ and A is finite then $\mathrm{CB}(a/A) \leq \mathrm{CB}(b/A)$.

The following lemma has a substantially easier proof if D is a group. We nevertheless give the proof in general.

LEMMA **3.3.** *Let D be a weakly minimal definable set, defined over the finite set B. Let A be a finite subset of D. Then only finitely many non-algebraic strong types over B are realized in $D \cap \mathrm{acl}(AB)$.*

Proof. For ease of notation we assume $B = \emptyset$. It is also easy to reduce to the case where A is $\{a\}$, $a \notin \mathrm{acl}(\emptyset)$.

Suppose by way of contradiction that $\{b_i: i < \omega\} \subset \mathrm{acl}(a) \cap D$, where the $q_i = \mathrm{stp}(b_i/\emptyset)$ $(=\mathrm{tp}(b_i/\mathrm{acl}(\emptyset)))$ are non-algebraic and pairwise distinct. Let

U be a non-principal ultrafilter on ω, and let $q = \prod_i q_i/U$, the ultraproduct with respect to U of the q_i. Thus q is a non-algebraic strong type over $\varnothing$ in D and

(i) for any b realizing q and equivalence relation E on D defined over $\varnothing$ and with finitely many classes, we have $\{i < \omega: E(b, b_i\} \in U$.

Let b realize q, with b independent from a over $\varnothing$. Let a' realize $\text{stp}(a/\varnothing)$ such that a' is independent with $\{a, b\}$ over $\varnothing$.

CLAIM. *There is* $b' \in \text{acl}(a, b, a') - \text{acl}(a, b)$ *such that* $\text{stp}(b') = q$.

Let p be a U-rank 1 modular strong type over $\varnothing$. As $\text{stp}(a'/a)$ is modular, it is non-weakly orthogonal to $p|a$, whereby there is c realizing p, such that c and a' are interalgebraic over a. Thus c is independent with b over $\varnothing$ and so realizes $p|b$. Again $p|b$ is non-weakly orthogonal to the modular type $q|b$, whereby there is b' realizing $q|b$, with b' and c interalgebraic over b. Clearly $b' \in \text{acl}(a, b, a') - \text{acl}(a, b)$. This ends the proof of the claim.

Now choose b_i' for $i < \omega$ such that $\text{stp}(a', b_i'/\varnothing) = \text{stp}(a, b_i/\varnothing)$. Thus

(ii) $\text{stp}(b_i') = q_i$ and b_i' is interalgebraic with a' over $\{a, b\}$ for all i.

From the claim and forking symmetry

(iii) b' is interalgebraic with b_i' over $\{a, b\}$ for all i.

By Fact 3.2, for some α

(iv) $\text{CB}(b'/a, b) = \text{CB}(b_i'/a, b) = \alpha$, for all i.

Let $\psi(x)$ be a formula over ab which isolates $\text{tp}(b'/a, b)$ among types over ab of CB rank greater than or equal to α. As $r = \text{tp}(b'/a, b)$ is implied by '$\text{stp}(x) = \text{stp}(b) \wedge x \notin \text{acl}(a, b)$' there is some finite equivalence relation E over $\varnothing$ such that '$E(x, b) \wedge x \notin \text{acl}(a, b)$' $\rightarrow \psi(x)$. By (i) for some $X \in U$, we have $\psi(b_i')$ for all $i \in X$. Thus, by (iv) $\text{tp}(b_i'/a, b) = r$, for all $i \in X$. But since $\text{stp}(b'/\varnothing) = \text{stp}(b/\varnothing)$ and $\text{tp}(b_i'/a, b) = \text{tp}(b'/a, b)$ for all $i \in X$, we have $\models E(b_i', b)$ for all $i \in X$ and all finite equivalence relations E over $\varnothing$. Thus $\text{stp}(b_i'/\varnothing) = q$ for all $i \in X$, contradicting the distinctness of the q_i. This contradiction proves the lemma. ■

Our hypotheses, together with Proposition 1.3, imply the existence of a weakly minimal abelian group G defined over $\text{acl}(\varnothing)$. *We will assume that G is defined over* $\varnothing$. (This is justified by the purpose of the current analysis: classifying countable models of T. There is no loss in adding an algebraic parameter to the language, although clearly we should not add infinitely many algebraic parameters.) G is not connected by finite, so there are $\emptyset$-definable subgroups of finite index $G_0 > G_1 > \cdots > G_n > \ldots$, such that $\cap \{G_i : i < \omega\} = G^0$ (the connected component of G). The general theory of stable groups tells us that for any set A, and $a, b \in G - \mathrm{acl}(A)$,

(*) $\mathrm{stp}(a/A) = \mathrm{stp}(b/A)$ iff $a - b \in G_i$ for all i.

It is not difficult to conclude from this also

(**) $\mathrm{tp}(a/A) = \mathrm{tp}(b/A)$ iff for every i, and A-definable set X which is a union of cosets of G_i, $a \in X$ iff $b \in X$.

G and the G_i will henceforth be as above.

COROLLARY **3.4.** *For any finite set A, $\mathrm{acl}(A)$ intersects only finitely many cosets of G^0 in G.*

Proof. Suppose otherwise. Let $a \notin G^0 - \mathrm{acl}(A)$. If $c, d \in (\mathrm{acl}(A) \cap G)$ and $c - d \notin G^0$, then $\mathrm{stp}(a + c/A) \neq \mathrm{stp}(a + d/A)$, and both strong types are non-algebraic. But then $\mathrm{acl}(Aa)$ contains realizations of infinitely many non-algebraic strong types over A in G, contradicting Lemma 3.3. ■

Let $D = D(G^0, \emptyset)$ be the division ring associated to G^0, introduced in section 5 of Chapter 4. We will use S_0 to denote $\mathrm{acl}(\emptyset) \cap G^0$. We consider D as the division ring of '$\mathrm{acl}(\emptyset)$-definable' endomorphisms of G^0/S_0.

We want to extend the action of any $\lambda \in D$ on G^0/S_0 to H/S_0 for some *definable* subgroup of G. We now show how any member λ of D can be considered as acting on some $\mathrm{acl}(\emptyset)$-definable subgroup $H > G^0$. Fix $\lambda \in D$. Let S_λ be the unique minimal subgroup of $G^0 \times G^0$, corresponding to λ. So S_λ is defined over $\mathrm{acl}(\emptyset)$, and for $a, b \in G^0$, $\lambda(a/S_0) = b/S_0$ iff there are a', $b' \in G^0$ such that $a - a' \in S_0$, $b - b' \in S_0$, and $(a', b') \in S_\lambda$.

Let now S' be a weakly minimal $\mathrm{acl}(\emptyset)$-definable subgroup of $G \times G$. We will say that S' *represents* λ if (i) and (ii) below hold:

(i) $S' \cap (G^0 \times G^0) = S_\lambda$ (equivalently S_λ is the connected component of S');

(ii) if $\lambda \neq 0$, then for all $x \in G$, $(x, 0) \in S'$ iff $(x, 0) \in S_\lambda$, and $(0, x) \in S'$ iff $(0, x) \in S_\lambda$. (So $\ker(S') = \ker(S_\lambda) \subseteq S_0$, and $\mathrm{coker}(S') = \mathrm{coker}(S_\lambda) \subseteq S_0$.) (We will observe below that there always is some such S'.)

It should be noted that if S', S'' both represent λ then also $S' \cap S''$ represents λ (so a choice of S_λ is unique up to commensurability).

Given a choice of S' to represent λ, we will extend the definition of λ as follows: for $x, y \in G$ we say $\lambda(x/S_0) = y/S_0$ if there is $(x', y') \in S'$ such that $x - x'$ and $y - y'$ are in S_0. Note that (as $S_0 \subseteq G^0$) in fact:

(iii) if $\lambda(x/S_0) = y/S_0$, then there is x' such that $x - x' \in S_0$ and $(x', y) \in S'$, and there is also y' such that $y - y' \in S_0$ and $(x, y') \in S'$.

For H an acl($\varnothing$)-definable subgroup (of finite index) of G, we will say that λ is defined on H if (with above notation):

(iv) for any $x \in H$, there is y such that $\lambda(x/S_0) = y/S_0$ (equivalently by (iii) there is y such that $(x, y) \in S'$).

Note that the projection of S' on the first coordinate will be an acl($\varnothing$)-definable subgroup H' of G (of finite index in G) and so to say that λ is defined on some H means exactly that $H < H'$. We may write H' as dom(S'), or even as dom(λ) given the choice of S'.

If also

(v) for all $x \in H$, $\lambda(x/S_0) \subset H/S_0$,

we say that H is closed under λ.

If H is closed under λ and τ then it is easy to see that

(vi) H is closed under $-\lambda$, $\lambda.\tau$, and $\lambda + \tau$, and, moreover, for any $x, y \in H$, $(\lambda + \tau)(x/S_0) = \lambda(x/S_0) + \tau(x/S_0)$, and $\lambda(x + y/S_0) = \lambda(x/S_0) + \lambda(y/S_0)$.

Fact 3.5. For any $\lambda \in D$ there is an acl($\varnothing$)-definable weakly minimal $S' < G^0 \times G^0$ which represents λ.

With respect to such a choice of S', we have: if $x \in G$, λ is non-zero and $\lambda(x/S_0) = y/S_0$, then

(a) $x \in G^0$ iff $y \in G^0$; and

(b) if $x \notin$ acl($\varnothing$) then $y \notin$ acl($\varnothing$) and, moreover, stp(y) is determined by x.

Proof. Now by 1.6.18, S_λ is the intersection of *definable* (over acl($\varnothing$)) subgroups of $G \times G$. By 1.6.21 these can be chosen to be weakly minimal. As the kernel and cokernel of S_λ are both finite, we can clearly find S' satisfying (i), (ii), and (iii) above.

(a) Suppose $\lambda(x/S_0) = y/S_0$, and $y \in G^0$. By (iii) (and the fact that $S_0 \subseteq G^0$) we may assume that $(x, y) \in S'$. Now there is $x' \in G^0$ with $(x', y) \in S_\lambda$. As $S_\lambda \subseteq S'$, also $(x', y) \in S'$. Thus (subtracting) we have that $(x - x', 0) \in S'$. By (ii), $x - x' \in S_0$, so $x \in G^0$. Similarly if $x \in G_0$ then $y \in G^0$.

(b) Suppose for a contradiction that $x \in H -$acl($\varnothing$), and $\lambda(x/S_0) \subset$ acl($\varnothing$). For S' as above, we may assume that $(x, y) \in S'$, with $y \in$ acl($\varnothing$). Let x' have the same strong type as x with x and x' independent over $\varnothing$. Then also

$(x', y) \in S'$. So $(x - x', 0) \in S'$. But $x - x' \in G^0 - S_0$, and $S' \cap (G^0 \times G^0) = S_\lambda$. Thus $\lambda = 0$, a contradiction.

Thus if $y/S_0 = \lambda(x/S_0)$, then $y \notin \mathrm{acl}(\varnothing)$. For any $z \in S_0$, $\mathrm{stp}(y + z) = \mathrm{stp}(y)$, and thus $\mathrm{stp}(y/\varnothing)$ is determined. ■

We will often be interested only in $\mathrm{stp}(y)$ for $y \in \lambda(x/S_0)$. So in this situation, we may, by 3.5(b), write $y = \lambda(x)$ (for $x \in \mathrm{dom}(\lambda) - \mathrm{acl}(\varnothing)$). (This depends on a choice of suitable S'). We may write S'_λ for S' to emphasize the dependence on λ.

LEMMA 3.6. *D is a locally finite field.*

Proof. By 3.4 we may assume that $\mathrm{acl}(\varnothing) \cap G = S_0$.

By a result of Jacobson, it is enough to show that for every non-zero $\lambda \in D$ there is $k > 0$ such that $\lambda^k = 1$.

Let us fix non-zero $\lambda \in D$. Let S' be chosen to represent λ. Clearly then the composition (as a relation) of S' with itself k times represents λ^k. Moreover, $\{(x, y)\colon (y, x) \in S'\}$ represents λ^{-1} etc.

CLAIM. *There is $a \in G$, $a \notin G^0$ such that* either *$\lambda^n(a)$ is defined for all $n < \omega$* or *$\lambda^{-n}(a)$ is defined for all $n < \omega$.*

Proof of claim. Let H_0 be $\mathrm{acl}(\varnothing)$-definable of finite index in G such that λ is defined on H_0. Let $H_n = H_0 \cap \lambda^{-1}(H_0) \cap \cdots \cap \lambda^{-n}(H_0)$. Note that H_n is $\mathrm{acl}(\varnothing)$-definable of finite index in G.

Case (*i*). For some n, $H_n = H_{n+1}$.
This means that H_n is closed under λ, so under λ^m for all m. Choose $a \in H$, $a \notin G^0$.

Case (*ii*). For all n, H_{n+1} is properly contained in H_n.
If $b \in H_n$, $b \notin H_{n+1}$, then $\lambda^n(b)$ is defined and in H_0 but not in H_1. By compactness there is $a \in H_0$, $a \notin H_1$, such that $\lambda^{-n}(a)$ is defined for all n. The claim is proved.

Let $a \in G - G^0$ be chosen to satisfy the claim. By our assumption about $\mathrm{acl}(\varnothing) \cap G$, we have that $a \notin \mathrm{acl}(\varnothing)$. Without loss of generality $\lambda^n(a)$ is defined for all n, that is $\lambda^n(a/S_0)$ is defined for all n. For any $b \in \lambda^n(a/S_0)$ clearly $b \in \mathrm{acl}(a)$. But by 3.4 only finitely many strong types over $\varnothing$ are realized by elements of $\mathrm{acl}(a)$. It clearly follows that for some $m < n$, $\lambda^m(a/S_0) = \lambda^n(a/S_0)$. By (v) $(\lambda^n - \lambda^m)(a/S_0) = 0/S_0$. By 3.5(b), $\lambda^n - \lambda^m = 0$ (working in D), and thus $\lambda^{n-m} = 1$ (in D). ■

COROLLARY 3.7. *Let F be a finite subfield of D. Then (with respect to some choice of weakly minimal subgroups representing the elements of F), there is an* acl($\emptyset$)*-definable subgroup H of finite index in G such that H is closed under every $\lambda \in F$. Moreover, given any* acl($\emptyset$) *definable subgroup H_1 of G (of finite index) such H can be found inside H_1.*

Proof. Let H_1 be a given acl($\emptyset$)-definable subgroup of finite index of G. By passing to a possibly smaller subgroup we may assume that each $\lambda \in F$ is defined on H_1. Let $H = \cap \{\lambda(H_1): \lambda \in F\}$. ■

We now begin an analysis, due to Newelski, of 1-types in G over finite sets, the eventual purpose of which is to prove that, if T has fewer than continuum many countable models then any non-isolated rank 1 type over a finite set must have finite multiplicity. Although we will be working essentially inside G, it is useful to state a number of facts relating isolation and multiplicity which hold more generally.

LEMMA 3.8. (i) *Let A be finite, and* stp(a/A) *be modular of rank* 1. *Then* tp(a/A) *has finite multiplicity.*

(ii) *Let $p \in S(A)$ be isolated and of rank* 1. *Then p has infinite multiplicity.*

Proof. (i) If not then tp(a/A) has 2^ω extensions $\{p_i: i < 2^\omega\}$ to strong types over A, all of which are modular. The p_i are thus weakly non-orthogonal, whereby they are all realized in acl(aA), which is impossible.

(ii) Let $\varphi(x)$ isolate p. If p had finite multiplicity then $\varphi(x)$ is a member of only finitely many non-algebraic strong types. Thus φ has Morley rank 1, contradicting our current assumptions. ■

Let $p \in S(A)$ be non-algebraic (and eventually rank 1) and $A \subseteq B$, where A, B are finite. $S_p(B)$ will denote the space $\{q \in S(B): q \text{ extends } p\}$, and $S_{p,na}(B)$ denotes $\{q \in S(B): q \text{ extends } p, \text{ and } q \text{ is non-algebraic}\}$. Both $S_p(B)$ and $S_{p,na}(B)$ are closed subspaces of the countable space $S(B)$. If $q \in S(B)$ and q extends p, we will say that *q is isolated over p*, if q is isolated in $S_p(B)$.

The next lemma is crucial.

LEMMA 3.9. (i) *Let $p \in S(A)$ be of U-rank* 1 *of infinite multiplicity. Let $A \subseteq B$, where A, B are finite. Let $q \in S(B)$ be non-algebraic, $q \supseteq p$. If q is isolated in $S_{p,na}(B)$, then q has infinite multiplicity, and, moreover, q is isolated over p.*

(ii) *Let p and A and B be as in* (i). *Suppose that p has some non-algebraic* *extension $q \in S(B)$ which has finite multiplicity. Then p has infinitely many extensions in $S(B)$ which are isolated over p and non-algebraic.*

(iii) *p, A, and B as above. If p is realized in* $\mathrm{acl}(B)$, *then some non-algebraic extension of p in $S(B)$ has finite multiplicity.*

Proof. (i) Let $\varphi(x)$ over B isolate q in $S_{p,na}(B)$. Let Ψ be the formulae over $\mathrm{acl}(A)$ (or equivalently the formulae defining classes of finite equivalence relations over A). Let a realize q. Then $\mathrm{stp}(a/B)$ is implied by $p(x) \cup$ '$x \notin \mathrm{acl}(B)$' $\cup \{\psi(x) : \psi \in \Psi \text{ and } \psi(a)\}$. By compactness there is a single $\psi \in \Psi$ which is true of a such that $p(x) \cup$ '$x \notin \mathrm{acl}(B)$' $\cup \{\psi(x)\} \models \varphi(x)$. Let $\psi_1(x)$ be the disjunction of the B-conjugates of ψ. Then clearly $\psi_1 \in q$, ψ_1 is almost over A, and ψ_1 isolates q in $S_{p,na}(B)$. As p has infinite multiplicity, clearly p extends to infinitely many strong types over A containing ψ_1. The non-forking extension over $\mathrm{acl}(B)$ of each such strong type contains q, whereby q has infinite multiplicity.

To conclude that q is isolated over p, it suffices to prove:

CLAIM. $p(x) \cup \{\psi_1(x)\} \models x \notin \mathrm{acl}(B)$.

Proof of claim. Suppose $b \in \mathrm{acl}(B)$ satisfies $p(x) \cup \{\psi_1(x)\}$. Let c realize the non-forking extension of $\mathrm{stp}(b/A)$ to B. Then $\mathrm{tp}(c/B) = q$. As q has infinite multiplicity there are c_i for $i < 2^\omega$ such that $\mathrm{tp}(c_i/B) = q$ and $\mathrm{stp}(c_i/B) \neq \mathrm{stp}(c_k/B)$ for $i \neq k$. As q does not fork over A, $\mathrm{stp}(c_i/A) \neq \mathrm{stp}(c_k/A)$ for $i \neq k$. For $i < 2^\omega$, let b_i be such that $\mathrm{tp}(b_i, c_i/B) = \mathrm{tp}(b, c/B)$. In particular $b_i \in \mathrm{acl}(B)$ for all i, and $\mathrm{stp}(b_i/A) \neq \mathrm{stp}(b_k/A)$ for $i \neq k$. Thus $\mathrm{acl}(B)$ is uncountable, which is a contradiction. ■

(ii) Let $q \in S(B)$ be a non-algebraic extension of p, with finite multiplicity. By part (i), q is not isolated in $S_{p,na}(B)$. Thus $S_{p,na}(B)$ contains infinitely many isolated points, all of which are, by (i) isolated in $S_p(B)$.

(iii) Let $a \in \mathrm{acl}(B)$ realize p. Let b realize $\mathrm{stp}(a/A)|B$. Then $\mathrm{tp}(b/Ba)$ is stationary, whereby clearly $\mathrm{tp}(b/B)$ has finite multiplicity. ■

In the case where we are working inside the weakly minimal group G, we obtain (by virtue of 'homogeneity') the following refinement of Lemma 3.9.

LEMMA 3.10. (i) *Let $p \in S(B)$ be where B is finite, $p(x) \models$ '$x \in G$'. Suppose, moreover, that there is $\varphi(x) \in p$ such that $\{\varphi(x)\} \cup$ '$x \notin \mathrm{acl}(B)$' $\models p(x)$. Then p is isolated by a formula which is a union of G_i-cosets for some i.*

(ii) *Let $p \in S(A)$ and $q \in S(B)$ an extension of p, with p, q non-algebraic and $p(x) \vDash x \in G$. Suppose p has infinite multiplicity and q is isolated in $S_{p,na}(p)$. Then q is isolated over p, by a formula over B which is a union of G_i-cosets for some i.*

Proof. (i) Like the proof of Lemma 3.9(i). We use the fact

($*$) for any $a \in G$, $a \notin \text{acl}(B)$, $\text{stp}(a/B)$ is determined by the cosets of the G_i which contain a.

It follows that we can choose $\varphi(x)$ to be a union of G_i-cosets for some i. As in the claim in the previous lemma, show that $\varphi(x) \cap \text{acl}(B)$ is empty.

(ii) Just as in Lemma 3.9(i). The main point is that we can restrict the set Ψ to those formulae which define cosets of the subgroups G_i, whereby the formula ψ_1 obtained there will be a union of G_i-cosets for some i. ■

To continue the Newelski analysis, we must first take more care in defining the sequence G_i of subgroups (whose cosets determine strong types in G).

As D is a countable, locally finite field, we can write $D = \cup\{F_i : i < \omega\}$, where $F_0 \subset F_1 \subset \ldots$ is a chain of finite subfields, such that, moreover, if $\lambda \in F_i$ then any conjugate of λ over $\varnothing$ is also in F_i. By Lemma 3.7 (and intersecting conjugates), we can find $\varnothing$-definable subgroups G_i of G, of finite index in G, such that

(i) $G_o \geq G_1 \geq \ldots$;

(ii) $\cap\{G_i : i < \omega\} = G^0$; and

(iii) G_i is closed under F_i for all i.

This will enable us to analyse algebraic dependence on G in terms of D.

For convenience we let $E_n(x, y)$ denote the equivalence relation '$x - y \in G_n$'. So E_n is $\varnothing$-definable with finitely many classes, and for any $a \in G$, and set A with $a \notin \text{acl}(A)$, $\text{stp}(a/A)$ is determined exactly by the equivalence classes of the E_n which contain a.

Definition 3.11. *Let $a \in G$. Let A be a finite set with $a \notin \text{acl}(A)$. Let $p = \text{tp}(a/A)$. We define a function $f: \omega \to \omega$ associated to p, and we write $f = f(p) = f(a/A)$, as follows:*

(i) *$f(0)$ = the number of G_0-cosets containing a realization of p;*

(ii) *$f(i + 1)$ = the number of G_{i+1}-cosets which contain a realization of p, and are contained in $a + G_i$.*

Remark 3.12. *Clearly the definition of $f(i + 1)$ depends only on p, not on a. We could also of course define $f(p)$ similarly for algebraic p.*

Let Min: $\omega \to \omega$ be the function $\mathrm{Min}(i) = 1$ for all i. Let Max: $\omega \to \omega$ be the function $\mathrm{Max}(0) = [G: G_0]$, $\mathrm{Max}(i+1) = [G_i: G_{i+1}]$.

The relations $\leq^*$ and $=^*$ between functions from ω to ω denote eventual domination, and eventual equality, respectively. We write $f <^* g$ if $f \leq^* g$ but *not* $f =^* g$.

Note that if $A \subseteq B$, then $f(a/B) \leq^* f(a/A)$, and that $\mathrm{Min} \leq^* f(a/A) \leq^* \mathrm{Max}$.

Lemma 3.13. *Let $a \in G$, $A \subseteq B$ be finite, and $a \notin \mathrm{acl}(B)$. Let $p = \mathrm{tp}(a/A)$ and $q = \mathrm{tp}(a/B)$ an extension of p. Then*

(i) $f(p) =^* \mathrm{Min}$ *iff p has finite multiplicity.*

(ii) $f(p) =^* \mathrm{Max}$ *iff p is isolated.*

(iii) *If* $\mathrm{mult}(p)$ *is infinite, then* $f(p) =^* f(q)$ *iff q is isolated over p.*

(iv) if $B \subseteq \mathrm{acl}(A)$ then $f(p) =^* f(q)$.

Proof. (i) Let $f(p) =^* \mathrm{Min}$, and suppose for all $n \geq i$, $f(p)(n) = 1$. It is then clear, by $(*)$, that any strong type over A extending p is determined by its E_i-class, so p has finite multiplicity. Conversely, if p has finite multiplicity, then there is some E_i such that the strong types over A extending p are distinguished by their E_i-classes. Clearly $f(p)(n) = 1$ for all $n \geq i$.

(ii) First suppose p to be isolated. By Lemma 3.10(i), p is isolated by some X which is a union of G_i-cosets for some i. Clearly $f(p)(n) = \mathrm{Max}(n)$ for all $n > i$.

Now suppose that $f(p) =^* \mathrm{Max}$. Let i be such that $f(p)(n) = \mathrm{Max}(n)$ for all $n \geq i$. Let X be the union of those G_i-cosets containing a realization of p. It follows that for every n, every G_n-coset contained in X contains a realization of p. On the other hand, $(*)$ says that, if $b, c \notin \mathrm{acl}(A)$ and b and c are in the same G_n-cosets for all $n \geq i$, then $\mathrm{stp}(b/A) = \mathrm{stp}(c/A)$. Thus, for any $b \in X - \mathrm{acl}(A)$, $\mathrm{tp}(b/A) = p$. By Lemma 3.10(i), p is isolated.

(iii) If q is isolated over p, then by Lemma 3.10(ii) (and the assumption that p has infinite multiplicity) this isolation is by a formula X which is a union of G_i-cosets for some i. Clearly $f(q)(n) = f(p)(n)$ for all $n > i$.

Conversely, suppose $f(p)(n) = f(q)(n)$ for all $n \geq i$. Let X be the union of those G_i-cosets containing a realization of q. (Note X is B-definable.) As in (ii) it follows that every $b \in X - \mathrm{acl}(B)$ which realizes p also realizes q. By Lemma 3.10(ii) q is isolated over p.

(iv) We may assume that p has infinite multiplicity, so (iii) applies. In fact it follows from the easy direction of (iii). For if $B \subseteq \mathrm{acl}(A)$, then clearly p has only finitely many extensions over B, all of which must thus be isolated over p. ■

LEMMA 3.14. *Let $a, b \in G - \mathrm{acl}(A)$, where A is finite. Suppose $a \notin \mathrm{acl}(Ab)$. Then $f(a/A) =^* f(a/A \cup \{b\})$ iff $f(b/A) =^* f(b/A \cup \{a\})$.*

Proof. Suppose that $f(a/A) =^* f(a/A \cup \{b\})$. Write $B = A \cup \{b\}$. Thus there is i, such that every $a' \in (a + G_i) - \mathrm{acl}(B)$ which realizes $\mathrm{tp}(a/A)$ realizes $\mathrm{tp}(a/B)$. Let c be a name for $a + G_i$. So $c \in \mathrm{acl}(\emptyset)$ and we see that $\mathrm{tp}(a/Ac)$ has a unique non-forking extension over Bc. Thus $\mathrm{tp}(b/A \cup \{c\})$ has a unique non-forking extension over $A \cup \{a, c\}$. Clearly then, $f(b/A \cup \{c\}) =^* f(b/A \cup \{a, c\})$. As $c \in \mathrm{acl}(\emptyset)$, by Lemma 3.10(iv), we conclude that $f(b/A) =^* f(b/A \cup \{a\})$. ■

LEMMA 3.15. *Let $a, b \in G - \mathrm{acl}(A)$,* and $a \in \mathrm{acl}(A \cup \{b\})$. *Then $f(a/A) \leq^* f(b/A)$.*

Proof. By Chapter 4, section 5, $\mathrm{stp}(a, b/A)$ is the generic type of some ∞-definable (over $\mathrm{acl}(A)$) coset X of a connected, minimal ∞-definable subgroup S of $G \times G$. As S is connected clearly $S \subseteq G^0 \times G^0$. So $S = S_\lambda$ for some $\lambda \in D$. Now our choice of the G_i means that there is some weakly minimal $\mathrm{acl}(\emptyset)$-definable representative S' of λ, and some $n_0 < \omega$ such that with respect to this choice of S', G_n is closed under λ and λ^{-1} for all $n \geq n_0$. Now S_λ is the connected component of S'. Thus X is contained in a unique translate Y of S', $\mathrm{tp}(a, b/\mathrm{acl}(A))$ is a generic type of Y, and clearly Y is definable over a finite part of $\mathrm{acl}(A)$. By Lemma 3.13 we may assume Y to be A-definable.

CLAIM. *Let (a_1, b_1) and (a_2, b_2) both realize* $\mathrm{tp}(a, b/A)$. *Then for any $n \geq n_0$, $a_1 - a_2 \in G_n$ iff $b_1 - b_2 \in G_n$.*

Proof of claim. So $(a_1, b_1), (a_2, b_2) \in Y$, whereby $(a_1 - a_2, b_1 - b_2) \in S'$. As G_n is closed under λ and λ^{-1}, $a_1 - a_2 \in G_n$ iff $b_1 - b_2 \in G_n$. ■

Now fix $n \geq n_0$. Let $f(a/A)(n + 1) = k$. Let $a_1, \ldots, a_k \in a + G_n$ be realizations of $p = \mathrm{tp}(a/A)$ which are in different G_{n+1}-cosets. For each $r = 1, \ldots, k$, let b_i be such that $\mathrm{tp}(a_i, b_i/A) = \mathrm{tp}(a, b/A)$. By the claim, each $b_i \in b + G_n$ and the b_i are in different G_{n+1}-cosets. Thus $f(b/A)(n + 1) \geq k$. The lemma is proved. ■

We conclude this section with the observation that inside G, non-algebraic isolated types are dense. This should be compared with Lemma 1.4.

LEMMA 3.16. *Let $\varphi(x)$ be a non-algebraic formula over a finite set A, such that $\varphi(x) \to x \in G$. Then there is an isolated non-algebraic type $p \in S(A)$ containing φ.*

Proof. Let $Q = \{p \in S(A)$: p contains φ and p is non-algebraic$\}$. Q is non-empty and countable, so contains an isolated point p. By Lemma 3.10(i), p is actually an isolated type (in $S(A)$). ■

4 Newelski's lemma

Assumption 3.1 is still in place. The weakly minimal group G is as in section 3. In this section we will prove the following result, which has gone under the name of 'Saffe's conjecture' or 'Saffe's hypothesis':

THEOREM **4.1.** *Let A be finite and let $p \in S(A)$ be of ∞-rank* 1, *non-isolated, and of infinite multiplicity, Then T has 2^ω countable models.*

Before getting into the proof, we mention a certain construction which gives 2^ω countable models, as well as some important consequences. Let A be a set and M a model containing A. A family $R = \{r_i: i \in I\} \subseteq S(A)$ is said to be *almost orthogonal in M*, if each r_i is realized in M, and, moreover, whenever $a_i \in M$ realizes r_i for all i, then $\{a_i: i \in I\}$ is A-independent. If we omit 'in M' we mean 'in the universe.'

LEMMA **4.2.** *Let A be finite and $A \subseteq M$. Suppose there is a family $\{q_i: i < \omega\}$ in $S(A)$ of non-isolated types, which is an almost orthogonal in M family. Then T has 2^ω countable models.*

Proof. Let $a_i \in M$ realize q_i for all $i < \omega$. For $X \subseteq \omega$, let $A_X = A \cup \{a_i: i \in X\}$. Let $k \in \omega - X$. Then every realization $a \in M$ of q_k is independent with A_X over A. We claim that there is no consistent formula $\varphi(x)$ over A_X such that $\models \varphi(x) \rightarrow q_k(x)$. For otherwise, let a satisfy $\varphi(x)$ in M. So a realizes q_k, and as q_k is non-isolated, the open mapping theorem implies that a forks with A_X over A, contradicting the almost orthogonality in M hypothesis. By the omitting types theorem, there is a model M_X containing A_X which omits q_k for all $k \notin X$. Thus we obtain 2^ω countable models over A. As A is finite we obtain 2^ω countable models. ■

COROLLARY **4.3.** *Let D_1 be a weakly minimal definable set, defined over $\varnothing$, say. Let A be finite. Suppose there is a family $Q = \{q_i: i < \omega\}$ of distinct types in $S(A)$ such that for all i, $\models q_i(x) \rightarrow x \in D_1$, and q_i is rank* 1 *and of finite multiplicity. Then T has 2^ω countable models.*

Proof. We will construct an infinite, almost orthogonal subfamily $\{r_i: i < \omega\}$ of Q, inductively. Suppose we have already got $r_1, \ldots, r_n$. The induction assumption, together with the fact that each r_i has finite multiplicity, implies that their are only finitely many types over A extending $r_1(x_1) \cup \cdots \cup r_n(x_n)$. For each realization $\mathbf{a} = (a_1, \ldots, a_n)$ of such a type, by Lemma 3.3, only finitely many non-algebraic strong types over A are realized in $\mathrm{acl}(A, \mathbf{a})$. Thus there is $q \in Q$ such that for all such $\mathbf{a}$, q is not realized in $\mathrm{acl}(A\mathbf{a})$. Put $r_{n+1} = q$. So we have our infinite, almost orthogonal subfamily of Q. But each $r \in Q$ is non-isolated (otherwise it would have Morley rank 1). By Lemma 4.2, T has 2^{ω} countable models. ■

Next we show that in order to prove Theorem 4.1, we may work inside G.

LEMMA **4.4.** *Suppose that A is finite, $p \in S(A)$ has U-rank* 1, *and is non-isolated and of infinite multiplicity. Then if a realizes p, there is $b \in G$ such that* $\mathrm{tp}(b/Aa)$ *is non-isolated and of infinite multiplicity.*

Proof. Let c realize $\mathrm{stp}(a/A)$ such that c is independent from a over A. So $\mathrm{tp}(c/aA)$ is modular, and hence c is interalgebraic over aA with some d realizing the generic type of G^0 over Aa. By compactness there is a formula $\psi(x)$ over $\mathrm{acl}(A)$ and in $\mathrm{stp}(a/A)$ such that whenever c' satisfies ψ and c' is independent with a over A, then there is $d' \in G$ interalgebraic with c' over Aa. Without loss of generality ψ is over A (otherwise augment A). By Lemma 3.9(ii) (as $\mathrm{tp}(c/Aa)$ is stationary) there is c' satisfying p such that $\mathrm{tp}(c'/Aa)$ is non-algebraic and isolated in p. By Lemma 3.9(i), $\mathrm{tp}(c'/Aa)$ has infinite multiplicity, and, moreover, being a non-forking extension of p, is also non-isolated. Let $b \in G$ be interalgebraic with c' over Aa. Clearly then $\mathrm{tp}(b/Aa)$ is non-isolated of infinite multiplicity. ■

The next Lemma proves Theorem 4.1 under some additional assumptions. We work back in G.

LEMMA **4.5.** *Suppose that A is finite and $q \in S(A)$ (necessarily of U-rank* 1) *is such that $\models q(x) \rightarrow x \in G$, and*

(A) *q is non-isolated and of infinite multiplicity, and* **whenever** *$b_0, \ldots, b_n$ are realizations of q with $\mathrm{tp}(b_i/b_0, \ldots, b_{i-1}, A)$ non-algebraic, and isolated over q for all $i \leq n$,* **then** *every complete extension q' of q over $A \cup \{b_0, \ldots, b_n\}$ of infinite multiplicity is isolated over q.*

Then T has 2^{ω} countable models.

Proof. The 2^{ω} countable models will be found using Lemma 4.2. Fix $q = \mathrm{tp}(a/A)$ as given by the assumptions. Let $B = A \cup \{a\}$. Then by Lemma 3.9(ii) q has infinitely many non-algebraic extensions over B which are isolated over q. Let R be an infinite family of such extensions. We will try to construct inductively an almost orthogonal subfamily $\{p_i: i < \omega\}$ of R, satisfying the additional condition:

(i) for any n, $P_n(\mathbf{x}) = p_0(x_0) \cup \cdots \cup p_n(x_n)$ has only finitely many extensions to complete types over B.

Suppose we have already constructed the almost orthogonal family, $\{p_0, \ldots, p_n\}$ satisfying (i). Note that

(ii) for all $(b_0, \ldots, b_n)$ satisfying $P_n(\mathbf{x})$, $\mathrm{tp}(b_i/b_0, \ldots, b_{i-1}, B)$ is isolated over q, for all $i \leq n$ (and also of course $\mathrm{tp}(a/A)$ is isolated over q, trivially).

Fix now some $\mathbf{b}$ realizing $P_n(\mathbf{x})$. By Corollary 4.3, we may assume that only finitely many $r \in R$ can have a non-algebraic extension over $B \cup \{\mathbf{b}\}$ of finite multiplicity. So by (ii) and our hypothesis (A) on q, for all but finitely many $r \in R$, every non-algebraic extension of r over $B\mathbf{b}$ is isolated over q. By compactness of $S_{r,na}(B\mathbf{b})$, any such r has only finitely many non-algebraic extensions over $B\mathbf{b}$. By Lemma 3.9(iii), every extension of r over $B\mathbf{b}$ is non-algebraic. We have shown that for all but finitely many $r \in R$, every extension of r over $B\mathbf{b}$ is non-algebraic and isolated over r (and thus over q, as r is isolated over q). By the inductive assumption (i), there is $r \in R$ such that for all $\mathbf{b}$ realizing $P_n(\mathbf{x})$, r has only finitely many extensions over $B\mathbf{b}$, all non-algebraic and isolated over q. Put p_{n+1} to be some such r. Clearly $\{p_0, \ldots, p_{n+1}\}$ is almost orthogonal, and, moreover, (i) holds for $n+1$ in place of n. Thus the almost orthogonal family $P = \{p_i: i < \omega\}$ can be constructed. Each p_i, being a non-forking extension of q, is non-isolated. By Lemma 4.2, T has 2^ω countable models. ■

Proof of Theorem. 4.1. By Lemma 4.4 there is finite A and $p_0 \in S(A)$ such that $p_0(x) \vDash x \in G$, and p_0 is of U-rank 1, non-isolated, and of infinite multiplicity. We may assume $A = \varnothing$. By Lemma 4.5 we may also assume that for every non-isolated U-rank 1 type $q \in S(A)$ of infinite multiplicity (A finite) in G, the hypothesis (A) of Lemma 4.5 fails. We then clearly obtain.

(B) There are realizations a_i of p_0 for $i < \omega$, and a strictly increasing sequence $\{n(k): k < \omega\}$ of natural numbers with $n(0) = 0$, such that, denoting $\{a_m: m < i\}$ by A_i, and $\mathrm{tp}(a_{n(k)}/A_{n(k)})$ by q_k,

(i) for all i, $\mathrm{tp}(a_i/A_i)$ is non-algebraic;

(ii) for all $k < \omega$, and i such that $n(k) \leq i < n(k+1)$, $\mathrm{tp}(a_i/A_i)$ extends q_k and is isolated over q_k.

(iii) for all k, q_{k+1} extends q_k, is not isolated over q_k, and has infinite multiplicity.

Note that every $tp(a_i/A_i)$ is non-isolated (being a non-forking extension of p_0), and has infinite multiplicity (by construction if $i = n(k)$, and by Lemma 3.9(i) otherwise).

Let $f_k = f(q_k)$. So by Lemma 3.13(iii), we also have

(iv) for all k, and i such that $n(k) \leq i < n(k+1)$, $f(a_i/A_i) =^* f_k$, and, moreover, $f_{k+1} <^* f_k$.

We will use the sequence $\{f_k\}$ to construct 2^ω countable models of T. In fact we will construct 2^ω countable models of $Th(G)$ (i.e. elementary substructures of G), but it is an easy exercise using definability of ∞-rank to show that any countable model M of $Th(G)$ is of the form G^N for some countable model N of T.

We now work inside G (or $Th(G)$). **All elements we talk about will be elements of G (not G^{eq}).**

REMARK **4.6.** *For any countable model M, and countable set X,* $acl(MX)$ *is a model.*

Proof. By the omitting types theorem there is a model M_1 containing $M \cup X$ such that for every $b \in M_1 - M$, b forks with X over M. (If $b' \notin M$ and $tp(b'/MX)$ does not fork over M, then $tp(b'/MX)$ is finitely satisfiable in M, so could not be isolated.) As every element has rank at most 1, $M_1 = acl(MX)$. ∎

Let A denote $\{a_i: i < \omega\}$. We first construct a base model M_0. In fact just choose M_0 to be a countable model which contains A and is almost atomic over A, that is for any tuple b from M_0, for all sufficiently inclusive finite sets $A_0 \subset A$, $tp(\mathbf{b}/A_0)$ is isolated. This is easy as T is small. Let B be a maximal A-independent set of (non-algebraic) elements in M_0. Then clearly by weak minimality of G, $M_0 = acl(A \cup B)$, and we remark:

Fact 4.7. For all $b_1, \ldots, b_n \in B$ and all finite $A_0 \subset A$, $tp(b_n/A_0 \cup \{b_1, \ldots, b_{n-1}\})$ is isolated and non-algebraic. In particular, by Lemma 3.13(ii),

$$f(b_n/A_0 \cup \{b_1, \ldots, b_{n-1}\}) =^* \text{Max}.$$

Now let $X \subseteq \omega$ be infinite. We construct a countable model M_X as follows.

Define inductively sets $B_k = \{b_{k,m}: m < \omega\}$ for $k \in X$, such that for all $k \in X$ and $m < \omega$, $tp(b_{k,m}/A_{n(k)}) = q_k$, and for all finite D such that $A_{n(k)} \subseteq D \subseteq A \cup B \cup \bigcup\{B_i: i < k, i \in X\} \cup \{b_{k,n}: n < m\}$, $tp(b_{k,m}/D)$ is non-algebraic and isolated over q_k.

This can easily be accomplished. Suppose we have already defined B_i for $i < k$ and $b_{k,n}$ for $n < m$. Write $A \cup B \cup \bigcup\{B_i: i < k\} \cup \{b_{k,n}: n < m\}$ as an increasing union of finite sets D_i, where $D_0 = A_{n(k)}$. Define an increasing sequence of types $r_i \in S(D_i)$, such that $r_0 = q_k$, and r_{i+1} is isolated over r_i. Given $r_i \in S(D_i)$ which by induction has infinite multiplicity, choose an extension r_{i+1} of r_i in $S(D_{i+1})$ which is isolated over r_i among non-algebraic types. (The space of non-algebraic extensions of r_i in $S(D_{i+1})$ is closed in $S(D_{i+1})$ and countable, as T is small, so has an isolated point). As r_i has infinite multiplicity, by Lemma 3.9(i), r_{i+1} is isolated over r_i. Thus r_i is isolated over $r_0 = q_k$. Let $b_{k,m}$ realize $\cup\{r_i: i < \omega\}$.

By Remark 4.6, $\mathrm{acl}(M_0 \cup \bigcup\{B_k: k \in X\})$ is a model M_X. Note that $M_X = \mathrm{acl}(A \cup B \cup \bigcup\{B_k: k \in X\})$. Let $C = A \cup B \cup \bigcup\{B_k: k \in X\}$. Note that, by the construction, C is independent over $\varnothing$. We put the following ordering on $\cup\{B_k: k \in X\}$: $b_{k,m} < b_{t,n}$ iff either $k < t$, or $k = t$ and $m < n$.

LEMMA **4.8.** *Let $b \in B_k$ $(k \in X)$. Let D be finite such that $A_{n(k)} \subseteq D \subseteq C - \{b\}$. Then $f(b/D) = {}^* f_k$; equivalently, by Lemma* 3.13, *$tp(b/D)$ is (non-algebraic and) isolated over q_k.*

Also for any $b \in B$ and finite $D \subseteq C - \{b\}$, $\mathrm{tp}(b/D)$ is isolated, so $f(b/D) =^* \mathrm{Max}$.

Proof. This is just an application of the 'symmetry lemma' 3.14. Let $D_1 = D \cap (A \cup B \cup \{B_i: i < k\} \cup \{c \in B_k: c < b\})$. Let $D - D_1 = \{d_1, \ldots, d_n\}$, where $d_1 < d_2 < \ldots < d_n$. We may assume that D (and thus D_1) includes $A_{n(r)}$, where r is maximum such that some $d_i \in B_r$. By construction $f(b/D_1) =^* f(b/A_{n(k)}) = f_k$. Also, by construction, for each $i = 1, \ldots, n$,

$$f(d_i/D_1 \cup \{d_1, \ldots, d_{i-1}, b\}) =^* f(d_i/D_1 \cup \{d_1, \ldots, d_{i-1}\})$$

$(=^* f_{t(i)}$ where $d_i \in B_{t(i)})$. By repeated applications of Lemma 3.14 (and the fact that C is independent over $\varnothing$), we conclude that $f(b/D) =^* f(b/D_1)$, which suffices by the above.

The second part of the lemma is proved similarly, using Fact 4.7, the construction of the B_k, and the symmetry lemma. ■

The final lemma shows we can recover X from M_X.

LEMMA **4.9.** *For any $k < \omega$, $k \in X$, if and only if in M_X (i.e. all elements are chosen from M_X)*

($*$) *there is a finite set A' such that for any finite set $B' \supseteq A'$, there is an element a realizing p_0 such that $[f(a/B') = {}^* f_k$, and for no $b_1, \ldots, b_n$ realizing p_0 do we have $a \in \mathrm{acl}(B', b_1, \ldots, b_n)$ and $f(b_i/B') <^* f_k$ for all $i = 1, \ldots, n$.]*

Proof. Suppose first that $k \in X$. Choose A' to be $A_{n(k)}$. Let $B' \supseteq A'$ be any finite set in M_X. As $M_X = \mathrm{acl}(C)$, we can find finite $B'' \subseteq C$ such that $A' \subseteq B''$

and $B' \subseteq \text{acl}(B'')$. Let $a \in B_k - B''$. By Lemma 4.8, $f(a/B'') =^* f_k$, and thus, by Lemma 3.13(iv), $f(a/B') =^* f_k$. Suppose, by way of contradiction, that there are $b_1, \ldots, b_n$ realizing p_0 in M_X, such that $a \in \text{acl}(B', b_1, \ldots, b_n)$ and $f(b_i/B') <^* f_k$ for all i. Then also $a \in \text{acl}(B'', b_1, \ldots, b_n)$ and $f(b_i/B'') <^* f_k$ for all i. Let $D \subseteq C - B''$ be finite and minimal such that $b_i \in \text{acl}(B'' \cup D)$ for all i. Thus $a \in \text{acl}(B'' \cup D)$. As C is independent over $\varnothing$, $a \in D$. Let $D_1 = D - \{a\}$. By minimality of D, for some $t \leq n$, $b_t \notin \text{acl}(B'' \cup D_1)$. By forking symmetry $a \in \text{acl}(B'' \cup D_1 \cup \{b_t\})$. By Lemma 3.15, $f(a/B'' \cup D_1) \leq^* f(b_t/B'' \cup D_1)$. But by 4.8, $f(a/B'' \cup D_1) =^* f_k$, and (as $\text{tp}(b_t/B'' \cup D_1)$ is an extension of $\text{tp}(b_t/B'')$), $f(b_t/B'' \cup D_1) =^* f_k$.

This contradiction proves that $(*)$ holds.

Conversely, suppose $k \notin X$. We must show that $(*)$ does not hold in M_X. Let A' be any finite set in M_X. Let $r \in X$ be minimal such that $r > k$. Choose finite $B'' \subseteq C$ such that $A' \subseteq \text{acl}(B'')$ and $B'' \supseteq A_{n(r)}$. Put $B' = B'' \cup A'$. Suppose a realizes p_0 in M_X and $f(a/B') =^* f_k$. We have to find $b_1, \ldots, b_n$ realizing p_0 in M_X, such that $f(b_i/B') <^* f_k$ for all i, and $a \in \text{acl}(B', b_1, \ldots, b_n)$.

Note that as B' is interalgebraic with B'' also $f(a/B'') =^* f_k$ (by 3.13(v)). Choose a finite minimal $D \subseteq C - B''$ such that $a \in \text{acl}(B'' \cup D)$.

Claim. *For every $b \in D$, either $b \in A$ or $b \in B_i$ for some $i \in X$ with $i \geq r$.*

Proof of claim. Suppose otherwise and let $b \in D$ be such that $b \in B$, or $b \in B_i$ for some $i \in X$ with $i < r$. Let $D_1 = D - \{b\}$. By minimal choice of D, $a \notin \text{acl}(B'' \cup D_1)$. By forking symmetry $b \in \text{acl}(B'' \cup D_1 \cup \{a\})$. By Lemma 3.15, $f(b/B'' \cup D_1) \leq^* f(a/B'' \cup D_1)$. Thus, as $f(a/B'' \cup D_1) \leq^* f(a/B'') =^* f_k$, we deduce that $f(b/B'' \cup D_1) \leq^* f_k$. On the other hand, as B'' contains $A_{n(i)}$, we see by Lemma 4.8 that $f(b/B'' \cup D_1) =^* f(b/D_1) =^* f_i$ or Max, depending on whether $b \in B_i$ or B. Thus $b \in B_i$ (where $i \in X$ and $i < r$) and $f_i \leq^* f_k$. But as $k \notin X$, and r was chosen minimal in X greater than k, we must have $i < k$, whereby $f_k <^* f_i$. This contradiction proves the claim. ■

By the claim and the construction, for every $b \in D$, $\text{tp}(b/B'')$ is an extension of q_r, and thus $f(b/B'') \leq^* f_r <^* f_k$. Now B' is interalgebraic with B'', and thus also $a \in \text{acl}(B' \cup D)$, and using 3.13(iv), for every $b \in D$, also $f(b/B') <^* f_k$. This completes the proof of the lemma. ■

By Lemma 4.9, if $X \neq Y$ are infinite subsets of ω, then M_X and M_Y are non-isomorphic countable models of $Th(G)$. As remarked earlier, this gives 2^ω countable models of T. The proof of Theorem 4.1 is complete.

An important consequence of Theorem 4.1 is

Corollary **4.10.** *Suppose that T has less than 2^ω countable models. Let D be* *a weakly minimal set defined over a finite set A. Then there are only finitely many non-isolated* $q \in S(A)$, *with* $\models q(x) \to x \in D$.

Proof. By Corollary 4.3 and Theorem 4.1. ■

5 Countable models of unidimensional theories

The hypotheses of the last two sections remain in place, but we also assume.

ASSUMPTION. *T has less than* 2^ω *countable models.*

We will give some information on the structure of the countable models of T. We work back in T^{eq}, unless otherwise stated. Theorem 4.1 together with 3.8 says, if A is finite and $p \in S(A)$ has U-rank 1, then p is non-isolated iff p has finite multiplicity.

p_0 will denote (at the right time) a fixed, stationary, modular rank 1 type over $\emptyset$. (By Proposition 1.1 there is a weakly minimal group G defined over a finite part e of $\mathrm{acl}(\emptyset)$. The generic type of the connected component G^0 of G is over e, stationary, and modular. As we are interested in counting countable models, we may name e.)

LEMMA **5.1.** *Let A be finite. Suppose* $\mathrm{tp}(b/A)$ *is isolated and* $\mathrm{tp}(a/A)$ *is non-isolated of U-rank* 1. *Then a is independent from b over A and, moreover,* $\mathrm{tp}(b/Aa)$ *is also isolated. If* $\mathrm{tp}(a/A)$ *is also stationary then* $\mathrm{tp}(b/A) \models \mathrm{tp}(b/Aa)$.

Proof. If $a \in \mathrm{acl}(Ab)$ then $\mathrm{tp}(a/A)$ would be isolated. Thus a is independent with b over A. Theorem 4.1, $\mathrm{tp}(a/A)$ has finite multiplicity. We have just seen that every extension of $\mathrm{tp}(a/A)$ over Ab is a non-forking extension; thus (as $\mathrm{tp}(a/A)$ has finite multiplicity) there are only finitely many extensions. So also $\mathrm{tp}(b/A)$ has only finitely many extensions over Aa, which must therefore all be isolated. So $\mathrm{tp}(b/Aa)$ is isolated.

If also $\mathrm{tp}(a/A)$ is stationary, then $\mathrm{tp}(a/A)$ has a unique extension over Ab. Thus clearly $\mathrm{tp}(b/A) \models \mathrm{tp}(b/Aa)$. ■

PROPOSITION **5.2.** *Let A be finite,* $\mathrm{tp}(b/A)$ *of U-rank* 1 *and isolated, and* $\mathrm{tp}(a/A, b)$ *of U-rank* 1 *and non-isolated. Then there is c such that* $U(c/A) = 1$, $\mathrm{tp}(c/A)$ *is non-isolated, c is independent from b over A, and a and c are interalgebraic over Ab.*

Proof. Without loss of generality $A = \emptyset$. Let $p = \mathrm{tp}(b/\emptyset)$.

CLAIM. *There are non-algebraic extensions* $p_i \in S(b)$ *of p, for* $i < \omega$, *such that whenever* b_i *realizes* p_i *for all i,* $\{b_i : i < \omega\}$ *is independent and atomic over b.*

Proof of claim. We make a construction like that in the proof of Lemma 4.5. First by Lemma 3.9, there is an infinite family $R \subseteq S(b)$ of non-algebraic extensions of p such that every $r \in R$ is isolated over p and thus isolated. Suppose we have already constructed $p_0, \ldots, p_n \in S(b)$ satisfying the requirements in the claim (i.e. whenever b_i realizes p_i for $i \leq n$ then $\{b_i : i \leq n\}$ is independent and atomic over b.) Thus there are only finitely many extensions of $P_n(\mathbf{x}) = p_0(x_0) \cup p_1(x_1) \cup \cdots \cup p_n(x_n)$ in $S(b)$. Fix $\mathbf{b}$ realizing $P_n(\mathbf{x})$. By Corollary 4.10 (working in the weakly minimal set p, which is definable as p is isolated), only finitely many $r \in R$ can have a non-isolated extension over $b\mathbf{b}$. By Lemma 3.9, for all but finitely many $r \in R$, every extension of r over $b\mathbf{b}$ is isolated and non-algebraic. Thus there is $r \in R$ such that for all $\mathbf{b}$ realizing $P_n(\mathbf{x})$ every extension of r over $b\mathbf{b}$ is isolated and non-algebraic. Put p_{n+1} to be such an r. The claim is proved. ■

Now fix b_i realizing p_i and let a_i be such that $\text{tp}(a_i, b_i) = \text{tp}(a, b)$. As $\text{tp}(a_i/b_i)$ is non-isolated of rank 1, and $\text{tp}(b/b_i)$ is isolated, Lemma 5.1 says that a_i is independent from b over b_i. As b is independent from b_i over $\varnothing$, it follows that $\{a_i, b_i\}$ is independent from b over $\varnothing$, whereby clearly $\text{tp}(a_i, b_i/b)$ is non-isolated. Let $q_i(x, y) = \text{tp}(a_i, b_i/b)$. By Lemma 4.2, $\{q_i : i < \omega\}$ is not an almost orthogonal family. Thus there are realizations, which we again call $(a_1, b_1), \ldots, (a_n, b_n)$ of $q_1, \ldots, q_n$ respectively, such that $a_n b_n$ forks with $\{a_1, b_1, \ldots, a_{n-1}, b_{n-1}, b\}$ over b, and we can assume n is minimal. Write $\mathbf{b}$ for $\{b_1, \ldots, b_{n-1}\}$ and $\mathbf{a}$ for $\{a_1, \ldots, a_{n-1}\}$. As $a_n b_n$ is independent from b over $\varnothing$, we see that $a_n b_n$ forks with $\mathbf{a}\mathbf{b}b$ over $\varnothing$. Now as the b_i realize the p_i,

(i) b_n is independent from $\mathbf{b}b$ over $\varnothing$, and $\text{tp}(b_n/\mathbf{b}b)$ is isolated.

By minimality of n, $\{(a_1, b_1), \ldots, (a_{n-1}, b_{n-1}), b\}$ is independent over $\varnothing$, and thus for each $i \leq n-1$,

(ii) $\text{tp}(a_i/a_1, \ldots, a_{i-1}, \mathbf{b}, b)$ is non-isolated (of rank 1).

From (i) and (ii) it follows, by induction on i and Lemma 5.1, that for all i, a_i is independent from b_n over $\{a_1, \ldots, a_{i-1}, \mathbf{b}, b\}$ and, moreover, $\text{tp}(b_n/a_1, \ldots, a_i, \mathbf{b}, b)$ is isolated. Thus

(iii) $\text{tp}(b_n/\mathbf{a}, \mathbf{b}, b)$ does not fork over $\mathbf{b}b$; and

(iv) $\text{tp}(b_n/\mathbf{a}, \mathbf{b}, b)$ is isolated.

By (i) and (iii) we see that

(v) b_n is independent from $\mathbf{a}\mathbf{b}b$ over $\varnothing$.

Thus $U(a_n b_n/\varnothing) = 2$, and $U(a_n b_n/\mathbf{a}\mathbf{b}b) = 1$. As T is 1-based

$$C = \text{Cb}(\text{stp}(a_n b_n/\mathbf{a}\mathbf{b}b)) \subseteq \text{acl}(a_n b_n).$$

Additivity of U-rank implies that $U(\text{tp}(C)) = 1$. Let $c \in C - \text{acl}(\varnothing)$. So $U(\text{tp}(c)) = 1$, b_n is independent from c over $\varnothing$ (as $c \in \text{acl}(\mathbf{ab}b)$ and by (v)), and so $a_n \in \text{acl}(b_n, c)$. As $\text{tp}(a_n, b_n/\varnothing) = \text{tp}(a, b/\varnothing)$ we have proved the proposition except for the non-isolation of $\text{tp}(c/\varnothing)$.

Suppose, by way of contradiction, that $\text{tp}(c/\varnothing)$ is isolated. Now by (iv), (v), and the fact that $c \in \text{acl}(\mathbf{ab}b)$, $\text{tp}(b_n/c)$ is isolated. Thus $\text{tp}(b_n, c)$ is isolated, so $\text{tp}(c/b_n)$ is isolated. But a_n forks with c over b_n, so we contradict Lemma 5.1. Thus $\text{tp}(c/\varnothing)$ is non-isolated. The proof of the proposition is complete. ■

Although the following corollary is not required for the proofs of our classification theorems, it does give some picture of the structure of the models of T, in particular strengthening the coordinatization lemma.

COROLLARY **5.3.** *For any a there are $b_1, \ldots, b_n \in \text{acl}(a)$ such that for all $i \leq n$ $\text{tp}(b_i/b_1, \ldots, b_{i-1})$ is of U-rank 1, and non-isolated of finite multiplicity, and $\text{tp}(a/b_1, \ldots, b_n)$ is isolated. Clearly $n \leq U(a)$.*

Proof. We prove the corollary by induction on $U(\text{tp}(a/\varnothing)) = m$. Remember we work in T^{eq}. Also by Theorem 4.1 every U-rank 1 non-isolated type over a finite set has finite multiplicity. Fix a, with $U(a/\varnothing) = m$. If $m = 1$ there is no problem ($\text{tp}(a/\varnothing)$ is either non-isolated or isolated). Suppose $m > 1$. By 1-basedness, Proposition 2.5.9, and additivity of U-rank, there are $a_1, \ldots, a_m \in \text{acl}(a)$, such that $U(a_i/a_1, \ldots, a_{i-1}) = 1$ for all $i \leq m$, and $a_m = a$.

For $i \leq m$, let A_i denote $(a_1, \ldots, a_{i-1})$.

Case (i). For all $i \leq m$, $\text{tp}(a_i/A_i)$ is non-isolated.
In this case, the a_i do the job.

Case (ii). $\text{tp}(a/A_m)$ is isolated.
Note that by addivitity of U-rank, $U(\text{tp}(A_m/\varnothing)) = m - 1$. So we can apply the induction hypothesis to A_m and find (for some n) $b_1, \ldots, b_n \in \text{acl}(A_m)$, with $\text{tp}(b_i/b_1, \ldots, b_{i-1})$ of U-rank 1, non-isolated, for all $i \leq n$, and with $\text{tp}(A_m/b_1, \ldots, b_n)$ isolated. Then the $b_i \in \text{acl}(a)$, and by the case assumption $\text{tp}(a/b_1, \ldots, b_n)$ is isolated.

Case (iii). Otherwise.
Let $i < m$ be maximal such that $\text{tp}(a_i/A_i)$ is isolated. Thus $\text{tp}(a_{i+1}/A_i, a_i)$ is non-isolated. By Proposition 5.2, there is c, with $U(c/A_i) = 1$, $\text{tp}(c/A_i)$ non-isolated, and c interalgebraic with a_{i+1} over A_i, a_i. Note that $U(a_i, a_{i+1}, c/A_i) = U(a_i, a_{i+1}/A_i) = U(a_i, c/A_i) = U(c, a_{i+1}/A_i) = 2$. Thus

(a) (a_i, a_{i+1}), (a_i, c), and (c, a_{i+1}) are pairwise interalgebraic over A_i.

By Lemma 5.1, $\text{tp}(a_i/A_i, c)$ is isolated. Thus by (a), we conclude

(b) $\text{tp}(a_{i+1}/A_i, c)$ is isolated.

We now modify the sequence of $a_1, \ldots, a_m$ by replacing a_i by c. Formally let $a'_k = a_k$ if $k \neq i$, and let $a'_i = c$. Clearly $U(a'_k/A'_k) = 1$, for all k, and by (a) for $k > i + 1$, $\text{tp}(a'_k/A'_k)$ is not isolated. Thus now by (b), $i + 1$ is maximal such that $\text{tp}(a'_{i+1}/A'_{i+1})$ is isolated. Continuing this way, we reach eventually the situation of Case (ii).
The Corollary is proved. ■

COROLLARY **5.4.** (*of Proposition* 5.2) *Let A be finite, and suppose* $\text{tp}(b/A)$ *is isolated,* $\text{tp}(a/A, b)$ *is non-isolated of U-rank* 1. *Then there is c interalgebraic with a over Ab such that* $U(c/A) = 1$, *c is independent with b over A, and* $\text{tp}(c/A)$ *is non-isolated.*

Proof. Assume $A = \varnothing$ for convenience. By Proposition 2.5.9 (and the fact that $U(\text{tp}(b))$ is finite), there are $b_1, \ldots, b_n \in \text{acl}(b)$, such that $b_n = b$, and for all $i \leq n$, $\text{tp}(b_i/\mathbf{b}_i)$ has U-rank 1 (where $\mathbf{b}_i = (b_1, \ldots, b_{i-1})$ and $\mathbf{b}_0 = \varnothing$). We may clearly assume that $b = (b_1, \ldots, b_n) = \mathbf{b}$, as b is interalgebraic with $\mathbf{b}$ and so $\text{tp}(a/\mathbf{b})$ still has U-rank 1, and is non-isolated. Moreover, $\text{tp}(\mathbf{b}/\varnothing)$ is isolated, as is $\text{tp}(b_i/\mathbf{b}_i)$ for all $i \leq n$. Using Proposition 5.2, find (downward inductively) $c_n, c_{n-1}, \ldots, c_1$, such that $U(c_i/\mathbf{b}_i) = 1$, $tp(c_i/\mathbf{b}_i)$ is non-isolated, c_i is independent from $\mathbf{b}$ over $\mathbf{b}_i$, and c_i is interalgebraic with a over $\mathbf{b}$: suppose we have already obtained c_i, with $i > 1$. By Proposition 5.2, find c_{i-1}, with $U(c_{i-1}/\mathbf{b}_{i-1}) = 1$, $\text{tp}(c_{i-1}/\mathbf{b}_{i-1})$ non-isolated, c_{i-1} interalgebraic with c_i over $\mathbf{b}_i$, and c_{i-1} independent from $\mathbf{b}_i$ over $\mathbf{b}_{i-1}$. Then using the inductive hypothesis, clearly c_{i-1} is independent from $\mathbf{b}$ over $\mathbf{b}_{i-1}$, and interalgebraic with a over $\mathbf{b}$.

Then $c = c_1$ satisfies the requirements of the corollary. ■

We now take the opportunity to introduce a certain dependence relation on types.

DEFINITION **5.5.** *Let A be any set and let Q be a family of rank* 1, *non-modular strong types over A. Let $q, q_1, \ldots, q_n \in Q$. We will say* $q \in \text{ACL}(q_1, \ldots, q_n)$ *if there are realizations $a_1, \ldots, a_n$ of $q_1, \ldots, q_n$ respectively such that q is realized in* $\text{acl}(A, a_1, \ldots, a_n)$.

We will make heavy use of Corollary 5.2.5 which we state again as a fact:

Fact 5.6. If q is a rank 1, non-modular strong type over A, and $B \supseteq A$ then q is realized in $\text{acl}(B)$ iff $q|B$ is modular.

LEMMA 5.7. *Let A and Q be as in 5.5. Let $q, q_1, \ldots, q_n \in Q$. Then the following are equivalent*

(i) $q \in \mathrm{ACL}(q_1, \ldots, q_n)$.

(ii) *Whenever a_i realizes q_i for $i = 1, \ldots, n$, q is realized in* $\mathrm{acl}(A, a_1, \ldots, a_n)$.

Proof. (ii) → (i) is immediate.

(i) → (ii). The proof is by induction on n. Let a realize q, a_i realize q_i for $i \leq n$ and $a \in \mathrm{acl}(A, a_1, \ldots, a_n)$. We may assume

(∗) for any proper subset Q_0 of $\{q_1, \ldots, q_n\}$, $q \notin \mathrm{ACL}(Q_0)$,

so $\{a_1, \ldots, a_n\}$ is A-independent.

CLAIM. *Whenever b_i realizes q_i for $i = 1, \ldots, n$, $\{b_1, \ldots, b_n\}$ is A-independent.*

Proof of claim. Suppose not. Let $b_1, \ldots, b_n$ be a counterexample, and without loss of generality $b_n \in \mathrm{acl}(b_1, \ldots, b_{n-1})$. Thus $q_n \in \mathrm{ACL}(q_1, \ldots, q_{n-1})$. By the induction assumption, and Fact 5.6, $\mathrm{stp}(a_n/A, a_1, \ldots, a_{n-1})$ is modular. But a and a_n are interalgebraic over $A, a_1, \ldots, a_{n-1}$. Thus $\mathrm{stp}(a/A, a_1, \ldots, a_{n-1})$ is modular. By Fact 5.6, q is realized in $\mathrm{acl}(A, a_1, \ldots, a_{n-1})$, contradicting (∗). The claim is proved. ■

By the claim $\mathrm{stp}(b_1, \ldots, b_n/A) = \mathrm{stp}(a_1, \ldots, a_n/A)$ whenever the b_i are realizations of the q_i. So by automorphism, for any such b_i, q is realized in $\mathrm{acl}(A, b_1, \ldots, b_n)$, giving (ii). ■

COROLLARY 5.8. (*A and Q as in 5.5, $q, q_1 \in Q$, Q_1 and Q_2 subsets of Q*) (i) *if $q \in \mathrm{ACL}(Q_1)$, and $Q_1 \subseteq \mathrm{ACL}(Q_2)$ then $q \in \mathrm{ACL}(Q_2)$.*

(ii) *If $q \in \mathrm{ACL}(Q_1 \cup \{q_1\}) - \mathrm{ACL}(Q_1)$, then $q_1 \in \mathrm{ACL}(Q_1 \cup \{q\}) - \mathrm{ACL}(Q_1)$.*

Proof. Immediate, by Lemma 5.7 and forking symmetry on rank 1 elements. ■

Thus, for any Q as in 5.5 we can speak of ACL-dim(Q), the cardinality of a maximal ACL-independent subset of Q.

LEMMA 5.9. *Let A be finite, and Q a family of U-rank* 1, *non-modular strong types over A, such that for each $q \in Q$, the type $r \in S(A)$ of which q is an extension has finite multiplicity. Then* ACL-dim(Q) *is finite.*

Proof. Let $R = \{r \in S(A)$: some extension of r is in $Q\}$. Let $Q' = \{q \in S(\mathrm{acl}(A))$: q extends some $r \in R\}$. Each $r \in R$ has finite multiplicity and is therefore non-isolated. By Lemma 4.2, R does not contain an infinite, almost orthogonal family. Thus, clearly there are $r_1, \ldots, r_k \in R$, such that for all $r \in R$ there are a_i realizing r_i for $i = 1, \ldots, k$ such that r is realized in $\operatorname{acl}(A, a_1, \ldots, a_k)$. Let $Q_0 = \{q \in Q'$: q extends r_i for some $i = 1, \ldots, k\}$. As each $r \in R$ has finite multiplicity, Q_0 is finite, say $Q_0 = \{q_1, \ldots, q_t\}$, and clearly for all $q \in Q'$, there are $b_1, \ldots, b_t$ realizing $q_1, \ldots, q_t$ respectively such that q is realized in $\operatorname{acl}(A, b_1, \ldots, b_t)$. This means that $Q' \subseteq \operatorname{ACL}\{q_1, \ldots, q_t\}$. Let $b_1, \ldots, b_t$ realize $q_1, \ldots, q_t$. Let $\mathbf{b} = (b_1, \ldots, b_t)$

By Lemma 5.7, each $q \in Q'$ is realized in $\operatorname{acl}(A, \mathbf{b})$. Let $B \subseteq \operatorname{acl}(A, \mathbf{b})$ consist of one realization of q for every $q \in Q'$. Let $C = \{b \in B$: $\operatorname{stp}(b/A) \in Q\}$. Let $\mathbf{c} \subseteq C$ be such that $\operatorname{tp}(\mathbf{b}/AC)$ does not fork over $A\mathbf{c}$. As $C \subseteq \operatorname{acl}(A\mathbf{b})$, it follows that $C \subseteq \operatorname{acl}(A, \mathbf{c})$. As C contains a realization of every $q \in Q$, we see that ACL-dim(Q) is finite. ■

We will apply Lemma 5.9 to the following families:

Definition **5.10.** *Let M be a model and A a finite subset of M. Let $Q_{M,A} = \{q \in S(\operatorname{acl}(A))$: q is rank 1 and non-modular, q is realized in M, and the type over A of which q is an extension has finite multiplicity$\}$.*

Lemma **5.11.** *Let M be a model. Let $\{q_1, \ldots, q_n\}$ be an* ACL*-basis of $Q_{M,\varnothing}$ (such a basis is finite by* 5.9*). Let $a_1, \ldots, a_n \in M$ realize $q_1, \ldots, q_n$ respectively. Let I be a p_0-basis of M (where p_0 is a fixed, modular, stationary rank* 1 *type over $\varnothing$). Then*

(i) *$I \cup \{a_1, \ldots, a_n\}$ is independent over $\varnothing$; and*

(ii) *for any $a \in M$ such that $U(a/\varnothing) = 1$ and $\operatorname{tp}(a/\varnothing)$ has finite multiplicity, $a \in \operatorname{acl}(I, a_1, \ldots, a_n)$.*

Proof. (i) As $\{q_1, \ldots, q_n\}$ is ACL-independent, by Lemma 5.7 and Fact 5.6, for each i, $\operatorname{tp}(a_i/a_1, \ldots, a_{i-1})$ is non-modular. Now we can prove for each i that $\operatorname{tp}(a_i/I, a_1, \ldots, a_{i-1})$ is non-algebraic (using 5.6). Thus $I \cup \{a_1, \ldots, a_n\}$ is independent over $\varnothing$.

(ii) If $\operatorname{stp}(a/\varnothing)$ is modular, then $a \in \operatorname{acl}(I)$. If not, then $\operatorname{stp}(a/\varnothing) \in Q_{M,\varnothing}$. So $\operatorname{stp}(a/\varnothing) \in \operatorname{ACL}(q_1, \ldots, q_n)$ and thus, by Lemma 5.7, is realized by some b in $\operatorname{acl}(a_1, \ldots, a_n)$. If $a \in \operatorname{acl}(b)$, fine. Otherwise $\operatorname{stp}(a/b)$ is modular and thus there is c realizing $p_0|b$ with $a \in \operatorname{acl}(b, c)$. But $c \in \operatorname{acl}(I)$. Thus we see that $a \in \operatorname{acl}(I, a_1, \ldots, a_n)$. ■

We can now give the structure theorem for the models of T in the weakly minimal case. Note that we are working in T^{eq}. Recall that to say T is weakly minimal means, in this case, that there is some sort S, say, such that S is weakly minimal and for any model M, $M = \operatorname{dcl}(S^M)$.

THEOREM 5.12. *Let T be weakly minimal. Let M be a model of T. Let $\{q_1, \ldots, q_n\}$ be an* ACL*-basis of $Q_{M,\varnothing}$, and $a_1, \ldots, a_n \in M$ realizations of $q_1, \ldots, q_n$ respectively. Let I be a p_0-basis of M. Then M is atomic over $I \cup \{\mathbf{a}\}$ (where $\mathbf{a} = (a_1, \ldots, a_n)$).*

Proof. As $M \subseteq \mathrm{dcl}(S^M)$, it is enough to prove that $\mathrm{tp}(\mathbf{c}/I \cup \mathbf{a})$ is isolated, where $\mathbf{c}$ is any tuple of elements in S^M.

So let $\mathbf{c}$ be such, say $\mathbf{c} = (c_1, \ldots, c_m) \in M$ with $c_i \in S$ for all i. We may clearly assume $\{c_1, \ldots, c_m\}$ to be (algebraically) independent over $\varnothing$. Let $\mathbf{c}'$ be a maximal subtuple of $\mathbf{c}$ such that $\mathrm{tp}(\mathbf{c}'/\varnothing)$ is isolated. Without loss $\mathbf{c}' = (c_1, \ldots, c_k)$ for some $k \leq m$. Thus for all i with $k < i \leq m$, $\mathrm{tp}(c_i/\mathbf{c}')$ is non-isolated (of rank 1). By Corollary 5.4 for each such i, there is d_i such that

(i) $\mathrm{tp}(d_i/\varnothing)$ is non-isolated of U-rank 1 and $c_i \in \mathrm{acl}(\mathbf{c}', d_i)$.

By Theorem 4.1, $\mathrm{tp}(d_i/\varnothing)$ has finite multiplicity, for all $k < i \leq m$, and thus by Lemma 5.11,

(ii) $d_i \in \mathrm{acl}(I \cup \mathbf{a})$.

On the other hand by Lemma 5.1 (applied iteratively) $\mathrm{tp}(\mathbf{c}'/\mathbf{a})$ is isolated. Again, by Lemma 5.1 and the fact that p_0 is stationary, we see that $\mathrm{tp}(\mathbf{c}'/\mathbf{a}) \models \mathrm{tp}(\mathbf{c}'/I \cup \mathbf{a})$, and thus

(iii) $\mathrm{tp}(\mathbf{c}'/I \cup \mathbf{a})$ is isolated.

By (i) and (ii) $\mathbf{c} \in \mathrm{acl}(\mathbf{c}' \cup I \cup \mathbf{a})$, and thus by (iii) $\mathrm{tp}(\mathbf{c}/I \cup \mathbf{a})$ is isolated. This proves the theorem. ■

COROLLARY 5.13. *Let T be weakly minimal. Then T has countably many countable models.*

Proof. Let M be countable. Let I and $\mathbf{a}$ be as in 5.12. By 5.12, M is atomic and thus prime over $I \cup \mathbf{a}$. So the isomorphism type of M is determined by $\mathrm{tp}(I, \mathbf{a})$. But I is a Morley sequence in the stationary type p_0, and by 5.11, I and $\mathbf{a}$ are independent over $\varnothing$. Thus $\mathrm{tp}(I, \mathbf{a})$ is determined by $\mathrm{tp}(\mathbf{a})$ and $|I|$. As T is small there are only countably many possibilities. Note that by the results of sections 1 and 2, $|I|$ can take all values (from 0 to ω, as M is countable). ■

The methods developed so far are not sufficient to eliminate the weakly minimal assumption from 5.13. However, if we restrict ourselves to 'finite-dimensional' models, they are. We call M finite dimensional if $\dim(p_0, M)$ is finite. Note that this is independent of the choice of p_0.

THEOREM **5.14.** *Any finite-dimensional model of T is atomic over a finite set. In particular T has only countably many countable finite-dimensional models.*

Proof. Let S be some sort such that $M = \mathrm{dcl}(S^M)$ for all models M of T. (For example, S could be the home sort.) Let $n < \omega$ be such that S has ∞-rank (or U-rank) n.

Let M be a finite-dimensional model of T. We define finite sets $A(1) \subseteq A(2) \subseteq \cdots \subseteq A(n)$ in M such that M will be atomic over $A(n)$.

Let first $A(1) = I \cup \mathbf{a}_1$ where I, $\mathbf{a}_1$ are as in 5.11. That is, I is a p_0-basis of M (necessarily finite by assumption) and $\mathbf{a}_1$ is a tuple realizing each strong type over $\varnothing$ from some ACL-basis of $Q_{M,\varnothing}$ (also finite by 5.9). Given $A(i)$ for $i < n$, let $\mathbf{a}_{i+1}$ be a finite tuple from M realizing each strong type over $A(i)$ from some ACL-basis of $Q_{M,A(i)}$ (also finite by 5.9). Let $A(i+1) = A(i) \cup \mathbf{a}_{i+1}$.

We will show by induction on $m \leq n$.

CLAIM. *If* $\mathbf{c}$ *is a tuple such that* $U(c) \leq m$ *for all* $c \in \mathbf{c}$ *then* $\mathrm{tp}(\mathbf{c}/A(m))$ *is isolated.*

Proof. For $m = 1$, this is precisely what we did *in the proof* of Theorem 5.12, so we do not repeat it.

Suppose the claim has been proven for m with $1 < m < n$. We prove it for $m + 1$. Let $\mathbf{c} = (c_1, \ldots, c_t)$ be our tuple. By coordinatization, or rather an argument similar to that given in the proof of coordinatization, we can find for each $i = 1, \ldots, t$, some $b_i \in \mathrm{acl}(c_i)$ such that $U(b_i) \leq m$ and $U(c_i/b_i) \leq 1$. (Note that we will have recourse to imaginaries and that $M^{\mathrm{eq}} \subseteq \mathrm{dcl}(M)$.) Let $\mathbf{b} = (b_1, \ldots, b_t)$. By the induction hypothesis,

(i) $\mathrm{tp}(\mathbf{b}/A(m))$ is isolated.

As for each $a \in A(m+1)$ $\mathrm{tp}(a/A(m))$ is non-isolated of U-rank 1, by Lemma 5.1 we see that $\mathrm{tp}(\mathbf{b}/A(m+1))$ is isolated.

We may clearly assume that $U(c_i/\mathbf{b}A(m)) = 1$ for $i = 1, \ldots, t$ and that, moreover, $\{c_1, \ldots, c_t\}$ is $\mathbf{b}A(m)$-independent.

Let $\mathbf{c}'$ be a maximal subtuple of $\mathbf{c}$ such that $\mathrm{tp}(\mathbf{c}'/\mathbf{b}A(m))$ is isolated. Thus

(ii) $\mathrm{tp}(\mathbf{c}'\mathbf{b}/A(m)$ is isolated.

Without loss $\mathbf{c}' = (c_1, \ldots, c_k)$ for some $k < t$. So for $i > k$, $\mathrm{tp}(c_i/\mathbf{c}'\mathbf{b}A(m))$ is non-isolated of U-rank 1. By Corollary 5.4 and (ii) for each $i >$ k there is d_i such that $\mathrm{tp}(d_i/A(m))$ is non-isolated of U-rank 1 and

(iii) $c_i \in \mathrm{acl}(d_i, \mathbf{c}', \mathbf{b}, A(m))$.

Fix *i*. By 4.1, $\mathrm{tp}(d_i/A(m))$ has finite multiplicity. As I, the p_0-basis of M is contained in $A(m)$, $q = \mathrm{stp}(d_i/A(m))$ is non-modular, and thus in $Q_{M,A(m)}$. Thus q is realized in $\mathrm{acl}(A(m+1))$ (by choice of $A(m+1)$), by some d. If d is independent from d_i over $A(m)$, then $\mathrm{stp}(d_i/dA(m))$ is modular so not weakly orthogonal to $p_0|dA(m)$. But again the p_0-basis of M is contained in $A(m)$. Thus d and d_i fork over $A(m)$, where $d_i \in \mathrm{acl}(A(m+1))$. We have shown

(iv) For all $i = k+1, \ldots, t$, $d_i \in \mathrm{acl}(A(m+1))$.

Now by (ii), and Lemma 5.1, $\mathrm{tp}(\mathbf{c}'\mathbf{b}/A(m+1))$ is isolated. By (iii) and (iv), $\mathbf{c} \in \mathrm{acl}(\mathbf{c}'\mathbf{b}, A(m+1))$. Thus $\mathrm{tp}(\mathbf{c}/A(m+1))$ is isolated. This proves the claim. ■

By the claim, if $\mathbf{c}$ is any tuple of elements, then $\mathrm{tp}(\mathbf{c}/A(n))$ is isolated. Thus M is atomic over $A(n)$. ■

Note that in particular Theorem 5.14 says that any elementary substructure of M_0, the p_0-envelope over $\varnothing$, is atomic over a finite set. (M_0 has p_0-dimension 0.)

7

Regular types

The purpose of this chapter is to generalize much of the preceding results on minimal or U-rank 1 types to regular types (in a stable theory). This includes the classification of the geometries attached to locally modular regular types. In the case of a minimal type p, dimension or U-rank was the crucial tool. By looking at canonical bases for example, we were forced into p^{eq}, that is into looking at types of the form $tp(a/A)$ where a is in the definable closure of some set of realizations of the minimal type p under consideration. It was fairly easy to extend the dimension theory to such types (by using U-rank, for example). The same problem arises when we study a regular type p. The main difference is that a forking extension of p need not be algebraic. It is not too difficult to develop a dimension theory for p^{eq} (extending the usual 'regular type' dimension theory for tuples of realizations of p). We nevertheless develop in section 1 a theory, that of p-simple types and p-weight, which goes beyond p^{eq}, but is quite elegant. Once the machinery is in place, much of the generalization of the minimal type (or finite rank) theory becomes routine, although for the benefit of the reader we will sometimes repeat arguments from earlier chapters.

In section 4 we introduce the notion of internality (which could have and maybe should have appeared earlier in the book), and among other things we prove the existence of definable automorphism groups under certain assumptions.

Throughout this chapter T is a stable theory and $\mathbf{C}$ is a big saturated model. As usual we work in $\mathbf{C}^{eq}$.

1 p-weight

Regular types were defined and introduced in section 4.5 of Chapter 1. The reader is advised to review that section. Regular types also appeared in Chapter 4. Of course minimal types are regular. In any case we repeat the definition: the stationary type $p \in S(A)$ is *regular* if for any $B \supseteq A$, and elements a, b such that $tp(a/B)$ is a non-forking extension of p and $tp(b/B)$ is a forking extension of p, a is independent from b over B.

As was pointed out in section 4.5 of Chapter 1, if $p(x) \in S(A)$ is a regular type, then $p^{\mathbf{C}}$, the set of realizations of p in the universe, has the structure of a pregeometry under the following 'closure' operation: if $B \subseteq p^{\mathbf{C}}$, then $\mathrm{cl}(B) = \{b \in p^{\mathbf{C}}: b$ forks with B over $A\}$. Of course cl(-) here depends on A. (In fact, as p has a unique non-forking extension over any $C \supseteq A$, this is a homogeneous pregeometry.) We point out that this feature *characterizes* regular types.

Remark 1.1. *Let $p \in S(A)$ be a stationary type, and that (for* cl(-) *as above) $(p^{\mathbf{C}}, \mathrm{cl})$ is a pregeometry. Then p is regular.*

Proof. Suppose that $(p^{\mathbf{C}}, \mathrm{cl})$ is a pregeometry. Let $B \supseteq A$, and a, b realizations of p, with a independent from B over A and b dependent with B over A. We wish to show that a is independent from b over B. By taking non-forking extensions, we may assume B to be an $|A|^+$-saturated model M. By Lemma 1.4.5.9,

$$(*) \qquad \mathrm{tp}(a, b/A \cup p^M) \models \mathrm{tp}(a, b/M).$$

As b forks with M over A, b forks with p^M over A, that is $b \in \mathrm{cl}(p^M)$. Also of course $a \notin \mathrm{cl}(p^M)$. Thus (as we assume $(p^{\mathbf{C}}, \mathrm{cl})$ to be a pregeometry) $a \notin \mathrm{cl}(p^M \cup \{b\})$. This means that a is independent from $p^M \cup \{b\}$ over A.

By $(*)$, a is independent from $M \cup \{b\}$ over A; thus a is independent from $M \cup \{b\}$ over M. ■

We now fix a stationary regular type $p \in S(A)$ where we assume $A = \mathrm{Cb(p)}$. So A has cardinality at most $|T|$, and A is contained in the algebraic closure of some subset A_0 of A of cardinality $< \kappa(T)$. We will again be using some 'dimension' notation, hopefully without confusion. If Y is a set of realizations of p, and B is any superset of A, then by a B-basis of Y we mean a maximal subset Z of Y such that for any $a \in Z$, a is independent from $B \cup (Z - \{a\})$ over A. Regularity of p (or alternatively Lemmas 1.4.5.3 and 1.4.4.1) implies that any two B-bases of Y have the same cardinality, which, for now, we call the dimension of Y over B or $\dim_B(Y)$, or $\dim_{p,\mathrm{B}}(Y)$ if we want to mention p. Quite soon we will be using notation for 'p-weight' instead. Remember that p is fixed for now.

Definition 1.2. *Let r and q be stationary types over maybe different sets. We say that q is hereditarily orthogonal to r if every extension of q is orthogonal to r.*

Note that if q is hereditarily orthogonal to r then so is any extension of q.

We will define a class of types, the p-simple types, for which an analogue to 'dimension' (or U-rank) is available.

Definition 1.3. *Let q be a strong type over a set X, where p is non-orthogonal to X. q is said to be p-simple if there is a set $B \supseteq A \cup X$, some realization a of $q|B$ and a set Y of realizations of p such that $\mathrm{stp}(a/BY)$ is hereditarily orthogonal to p.*

An obvious example of a p-simple type is $\mathrm{stp}(a_1, \ldots, a_m/A)$ where each a_i realizes p.

Remark 1.3′. (i) *First note if q is p-simple then by definition q is a strong type over a set X such that p is non-orthogonal to X. Thus if $\mathrm{stp}(a/X)$ does not fork over Y and p is non-orthogonal to Y then $\mathrm{stp}(a/X)$ is p-simple iff $\mathrm{stp}(a/Y)$ is p-simple.*

(ii) *Suppose p is non-orthogonal to Y, $Y \subseteq X$, q is a strong type over Y, and q' is its non-forking extension over X. Then q is hereditarily orthogonal to p iff q' is hereditarily orthogonal to p.*

Proof. Only (ii) requires comment. Suppose q is not hereditarily orthogonal to p. So there is $Z \supseteq Y$ and some stationary extension q_1 of q over Z such that q_1 is non-orthogonal to p. Let Z' realize $\mathrm{stp}(Z/Y)$ with Z' independent from X over Y. Let q_1' be the copy of q_1 over Z'. Let A' be the image of A under an automorphism fixing $\mathrm{acl}(Y)$ and taking Z to Z'. Let p' be the copy of p over A'. Then clearly q_1' is non-orthogonal to p'. By 1.4.3.3, 1.4.4.2, and 1.4.5.3, p is domination equivalent to p'. Thus q_1' is non-orthogonal to p. Let q_1'' be the non-forking extension of q_1' over $\mathrm{acl}(Z' \cup X)$, realized by c, say. As Z' is independent from X over Y, it follows that $\mathrm{stp}(c/X)$ does not fork over Y, whereby $\mathrm{stp}(c/X) = q'$. Thus q_1'' is an extension of q' which is non-orthogonal to p, showing that q' is not hereditarily orthogonal to p. ■

The following is almost immediate:

Lemma 1.4. (i) *Suppose that $\mathrm{stp}(a/X)$, $\mathrm{stp}(b/X)$ are p-simple. Then so is $\mathrm{stp}(a, b/X)$.*

(ii) *If $\mathrm{stp}(a/X)$ is p-simple and $b \in \mathrm{acl}(aX)$ then $\mathrm{stp}(b/X)$ is p-simple.*

Proof. (i) Let B_1, Y_1 witness the p-simplicity of $\mathrm{stp}(a/X)$, and B_2, Y_2 witness the p-simplicity of $\mathrm{stp}(b/X)$. Let $a_1 b_1$ realize $\mathrm{stp}(ab/X)|(B_1 B_2)$. Without loss of generality both $\mathrm{stp}(a_1/B_1 Y_1)$ and $\mathrm{stp}(b_1/B_2 Y_2)$ are hereditarily orthogonal to p. Thus so is $\mathrm{stp}(a_1 b_1/B_1 B_2 Y_1 Y_2)$.

(ii) is left to the reader. ■

LEMMA 1.5. *Let $B \supseteq A$ and let Y be a set of realizations of p such that* $\mathrm{stp}(a/B \cup Y)$ *is hereditarily orthogonal to p. Let I be a B-basis of Y. Then* $\mathrm{stp}(a/B \cup I)$ *is hereditarily orthogonal to p.*

Proof. Let C contain $B \cup I$, and let b realize $p|C$. As each $y \in Y$ forks with $B \cup I$ over A, regularity of p implies that b realizes $p|C \cup Y$. Thus b is independent with a over $C \cup Y$, whereby b is independent with a over C. ■

DEFINITION 1.6. *Let $q = \mathrm{stp}(a/X)$ be p-simple. Then the p-weight of q, $w_p(q)$, is defined to be* $\min\{\kappa$: *there is $B \supseteq A \cup X$, there is a' realizing $q|B$, and there is I, an independent set of realizations of $p|B$, such that* $\mathrm{stp}(a'/BI)$ *is hereditarily orthogonal to p and $|I| = \kappa\}$.*

By Lemma 1.5, clearly $w_p(q)$ exists, is at most $|T|$, and is finite if T is superstable. (Suppose B, I are as in Definition 1.6. Let I' be a maximal subset of I such that a' is independent from I over B. So we may incorporate I' into B, and thus we may assume that a' forks with c over B for each $c \in I$. Now use section 4.4 of Chapter 1.)

Moreover, again, standard forking considerations show that $w_p(q) = w_p(q_1)$ if q_1 is parallel to q. Note also that if $\mathrm{stp}(a/X)$ is p-simple and $b \in \mathrm{acl}(a, X)$ then $w_p(\mathrm{stp}(b/X)) \leq w_p(\mathrm{stp}(a/X))$.

We will write $w_p(a/X)$ in place of $w_p(\mathrm{stp}(a/X))$ when it is convenient. It is also clear (using Remark 1.3′(ii)) that if q is p-simple, then $w_p(q) = 0$ iff q is hereditarily orthogonal to p.

As remarked above there is a subset A_0 of A of cardinality $< \kappa(T)$ such that $A \subseteq \mathrm{acl}(A_0)$. As we are always concerned with strong types over given sets, the reader can assume below that $A = A_0$. (This explains why we are able to work inside a-models below rather than in $|T|^+$-saturated models.)

LEMMA 1.7. *Let $q = \mathrm{stp}(a/X)$ be p-simple. Let M be an a-model containing $A \cup X$ with* $\mathrm{stp}(a/M) = q|M$. *Let M_1 be any a-model containing $M \cup \{a\}$, and let $Y = \{b \in M_1$: b realizes $p|M$ and b forks with a over $M\}$. Then* $\mathrm{stp}(a/MY)$ *is hereditarily orthogonal to p and* $w_p(q) = \dim_{p,M}(Y)$.

Proof. As $w_p(\text{-})$ is a parallelism invariant we may suppose that $|X| < \kappa(T)$. Let $B \supseteq A \cup X$ and let I be a B-independent set of realizations of $p|B$ such that $\mathrm{stp}(a/B) = q|B$, $\mathrm{stp}(a/BI)$ is hereditarily orthogonal to p and $w_p(q) = \kappa = |I|$. We may assume that

(*) $\quad$ a forks with b over B for every $b \in I$.

(For let I_0 be a maximal subset of I such that a is independent with I_0 over B, and replace B by $B \cup I_0$, and I by $I - I_0$.) Let B_0 be such that $A \cup X \subseteq B_0 \subseteq B$, $\operatorname{stp}(a, I/B)$ does not fork over B_0, and $|B_0| < \kappa(T)$. Then $\operatorname{stp}(a/B_0 I)$ is hereditarily orthogonal to p. As both M and M_1 are a-models, we may assume that $B_0 \subseteq M$ and $I \subseteq M_1$. Let J be an M-basis of I. Then by Lemma 1.5, $\operatorname{stp}(a/MJ)$ is hereditarily orthogonal to p, $w_p(q) = |J|$ and by $(*)$, a forks with b over M for all $b \in J$. Thus $J \subseteq Y$, and clearly J is an M-basis of Y. This proves the lemma. ■

COROLLARY **1.8.** *Let q and M be as in the hypothesis of Lemma* 1.7. *Let M_1 be an a-prime model over $M \cup \{a\}$. Let I be an M-basis of $\{b \in M_1$: b realizes $p|M\}$. Then* $\operatorname{stp}(a/MI)$ *is hereditarily orthogonal to p and $w_p(q) = |I|$.*

Proof. Immediate from 1.7 noting that (as Ma dominates M_1 over M), b forks with a over M, for all $b \in M_1 - M$. ■

An advantage to the approach and definitions outlined above is that the notions (of p-simplicity and p-weight) are invariant under replacing p by another regular type non-orthogonal to p:

LEMMA **1.9.** *Let p' be a regular type non-orthogonal to p. Let q be p-simple. Then q is also p'-simple and, moreover, $w_p(q) = w_{p'}(q)$.*

Proof. Note that p' is non-orthogonal to the domain of q (as p is). Let M be an a-model containing A, the domain of q, and also the domain of p'. Let a realize $q|M$. Let M_1 be an a-prime model over $M \cup a$. Let I be a $p|M$ basis of M_1 and J a $p'|M$ basis of M_1. Then, as p and p' are non-orthogonal, $|I| = |J|$. Note that by Corollary 1.8, we have

$(*)$ $\qquad$ $\operatorname{stp}(a/MI)$ is hereditarily orthogonal to p.

CLAIM. $\operatorname{stp}(a/MJ)$ *is hereditarily orthogonal to p'.*

Proof of claim. Let B contain MJ, and b realize $p'|B$. We must show that a is independent with b over B. Suppose by way of contradiction that a forks with b over B. We can clearly find c realizing $p|B$ such that b forks with c over B. As p' is regular a forks with c over B. Note that each $d \in I$ forks with J over M and thus forks with B over M. By regularity of p, c is independent from BI over B and thus realizes $p|BI$. But still a forks with c over BI. This contradicts $(*)$, proving the claim. ■

By the claim, q is p'-simple, and by 1.8, $w_{p'}(q) = |J| = |I| = w_p(q)$. ■

LEMMA **1.10.** *Let* $\mathrm{stp}(a/X)$ *be p-simple and* $X \subseteq Y$. *Then* $\mathrm{stp}(a/Y)$ *is p-simple and* $w_p(\mathrm{stp}(a/Y)) \leq w_p(\mathrm{stp}(a/X))$.

Proof. Let A' be such that $\mathrm{stp}(A'/X) = \mathrm{stp}(A/X)$ and A' is independent from aYA over X. Let $p' \in S(A')$ be the copy of p over A'. As p is non-orthogonal to X, by Lemma 1.4.3.3(iii), p' is non-orthogonal to p, and thus by Lemma 1.9, we may work with p' in place of p. Let M_1 be an a-model containing $X \cup A'$ such that M_1 is independent from aY over XA'. Then

(i) $\mathrm{tp}(a/M_1)$ does not fork over X; and
(ii) $\mathrm{tp}(a/M_1 Y)$ does not fork over Y.

By (ii) let M_2 be an a-model containing $M_1 Y$ such that $\mathrm{tp}(a/M_2)$ does not fork over Y.

Let, by Lemma 1.7 and (i), I be an independent set of realizations of $p'|M_1$ such that $\mathrm{stp}(a/M_1 I)$ is hereditarily orthogonal to p', and $w_{p'}(\mathrm{stp}(a/X)) = |I|$. Then $\mathrm{stp}(a/M_2 I)$ is hereditarily orthogonal to p'; thus by choice of M_2, $\mathrm{stp}(a/Y)$ is p'-simple. Moreover, by Lemma 1.5, $w_{p'}(\mathrm{stp}(a/Y)) \leq |I|$.

It should be remarked that the theory would break down if in the definition of $q = \mathrm{stp}(a/X)$ being p-simple (Definition 1.3) we waived the requirement that p be non-orthogonal to X. Consider for example the theory T of an equivalence relation E with infinitely many infinite classes. T is ω-categorical and ω-stable of Morley rank 2. Let a_0 be any element in a model, and let p be the complete stationary type over a_0 isolated by the formula $E(x, a_0)$. p is orthogonal to $\varnothing$. Let $q(x)$ be the unique 1-type over $\varnothing$. The non-forking extension of q over $\{a_0\}$ is hereditarily orthogonal to p, but q itself is not hereditarily orthogonal to p (as p is an extension of q).

LEMMA **1.11.** *Let* $\mathrm{stp}(a/X)$ *and* $\mathrm{stp}(b/X)$ *be p-simple. Then* $w_p(a, b/X) = w_p(a/X) + w_p(b/a, X)$.

Proof. Note by Lemma 1.4(i) that $\mathrm{stp}(a, b/X)$ is p-simple. Let M be an a-model containing X such that ab is independent with M over X. Replacing p by a suitable conjugate over X, we may, using 1.9, assume that $A \subseteq M$. Let M_1 be a-prime over $M \cup \{a\}$ such that b is independent with M_1 over aM (and so over aX), and let M_2 be any a-model containing $M_1 \cup \{b\}$. Let I be an M-basis for $\{b \in M_1\colon b \text{ realizes } p|M\}$. Then by Corollary 1.8, $w_p(a/X) = |I|$. Let $J \subseteq M_2$ be a maximal set of realizations of $p|M$ in M_2

such that $I \cup J$ is M-independent and every $c \in J$ forks with ab over M. Then by Lemma 1.7, $w_p(a, b/X) = |I \cup J|$, and $\text{stp}(a, b/MIJ)$ is hereditarily orthogonal to p. So to prove the Lemma it suffices to show that $|J| = w_p(b/aX)$ $(= w_p(b/M_1))$.

Regularity of p implies that J is an M_1-independent set of realizations of $p|M_1$, each element of which forks with b over M_1 (see the proof of 1.4.5.10). Thus by Lemma 1.7 it suffices to show

CLAIM. *If c realizes $p|M_1J$ then c is independent from b over M_1.*

But $\text{stp}(ab/MIJ)$ is hereditarily orthogonal to p, whereby $\text{stp}(ab/M_1J)$ and thus $\text{stp}(b/M_1J)$ is orthogonal to p, which yields the claim. ■

We will be frequently using the following consequence of Lemma 1.4(i), Lemma 1.10, and Lemma 1.11: if $\text{stp}(a_i/X_i)$ is p-simple of finite p-weight, for $i = 1, \ldots, n$, and Y contains $\bigcup_i X_i$, then $\text{stp}(a_1, \ldots, a_n/Y)$ is p-simple of finite p-weight.

The general theory of p-simple types is of course relevant to the finite rank theory too. A typical example of a type which is *not* p-simple, but nevertheless has (ordinary) weight 1 and is non-orthogonal to p, is the following. Let $T = Th((\mathbf{Z}/4\mathbf{Z})^{(\omega)})$ (as a commutative group). This theory is totally categorical of U-rank 2. Let G be a big model. Let q be the generic type of G (over $\emptyset$). $2.G$, or what amounts to the same thing, $\{x \in G: 2.x = 0\}$, is strongly minimal. Let p be its generic type. Then p is regular. As T is totally categorical, q is non-orthogonal to p. Let M be a model, and let a realize $q|M$. Then as is easily seen the (a-)prime model M_1 over $M \cup \{a\}$ is exactly $M \oplus \langle a \rangle$. Then $2a$ realizes $p|M$ and is a basis for $p|M$ in M_1. By unidimensionality, q has weight 1. On the other hand $\text{tp}(a/M \cup \{2a\})$ is not algebraic, and hence is also non-orthogonal to p. Thus (by 1.8), q is not p-simple.

We now work with p-simple types which have finite p-weight.

DEFINITION **1.12.** *The strong type q is said to be* p-semi-regular *if q is p-simple and (eventually) domination equivalent to some non-zero power $p^{(n)}$ of p (where remember $p^{(n)}$ is the type over A of an A-independent sequence of realizations of p).*

REMARK **1.13.** *We recall Lemma* 1.4.5.4 *which implies that q being domination equivalent to $p^{(n)}$ (p regular) means exactly*

(∗) *for some a-model M containing A and the domain of q, there are b realizing $q|M$ and an M-independent set I consisting of n realizations of p, such that $M[I] = M[b]$.*

Note (∗) *implies also that b dominates I over M and I dominates b over M.*

If q is domination equivalent to $p^{(n)}$ and q is p-simple (so q is p-semi-regular) then Corollary 1.8 implies that $w_p(q)$ is exactly n. Note also that the notion 'p-semi-regular' is invariant under replacing p by some regular p' non-orthogonal to p.

Lemma 1.14. *Let* $\text{stp}(a/X)$ *be p-semi-regular, and $Y \supseteq X$. Then a is independent from Y over X iff $w_p(a/X) = w_p(a/Y)$.*

Proof. We have already remarked that $w_p(\text{-})$ is invariant under non-forking extensions.

Suppose conversely that a forks with Y over X. Let M be an a-model containing X such that aY is independent with M over X. (We may suppose again p to be over a subset of A.) Suppose $w_p(a/X) = n$, and let I be an independent set of n realizations of $p|M$ as in Remark 1.13. Then $\text{stp}(a/IYM)$ is hereditarily orthogonal to p. But as I dominates a over M and a forks with Y over M, I forks with Y over M, and thus if $J \subseteq I$ is an MY-basis of I, we have $|J| < n$. By Lemma 1.5, $\text{stp}(a/JYM)$ is hereditarily orthogonal to p, so $w_p(a/Y) = w_p(a/YM) < n$. ■

Corollary 1.15. *Let r be any stationary type. The following are equivalent:*

(i) *r is regular and non-orthogonal to p;*

(ii) *r is p-semi-regular with $w_p(r) = 1$.*

Proof. (i) → (ii) is immediate (using Lemma 1.9).
(ii) → (i). Suppose r, a stationary type over X, is p-semi-regular with $w_p(r) = 1$. Let r' be a forking extension of r. By Lemma 1.14 (r' is p-simple and) $w_p(r') = 0$. Thus r' is hereditarily orthogonal to p, in particular orthogonal to p. As r is domination equivalent to p, r' is orthogonal to r.

Thus r is regular. ■

Finally in this section we consider the problem of the existence of enough p-simple and p-semi-regular types. Typically, if $\text{stp}(a/X)$ is non-orthogonal to p, we will be able to find some a' in $\text{acl}(aX)$ with $\text{stp}(a'/X)$ p-simple and sometimes p-semi-regular (and of non-zero p-weight). For various applications and also for just aesthetic reasons we would like to obtain such a' in $\text{dcl}(aX)$. To prove this it is convenient, although not essential, to give a generalization of canonical bases to the case of non-stationary types. Recall that if $q(x)$ is a stationary type, say $q = \text{stp}(a/X)$ (or if you wish $\text{tp}(a/\text{acl}(X))$), then the canonical base of q is just the set of φ-definitions of q (or equivalently of the non-forking extension $\mathbf{q}$ of q to $\mathbf{C}$) for $\varphi(x, y)$ in L. An automorphism of **C** fixes **q** iff it fixes $\mathrm{Cb}(q)$ pointwise, and other properties are given in Lemmas 1.2.26 and 1.2.28.

Suppose now that $q(x) \in S(X)$ is not necessarily stationary. Let $Q = \{\mathbf{q}_i : i \in I\}$ be the set of non-forking extensions of q over **C**. From section 2 of Chapter 1, we know that for each $\varphi(x, y)$ in L, there are only finitely many distinct complete φ-types among the $\mathbf{q}_i$.

Let $\psi_{\varphi,1}(y), \psi_{\varphi,2}(y), \ldots, \psi_{\varphi.k(\varphi)}(y)$ be the φ-definitions of these types, and let $c_\varphi = \{\psi_{\varphi,j}(y) : j = 1, \ldots, k(\varphi)\}$, as an element of $\mathbf{C}^{\mathrm{eq}}$. Then by $\mathrm{Cb}_1(q)$, we mean $\{c_\varphi : \varphi \in L\}$. We sum up the pertinent facts in the next remark. In (i)–(v), $q \in S(X)$ and $Q = \{\mathbf{q}_i : i \in I\}$ is the set of non-forking extensions of q over **C**.

Remark 1.16. (i) $\mathrm{Cb}_1(q) \in \mathrm{dcl}(X)$.

(ii) *An automorphism of* **C** *fixes* Q *setwise iff it fixes* $\mathrm{Cb}_1(q)$ *pointwise.*

(iii) *Let* $Y \subseteq X$. *Then* q *does not fork over* Y *iff* $\mathrm{Cb}_1(\mathrm{q}) \subseteq \mathrm{acl}(Y)$.

(iv) *Let* J_i *be a Morley sequence in* $q_i|\mathrm{acl}(X)$ *for* $i \in I$. *Then* $\mathrm{Cb}_1(q) \subseteq \mathrm{dcl}(\bigcup_i J_i)$.

(v) q *is the unique non-forking extension over* X *of the restriction of* q *to* $\mathrm{Cb}_1(q)$.

(vi) *Let* $q = \mathrm{tp}(a/bc)$ *where* $c \in \mathrm{dcl}(ab)$. *Then* $c \in \mathrm{dcl}(\mathrm{Cb}_1(q), b)$.

Proof. (i)–(iv) are left as exercises.

(v) Let q^0 denote the restriction of q to $\mathrm{Cb}_1(q)$. Suppose by way of contradiction that $r \in S(X)$ is a non-forking extension of q^0, and $r \neq q$. Let $q_1 \in S(\mathbf{C})$ be a non-forking extension of q, and $r_1 \in S(\mathbf{C})$ a non-forking extension of r. So both q_1 and r_1 are non-forking extensions of q^0. Thus there is an automorphism of **C** which fixes $\mathrm{Cb}_1(q)$ pointwise and takes q_1 to r_1. But $q_1 \in Q$ and $r_1 \notin Q$. So we contradict (ii).

(vi) Let C denote $\mathrm{Cb}_1(q)$. Suppose by way of contradiction that there is c_1 such that $\mathrm{tp}(c_1/\mathrm{Cb}) = \mathrm{tp}(c/\mathrm{Cb})$, but $c_1 \neq c$. Let f be a $\varnothing$-definable function such that $c = f(a, b)$. Let $q' \in S(bc_1)$ be the image of q under an automorphism which fixes Cb and takes c to c_1 (note also $C \subseteq \mathrm{dcl}(bc_1)$). Thus $q(x) \vDash f(x, b) = c$, and $q'(x) \vDash f(x, b) = c_1$. Let q'' be a non-forking extension of q' over bc_1c, and q''' the restriction of q'' to bc. Then $q'''(x) \vDash f(x, b) \neq c$. But both q and q''' are non-forking extensions over bc of $q|C$. This contradicts (v). ■

Now we continue.

Lemma 1.17. *Let X be algebraically closed and $\operatorname{tp}(a/X)$ be non-orthogonal to p. Then there is $e \in \operatorname{dcl}(aX)$ such that $\operatorname{stp}(e/X)$ is p-simple with finite and non-zero p-weight. Moreover, e has the following feature: there is $B \supseteq X$ with e independent from B over X, and there are $c_1, \ldots, c_n$ such that for all i, $\operatorname{tp}(c_i/B)$ is an extension of an X-conjugate of p, and $e \in \operatorname{dcl}(Bc_1 \cdots c_n)$.*

Proof. We may change A to an X-conjugate at the outset so that $\operatorname{tp}(a/X \cup A)$ does not fork over X. Let $A \cup X \subseteq Y$ be such that a is independent from Y over X and for some c realizing $p|Y$, a forks with c over Y. In particular a forks with cY over X and thus $\operatorname{tp}(cY/aX)$ forks over X. Let $E = \operatorname{Cb}_1(\operatorname{tp}(cY/aX))$. Then, by 1.16(iii), E is not contained in X, and it is easy to see that c forks with E over Y (for otherwise $\operatorname{tp}(c/EY)$ does not fork over Y; but $\operatorname{tp}(Y/EX)$ does not fork over X, which yields that $\operatorname{tp}(cY/EX)$ does not fork over X, a contradiction). So let $e \in \operatorname{Cb}_1(\operatorname{tp}(cY/aX))$ be such that $e \notin X$ $(= \operatorname{acl}(X))$ and c forks with e over Y. By 1.16(i), $e \in \operatorname{dcl}(aX)$. Also by 1.16(iv), $e \in \operatorname{dcl}(c_1 Y_1, \ldots, c_n Y_n)$ where $\{c_1 Y_1, \ldots, c_n Y_n\}$ is some aX-independent set of realizations of $\operatorname{tp}(cY/aX)$, and without loss of generality $c_1 Y_1 = cY$. As p is non-orthogonal to $X = \operatorname{acl}(X)$ and $\operatorname{tp}(Y_i/X) = \operatorname{tp}(Y/X)$, $\operatorname{stp}(c_i/Y_i)$ is regular, non-orthogonal to p, and thus in particular p-simple of p-weight 1 by 1.15. Thus $\operatorname{stp}(c_1, \ldots, c_n/Y_1 \cdots Y_n)$ is p-simple of finite p-weight, whereby also $\operatorname{stp}(e/Y_1 \cdots Y_n)$ is p-simple of finite p-weight. But $e \in \operatorname{dcl}(aX)$ and as a is independent with Y over X, also a is independent with $Y_1 \cdots Y_n$ over X. Thus e is independent with $Y_1 Y_2 \cdots Y_n$ over X, whereby $\operatorname{stp}(e/X)$ is p-simple with finite p-weight. As e forks with c over Y and c realizes $p|Y$, $w_p(e/Y) > 0$. As $\operatorname{tp}(e/Y)$ does not fork over X, $w_p(e/X) > 0$. The moreover clause in the lemma is clear: take $B = Y_1 Y_2 \cdots Y_n$. ■

Before giving the existence theorem for p-semi-regular types we give a useful criterion for a p-simple type to be p-semi-regular.

Lemma 1.18. *Let X be algebraically closed, and $\operatorname{tp}(a/X)$ be p-simple. Then the following are equivalent:*

(i) $\operatorname{tp}(a/X)$ *is p-semi-regular;*

(ii) *for any $e \in \operatorname{acl}(aX) - X$, $w_p(e/X) \geq 1$;*

(iii) *for any $e \in \operatorname{dcl}(aX) - X$, $w_p(e/X) \geq 1$.*

The same equivalence holds with Y in place of X, and $\operatorname{stp}(a/Y)$ in place of $\operatorname{tp}(a/X)$ as long as Y contains some algebraically closed set Y_0 to which p is non-orthogonal.

Proof. Clearly (ii) $\rightarrow$ (iii).

(i) $\rightarrow$ (ii). Suppose $\operatorname{tp}(a/X)$ to be p-semi-regular and let $e \in \operatorname{acl}(aX) - X$. Then a forks with e over X. By 1.14, $w_p(a/Xe) < w_p(a/X)$. Clearly $w_p(e/aX) = 0$. By Lemma 1.11, $w_p(e/X) > 0$.

(iii) $\rightarrow$ (i). Assume (iii). Let M be an a-model containing $X \cup A$ with a independent from M over X. (Again by replacing A by a suitable X-conjugate.) Let I be a maximal independent set of realizations of $p|M$ in M_1, the a-prime model over $M \cup \{a\}$. So I is finite, a dominates I over M and by Corollary 1.8, $\operatorname{stp}(a/MI)$ is hereditarily orthogonal to p. We will show that I dominates a over M. Suppose not. So let c be such that c is independent with I over M but c forks with a over M. Let $E = \operatorname{Cb}(\operatorname{stp}(aI/cM))$. Then $E \subseteq \operatorname{acl}(cM) - M$, and a forks with E over M. Let $e \in E - M$ be such that e forks with a over M. As $e \in \operatorname{dcl}(a_1I_1, \ldots, a_nI_n)$ for some Morley sequence $\{a_iI_i\}$ in $\operatorname{stp}(aI/cM)$, and $\operatorname{stp}(a_i/I_iM)$ is p-simple of p-weight 0 for all i, it follows, by 1.11, that $\operatorname{stp}(e/M \cup \bigcup_i I_i)$ is p-simple with p-weight 0. But clearly c, and thus e, is independent with $\bigcup_i I_i$ over M, whereby $\operatorname{tp}(e/M)$ is p-simple of p-weight 0. Now let $E' = \operatorname{Cb}_1(\operatorname{tp}(eM/aX))$. A similar argument, using an aX-independent set of realizations of $\operatorname{tp}(eM/aX)$ gives us (using 1.16) $e' \in E' - X$ such that $w_p(e'/X) = 0$ and $e' \in \operatorname{dcl}(aX)$. This contradicts (iii). Thus I does dominate a over M, so $\operatorname{tp}(a/M)$ and also $\operatorname{tp}(a/X)$ are p-semi-regular.

The additional remark in the lemma follows by the same proof as above (if for example some strong type q is non-orthogonal to p, then any Y-conjugate of q will in particular by a Y_0-conjugate, so will also be non-orthogonal to p). ■

Corollary 1.19. *Suppose X is algebraically closed, and that $\operatorname{tp}(a/X)$ is non-orthogonal to p but orthogonal to every forking extension of p. Then there is $e \in \operatorname{dcl}(aX)$ such that $\operatorname{tp}(e/X)$ is p-semi-regular with non-zero p-weight.*

Proof. First note that (by automorphism)

(*) if p' is an X-conjugate of p, then $\operatorname{tp}(a/X)$ is orthogonal to every forking extension of p'.

Let e be as given by Lemma 1.17, and let $B, c_1, \ldots, c_n$ be as given in the moreover clause there. We claim that $\operatorname{tp}(e/X)$ is p-semi-regular. We use Lemma 1.18. So let $d \in \operatorname{acl}(eX) - X$. We need to show that $\operatorname{tp}(d/X)$ has non-zero p-weight, equivalently is non-orthogonal to p. Note that d is independent with B over X, and $d \in \operatorname{acl}(B, c_1, \ldots, c_n)$. By reordering the c_i, we may assume that d is independent with $\{c_1, \ldots, c_k\}$ over B and d forks with c_i over $B \cup \{c_1, \ldots, c_k\}$ for $i = k+1, \ldots, n$ (with $k < n$). Now as

$\mathrm{tp}(a/X)$ is orthogonal to every forking extension of an X-conjugate of p, and $d \in \mathrm{acl}(aX)$, the same is true of $\mathrm{tp}(d/X)$. By the properties of the c_i it follows that $\mathrm{stp}(c_{k+1}/Bc_1, \ldots, c_k)$ must be a *non-forking* extension of an X-conjugate of p, in particular (as p is non-orthogonal to X and X is algebraically closed) $\mathrm{stp}(c_{k+1}/Bc_1, \ldots, c_k)$ is regular and non-orthogonal to p. Thus $\mathrm{tp}(d/X)$ is non-orthogonal to p, as desired. ■

COROLLARY **1.20.** (*of Lemma* 1.18) *Let X be algebraically closed and suppose* $\mathrm{tp}(a/X)$ *is p-simple with* $w_p(a/X) = n$. *Let* $Y = \{b \in \mathrm{dcl}(aX)\colon w_p(b/X) = 0\}$. *Then* $\mathrm{stp}(a/Y)$ *is p-semi-regular with p-weight n.*

Proof. We use the remark at the end of Lemma 1.18. It suffices to remark that

(i) if $c \in \mathrm{dcl}(aY)$ and $\mathrm{tp}(c/X)$ is p-simple with $w_p(c/Y) = 0$ (where X, Y are as above), then $w_p(c/X) = 0$ (so $c \in Y$).

(ii) $w_p(a/Y) = n$.

The point here is that Y may be infinite. To see (i), suppose by way of contradiction that $B \supseteq X \cup A$ is independent with c over X, and c forks with d over B for some d realizing $p|B$. As $\mathrm{tp}(y/X)$ is hereditarily orthogonal to p for all $y \in Y$, d is independent with BY over B, so d forks with c over BY, contradicting the fact that $\mathrm{stp}(c/Y)$ is hereditarily orthogonal to p. (ii) is similar. ■

We write down for future reference the following obvious special case of p-weight (which connects dimension and p-weight):

Fact 1.21. Let $\mathbf{a}$ be a tuple of realizations of p, and let $A \subseteq X$. Then $w_p(\mathbf{a}/X)$ is precisely $\dim_X(\mathbf{a})$, which, remember, is the cardinality of a maximal subtuple $\mathbf{b}$ of $\mathbf{a}$ such that $\mathbf{b}$ is an X-independent set of realizations of $p|X$.

2 Local modularity

The aim in this section is to characterize and give properties of locally modular regular types, as was done in Chapter 2 for minimal types.

We fix a stationary regular type $p \in S(A_0)$, where $A_0 = \mathrm{Cb}(p)$.

So far we have talked about finitary types (i.e. types of finite tuples) being p-simple. For B an infinite tuple (i.e. an enumerated infinite set) we say that $\mathrm{stp}(B/X)$ is p-simple if for any finite tuple b from B $\mathrm{stp}(b/X)$ is p-simple. For such B we say that $w_p(B/X) = n$, if $n = \max\{w_p(b/X)\colon b \text{ a finite tuple from } B\}$. If b is a finite tuple from B such that $w_p(b/X) = w_p(B/X) = n$, then by Lemma 1.11, $w_p(d/bX) = 0$ for all finite tuples d from B. In fact it is easy to see that all the results of section 1 hold for infinitary p-simple types. It will be convenient to call $\{b: \text{stp}(b/X) \text{ is } p\text{-simple and } w_p(b/X) = 0\}$ the *p-closure of X*, denoted $\text{cl}_p(X)$.

We begin with some observations on canonical bases.

LEMMA **2.1.** *Let* $\text{stp}(a/X)$ *be p-simple of finite p-weight. Let* $X \subseteq Y$, *and suppose* $w_p(a/X) = w_p(a/Y) = n$. *Let* $C = \text{Cb}(\text{stp}(a/Y))$. *Then* $C \subseteq \text{cl}_p(X)$.

Proof. Let $w_p(a/Y) = m$. Let $a = a_0, a_1, a_2, \ldots$ be a Morley sequence in $\text{stp}(a/Y)$. So $C \subseteq \text{dcl}(\{a_i: i < \omega\}) \cap \text{acl}(Y)$, thus $\text{stp}(C/X)$ is p-simple, by section 1.

It is also clear that for all i,

(*) $w_p(a_i/a_0, \ldots, a_{i-1}, X) = w_p(a_i/a_0, \ldots, a_{i-1}, X, c) = n$, for all tuples c from C.

Let c be a tuple from C, and suppose $c \in \text{dcl}(a_0, \ldots, a_m)$. So

$$w_p(c/a_0, \ldots, a_m, X) = 0.$$

Now

$$\begin{aligned} w_p(a_0, \ldots, a_m, c/X) &= w_p(a_0, \ldots, a_m/c, X) + w_p(c/X) \\ &= w_p(c/a_0, \ldots, a_m, X) + w_p(a_0, \ldots, a_m/X), \end{aligned}$$

by Lemma 1.11. It follows from (*) and 1.11 that $w_p(c/X) = 0$. Thus $C \subseteq \text{cl}_p(X)$. ∎

LEMMA **2.2.** *Let* $\text{stp}(a/X)$ *be p-simple of finite p-weight. Let* $X \subseteq Y$. *Then there is a finite* $Y_0 \subseteq Y$ *such that* $w_p(a/Y) = w_p(a/XY_0)$.

Proof. As in the proof of Lemma 1.10, we may assume that aY is independent with A_0 $(= \text{Cb}(p))$ over X. Suppose $w_p(a/X) = n$, and $w_p(a/Y) = m \leq n$. Let M be an a-model containing $X \cup A_0$, which is independent from aY over $X \cup A_0$ (and so also over X) and let I be an M-independent set of realizations of $p|M$ such that $|I| = n$ and $w_p(a/MI) = w_p(I/aM) = 0$.

CLAIM. $w_p(I/MY) = m$ $(= w_p(a/Y))$.

Proof of claim. First, as a is independent from MY over Y, $w_p(a/MY) = w_p(a/Y)$. Now, using Lemma 1.11, $m = w_p(a/MY) = w_p(aI/MY) = w_p(I/MY)$. ∎

On the other hand, by Fact 1.21, $w_p(I/MY) = |J|$ where J is an MY-basis of I. Let Y_0 be a finite subset of Y such that for all $b \in I - J$, $\mathrm{tp}(b/MY_0J)$ is a forking extension of p. Thus J is an MY_0-basis of I, whereby again $w_p(I/MY_0) = |J| = m$. The same application of 1.11 shows that $w_p(a/MY_0) = m$. As aY_0 is independent with M over X we conclude that $w_p(a/XY_0) = m$. ■

LEMMA 2.3. *Let* $\mathrm{stp}(a/X)$ *be p-simple of finite p-weight. Let* $X \subseteq Y$ *and let* $C = \mathrm{Cb}(\mathrm{stp}(a/Y))$. *Then* $\mathrm{stp}(C/X)$ *is p-simple of finite p-weight.*

Proof. The fact that $\mathrm{stp}(C/X)$ is p-simple follows as in Lemma 2.1. Let J be a Morley sequence in $\mathrm{stp}(a/Y)$ such that a is independent from J over Y. Then $\mathrm{stp}(a/YJ)$ does not fork over Y, and also does not fork over XJ. So $w_p(a/Y) = w_p(a/XJ) = m$, say. By Lemma 2.2, there is finite $J_0 \subseteq J$ such that $w_p(a/XJ_0) = m$. But J_0 is a finite set of realizations of $\mathrm{stp}(a/X)$; thus $w_p(J_0/X)$ is finite, say $= k$. By Lemma 2.1, $C \subseteq \mathrm{cl}_p(XJ_0)$. Thus clearly for all tuples c from C, $w_p(c/X) \leq k$, so $w_p(C/X)$ is finite. ■

We now consider in more detail the pregeometry of p, and discuss notation. We have already remarked (in 1.1) that $p^{\mathbf{C}}$ (= the set of realizations of p in the big model $\mathbf{C}$) is a pregeometry under the closure operation $\mathrm{cl}(B) = \{a \in p^{\mathbf{C}}: a \text{ forks with } B \text{ over } A_0\}$. Note that $\mathrm{cl}(B)$ will be in general a 'large' subset of $\mathbf{C}$. We say that p is modular or locally modular, if this pregeometry is. Note that $\mathrm{cl}(B)$ is precisely $\mathrm{cl}_p(\mathrm{B} \cup A_0) \cap p^{\mathbf{C}}$. Moreover, for $B \subseteq p^{\mathbf{C}}$, $\dim(B)$ is, if finite, precisely $w_p(\mathrm{stp}(B/A_0))$ (by Fact 1.21). Thus modularity of p is expressed by either

(i) if X, Y are 'p-closed' subsets of $p^{\mathbf{C}}$ (i.e. $\mathrm{cl}_p(XA_0) \cap p^N = X$ and $\mathrm{cl}_p(YA_0) \cap p^{\mathbf{C}} = Y$), and $w_p(X/A_0)$, $w_p(Y/A_0)$ are finite, then $w_p(X/A_0) + w_p(Y/A_0) = w_p(X \cup Y/A_0) + w_p(X \cap Y/A_0)$;

or

(ii) if $\{a, b\} \cup X \subseteq p^{\mathbf{C}}$, and $a \in \mathrm{cl}_p(X \cup \{b\} \cup A_0)$ then there is $c \in \mathrm{cl}_p(X)$ such that $a \in \mathrm{cl}_p(\{c, b\} \cup A_0)$.

We have discussed in section 1 of Chapter 2 the localization of a pregeometry (S, cl) at a subset A of S, where we pointed out that both modularity and local modularity are preserved under such localizations. In the case of the pregeometry $p^{\mathbf{C}}$, its localization at a set $A \subset p^{\mathbf{C}}$ is the pregeometry with underlying set $p^{\mathbf{C}}$ and closure operation: $\mathrm{cl}(B) = \mathrm{cl}_p(B \cup A \cup A_0)$. If we restrict this pregeometry to $p^{\mathbf{C}} - \mathrm{cl}_p(A \cup A_0)$ = the set of realizations of $p|(A \cup A_0)$ (the non-forking extension of p to $A \cup A_0$), we clearly obtain the same geometry. Thus we see that if p is modular, so is $p|(A \cup A_0)$ where A is any (small) set of realizations of p. Now suppose A to be *any* small set containing A_0. We can again consider the 'localization' of the pregeometry p^N to A: for $\{a\} \cup B \subset p^N$ $a \in \text{cl}(B)$ iff $a \in \text{cl}_p(\text{B} \cup A)$ (iff $\text{tp}(a/B \cup A)$ is a forking extension of p). Again this yields the same geometry as that coming from the regular type $p|A$. However, such a geometry is essentially the same as that arising from localizing at a subset of $p^{\mathbb{C}}$, due to stability. Let $X \subseteq p^{\mathbb{C}}$, and let M be a saturated model containing A such that X is independent with M over A. Let $Y = p^M$. Lemma 1.4.5.9 tells us that $\text{tp}(X/A_0 \cup Y) \models \text{tp}(X/M)$. It follows that if $\{x\} \cup X_1 \subseteq X$ then x is independent from MX_1 over A_0X_1Y, whence $w_p(x/A_0X_1Y) = w_p(x/MX_1)$, whence $w_p(x/X_1 \cup A) = w_p(x/A_0X_1Y)$. Thus, for example, $p|A$ is (locally) modular iff $p|(A_0 \cup Y)$ is (locally) modular. In particular if p is modular, so is any non-forking extension of p.

We also obtain associated 'geometries' (although these will not satisfy the exchange axiom) by looking at p-simple types. Let A be any set such that p is non-orthogonal to A. Let $D(p, A) = \{a$: $\text{stp}(a/A)$ is p-simple of finite p-weight$\}$. We can equip $D(p, A)$ with the closure operation: a is in the p-closure of B if $a \in \text{cl}_p(B \cup A)$. We say that $D(p, A)$ is modular just if for finite-dimensional p-closed subsets X, Y of $D(p, A)$, $w_p(X/A) + w_p(Y/A) = w_p(X \cup Y/A) + w_p(X \cap Y/A)$, or equivalently

$$w_p(X/Y \cup A) = w_p(X/(X \cap Y) \cup A).$$

The following gives an analogous characterization of local modularity to that obtained in Chapter 2 for minimal types.

PROPOSITION 2.4. (*Remember* $p \in S(A_0)$ *is a fixed, stationary regular type*). *The following are equivalent*:

(i) *p is locally modular*;

(ii) *for some* $A \supseteq A_0$, $p|A$ is modular;

(iii) *for every A such that p is non-orthogonal to A, $D(p, A)$ is modular*;

(iv) *suppose a_1, a_2 are realizations of p, $A \supseteq A_0$, and* $\text{stp}(a_1, a_2/A)$ *is p-semi-regular of p-weight* 1. *Let* $C = \text{Cb}(\text{stp}(a_1, a_2/A))$. *Then* $w_p(C/A_0) \leq 1$.

Proof. Note: Let $q = \text{stp}(a_1, a_2/A)$ for a_1, a_2, A as in (iv). We can again think of (the parallelism class of) q as a plane curve over p. $w_p(C/A_0)$ measures the 'dimension' of the family of A_0-conjugates of q. So (iv) states that every definable family of plane curves over p is at most 1-dimensional.

(i) → (ii) is immediate.

(ii) → (iii). By (ii) and previous remarks, $p|M$ is modular for all sufficiently saturated models containing A_0. Fix A with p non-orthogonal to A. Suppose by way of contradiction that $D(p, A)$ is not modular. Then clearly there are finite tuples a, b, c from $D(p, A)$, with $c \subseteq a$, $c \subseteq b$, $w_p(a/b, A) < w_p(a/c, A)$, and

$$(*) \qquad \mathrm{cl}_p(c, A) = \mathrm{cl}_p(a, A) \cap \mathrm{cl}_p(b, A).$$

We may assume that abc is independent from A_0 over A. Let M be an a-model containing $A_0 \cup A \cup c$ such that ab is independent with M over Ac, and such that, moreover, $p|M$ is modular. We then clearly have

$(**)$ $w_p(a/bM) = w_p(a/bA) < w_p(a/cA) = w_p(a/M)$, and

$(***)$ Ma is independent with ab over Aa, and Mb is independent with ab over Ab.

Step I. *There is d realizing $p|M$ with $d \in \mathrm{cl}_p(aM) \cap \mathrm{cl}_p(bM)$.*

Proof. Let J_1, J_2 each be an M-independent set of realizations of $p|M$ such that $w_p(a/J_1M) = w_p(J_1/aM) = 0$, and $w_p(b/J_2M) = w_p(J_2/bM) = 0$.

By $(**)$ $w_p(J_1/J_2M) < w_p(J_1/M)$. As $p|M$ is modular, there is d realizing $p|M$ with $d \in \mathrm{cl}_p(J_1M) \cap \mathrm{cl}_p(J_2M)$. But then $d \in \mathrm{cl}_p(aM) \cap \mathrm{cl}_p(bM)$. ■

Step II. *There are $e \in D(p, A)$ and $m \in D(p, A) \cap M$ such that*

$$e \in \mathrm{cl}_p(a, m, A) \cap \mathrm{cl}_p(b, m, A)$$

and $w_p(e/M) = 1$.

Proof. Note that, by Step I, $w_p(ab/Md) < w_p(ab/M)$. Let

$$E = \mathrm{Cb}(\mathrm{stp}(ab/Md)).$$

By Lemma 2.3, $\mathrm{stp}(E/A)$ is p-simple of finite p-weight, and *clearly* $w_p(E/M) = 1$. Let e be a finite tuple in E such that $w_p(e/M) = 1$. As $e \in \mathrm{acl}(Md)$, also $e \in \mathrm{cl}_p(aM) \cap \mathrm{cl}_p(bM)$, by Step I. Now, as d forks with E over M, we can choose m (by 2.3 again) to be a suitable tuple from $\mathrm{Cb}(\mathrm{stp}(e, a, b/M))$. ■

Now by $(***)$ $\{a, m\}$ is independent with ab over Aa. Thus $w_p(a, m/abA) = w_p(a, m/aA)$, and it easily follows that $w_p(B/abA) = w_p(B/aA)$ for any $B \subseteq \mathrm{cl}_p(a, m, A)$. Similarly, using the fact that $\{b, m\}$ is independent with ab over Ab, we see that $w_p(B/abA) = w_p(B/bA)$ for any $B \subseteq \mathrm{cl}_p(b, m, A)$. But by Step II, $\{e, m\} \subseteq \mathrm{cl}_p(a, m, A) \cap \mathrm{cl}_p(b, m, A)$. Thus we conclude that

Step III. $w_p(e, m/abA) = w_p(e, m/aA) = w_p(e, m/bA)$.

By Lemma 2.1 and Step III, $C = \text{Cb}(\text{stp}(e, m/abA)) \subseteq \text{cl}_p(aA) \cap \text{cl}_p(bA)$ and by (*) the latter is contained in $\text{cl}_p(M)$. Now $w_p(e/m, abA) = w_p(e/m, C)$. But the former is 0, and the latter is ≥ 1, by Step II, a contradiction. This completes the proof of (ii) $\rightarrow$ (iii).

(iii) $\rightarrow$ (iv). Let a_1, a_2, and C be as in the hypothesis of (iv). By (iii) $D(p, A_0)$ is modular. By 2.3, $\text{stp}(C/A_0)$ is p-simple of finite p-weight. Thus $\{a_1, a_2\} \cup C \subseteq D(p, A_0)$. If $w_p(a_1, a_2/A_0) = 1$, then by 2.1, $w_p(C/A_0) = 0$. Otherwise $w_p(a_1, a_2/A_0) = 2$. As $D(p, A_0)$ is modular $w_p(a_1, a_2/CA_0) = w_p(a_1, a_2/EA_0)$ ($= 1$) where $E = \text{cl}_p(a_1, a_2, A_0) \cap \text{cl}_p(C, A_0)$. As

$$w_p(E/CA_0) = 0, \; w_p(a_1, a_2/ECA_0) = w_p(a_1, a_2/CA_0) = 1.$$

By semi-regularity of $\text{stp}(a_1, a_2/CA_0)$ and 1.14, (a_1, a_2) is independent from E over CA_0. So $C = \text{Cb}(\text{stp}(a_1, a_2/ECA_0))$. Finally by Lemma 2.1, and the fact that $w_p(a_1, a_2/EA_0) = 1$, we see that $C \subseteq \text{cl}_p(EA_0)$. Thus $w_p(C/A_0) \leq w_p(E/A_0)$. But clearly by 1.11, $w_p(E/A_0) = 1$. This proves (iv).

(iv) $\rightarrow$ (i). (This is a direct translation of the proof of (ii) $\rightarrow$ (iii) in Proposition 2.2.6 so we are brief.) Let e realize p. We want to show that the localization of p at e is modular. So let $\{a_1, a_2\} \cup B \subseteq p^{\mathbf{C}}$ be such that $a_2 \in \text{cl}_p(a_1, B, e, A_0)$. We want to find $d \in \text{cl}_p(B, e, A_0)$ with $a_2 \in \text{cl}_p(a_1, d, e, A_0)$. We may clearly assume that $a_1 \notin \text{cl}_p(B, e, A_0)$ and $a_2 \notin \text{cl}_p(a_1, e, A_0)$. Thus $\{a_1, a_2, e\}$ is A_0-independent, and a_i realizes $p|(B, e, A_0)$ for $i = 1, 2$. Let $B_1 = \text{cl}_p(B, e, A_0) \cap \text{dcl}(a_1, a_2, BeA_0)$. Then by Corollary 1.20, $\text{stp}(a_1, a_2/B_1)$ is p-semi-regular of p-weight 1. Let $C = \text{Cb}(\text{stp}(a_1, a_2/B_1))$. By (iv) $w_p(C/A_0) = 1$, and as

$$w_p(a_1, a_2/CA_0) + w_p(C/A_0) = w_p(C/a_1, a_2, A_0) + w_p(a_1, a_2/A_0),$$

we see that $w_p(C/a_1, a_2, A_0) = 0$. As e realizes $p|(a_1, a_2, A_0)$, e is independent with C over A_0. For similar reasons a_1 is independent with C over A_0. Thus $\text{stp}(a_1/CA_0) = \text{stp}(e/CA_0)$, so we can find d such that $\text{tp}(e, d/CA_0) = \text{tp}(a_1, a_2/CA_0)$. It follows that $d \in \text{cl}_p(B, e, A_0) \cap \text{cl}_p(a_1, a_2, e, A_0)$, and easily $a_2 \in \text{cl}_p(a_1, d, e, A_0)$ as required. ∎

A trivial consequence of the above proposition is that if p is locally modular and p' is a regular type non-orthogonal to p, then p' is locally modular. (For $D(p, A) = D(p', A)$.)

Corollary 2.5. *Let T be 1-based. Then every regular type is locally modular.*

Proof. If T is 1-based then for any algebraically closed sets X and Y, X is independent with Y over $X \cap Y$. This is, moreover, also true after naming any parameters. Thus for any regular type p non-orthogonal to A and

p-closed subsets X, Y of $D(p, A)$, $w_p(X/Y) = w_p(X/X \cap Y)$, that is $D(p, A)$ is modular. ■

In Chapter 2 we deduced the global property of 1-basedness from the assumption that T is a theory of finite U-rank all of whose minimal types are locally modular. Property (iii) in Proposition 2.4 can be considered as a kind of global consequence of local modularity of the regular type p. We mention some further analogues.

LEMMA 2.6. *Suppose p is regular and locally modular.*

(i) *Suppose* $\mathrm{stp}(a/A)$ *is p-simple of finite p-weight, $B \supseteq A$ and* $\mathrm{stp}(a/B)$ *is p-semi-regular. Then* $\mathrm{Cb}(\mathrm{stp}(a/B)) \subseteq \mathrm{cl}_p(aA)$.

(ii) *Suppose* $\mathrm{stp}(a/A)$ *and* $\mathrm{stp}(b/A)$ *are p-simple of finite p-weight and that both* $\mathrm{stp}(a/bA)$ *and* $\mathrm{stp}(b/aA)$ *are p-semi-regular. Then* $\mathrm{Cb}(\mathrm{stp}(a/bA)) \subseteq \mathrm{acl}(aA)$.

Proof. (i) Let $C = \mathrm{Cb}(\mathrm{stp}(a/B))$. Then by 2.3, $\mathrm{stp}(C/A)$ is p-simple of finite p-weight. Let $E = \mathrm{cl}_p(aA) \cap \mathrm{cl}_p(CA)$. By 2.4 and 1.11, $w_p(a/CA) = w_p(a/EA) = w_p(a/ECA)$. By p-semi-regularity of $\mathrm{stp}(a/CA)$, and 1.14, $C = \mathrm{Cb}(\mathrm{stp}(a/ECA))$. By Lemma 2.1, $C \subseteq \mathrm{cl}_p(EA)$. Thus $C \subseteq \mathrm{cl}_p(aA)$, as required.

(ii) Let $C = \mathrm{Cb}(\mathrm{stp}(a/bA))$. By (i) $C \subseteq \mathrm{cl}_p(aA) \cap \mathrm{acl}(bA)$. As $\mathrm{stp}(b/Aa)$ is p-semi-regular, by 1.18 we have $C \subseteq \mathrm{acl}(aA)$. ■

LEMMA 2.7. *Let* $\mathrm{stp}(a/A)$ *be non-orthogonal to the regular locally modular type p. Then there is $c \in \mathrm{cl}_p(aA)$ with* $\mathrm{stp}(c/A)$ *p-simple of p-weight* 1 *(so* $\mathrm{stp}(c/\mathrm{cl}_p(A))$ *is regular).*

Proof. By Lemma 1.17 there is $e \in \mathrm{acl}(aA)$ such that $\mathrm{stp}(e/A)$ is p-simple with $w_p(e/A) = n$ for some $n > 0$. It is easy to find some $B \supseteq A$ such that $w_p(e/B) = n - 1$ and $\mathrm{stp}(e/B)$ is p-semi-regular. Let $C = \mathrm{Cb}(e/B)$. By 2.6(i) $w_p(C/eA) = 0$. Thus 1.11 implies that $w_p(C/A) = 1$. Let $c \in C$ with $w_p(c/A) = 1$. ■

LEMMA 2.8. *Suppose $p \in S(A)$ is regular and modular. Let* $\mathrm{stp}(a/A)$ *be p-simple of finite p-weight and let B be a finite set of realizations of p. Then* $w_p(a/BA) = w_p(a/((\mathrm{cl}_p(BA) \cap \mathrm{cl}_p(aA) \cap p^{\mathrm{C}}) \cup A))$.

Proof. Let $C = \mathrm{cl}_p(BA) \cap \mathrm{cl}_p(aA)$, and let $D = \mathrm{cl}_p(BA) \cap \mathrm{cl}_p(aA) \cap p^{\mathrm{C}}$. By Proposition 2.4, $w_p(a/BA) = w_p(a/CA)$. So it sufficies to show that $C \subseteq \mathrm{cl}_p(DA)$. Let B_1 be such that $\mathrm{stp}(B_1/CA) = \mathrm{stp}(B/CA)$ and B_1 is independent with B over CA. Thus $C = \mathrm{cl}_p(BA) \cap \mathrm{cl}_p(B_1A)$. On the other hand $w_p(B_1/BA) = w_p(B_1/\mathrm{cl}_p(BA) \cap \mathrm{cl}_p(aA) \cap p^{\mathrm{C}}, A)$ $(= w_p(B_1/DA))$ by modularity of p. It easily follows that $w_p(C/A) = w_p(D/A)$. As $D \subseteq C$, it follows that $C \subseteq \mathrm{cl}_p(DA)$ as required. ∎

As in Corollary 2.5.5, we can conclude:

COROLLARY **2.9.** *Let p, q be regular modular types over A. If p and q are non-orthogonal then they are non-weakly orthogonal.*

We end this section with some examples of (superstable) theories all of whose regular types are locally modular but in which 1-basedness fails to differing degrees. Surprisingly, a lot of 'pathology' can be found in theories all of whose regular types are trivial.

Example 2.10. (The free pseudoplane) This has already been met in Example 4.6.1. We recall that T is the theory of a symmetric irreflexive binary relation R, which satisfies

(i) for any x there are infinitely many y such that $R(x, y)$; and

(ii) there are no R-loops, that is for all $n > 1$ there do not exist distinct $x_0, \ldots, x_n$ with $x_0 = x_n$ and $R(x_i, x_{i+1})$ for all $i < n$.

T is ω-stable.

Let M be a model of T. The regular types over M are (up to non-orthogonality)

(i) $p_0(x) = \{\neg R_n(a, x): a \in M, n < \omega\}$;

(ii) for $a \in M$, $p_a(x) = \{R(a, x)\} \cup \{x \neq b: b \in M\}$.

p_0 is definable over $\varnothing$, and p_a is definable over $\{a\}$.

These types are all *trivial* (where remember that a regular type is said to be trivial if its pregeometry is trivial), and thus locally modular. As pointed out in Chapter 4, T is not 1-based (the relation R is a type-definable pseudoplane). T has NDOP but infinite depth.

Example 2.11. Let G be the free group on an infinite set A of generators. Let B be a set on which G acts fixed point freely (i.e. if $g \in G$ fixes a point in B then g is the identity).

Let M be the 2-sorted structure (A, B, R) where $R \subseteq A \times B \times B$ is the graph of the action of A on B. M is ω-stable. There are, up to non-orthogonality, two regular types p, q both defined over $\varnothing$. $p(x) =$ '$x \in A$' has Morley rank 1. $q(x) =$ '$x \in B$' has Morley rank ω. Both types are trivial. $Th(M)$ is not 1-based, as the relation R_1 on B: $R_1(x, y)$ iff there is $z \in A$, $R(z, x, y)$, is a type-definable pseudoplane.

Example 2.12. This a vector space version of 2.11. Let A be an infinite set, and let $K = \mathbb{Q}(X_a: a \in A)$, the field of rational functions over $\mathbb{Q}$ in the infinitely many indeterminates X_a. Let V be a K-vector space. Let $M = (V, +, A, R)$, where $R \subseteq A \times V \times V$ and $R(a, v, w)$ iff $X_a.v = w$. M is ω-stable, with two regular types, p, the generic type of V, and $q =$ '$x \in A$'. p is modular, as for $b, c_1, \ldots, c_n$ realizing p, b forks with $\{c_1, \ldots, c_n\}$ iff b is in the K-linear span of $\{c_1, \ldots, c_n\}$. q is trivial (as A is an indiscernible set). The relation v realizes p, w realizes p, and there is $a \in A$ such that $R(a, v, w)$, is a type-definable pseudoplane, so $Th(M)$ is not 1-based.

Note that if a stable theory is 1-based, and some regular type $p \in S(A)$ is, say, non-orthogonal to $\varnothing$, then there is some regular $q \in S(\mathrm{acl}(\varnothing))$, which is non-orthogonal to p. For, without loss A is a big model M, whereby we can find a realizing p and b such that $\mathrm{tp}(b/M)$ does not fork over $\varnothing$ and such that a forks with b over M. Let $c \in \mathrm{Cb}(\mathrm{stp}(b/aM)) - M$. By 1-basedness $c \in \mathrm{acl}(b)$, whereby $\mathrm{tp}(c/M)$ does not fork over $\varnothing$. On the other hand, as $c \in \mathrm{acl}(aM)$, $\mathrm{tp}(c/M)$ is regular and non-orthogonal to p. So $\mathrm{stp}(c/\varnothing)$ is the required type.

The next example. which is a kind of 'fibred space' version of Example 2.11, shows that this property can fail, even if all regular types are locally modular (in fact trivial).

Example 2.13. The model will be $M = (B, I, \pi, \sigma)$, where B and I are infinite sets, $\pi: B \to I$ is a surjection, and σ is a partial function from $I \times I \times B \to B$ satisfying certain axioms, which shall be given in a moment.

Let us denote $\pi^{-1}((i\})$ by B_i. $\sigma(i, j, b)$ will be defined iff $i \neq j$ and $b \in B_i$, in which case $\sigma(i, j, b) \in B_j$.

Let us denote by σ_{ij} the map $\sigma(i, j, \text{-})$ from B_i to B_j. We think of σ_{ij} as acting on the right. By a word w we mean a sequence $\sigma_{i(1)i(2)}.\sigma_{i(2)i(3)} \cdot \ldots \cdot \sigma_{i(n-1)i(n)}$. Note that such a word defines a map from $B_{i(1)}$ into $B_{i(n)}$. Let X be a subset of I. The word w will be said to have support in X if for every σ_{ij} in w, both i and j are in X. A word w is said to be reduced if w contains no subword $\sigma_{ij}.\sigma_{ji}$. The axioms are:

(i) for all $i \neq j$ in I, σ_{ij} is a bijection from B_i onto B_j.

(ii) $\sigma_{ji} = \sigma_{ij}^{-1}$.

(iii) ('Free action') If $w = \sigma_{ij} \cdot \ldots \cdot \sigma_{st}$, and $u = \sigma_{ik} \cdot \ldots \cdot \sigma_{rt}$ are reduced words such that for some $b \in B_i$, $(b)u = (b)u$, then $w = u$.

It can be checked that the axioms are consistent, and give rise to a complete theory $T = Th(M)$ which is ω-stable, and has quantifier elimination (although some care should be exercised in totalizing the partial maps). We assume M to be saturated. We list some properties of T:

I. I is an indiscernible set, so the type $p_0(x) =$ '$x \in I$' has Morley rank 1. For $i \in I$, $p_i(x) =$ '$x \in B_i$' is regular with U-rank and Morley rank ω. p_j is non-orthogonal to p_i for $i \neq j$ (via σ_{ij}). All these types are trivial. In fact these are the only regular types up to non-orthogonality, so T is 2-dimensional.

II. Let $a \in M^{eq}$. Then $\text{stp}(a/\emptyset)$ is hereditarily orthogonal to p_i (for some, or all i) iff $\text{stp}(a/\emptyset)$ has finite U-rank (or Morley rank) iff a is interalgebraic with a finite subset of I.

III. Let $b, b_1, \ldots, b_n \in B$, and $X \subseteq I$. Then $b \in \text{acl}(X, b_1, \ldots, b_n)$ iff either $b = b_i$ for some i, or for some i and reduced word w with support in X, $b = (b_i)w$.

Clearly each p_i is non-orthogonal to $\emptyset$. We claim, however, that there is no regular type $p \in S(\text{acl}(\emptyset))$ with p non-orthogonal to some (or all) p_i. For suppose by way of contradiction that there is such a type p. Let a realize p. Then a is independent with I over $\emptyset$. Fix $i(0) \in I$. As $p|i(0)$ is non-orthogonal to $p_{i(0)}$ and both are trivial, by Lemma 2.9, they are non-weakly orthogonal, that is there is $b_0 \in B_{i(0)}$ such that b_0 forks with a over $i(0)$. Clearly $\text{stp}(ab_0/i(0))$ is p-simple with p-weight 1. Note that $\text{stp}(i(0)/\emptyset)$ is p-simple of p-weight 0. By Corollary 1.20 and ω-stability we may 'enlarge' $i(0)$ to some $c \in M^{eq}$ with $\text{stp}(c/\emptyset)$ hereditarily orthogonal to p, and such that $\text{stp}(ab_0/c)$ is regular, of p-weight 1. By II above we may assume c to be a finite subset X_0 of I (containing i_0). Moreover, we claim

IV. $b_0 \in \text{acl}(a, X_0)$.

For if not, then $\text{stp}(b_0/aX_0)$ has finite, non-zero Morley rank, so is non-orthogonal to p_0. As p_0 is trivial, one can easily find $j \in I - \text{acl}(aX_0)$ such that $j \in \text{acl}(b_0, a, X_0)$. But this contradicts the fact that $\text{stp}(ab_0/X_0)$ is regular and non-orthogonal to p, whereas p_0 is orthogonal to p.

Now let $\{b_0X_0, b_1X_1, \ldots\}$ be a Morley sequence in $\text{stp}(b_0X_0/a)$. As X_0 is independent with a over $\emptyset$, $\{X_n: n < \omega\}$ is independent over $\emptyset$, and thus the X_n are pairwise disjoint. Moreover, as $\pi(b_0) = i_0 \in X_0$, also $\pi(b_n) \in X_n$. Let (by superstability) $n < \omega$ be such that $\text{stp}(b_nX_n/ab_0X_0 \ldots b_{n-1}X_{n-1})$ is definable over $\{b_0X_0, \ldots, b_{n-1}X_{n-1}\}$. Thus

$$b_n \in \text{acl}(b_0, \ldots, b_{n-1}, X_0, \ldots, X_{n-1}, X_n).$$

By III above, without loss of generality, $b_n \in \text{acl}(b_0, X_0, \ldots, X_{n-1}, X_n)$, and in fact there is a reduced word w_0 over $X_0 \cup \cdots \cup X_n$ such that $b_n = (b_0)w_0$. Let $b^0 = b_0$, $b^1 = b_n$, and $b^2 = b_{2n}$, and $X^0 = X_0 \cup \cdots \cup X_{n-1}$, $X^1 = X_n \cup \cdots \cup X_{2n-1}$, and $X^2 = X_{2n} \cup \cdots \cup X_{3n-1}$. By indiscernibility, $\text{tp}(b^0X^0, b^1X^1) = tp(b^1X^1, b^2X^2) = \text{tp}(b^0X^0, b^2X^2)$. Let w_1 be the image of w_0 under the isomorphism taking $b^0b^1X^0X^1$ to $b^1b^2X^1X^2$, and let w_3 be the image of w_0 under the isomorphism taking $b^0b^1X^0X^1$ to $b^0b^2X^0X^2$. Then w_1 and w_2 are reduced words over X^0X^1 and X^0X^2 respectively, and we have $b^1 = (b^0)w_0$, $b^2 = (b^1)w_1$, and $b^2 = (b^0)w_2$.

Thus $(b^0)w_0.w_1 = b^2 = (b^0).w_2$.

Let w be the reduced form of the word $w_0.w_1$. By axiom (iii) $w = w_2$. But as X^0, X^1, X^2 are disjoint it is clear that w must contain some symbols with indices from X^1. So $w \neq w_2$, a contradiction. Thus there is no regular type over acl($\varnothing$) which is non-orthogonal to p_i.

One can again cook up a vector space version of the previous example in which the fibres B_i are replaced by, say, $\mathbb{Q}$-vector spaces, and the bijections σ_{ij} by isomorphisms.

Our final example is again ω-stable and 2-dimensional, but one of the dimensions is non-locally modular.

Example 2.14. Let $M = (V, +, K, +^*, .^*, f)$, where $(K, +^*, .^*)$ is an algebraically closed field, V is a K-vector space, and $f: K \times V \to V$ gives the action of K on V ($f(k, a) = k.a$). $Th(M)$ is ω-stable. There are two regular types up to non-orthogonality: p_0 the generic type of K, which has Morley rank 1, and p_1 the generic type of V which has Morley rank and U-rank ω. p_1 is modular, for forking on realizations of p_1 is precisely K-linear dependence. Now suppose M is saturated, and let a, b (in the monster model $\mathbf{C}$) be M-independent generic elements of V over M. Let X be the definable set $\{k_1.a + k_2.b: k_1, k_2 \in K^{\mathbf{C}}\}$, and let c be X as an object in $\mathbf{C}^{\mathrm{eq}}$. Then easily tp(c/M) is p_1-semi-regular of p_1-weight 2 (as $c \in$ dcl(a, b), stp($a, b/M$) is p_1-semi-regular, and each of a and b forks with c over M). *However, there is no* $d \in$ acl(c, M) *such that the* p_1*-weight of* tp(d/M) = 1. For suppose such d existed. Then tp(d/M) is non-orthogonal to p_1, and thus (as M is saturated) there is e realizing $p_1|M$ such that e forks with d over M. Thus e forks with $\{a, b\}$ over M. By the above characterization of forking on p_1, there is $e' \in X$ such that e' forks with e over M. Thus e' forks with d over M. But we can find $e'' \in X$ such that $\{e', e''\}$ is a K^N-basis of X. In particular e' and e'' are independent over M and one can check that e' and e'' have the same strong type over Mc. In particular both e' and e'' fork with d over M, which is a contradiction to the p_1-weight of tp(d/M) being 1.

I do not know of an example of a superstable theory T in which all regular types are locally modular, with an element a in a model of T such that there is no $c \in$ acl(a) with stp($c/\varnothing$) regular.

3 Group existence theorems

In this section we state some results of Chapter 5 in the context of regular types.

THEOREM 3.1. *Let p be a locally modular non-trivial regular type, which is non-orthogonal to $\varnothing$. Then there is some set E, with $w_p(E/\varnothing) = 1$, and there is an ∞-definable connected group G defined over E such that the generic type of G is regular and non-orthogonal to p.*

Proof. The proof is almost a direct copy of that of Theorem 5.1.1, which the reader should look back at. Because p-closure is not the same as algebraic closure, there are a few additional things to be said, which we proceed to do. The first main point is:

LEMMA 3.2. *There are a, b, c, and E in $D(p, \varnothing)$, such that $w_p(E/\varnothing) = w_p(a/E) = w_p(b/E) = w_p(c/E) = 1$, $\{a, b, c\}$ is pairwise independent over E, each of* $\mathrm{stp}(a/E)$, $\mathrm{stp}(b/E)$, *and* $\mathrm{stp}(c/E)$ *is regular (so non-orthogonal to p), and a is interdefinable with b over Ec.*

Proof. We will work over $\mathrm{cl}_p(\varnothing)$, in the sense that at any given time we will allow ourselves to name a small set of parameters from $\mathrm{cl}_p(\varnothing)$: technically, if we have for example an element a such that $\mathrm{stp}(a/\varnothing)$ is p-simple, of p-weight n, then by Corollary 1.20 there is a small set $Y \subseteq \mathrm{cl}_p(\varnothing)$ such that $\mathrm{stp}(a/Y)$ is p-semi-regular with p-weight n, and we will name the elements of Y. This is fine, because at the end we can incorporate all these added parameters into E, and the conclusion will remain valid over the empty set.

First, bearing in mind the above remarks, we may assume p is over $\varnothing$. (For by the local modularity of p, the fact that it is non-orthogonal to $\varnothing$, and Lemma 2.7, there is some strong type over $\varnothing$ which is p-simple of p-weight 1. By 1.20 there is a regular type over a small part of $\mathrm{cl}_p(\varnothing)$ which is non-orthogonal to p, so we may replace p by this type and name the relevant small part of $\mathrm{cl}_p(\varnothing)$.) Now p is non-trivial, so for some minimal n there is a set $\{a_1, \ldots, a_n\}$ of realizations of p which is pairwise independent but not independent (over $\varnothing$). Thus $w_p(a_1, a_2/\varnothing) = 2$, $w_p(a_i/a_3, \ldots, a_n) = 1$ for $i = 1$, 2, and $w_p(a_1, a_2/a_3, \ldots, a_n) = 1$. By modularity of $D(p, \varnothing)$ (Proposition 2.4) there is $e \in \mathrm{cl}_p(a_3, \ldots, a_n) \cap \mathrm{cl}_p(a_1, a_2)$ such that $w_p(a_1, a_2/e) = 1$. It easily follows that $w_p(e/\varnothing) = 1$, and that $\{a_1, a_2, e\}$ is pairwise independent, but $w_p(a_1, a_2, e) = 2$. Replacing e by $E = \mathrm{cl}_p(e) \cap \mathrm{dcl}(a_1, a_2)$, everything remains true but now $\mathrm{stp}(a_1, a_2/E)$ is regular. Let (b_1, b_2) realize $\mathrm{stp}(a_1, a_2/E)$ such that $b_1 b_2$ is independent with $a_1 a_2$ over E. Then $\{a_1, a_2, b_1, b_2\}$ is a set of realizations of p, which is 3-wise independent, and $w_p(a_1, a_2, b_1, b_2/\emptyset) = 3$. By modularity of $D(p, \emptyset)$, there is $d \in \mathrm{cl}_p(a_1, b_2) \cap \mathrm{cl}_p(a_2, b_1)$, $d \notin \mathrm{cl}_p(\emptyset)$. Clearly $w_p(d/\emptyset) = 1$, and we may (by adding some p-weight 0 parameters as mentioned above) assume that $\mathrm{stp}(d/\emptyset)$ is regular. In exactly the same way, find $d_1 \in \mathrm{cl}_p(a_1, b_1) \cap \mathrm{cl}_p(E, d)$ with $w_p(d_1/\emptyset) = 1$ and $\mathrm{stp}(d_1/\emptyset)$ regular. We are now exactly in the situation described by the diagram in the proof of Theorem 5.11, which the reader should refer to:

(i) any point has p-weight 1;

(ii) the whole diagram has p-weight 3;

(iii) any line has p-weight 2;

(iv) any set of three non-collinear points, two of which are collinear, is independent (and so contains in its p-closure all the other points).

By augmenting E we may assume also that $\mathrm{stp}(d, d_1/E)$ is regular (note by construction $\mathrm{stp}(a_1, a_2/E)$ and $\mathrm{stp}(b_1, b_2/E)$ are already regular). What is missing again is the interdefinability of (a_1, a_2) and (b_1, b_2) over $E \cup (d, d_1)$. We do not even have interalgebraically. In fact we will directly obtain interdefinability by a use of the generalized canonical bases introduced in section 1.

Let $C = \mathrm{Cb}_1(\mathrm{tp}(a_1/dd_1b_1b_2))$. By Remark 1.16, $C \subseteq \mathrm{dcl}(dd_1b_1b_2)$, but C is also in the definable closure of some set of realizations of $\mathrm{tp}(a_1/dd_1b_1b_2)$. As in the proof of the claim in Lemma 5.1.2, using now p-weight in place of dimension, we see that if $\mathrm{tp}(a_1'/dd_1b_1b_2) = \mathrm{tp}(a_1/dd_1b_1b_2)$ then $a_1' \in \mathrm{cl}_p(a_1)$. It follows that $w_p(C) \leq 1$. On the other hand $\mathrm{tp}(a_1/dd_1b_1b_2)$ does not fork over C, whereby clearly $w_p(C) = 1$. Replace a_1 by some element of C which has p-weight 1. The new a_1 is in $\mathrm{dcl}(dd_1b_1b_2)$ and is in the p-closure of the old a_1, so properties (i)–(iv) above still hold. We can if we wish assume the new a_1 to have regular type over $\emptyset$, by adding some parameters in $\mathrm{cl}_p(\emptyset)$. The same thing can be done with a_2 (by considering $\mathrm{tp}(a_2/da_1b_1E)$). So now we have (i)–(iv) above still true, but with $(a_1, a_2) \in \mathrm{dcl}(edd_1b_1b_2)$.

The final step will destroy the diagram but give us Lemma 3.2. Let $a = (a_1, a_2)$, and $c = (d, d_1)$. Let $B = \mathrm{Cb}_1(\mathrm{tp}((b_1, b_2)/acE))$. Then $B \subseteq \mathrm{dcl}(acE)$, and B is also in the definable closure of a set of realizations of $\mathrm{tp}((b_1, b_2)/acE)$. But if (b_1', b_2') is such a realization, then a p-weight computation shows as before that $b_1' \in \mathrm{cl}_p(b_1)$, and note that $b_2' \in \mathrm{cl}_p(\mathrm{E}, b_1')$. Thus $B \subseteq \mathrm{cl}_p(b_1, E)$. Now $w_p(b_1, b_2) = 2$, and $(b_1, b_2) \in \mathrm{cl}_p(B)$. Thus $w_p(B) \geq 2$ from which it follows that $w_p(B/E) = 1$. Also $w_p(Ba/E) = w_p(Bc/E) = 2$. As $a \in \mathrm{dcl}(b_1b_2cE)$, by Remark 1.16(vi), $a \in \mathrm{dcl}(BcE)$. Thus we can choose b from B such that $w_p(b/E) = 1$ and $a \in \mathrm{dcl}(bcE)$ (and remember $b \in \mathrm{dcl}(acE)$). By adding some parameters in $\mathrm{cl}_p(E)$ to E, we may assume that each of $\mathrm{stp}(a/E)$, $\mathrm{stp}(b/E)$, and $\mathrm{stp}(c/E)$ is regular, and thus $\{a, b, c\}$ is pairwise E-independent. This completes the proof of Lemma 3.2. ■

We now complete the proof of Theorem 3.1 by finding the group G. This is again just as in the proof of Theorem 5.1.1. We now work over the set E, which we may assume to be also algebraically closed (i.e. assume $E = \mathrm{acl}(\varnothing) = \mathrm{dcl}(\varnothing)$). First let $\sigma \in \mathrm{Cb}(\mathrm{stp}(a, b/c))$ be such that a and b are interdefinable over σ. Then $\sigma \in \mathrm{dcl}(c)$, and clearly $\mathrm{stp}(\sigma)$ is regular and non-orthogonal to p (using Lemma 1.18). We will write $\sigma'.a' = b'$ and $\sigma'^{-1}.b' = a'$ iff $\mathrm{stp}(\sigma', a', b') = \mathrm{stp}(\sigma, a, b)$. σ is the 'germ' of a definable invertible function from $\mathrm{stp}(a)$ to $\mathrm{stp}(b)$, in the sense that

(i) for a' realizing $\mathrm{stp}(a)|\sigma$ there is (unique) b' realizing $\mathrm{stp}(b)$ such that $\sigma.a' = b'$, and, moreover, b' is independent with σ over $\varnothing$; and
(ii) if $\mathrm{stp}(\sigma) = \mathrm{stp}(\sigma')$, a' realizes $\mathrm{stp}(a)|\{\sigma, \sigma'\}$ and $\sigma.a = \sigma'.a$ then $\sigma = \sigma'$.

Let $q = \mathrm{stp}(a/\varnothing)$, and $s = \mathrm{stp}(\sigma/\varnothing)$. Let τ realize s such that $\{\sigma, \tau, a\}$ is independent over $\varnothing$. Let $a' = \tau^{-1}.\sigma.a$. Then a' realizes $q|\{\sigma, \tau\}$. Let $\upsilon \in \mathrm{Cb}(\mathrm{stp}(a, a'/\sigma, \tau))$ be such that a and a' are interdefinable over υ. By Proposition 2.4 and local modularity of q, $w_p(\upsilon) = 1$, and moreover, (again using 1.18), as $\upsilon \in \mathrm{dcl}(\sigma, \tau)$ and σ, τ are independent realizations of the regular type s which is non-orthogonal to p, υ is regular. We write $\upsilon = \tau^{-1}.\sigma$, $a' = \upsilon.a$, and $a = \upsilon^{-1}.a'$. υ 'is' the germ of an invertible function from q to q. Let $s' = \mathrm{stp}(\upsilon)$. Just as in the proofs of 5.1.6, 5.1.7 and 5.1.8, one shows that the realizations of s' are closed under 'generic' composition, and that the semi-group G of germs of invertible functions from q to q generated by the realizations of s' is a connected ∞-definable group (over $\varnothing$) with unique generic type s'. The generic action of G on q^{C} is $\varnothing$-definable and 'generically' regular. The proof of Theorem 3.1 is complete. ■

For future use it will be useful to write down the following version of Theorem 3.1.

THEOREM **3.3.** *Let p be a stationary regular type over $\varnothing$. Suppose p is locally modular and non-trivial.* Then, *there is a set E with $w_p(E/\varnothing) = 1$, there is a stationary regular type $q \in S(E)$ such that q is not weakly orthogonal to $p|E$, and there is a connected ∞-definable regular group G, defined over E, of germs of definable finctions on q.*

Proof. This is given by the proof of Theorem 3.1. We began there with regular p over $\varnothing$, found a_1 realizing p, E independent with a_1, and a such that $\mathrm{stp}(a/E) = q$ is regular, $a \in \mathrm{cl}_p(a_1, E)$, and the E-definable group G is a group of germs of invertible functions from q to q, everything being E-definable. ■

We end this section by stating the regular type version of the group configuration theorem of Chapter 5. The reason we cannot formally deduce Theorem 3.1 from the group configuration theorem is that in 3.1 we require the group to be defined over a set of p-weight at most 1. In any case the generalization of Proposition 5.4.5 to the regular type context is straightforward if one has understood the p-weight and generalized canonical base arguments in the proof above.

DEFINITION 3.4. *Let G be an ∞-definable group acting definably and transitively on an ∞-definable set X. Let p be a regular type. We say that G is p-simple, p-semi-regular, regular according to whether some (or equivalently all) generic types of G are. Similarly for X.*

The proof of the following is left as an exercise for the reader. We refer the reader to the diagram following Definition 5.4.3.

PROPOSITION 3.5. *Let p be a regular type over a saturated model M. Let a, b, c, x, y, z be elements whose types over M are p-semi-regular, such that*

(i) *each pair from a, b, c, x, y, z is M-independent;*

(ii) *each of a, b, c is in the p-closure of M together with the other two;*

(iii) $y \in \mathrm{cl}_p(a, x, M) \cap \mathrm{cl}_p(b, z, M)$, $x \in \mathrm{cl}_p(a, y, M) \cap \mathrm{cl}_p(c, z, M)$, *and* $z \in \mathrm{cl}_p(b, y, M) \cap \mathrm{cl}_p(c, x, M)$;

(iv) $a \in \mathrm{cl}_p(\mathrm{Cb}(\mathrm{stp}(x, y/M, a)), M)$, $b \in \mathrm{cl}_p(\mathrm{Cb}(\mathrm{stp}(y, z/M, b)), M)$,

$$c \in \mathrm{cl}_p(\mathrm{Cb}(\mathrm{stp}(x, z/M, c)), M).$$

Then there is an ∞-definable connected group G acting definably and transitively on an ∞-definable set X, where everything is defined over M, and both G and X are p-semi-regular. Moreover, there are generic elements a', b', c' of G over M and generic elements x', y', z' of X over M, such that $a'.x' = y'$, $b'.y' = z'$, $c'.x' = z'$, $c' = b'.a'$, and $\mathrm{cl}_p(a, M) = \mathrm{cl}_p(a', M)$, $\mathrm{cl}_p(b, M) = \mathrm{cl}_p(b', M)$, $\mathrm{cl}_p(c, M) = \mathrm{cl}_p(c', M)$, $\mathrm{cl}_p(x, M) = \mathrm{cl}_p(x', M)$, $\mathrm{cl}_p(y, M) = \mathrm{cl}_p(y', M)$, and $\mathrm{cl}_p(z, M) = \mathrm{cl}_p(z', M)$.

4 Internality

It is convenient at this juncture to introduce a notion which has been already used implicitly, namely internality. We also state and prove in full generality a group existence theorem in the context of internality: a special case of such a result was proved in Proposition 5.2.3.

Let **P** be a family of partial types (over various small subsets of the monster model **C**). For E a subset of **C** we will say that **P** is E-invariant (or based on E) if for every $\Psi \in \mathbf{P}$, and every E-automorphism f, the partial type $f(\Psi)$ is also in **P**. We say that a is in **P** over B or a realizes $\mathbf{P}|B$ if $\mathrm{tp}(a/B)$ is an extension of some $\Psi \in \mathbf{P}$.

DEFINITION **4.1.** (i) *Let $q \in S(A)$ be a stationary complete type. We say that q is foreign to* **P**, *if for every $B \supseteq A$, a realizing $q|B$, and c realizing $\mathbf{P}|B$, a is independent with c over B (i.e. q is orthogonal to every complete type extending some member of Ψ).*

(ii) *Let $q \in S(A)$ be stationary. We say that q is* **P***-internal, if there are $B \supseteq A$, a realizing $q|B$, and $c_1, \ldots, c_k$ realizing $\mathbf{P}|B$* such that $a \in \mathrm{dcl}(B, c_1, \ldots, c_k)$.

It is easy to see that if A, is, say, algebraically closed, **P** is A-invariant, and both $\mathrm{tp}(a/A)$ and $\mathrm{tp}(b/A)$ are **P**-internal, then $\mathrm{tp}(a, b/A)$ is **P**-internal. Also any algebraic type is **P**-internal (any **P**).

LEMMA **4.2.** (i) *Let $q \in S(A)$ be* **P***-internal and suppose* **P** *to be A-invariant. Then there is some $C \supseteq A$ such that for* any *a realizing q, there are $c_1, \ldots, c_n$ realizing $\mathbf{P}|C$ such that $a \in \mathrm{dcl}(C, c_1, \ldots, c_n)$.*

(ii) *Suppose* **P** *is a family of partial types over A, and q is a* **P***-internal stationary type over A. Then there exist a partial A-definable function $f(y_1, \ldots, y_m, z_1, \ldots, z_n)$, a sequence $a_1, \ldots, a_m$ of realizations of q, and a sequence $\Psi_1, \ldots, \Psi_n$ of partial types in* **P**, *such that for* any *a realizing q, there are c_1 realizing Ψ_1, c_2 realizing $\Psi_2, \ldots, c_n$ realizing Ψ_n, such that $a = f(a_1, \ldots, a_m, c_1, \ldots, c_n)$.*

Proof. (i) Let $B \supseteq A$, b realize $q|B$, and $d_1, \ldots, d_n$ realize $\mathbf{P}|B$ such that $b \in \mathrm{dcl}(B, d_1, \ldots, d_n)$. Let now M be some sufficiently saturated model containing A. Let a be any realization of q. Let $B' \supseteq A$ be such that $\mathrm{tp}(B'/A) = \mathrm{tp}(B/A)$ and B' is independent from aM over A. Then $\mathrm{tp}(aB'/A) = \mathrm{tp}(bB/A)$. Also $\mathrm{tp}(a/MB')$ does not fork over M, and so is definable over some small A' in M which we may suppose contains A. Saturation of M implies that we can find C in M such that $\mathrm{tp}(C/A') = \mathrm{tp}(B'/A')$. Thus $\mathrm{tp}(aC/A) = \mathrm{tp}(aB'/A) = \mathrm{tp}(bB/A)$. Let $c_1, \ldots, c_n$ be the images of $d_1, \ldots, d_n$ under an A-automorphism which takes bB to aC. Then $a \in \mathrm{dcl}(C, c_1, \ldots, c_n)$. As **P** is A-invariant each c_i realizes $\mathbf{P}|C$. In particular each c_i realizes $\mathbf{P}|M$ and $a \in \mathrm{dcl}(M, c_1, \ldots, c_n)$.

(ii) Let $B \supseteq A$, a realize $q|B$, and let $\mathbf{c}$ be a sequence of realizations of partial types in **P** such that $a \in \mathrm{dcl}(B, \mathbf{c})$. Let M be a big saturated model containing B such that $(a, \mathbf{c})$ is independent from M over B. Let $(a_i, \mathbf{c}_i)_{i<\omega}$

be a Morley sequence in $\text{stp}(a, \mathbf{c}/B)$ which is contained in M. Thus (from 1.2.28) $\text{tp}(a, \mathbf{c}/M)$ is definable over $A \cup \{a_i, \mathbf{c}_i : i < \omega\}$. In particular $a \in \text{dcl}(a_1, \ldots, a_m, \mathbf{c}_1, \ldots, \mathbf{c}_m, \mathbf{c}, A)$ for some m. Note that the $\mathbf{c}_i$ are sequences of elements realizing partial types from $\mathbf{P}$, and that a is independent from $(a_1, \ldots, a_m)$ over A. Let $\mathbf{d} = (\mathbf{c}_1, \ldots, \mathbf{c}_m, \mathbf{c})$, and $\mathbf{a} = (a_1, \ldots, a_m)$. Write $a = g(\mathbf{a}, \mathbf{d})$ for some A-definable function g.

So far we have shown

($*$) for (any) a' realizing $q|A\mathbf{a}$ there is a sequence $\mathbf{d}'$ of realizations of partial types from $\mathbf{P}$, such that $a' = g(\mathbf{a}, \mathbf{d}')$.

Now let a'' be an arbitrary realization of q. Let $\mathbf{a}'$ realize the non-forking extension of $\text{stp}(\mathbf{a}/A)$ over $A \cup \{a''\}$. So $\mathbf{a}' = (a_1', \ldots, a_m')$. By ($*$), for each i, there is $\mathbf{d}_i'$ (a sequence of realizations of $\mathbf{P}$) such that $a_i' = g(\mathbf{a}, \mathbf{d}_i')$. On the other hand, by automorphism, there is $\mathbf{d}''$ such that $\text{tp}(a'', \mathbf{a}', \mathbf{d}''/A) = \text{tp}(a, \mathbf{a}, \mathbf{d}/A)$, so $a'' = g(\mathbf{a}', \mathbf{d}'')$. Putting it all together we have some partial A-definable function f, such that $a'' = f(\mathbf{a}, \mathbf{d}_1', \ldots, \mathbf{d}_m', \mathbf{d}'')$. Note that the sequence of partial types in $\mathbf{P}$ realized by $(\mathbf{d}_1', \ldots, \mathbf{d}_m', \mathbf{d}'')$ does not depend on a''. We have proved (ii). ■

REMARK 4.3. *Let* $\mathbf{P}$ *be* A*-invariant. Assume* A *algebraically closed. Let* $\mathbf{Q} = \{q \in S(A)$: q *is* $\mathbf{P}$*-internal*$\}$. *Then* $\mathbf{Q}$ *is internally closed, in the sense that if* $r \in S(A)$ *is* $\mathbf{Q}$*-internal then* $r \in \mathbf{Q}$.

Proof. Let $r \in S(A)$ be $\mathbf{Q}$-internal. Let $B \supseteq A$, let a realize $r|B$, and let $c_1, \ldots, c_n$ be realizations of $\mathbf{Q}|B$ such that $a \in \text{dcl}(B, c_1, \ldots, c_n)$. Let M be a sufficiently saturated model containing A and independent with $B \cup \{a, c_1, \ldots, c_n\}$ over A. So each c_i realizes the non-forking extension of some $q_i \in \mathbf{Q}$ over M. But each q_i is $\mathbf{P}$-internal. So by saturation of M and A-invariance of $\mathbf{P}$, for each i there is a set C_i of realizations of $\mathbf{P}|M$ such that $c_i \in \text{dcl}(M, C_i)$. Thus $a \in \text{dcl}(MBC_1, \ldots, C_n)$. As a realizes $r|BM$, it follows that r is $\mathbf{P}$-internal, so $r \in \mathbf{Q}$. ■

Example 4.4. (i) Let $\mathbf{P}$ consist of a single partial type Ψ over, say, $\varnothing$. Then $\mathbf{P}$ is $\varnothing$-invariant. We say foreign (internal) to Ψ, instead of foreign (internal) to $\mathbf{P}$. The set $\{a \in \mathbf{C}: \text{stp}(a/\varnothing)$ is Ψ-internal$\}$ is a natural generalization of $\Psi^{\text{eq}} = \{a \in \mathbf{C}: a$ is definable over some elements satisfying $\Psi\}$. If $q \in S(\varnothing)$ is Ψ-internal, then Lemma 4.2(ii) shows that there is a partial (not necessarily $\varnothing$-)definable function from $(\Psi^{\mathbf{C}})^n$ onto $q^{\mathbf{C}}$, for some n.

If $\mathbf{P}$ consists of a single complete stationary type $p \in S(A)$ then to say that $q \in S(B)$ is foreign to p (i.e. to $\mathbf{P} = \{p\}$) means that p is hereditarily orthogonal to q.

(ii) Let $p \in S(A)$ be some stationary regular type which is, say, non-orthogonal to $\varnothing$. Let $\mathbf{P}$ = the set of conjugates of p over $\mathrm{acl}(\varnothing)$. So $\mathbf{P}$ is $\mathrm{acl}(\varnothing)$-invariant. If some stationary type q is $\mathbf{P}$-internal then clearly q is p-simple. The existence theorems (1.17 and 1.19) for p-simple and p-semi-regular types give directly $\mathbf{P}$-internal types.

(iii) Let again p be a stationary regular type, non-orthogonal to $\varnothing$. Let $\mathbf{P} = \{\mathrm{tp}(b/B)$: B is algebraically closed and $b \in \mathrm{cl}_p(B)\}$. Then $\mathbf{P}$ is $\mathrm{acl}(\varnothing)$-invariant and is, moreover, 'internally closed': if C is algebraically closed and $\mathrm{tp}(c/C)$ is $\mathbf{P}$-internal then $\mathrm{tp}(c/C) \in \mathbf{P}$. (Left to the reader).

LEMMA **4.5.** *Suppose* $\mathbf{P}$ *is A-invariant. Let* $B \supseteq A$, *let a be independent with B over A, and let E be a set of realizations of* $\mathbf{P}|B$. *Let* $C = \{c \in \mathrm{dcl}(aA)$: $\mathrm{stp}(c/A)$ *is* $\mathbf{P}$*-internal*$\}$. *Then BE is independent with aA over C.*

Proof. (Like the proof of Lemma 1.17) Let $D = \mathrm{Cb}_1(\mathrm{tp}(BE/aA))$. Then $D \subseteq \mathrm{dcl}(aA)$ and BE is independent with aA over D, so it suffices to show that $D \subseteq C$.

By 1.16, any $d \in D$ is in the definable closure of some aA-independent set $\{B_1E_1, \ldots, B_nE_n\}$ of realizations of $\mathrm{tp}(BE/aA)$. But then $B' = B_1 \cup \cdots \cup B_n$ is independent with a over A, and thus independent with d over A. As $\mathbf{P}$ is A-invariant, each E_i is a set of realizations of $\mathbf{P}|B'$. Thus $\mathrm{stp}(d/A)$ is $\mathbf{P}$-internal, whereby $d \in C$, as required. ■

COROLLARY **4.6.** *Suppose* $\mathbf{P}$ *is A-invariant and* $\mathrm{stp}(a/A)$ *is not foreign to* $\mathbf{P}$. *Then there is* $c \in \mathrm{dcl}(aA) - \mathrm{acl}(A)$ *such that* $\mathrm{stp}(c/A)$ *is* $\mathbf{P}$*-internal.*

Proof. Let $B \supseteq A$ with a independent from B over A, and suppose e realizes $\mathbf{P}|B$ and a forks with e over B. Then a forks with Be over A, so Be forks with aA over A. By Lemma 4.5, $C = \{c \in \mathrm{dcl}(aA)$: $\mathrm{stp}(c/A)$ is $\mathbf{P}$-internal$\}$ is not contained in $\mathrm{acl}(A)$. ■

Some further notions and results connected naturally with internality will be discussed later. For now we just wish to point out a nice stable group version of Corollary 4.6.

LEMMA **4.7.** *Let G be a connected* ∞*-definable group, defined over* $A = \mathrm{acl}(A)$. *Let* $q \in S(A)$ *be the generic type of G.*

(i) *Suppose q is not foreign to* $\mathbf{P}$, *where* $\mathbf{P}$ *is A-invariant. Then there is a proper A-definable (see below) normal subgroup N of G, such that the generic type* $q' \in S(A)$ *of* G/N *is* $\mathbf{P}$*-internal.*

(ii) *Suppose* p *is a regular type non-orthogonal to* q. *Let* **P** *be the set of conjugates of* p *over* A (*so* **P**-*internality automatically implies* p-*simplicity*). *Then there is an* A-*definable normal subgroup* N *of* G *such that the generic type of* G/N *is not only* **P**-*internal, but also* p-*simple* of non-zero p-weight.

Proof. For now, we call a subgroup N of G *A-definable* if there is single formula φ over A such that $N = \{a \in G: \varphi(a)\}$. (Earlier this was called relatively A-definable.) If N is such, then the coset space G/N is clearly ∞-definable.

(i) Let a realize q. First by Corollary 4.6 there is $c \in \mathrm{dcl}(aA) - \mathrm{acl}(A)$ such that $\mathrm{stp}(c/A)$ is **P**-internal. So $f(a) = c$ for some A-definable function $f(\text{-})$ which is defined on q.

Let $H = \{b \in G$: for a' realizing $q|(bA), f(a') = f(b.a')\}$. Note that if $b \in G$ and a' realizes $q|(bA)$ then $b.a'$ realizes q, so $f(b.a')$ is well defined and realizes $\mathrm{tp}(c/A)$. Moreover, it is easy to see that H is an A-definable subgroup of G (using definability of the type q).

CLAIM I. *Let* $b_1, b_2 \in G$ *and let* a' *realize* $q|(b_1, b_2, A)$. *Then* $f(b_1.a') = f(b_2.a')$ *iff* $b_1.b_2^{-1} \in H$.

Proof. $b_2^{-1}.a'$ realizes $q|(b_1, b_2, A)$ (as q is the unique generic type of G), and thus $\mathrm{tp}(b_1, b_2, b_2^{-1}.a'/A) = \mathrm{tp}(b_1, b_2, a'/A)$. Thus $f(b_1.a') = f(b_2.a')$ iff $f(b_1.(b_2^{-1}.a')) = f(b_2.(b_2^{-1}.a'))$ iff $f((b_1.b_2^{-1}).a') = f(a')$ iff $b_1.b_2^{-1} \in H$ (as a' realizes $q|(b_1.b_2^{-1}, A)$). ■

CLAIM II. *H is a proper subgroup of G.*

Proof. If by way of contradiction $H = G$, then if a_1, a_2 are independent realizations of q, we will have that $f(a_1) = f(a_2)$ (for if $b = a_1.a_2^{-1}$, then a_2 is independent with b over A, and so $f(a_2) = f((a_1.a_2^{-1}).a_2) = f(a_1)$). But then f is constant on q (if a_1, a_2 realize q choose a_3 realizing $q|(a_1, a_2, A)$ and by what we have just said $f(a_1) = f(a_3) = f(a_2)$). Thus $c \in \mathrm{dcl}(A)$, contradiction. ■

CLAIM III. *The right coset space* G/H *is* **P**-*internal, that is for any* $a' \in G$, $\mathrm{tp}((a'/H)/A)$ *is* **P**-*internal.*

Proof. Fix $a' \in G$. Let $X = \{a_i: i < |T|^+\}$ be a Morley sequence in $q|(a', A)$ and let $Y = \{f(a'.a_i): i < |T|^+\}$. We claim that $a'/H \in \mathrm{dcl}(X, Y, A)$. For let $a'' \in G$ be such that $\mathrm{tp}(a''/X, Y, A) = \mathrm{tp}(a'/X, Y, A)$. By stability there is $a_i \in X$ such that a_i is independent with $\{a', a''\}$ over A. But $f(a''.a_i) = f(a'.a_i)$ ($\in Y$), so by Claim I, $a''/H = a'/H$. As a'/H is independent with X over A, we have shown that $\mathrm{tp}((a'/H)/A)$ is $\mathrm{tp}(c/A)$-internal. By Remark 4.3, $\mathrm{tp}((a'/H)/A)$ is **P**-internal. This proves Claim III. ■

Now by stability $N = \cap\{H^g : g \in G\}$ is equal to a finite subintersection, say $N = H^{g(1)} \cap \cdots \cap H^{g(n)}$. Thus N is a normal A-definable subgroup of G. Let a realize $q|(A, g(1), \ldots, g(n))$; then a/N realizes the generic type of G/N over $(A, g(1), \ldots, g(n))$. Moreover, a/N is interdefinable with $(a^{g(1)^{-1}}/H, \ldots, a^{g(n)^{-1}}/H)$ over $A \cup \{g(1), \ldots, g(n)\}$. By Remark 4.3, $\mathrm{tp}((a/N)/A, g(1), \ldots, g(n))$ and thus $\mathrm{tp}((a/N)/A)$ is **P**-internal. This completes the proof of (i).

(ii) This proceeds just as in part (i), except that to begin with, for a realizing q, $c \in \mathrm{dcl}(aA)$ is chosen (by Lemma 1.17) to be not only **P**-internal but also such that $w_p(c/A) > 0$. Let H and N be as found in the proof of (i). It remains to be shown that the generic type of G/N has non-zero p-weight. Now, as before, $\mathrm{tp}((a/N)/A)$ is the generic type over A of G/N (where a realizes q). Suppose by way of contradiction that $w_p((a/N)/A) = 0$. But $a/H \in \mathrm{dcl}((a/N), A)$; thus $w_p((a/H)/A) = 0$.

An argument as in the proof of Claim II above shows that $c \in \mathrm{dcl}((a/H), A)$. But then $w_p(c/A) = 0$, a contradiction. ■

Connectedness of G is not essential in Lemma 4.7. We leave it to the reader to check that the lemma holds for G not connected and q some generic type of G, the conclusion being that there is A-definable normal N in G such that N has infinite index in G and some (all) generics of G/N are **P**-internal (or p-simple of non-zero p-weight).

Finally we show another important existence theorem for definable groups. The group here arises as an *automorphism* (or 'Galois') group. We work over $\emptyset$ here, but the same holds over any set A. The proof is really identical to that of Proposition 5.2.3, but there is some greater generality here, and we also give a more (and possibly too) detailed treatement.

THEOREM **4.8.** *Suppose* **P** *is a family of partial types over* $\emptyset$. *Let q be a strong type over* $\emptyset$ *which is* **P**-*internal. Let Q be the set of realizations of q in* **C**, *and P the union of the sets of all realizations of the partial types in* **P**. *Let G be the group of permutations of Q arising from elementary maps (or even automorphisms of* **C**) *which fix* $\mathrm{acl}(\emptyset) \cup P$ *pointwise. Then both G and its action on Q are ∞-definable over* $\mathrm{acl}(\emptyset)$; *that is, there is an ∞-definable group G_1 over* $\mathrm{acl}(\emptyset)$ *and an* $\mathrm{acl}(\emptyset)$-*definable action of G_1 on Q, such that this action is isomorphic to the action of G on Q.*

Proof. We should mention to begin that G need not be infinite. It will be infinite just if q is not realized in $\mathrm{acl}(P)$. Also the action of G on Q need not be transitive (i.e. Q need not be a homogeneous space for G), but it *will* be if some (any) realization of q is independent from P over $\emptyset$. On the other hand we will see that the action of G on Q always 'comes from' a certain canonical transitive definable action. We use Lemma 4.2(ii). Let f, $\mathbf{a}$, $\Psi_1, \ldots, \Psi_n$ be as there. That is, f is a partial function defined over $\emptyset$, $\mathbf{a}$ is some tuple of realizations of q, $\Psi_i \in \mathbf{P}$ for each $i = 1, \ldots, n$, and for any realization a of q there are c_i realizing Ψ_i for $i = 1, \ldots, m$, such that $a = f(\mathbf{a}, c_1, \ldots, c_n)$.

We will work over $\mathrm{acl}(\emptyset)$.

Let G be the the group of elementary permutations of Q over P, that is permutations σ of Q such that there is some *elementary* map σ' defined on $Q \cup P$ such that σ' fixes P pointwise, and $\sigma' | Q = \sigma$. We may clearly assume that $\mathrm{dcl}(P) = P$ (in $\mathbf{C}^{\mathrm{eq}}$), as G is unchanged. Let $r = \mathrm{tp}(\mathbf{a}/\emptyset)$ ($=\mathrm{stp}(\mathbf{a}/\emptyset)$, as we work over $\mathrm{acl}(\emptyset)$).

Let $r_{\mathbf{a}} = \mathrm{tp}(\mathbf{a}/P)$. As $\mathbf{a}$ is a sequence of realizations of q, G also acts on the set of realizations of $r_{\mathbf{a}}$. The first observation is

Claim I. *The map from G to $(r_{\mathbf{a}})^{\mathbf{C}}$ which takes σ to $\sigma(\mathbf{a})$ is a bijection.*

Proof. Clearly $\sigma_1(\mathbf{a}) \neq \sigma_2(\mathbf{a})$ iff $\sigma_1 \neq \sigma_2$. To show that the map is onto, suppose $\mathbf{b}$ realizes $r_{\mathbf{a}}$. So $\mathrm{tp}(\mathbf{a}/P) = \mathrm{tp}(\mathbf{b}/P)$. Any element of Q is of the form $f(\mathbf{a}, \mathbf{c})$ for some tuple $\mathbf{c}$ from P. As $\mathrm{tp}(\mathbf{a}/\emptyset) = \mathrm{tp}(\mathbf{b}/\emptyset)$, the same is true for $\mathbf{b}$ in place of $\mathbf{a}$. The map σ defined by $\sigma(f(\mathbf{a}, \mathbf{c})) = f(\mathbf{b}, \mathbf{c})$) (where $\mathbf{c}$ is any tuple from P such that $f(\mathbf{a}, \mathbf{c})$ is defined) is then well defined and in G. ■

The map above of course depends on $\mathbf{a}$. Claim I also holds for any realization $\mathbf{b}$ of r, in place of $\mathbf{a}$. In any case we can now identify G with the set of realizations of the type $r_{\mathbf{a}}$. However, $r_{\mathbf{a}}$ is a type over a 'large' set. Let $s_{\mathbf{a}} = r_{\mathbf{a}} \upharpoonright (\mathrm{dcl}(\mathbf{a}) \cup P)$.

Claim II. $s_{\mathbf{a}} \models r_{\mathbf{a}}$.

Proof. Let $C = \mathrm{Cb}_1(r_{\mathbf{a}})$ (see section 1). Let τ be an automorphism of $\mathbf{C}$ which fixes $\mathbf{a}$. As τ fixes P setwise, τ fixes setwise the set of non-forking extensions of r to $\mathbf{C}$. Thus τ fixes C. So $C \subseteq P \cap \mathrm{dcl}(\mathbf{a})$ (in fact $C = P \cap \mathrm{dcl}(\mathbf{a})$). Now let $\mathbf{b}$ realize $s_{\mathbf{a}}$. So $\mathrm{dcl}(\mathbf{b}) \cap P = \mathrm{dcl}(\mathbf{a}) \cap P$; also $\mathrm{tp}(\mathbf{b}/\emptyset) = \mathrm{tp}(\mathbf{a}/\emptyset) = r$. Thus also $C = \mathrm{Cb}_1(\mathrm{tp}(\mathbf{b}/P))$. $\mathrm{tp}(\mathbf{a}/P)$ and $\mathrm{tp}(\mathbf{b}/P)$ thus have the same non-forking extensions over $\mathbf{C}$, so in particular $\mathrm{tp}(\mathbf{b}/P) = \mathrm{tp}(\mathbf{a}/P)$. ■

$C_{\mathbf{a}}$ will denote $\mathrm{dcl}(\mathbf{a}) \cap P$. So $C_{\mathbf{a}}$ depends on $r_{\mathbf{a}}$. Of course Claim II yields $s_{\mathbf{b}} \models r_{\mathbf{b}}$ for any $\mathbf{b}$ realizing r, where $s_{\mathbf{b}} = r_{\mathbf{b}} \upharpoonright C_{\mathbf{b}}$.

Write $C_{\mathbf{a}}$ as $\{\lambda_i(\mathbf{a}): i \in I\}$ where λ_i are all the partial functions, defined over $\varnothing$, and defined at $\mathbf{a}$, with value in P. So λ_i is defined at any realization of r. So we have:

CLAIM III. *For any realizations* $\mathbf{a}$, $\mathbf{b}$ *of* r, $r_{\mathbf{a}}$ $(=\text{tp}(\mathbf{a}/P)) = r_{\mathbf{b}}$ $(=\text{tp}(\mathbf{b}/P))$ *if and only if* $\lambda_i(\mathbf{a}) = \lambda_i(\mathbf{b})$ *for all* $i \in I$.

CLAIM IV. *There is a formula* $\phi(\mathbf{x}_1, \mathbf{x}_2, y, z)$ *over* $\varnothing$, *such that for any* $\sigma \in G$, *any* $\mathbf{b}$ *realizing* r, *and any* $a' \in Q$, $\models \phi(\mathbf{b}, \sigma(\mathbf{b}), a', z)$ *iff* $z = \sigma(a')$.

Proof. Assume f is of the form $f(\mathbf{x}, w_1, \ldots, w_n)$. Let $\Psi(w_1, \ldots, w_n)$ be the partial type $\Psi_1(w_1) \cup \cdots \cup \Psi_n(w_n)$. Write $\mathbf{w}$ for $(w_1, \ldots, w_n)$.

Note by Claim III that

(∗) *if* $\mathbf{a}^1, \mathbf{a}^2$ realize r, $\lambda_i(\mathbf{a}^1) = \lambda_i(\mathbf{a}^2)$ for all $i \in I$, and $\mathbf{c}^1, \mathbf{c}^2$ realize $\Psi(\mathbf{w})$, *then* $[f(\mathbf{a}^1, \mathbf{c}^1) = f(\mathbf{a}^1, \mathbf{c}^2)$ iff $f(\mathbf{a}^2, \mathbf{c}^1) = f(\mathbf{a}^2, \mathbf{c}^2)]$.

By the compactness theorem, there is some formula $\psi(\mathbf{w})$, and also some finite subset J of I, such that (∗) is true with Ψ replaced by ψ and I by J.

Let $\phi(\mathbf{x}_1, \mathbf{x}_2, y, z)$ be the formula $(\exists \mathbf{w})(f(\mathbf{x}_1, \mathbf{w}) = y \wedge f(\mathbf{x}_2, \mathbf{w}) = z \wedge \psi(\mathbf{w}))$ (and if you want, add that $\lambda_i(\mathbf{x}_1) = \lambda_i(\mathbf{x}_2)$ for $i \in J$). This formula works, as the reader can check. ■

Let $X = \{(\mathbf{a}^1, \mathbf{a}^2): \text{tp}(\mathbf{a}^1/P) = \text{tp}(\mathbf{a}^2/P)\}$. By Claim III, $X = \{(\mathbf{a}^1, \mathbf{a}^2): \mathbf{a}^1, \mathbf{a}^2$ realize r and $\lambda_i(\mathbf{a}^1) = \lambda_i(\mathbf{a}^2)$ for each $i \in I\}$. Thus X is a set which is ∞-definable over $\varnothing$. By Claim I, if $(\mathbf{a}^1, \mathbf{a}^2) \in X$, then there is a *unique* $\sigma \in G$ such that $\mathbf{a}^2 = \sigma(\mathbf{a}^1)$. Also for ever $\sigma \in G$, and $\mathbf{a}^1$ realizing r, $(\mathbf{a}^1, \sigma(\mathbf{a}^1)) \in X$. Let E be the equivalence relation on X: $(\mathbf{a}^1, \mathbf{a}^2)$ $E(\mathbf{b}^1, \mathbf{b}^2)$ iff for some (unique) $\sigma \in G$, $\mathbf{a}^2 = \sigma(\mathbf{a}^1)$ and $\mathbf{b}^2 = \sigma(\mathbf{b}^1)$.

CLAIM V. E *is* $\varnothing$-*definable. In fact (for* $(\mathbf{a}^1, \mathbf{a}^2)$, $(\mathbf{b}^1, \mathbf{b}^2)$ *in* X), $(\mathbf{a}^1, \mathbf{a}^2)$ E $(\mathbf{b}^1, \mathbf{b}^2)$ *iff 'for* a' *realizing* $q|(\mathbf{a}^1, \mathbf{a}^2, \mathbf{b}^1, \mathbf{b}^2)$, $(\forall z)(\phi(\mathbf{a}^1, \mathbf{a}^2, a', z) \leftrightarrow \phi(\mathbf{b}^1, \mathbf{b}^2, a', z))$'. (*This is seen to be a formula over* $\varnothing$, *using the fact that the stationary type* q *is definable over* $\varnothing$.)

Proof. Let $\mathbf{a}^2 = \sigma(\mathbf{a}^1)$, $\mathbf{b}^2 = \tau(\mathbf{b}^1)$. Suppose the condition holds. Choose $\mathbf{a}^3$ realizing r, such that $\mathbf{a}^3$ is *independent* from $\{\mathbf{a}^1, \mathbf{a}^2, \mathbf{b}^1, \mathbf{b}^2\}$ over $\varnothing$. Then $\sigma(\mathbf{a}^3) = \tau(\mathbf{a}^3)$ (as $\mathbf{a}^3$ is a tuple of realizations of q, each member of which also satisfies the independence requirement). Let a'' be *any* element of Q. Then, as $\text{tp}(\mathbf{a}^3) = r$, $a'' = f(\mathbf{a}^3, \mathbf{c})$ for some $\mathbf{c}$ from P. But then (as σ, τ come from elementary maps which fix P pointwise), $\sigma(a'') = f(\sigma(\mathbf{a}^3), \mathbf{c}) = f(\tau(\mathbf{a}^3), \mathbf{c}) = \tau(a'')$. This shows that $\sigma = \tau$, as required. ■

So by Claim V, we have a canonical bijection $h: G \to X/E$ given by $h(\sigma) = (\mathbf{a}^1, \sigma(\mathbf{a}^1))/E$ ($\mathbf{a}^1$ any realization of r). Note that X/E is ∞-definable over $\emptyset$.

As in Claim V, the induced group operation on X/E is also $\emptyset$-definable.

CLAIM **VI.** *Let σ, τ, ν, be in G. Let $\mathbf{a}$, $\mathbf{b}$, $\mathbf{d}$ realize r. Then $\sigma.\tau = \nu$ if and only if 'for a' realizing $q|\{\mathbf{a}, \sigma(\mathbf{a}), \mathbf{b}, \tau(\mathbf{b}), \mathbf{d}, \nu(\mathbf{d})\}$, $\forall z(\phi(\mathbf{d}, \nu(\mathbf{d}), a', z) \leftrightarrow (\exists u)(\phi(\mathbf{b}, \tau(\mathbf{b}), a', u) \wedge \phi(\mathbf{a}, \sigma(\mathbf{a}), u, z)))$'.*

Proof. Like Claim V, and left to the reader. ■

Taking G_1 to be X/E, and by Claims IV, V, and VI, the theorem is proved. ■

REMARK **4.9.** *The group G_1 found in Theorem* 4.8 *clearly lives in Q^{eq}. Thus, as Q is* **P**-*internal G_1 is also* **P**-*internal. It follows that G_1 is definably isomorphic to a group living in* $\mathbf{P}^{\mathrm{eq}}$.

(Sketch proof: As G_1 is **P**-internal, there is, by 4.2(ii) and compactness, some $\emptyset$-definable function $f(\mathbf{x}, \mathbf{y})$ and some $\mathbf{b}$ such that for every $a \in G_1$ there is a tuple $\mathbf{c}$ of realizations of **P** such that $a = f(\mathbf{b}, \mathbf{c})$. Let X be the set of tuples $\mathbf{c}$ from P such that $f(\mathbf{b}, \mathbf{c}) \in G$, and let X' be X/E where $\mathbf{c}_1\, E\, \mathbf{c}_2$ if $f(\mathbf{b}, \mathbf{c}_1) = f(\mathbf{b}, \mathbf{c}_2)$. Then X'/E with the group operation $*$ induced from that of G_1 is $\mathbf{b}$-definably isomorphic to G_1. But by separation of parameters, X, E, as well as $*$, can be defined over parameters in **P**. Thus $(X', *)$ is a (∞-definable) group living in $\mathbf{P}^{\mathrm{eq}}$ which is $\mathbf{b}$-definably isomorphic to G_1.)

5 p-simple groups

The aim of this section is to continue the interpretations of p-weight, p-semi-regularity, etc., in the stable group context, and then look at what happens when p is locally modular.

G will in general be an ∞-definable group, defined over some base set A which we will assume to be algebraically closed. G is sometimes taken to be connected. p will be a regular type non-orthogonal to A. We will say that G has a certain property of its generic types have that property. We remark that if q_1, q_2 are generic types of G over A, and M is a saturated model containing A, then there is $c \in G^M$ such that $c.q_1' = q_2'$, where q_1', q_2' are the non-forking extensions of q_1, q_2 respectively to M. Thus if some generic type is p-internal, p-simple, or p-semi-regular, then all generic types have the corresponding property. As mentioned in section 3, we call G p-internal, p-simple, or p-semi-regular if some (so every) generic type of G has the corresponding property. Note, however, that as any element of G is a product of generic elements, if G is p-internal (p-simple) then for every $c \in G$ and $B \supseteq A$, $\mathrm{stp}(c/B)$ is p-internal (p-simple). Also if G is p-internal (p-simple) then for any definable normal subgroup N of G, G/N is p-internal (p-simple). Similarly for p-simplicity. If G is p-simple, then $w_p(G)$ is defined to be $w_p(q)$ where q is some generic type $q \in S(A)$ of G. By the above remarks this is well defined. Alll of this also applies to homogeneous spaces, that is objects of the form G/H (left or right coset space) acted on by G (on the left or right respectively). Namely, $w_p(G/H)$ = the p-weight of the type of some generic type of G/H.

LEMMA 5.1. (i) *Let G be p-simple with $w_p(G) = n$. Then for all $c \in G$, $w_p(c/A) \leq n$.*

(ii) *Let G be p-semi-regular with $w_p(G) = n$. Let $B \supseteq A$ and $a \in G$. Then* $\mathrm{stp}(a/B)$ *is a generic type of G (over B) iff $w_p(a/B) = n$.*

(iii) *Similarly for homogeneous spaces G/H, where H is a definable subgroup of G.*

Proof. (i) Let $c \in G$ and let a realize a generic type of G over A such that a is independent with c over A. Then $\mathrm{tp}(c.a/A)$ is a generic of G and so $w_p(c.a/A) = n$. On the other hand $w_p(c/A) = w_p(c/A, a) = w_p(c.a/A, a) \leq w_p(c.a/A) = n$.

(ii) Left to right is clear by definition.

Now suppose that $w_p(a/B) = n$. To show that $\mathrm{stp}(a/B)$ is generic it is enough to show that whenever $\mathrm{stp}(b/B)$ is generic and b is independent with a over B then $b.a$ is independent with b over B. So choose such b. Then $\mathrm{stp}(b.a/B)$ is generic, so p-semi-regular of p-weight n. On the other hand $w_p(b.a/B, b) = w_p(a/B, b) = n$. By Lemma 1.14, $b.a$ is independent with b over B.

(iii) has the same proof. ∎

LEMMA 5.2. *Let G be p-simple of finite p-weight. Then G is p-semi-regular iff there is no A-definable normal subgroup N of G with infinite index in G such that $w_p(G/N) = 0$.*

Proof. Let a realize some generic type of G over A. Let N be a normal A-definable subgroup of infinite index in G. Then a/N realizes a generic type of G/N over A and $a/N \notin \mathrm{acl}(A)$. Now $a/N \in \mathrm{dcl}(a, A)$. So by Lemma 1.18, if $\mathrm{tp}(a/A)$ is p-semi-regular, then $w_p((a/N)/A) > 0$. On the other hand, suppose $\mathrm{tp}(a/A)$ is not p-semi-regular. By Lemma 1.18 again there is $c \in \mathrm{dcl}(aA) - \mathrm{acl}(A)$ such that $w_p(c/A) = 0$. By Lemma 4.7 there is some A-definable normal N in G such that G/N is **Q**-internal, where $\mathbf{Q} = \{q \in S(A)$: q is p-simple of p-weight $0\}$. But then also $w_p(G/N) = 0$. ∎

Lemma 5.3. *Let G be p-simple (with finite p-weight). Let S be a definable subgroup of G. Then $w_p(G) = w_p(G/S) + w_p(S)$.*

Proof. Assume S to be B-definable for $B \supseteq A$. Let a be a generic element of G over B. Then a/S is a generic element of G/S over B. Moreover, a is a generic element of the coset aS over $B \cup \{a/S\}$, whereby $w_p(a/B \cup \{a/S\}) = w_p(S)$.

Now $a/S \in \mathrm{dcl}(a, A)$. So by Lemma 1.11, $w_p(a/B) = w_p(a, (a/S)/B) = w_p((a/S)/B) + w_p(a/(a/S), B)$, which is what we want. ■

Now suppose S to be an ∞-definable subgroup of the ∞-definable group G. Then G/S can no longer be considered as an ∞-definable set. However, it can be considered as an ∞-definable set of infinite tuples, as follows. By the proof of Lemma 4.4.2, we can find (relatively) definable subgroups S_i of G (for $i \in I$ with $|I| \leq |T|$), such that $S = \cap \{S_i: i \in I\}$, and $\{S_i: i \in I\}$ is closed under finite intersections, *and* for any A-automorphism f of $\mathbf{C}$, f fixes S setwise iff f fixes each S_i setwise. Let $g \in G$. Then g/S is on the one hand an ∞-definable set, which we denote by $g.S$. On the other hand we can consider the *element* g/S to be the infinite tuple $((g/S_i): i \in I)$, whereby G/S becomes the set of all such infinite tuples, which can thus be considered as an ∞-definable set of infinite tuples. Note also that g/S as defined here is (modulo A) a canonical set of defining parameters for $g.S$, in the sense that an A-automorphism will fix $g.S$ setwise iff it fixes g/S (i.e. fixes $\{g/S_i: i \in I\}$ pointwise). The theory of generics goes through for G/S (as a G-space) in the obvious way, and the reader is left to work out details of this as well as of Lemma 5.1(iii). In particular, if S is B-definable, g/S is a generic of G/S over B iff g/S_i is a generic of G/S_i for all i. Suppose now that G is p-simple of finite p-weight. Then for any $g \in G$, $w_p((g/S)/B)$ is well defined. This is because clearly $\{w_p((g/S_i)/B): i \in I\}$ has a maximum value m, and if $i(0) \in I$ is such that $w_p((g/S_{i(0)})/B) = m$, then $(g/S_i) \in \mathrm{cl}_p((g/S_{i(0)}), B)$ for all $i \in I$, whereby $w_p((g/S)/B) = m$. Thus we see also that $w_p(G/S) = \max\{w_p(G/S_i): i \in I\}$.

Lemma 5.4. *Let $S = \bigcap_{i \in I} S_i$ be an ∞-definable subgroup of G as above (where G is still p-simple of finite p-weight). Then for some $i(0) \in I$, $w_p(S) = w_p(S_{i(0)})$. Moreover,* Lemma 5.3 *holds for S.*

Proof. Let $w_p(G/S) = w_p(G/S_{i(0)})$ with $i(0) \in I$, as defined above. Let the S_i be all defined over $B \supseteq A$, and let g be generic of G over B. Then g/S_i is a generic of G/S_i over B for all i. As remarked above, for all $i \in I$, $g/S_i \in \mathrm{cl}_p(g/S_{i(0)}, B)$.

Also the coset $g.S$ is definable over $B \cup \{g/S_i: i \in I\}$. Thus

(∗) $\quad w_p(S) \geq w_p(g/B \cup \{g/S_i: i \in I\}) = w_p(g/B \cup \{g/S_{i(0)}\}) = w_p(S_{i(0)})$.

But $S < S_{i(0)}$, so by 5.1, $w_p(S) \leq w_p(S_{i(0)})$. Thus $w_p(S) = w_p(S_{i(0)})$. Now applying Lemma 5.3 to $S_{i(0)}$ gives the moreover clause. ■

LEMMA **5.5.** *Let G be p-simple of p-weight n. Let K be the intersection of all normal A-definable subgroups N of G such that* $w_p(G/N) = 0$. *Then K is p-semi-regular of p-weight n, and is, moreover, the unique* ∞*-definable connected subgroup of G which is p-semi-regular of p-weight n. We call K the p-connected component of G.*

Proof. Note that K is ∞-definable over A, and is, moreover, connected and normal in G. By Remark 5.4, $w_p(K) = n$. If K is not p-semi-regular, then by Lemma 5.2 there is a proper normal A-definable subgroup of K, say N, with $w_p(K/N) = 0$. Replacing N by $\cap \{N^g : g \in G\}$ (which is a finite subintersection), we may assume that N is normal in G. By Lemma 5.4, $w_p(G/N) = 0$. Now N is only an ∞-definable subgroup of G. So $N = \cap \{N_i : i \in I\}$, where the latter set is closed under finite intersections, and each N_i is a normal definable subgroup of G. Thus we must have $w_p(G/N_i) = 0$ for all $i \in I$. But then $K < N_i$ for all i, whereby $K = N$, a contradiction. Uniqueness of K is left to the reader. ■

PROPOSITION **5.6.** *Suppose G is p-simple of finite p-weight. Suppose p is locally modular. Let* $a \in G$, *and suppose* $q = \text{stp}(a/B)$ *is p-semi-regular. Then q is the generic type of some coset of some connected* ∞*-definable p-semi-regular subgroup S of G where S is definable over* $\text{acl}(B) \cap \text{cl}_p(A)$. *(Remember A is the base set over which G is defined and* $B \supseteq A$.)

Proof. (This is a somewhat 'simpler' version of Proposition 4.4.5 on 1-based groups.) Let S be the left stabilizer of q ($= \{b \in G$: for a realizing $q|(B, b)$, $b.a$ realizes $q|(B, b)\}$). So S is ∞-definable over $\text{acl}(B)$. Let $w_p(G) = n$, and $w_p(q) = m$. Let a realize q and let b be a generic of G over $B \cup \{a\}$. Then $b.a$ is a generic of G over B, so $w_p(b.a/B) = n$. Also $\text{stp}(b.a/B, b)$ is p-semi-regular with p-weight m (as $\text{stp}(a/B, b) = q|(B, b)$ is p-semi-regular with p-weight m). Let $C = \text{Cb}(\text{stp}(b.a/B, b))$. So $w_p(b.a/C, B) = m$ and by Lemma 2.6(i) $C \subseteq \text{cl}_p(b.a, B)$. Thus $n = w_p(b.a/B) = w_p(b.a, C/B) = w_p(b.a/C, B) + w_p(C/B)$, whereby

$$(*) \qquad w_p(C/B) = n - m.$$

CLAIM **I.** b/S *is interdefinable with C over B.*

Proof. Let M be a saturated model containing B, b such that a is independent with M over B. Then a realizes $q|M$ and $b.a$ is independent with M over B, b. Let f be a B-automorphism of M, and let $b' = f(b)$. Then

f fixes C pointwise iff $f(\text{tp}(b.a/M)) = \text{tp}(b.a/M)$ iff $\text{tp}(b'.a/M) = \text{tp}(b.a/M)$ iff $\text{tp}(b^{-1}.b'.a/M) = \text{tp}(a/M)$ iff $b^{-1}.b' \in \text{Stab}(q|M) = S$ iff $f(b.S) = b.S$ iff $f(b/S) = b/S$, as required. ■

By Claim I and (*) we see that $w_p((b/S)/B) = n - m$. As b/S is a generic of G/S over B, $w_p(G/S) = n - m$ and thus by Lemma 5.3, we have

CLAIM **II.** $w_p(S) = m$.

We can now see that S is connected, p-semi-regular, and q is a right translate of the generic type of S. For let S_1 be the p-connected component of S. S_1 is defined over $\text{acl}(B)$ and is connected and p-semi-regular with p-weight m (by Claim II and Lemma 5.5). Let $q_1 \in S(\text{acl}(B))$ be the generic type of S_1. Let a realize q and let c realize $q_1|(B, a)$. As $S_1 \subseteq S = \text{Stab}(q)$, $c.a$ realizes q. On the other hand $w_p(c.a/B, a) = w_p(c/B, a) = w_p(c/B) = m$ $(= w_p(c.a/B))$. As q is p-semi-regular, Lemma 1.14 implies that $c.a$ realizes $q|(B, a)$. But $c.a \in S_1.a$, and $w_p(S_1.a) = m$. Thus $c.a$ realizes over $\text{acl}(B, a)$ the generic type $X = S_1.a$. This shows that $q|(B, a)$ is the generic type of X. The same is then clearly true of q (showing in particular that X is defined over $\text{acl}(B)$). In fact S_1 must be equal to S. For if not, let $c \in S - S_1$, and let a realize $q|(B, c)$. Then $c.a \notin X$, so $c.a.$ does not realize q, contradicting the fact that S is the left stabilizer of q.

Finally we must show that S is $\text{cl}_p(A)$-definable (this is enough as we know that S has a canonical base $(\subseteq \text{acl}(B))$ and $\text{cl}_p(A)$ is definably closed). Let again b realize some generic type of G over B, and let a realize $q|(B, b)$. Then $a.b$ is a generic of G over B, and thus $a.b$ is independent with B over A. On the other hand $S = \text{(left)Stab}(\text{stp}(a/B, b)) = \text{(left)Stab}(\text{stp}(a.b/B, b))$. It should then be clear that S is definable over $C = \text{Cb}(\text{stp}(a.b/B, b))$ (an automorphism of the universe which fixes C pointwise fixes the global non-forking extension r of $\text{stp}(a.b/B, b)$ and thus fixes setwise $S = \text{Stab}(r)$). But $\text{stp}(a.b/B, b)$ is p-semi-regular (being a translate of $q|(B, b)$), and $\text{stp}(a.b/A)$ is p-simple, so by Lemma 2.6 (i), $C \subseteq \text{cl}_p(a.b, A)$.

To sum up what we have so far, S is definable over $\text{cl}_p(a.b, A)$ and also over $\text{acl}(B)$ and, moreover, $a.b$ is independent with B over A. Let C' be a set of canonical defining parameters for S and let $x \in C'$. Then $x \in \text{cl}_p(a.b, A) \cap \text{acl}(B)$. But $a.b$ is independent from B over A; hence $w_p(a.b/B) = w_p(a.b/Ax) = w_p(a/A)$. But $w_p(a.b, x/A) = w_p(a.b/x, A) + w_p(x/A)$. Thus $w_p(x/A) = 0$. Thus S is definable over $\text{cl}_p(A)$ as required. ■

COROLLARY **5.7.** *Let S be a ∞-definable connected p-semi-regular subgroup of the p-simple group G, where G is defined over A, and p is locally modular. Then S is defined over* $\text{cl}_p(A)$.

Proof. S is the stabilizer of its own generic type. ■

COROLLARY **5.8.** *Let G be a connected p-semi-regular group, where p is locally modular. Then G is abelian.*

Proof. Like the proof of Proposition 4.4.7.

In the present situation, for $g \in G$, the graph of Inn_g is a p-semi-regular subgroup of $G \times G$, so by 5.7 defined over $\mathrm{cl}_p(A)$ (assuming G defined over A). If g, g' are each generic in G over A, then (by semi-regularity of G), $\mathrm{tp}(g/\mathrm{cl}_p(A)) = \mathrm{tp}(g'/\mathrm{cl}_p(A))$. Thus $\mathrm{Inn}_g = \mathrm{Inn}_{g'}$. As in 4.4.7, G will be abelian. ■

Finally in this section we characterize p-closure on the generic type p of a locally modular regular group as linear dependence with respect to the division ring of $\mathrm{cl}_p(\emptyset)$-definable quasi-endomorphisms of G. This again is a copy into the regular type context of what was done for locally modular minimal groups in section 5 of Chapter 4. We nevertheless repeat some details, although we are brief.

Let us fix a connected group G, defined over $\emptyset$, say, whose generic type p is regular and locally modular. G is abelian (simply because G is connected and with regular generic), and so we write the group operation as $+$. Let $D = D(G) = \{S \subset G \times G$: S is a connected p-semi-regular p-weight 1 subgroup of G and $S \neq \{0\} \times G\}$. *By* 5.8 *each* $S \in D$ *is defined over* $\mathrm{cl}_p(\emptyset)$. Let $G_0 = \mathrm{cl}_p(\emptyset) \cap G$. Note that the definition of G_0 depends on the base set (in this case $\emptyset$), but the definition of D does not. $D \times \{0\}$ will be considered as the 0 element of D.

For $E, F \in D$, let $E \oplus F = \{(a, b + c)\colon (a, b) \in E$ and $(a, c) \in F\}$, and let $F \otimes E = \{(a, b)\colon$ for some c, $(a, c) \in E$ and $(c, b) \in F\}$. Let $E + F$ be the p-connected component of $E \oplus F$ and $F.E$ the p-connected component of $F \otimes E$. For E non-zero, let $E^{-1} = \{(b, a)\colon (a, b) \in E\}$. The kernel of E is $\{a\colon (a, 0) \in E\}$ and the cokernel of E is $\{b\colon (0, b) \in E\}$.

LEMMA **5.9.** (i) *If $E \in D$ then the projection of E on the first coordinate is G. The same is true of the projection of E on the second coordinate, if E is non-zero.*

(ii) *The cokernel of E is contained in G_0, and so is the kernel of E if E is non-zero.*

(iii) *Let $E \in D$.* Then $\{(a + G_0, b + G_0)\colon (a, b) \in E\}$ *is the graph of an endomorphism of G/G_0. We call this endomorphism f_E.*

(iv) *Let $E, F \in D$. Then $E + F$ and $F.E$ are in D. If E is non-zero $E^{-1} \in D$.*

(v) *If E is zero so is f_E. If E is non-zero then f_E is an isomorphism of G/G_0, and $f_{(E^{-1})} = (f_E)^{-1}$.*

(vi) *Let* $E, F \in D$. *Then* $f_{E+F} = f_E + f_F$, *and* $f_{F.E} = f_F.f_E$ *(composition).*

(vii) $(D, +, .)$ *is a division ring.*

Proof. (i) Let $A \subseteq \mathrm{cl}_p(\varnothing)$ be such that E is defined over A. Let (a, b) be generic in E over A. Then $\mathrm{stp}(a, b/A)$ is p-semi-regular (of p-weight 1), so not both $a, b \in \mathrm{cl}_p(A)$ $(= \mathrm{cl}_p(\varnothing))$. For suppose by way of contradiction that $a \in \mathrm{cl}_p(A)$. Then by Lemma 1.18, $a \in \mathrm{acl}(A)$. As every element of E is a sum of generics it follows that E_1, the projection of E on the first coordinate, is contained in $\mathrm{acl}(A)$ and is thus finite. But E_1 must also be connected, so $E_1 = \{0\}$. Moreover, b must be generic in G over A. Thus E_2, the projection of E on the second coordinate, equals G. Thus $E = \{0\} \times G$, contradicting the definition of G. Thus a must be generic in G over A, whereby $E_1 = G$.

The same argument shows that if E is non-zero then also $E_2 = G$.

(ii) The cokernel of E is an ∞-definable subgroup of G, defined over A. Suppose that for some generic $g \in G$ over A, $(0, g) \in E$. By (i) there is $(a, b) \in E$ with a generic in G over A. We may assume g is independent with $\{a, b\}$ over A. But then $(a, b + g) \in E$, and both a, $b + g$ are generic independent (over A) elements of G, so $w_p(E) \geq 2$, a contradiction. Thus $\mathrm{coker}(E)$ is contained in $\mathrm{cl}_p(A)$. As $A \subseteq \mathrm{cl}_p(\varnothing)$, $\mathrm{coker}(E) \subseteq \mathrm{cl}_p(\varnothing)$. A similar argument shows $\ker(E) \subseteq \mathrm{cl}_p(\varnothing)$, if E is non-zero.

(iii) It is enough to show that if $(a, b) \in E$ and $a \in G_0$ then $b \in G_0$. If not then b is generic in G over $\varnothing$ and thus generic in G over A. Thus $w_p(a, b/A) = 1$ so $\mathrm{stp}(a, b/A)$ is the generic type of E over A, by Lemma 5.1. But we have seen in (i) that this implies a is generic in G over A, a contradiction.

(iv) It is clear that if E is non-zero then $E^{-1} \in D$. For the rest it suffices to show that if $E, F \in D$, then both $E \oplus F$ and $F \otimes E$ have p-weight 1. Let a be generic in G over A, where both E, F are defined over A. If $(a, b) \in E$ and $(b, c) \in F$, then clearly $b, c \in \mathrm{cl}_p(A, a)$, so $b + c \in \mathrm{cl}_p(A, a)$, so $w_p(a, b + c/A) = 1$. This is enough to show that $w_p(E \oplus F) = 1$. A similar argument goes for $F \otimes E$.

(v) is clear.

(vi) Suppose $f_{E+F}(a + G_0) = b + G_0$. So there is $(a', b') \in E + F$ with $a - a' \in G_0$ and $b - b' \in G_0$. As $E + F \subseteq E \oplus F$ there are c, d with $c + d = b'$, $(a', c) \in E$ and $(a', d) \in F$. But then $f_E(a + G_0) = c + G_0$ and $f_F(a + G_0) = d + G_0$. So $f_E(a + G_0) + f_F(a + G_0) = b + G_0$. Thus $f_{E+F} = f_E + f_F$. Similarly, $f_{F.E} = f_F.f_E$. ■

We will, by virtue of Lemma 5.9, identify $D = D(G)$ with the ring of endomorphisms $\{f_E : E \in D\}$ of G/G_0. Thus G/G_0 is a vector space over D.

Proposition 5.10. *Let $a, b_1, \ldots, b_n$ be realizations of p (the generic type of the connected locally modular group G, defined over $\varnothing$). Then a forks with $\{b_1, \ldots, b_n\}$ over $\varnothing$ (or equivalently $a \in \mathrm{cl}_p(b_1, \ldots, b_n)$) iff there are $\sigma_1, \ldots, \sigma_n \in D$ such that $a + G_0 = \sigma_1(b_1 + G_0) + \cdots + \sigma_n(b_n + G_0)$ (i.e. $a + G_0$ is in the D-linear span of $\{b_1 + G_0, \ldots, b_n + G_0\}$).*

Proof. Suppose first the right-hand side to hold. Let $E(i)$ be the regular subgroup of $G \times G$ such that $\sigma_i = f_{E(i)}$, for $i = 1, \ldots, n$. Let $A \subseteq \mathrm{cl}_p(\varnothing)$ be such that the $E(i)$ are all defined over A. Then each b_i realizes $p|A$. Let $(b_i', c_i') \in E(i)$ be such that $b_i - b_i' \in G_0$. Then b_i' realizes $p|A$, and $c_i' \in \mathrm{cl}_p(A, b_i') = \mathrm{cl}_p(A, b_i)$ for all i. Moreover, $a - (c_1' + \cdots + c_n') \in G_0$, and thus $a \in \mathrm{cl}_p(b_1, \ldots, b_n, A)$. As $A \subset \mathrm{cl}_p(\varnothing)$, $a \in \mathrm{cl}_p(b_1, \ldots, b_n)$, that is a forks with $\{b_1, \ldots, b_n\}$ over $\varnothing$.

Conversely suppose a forks with $\{b_1, \ldots, b_n\}$ over $\varnothing$, where everything realizes p. We may assume $\{b_1, \ldots, b_n\}$ to be independent and generic over $\varnothing$. Let $A = \mathrm{dcl}(a, b_1, \ldots, b_n) \cap \mathrm{cl}_p(\varnothing)$. Then each of a, $b_1, \ldots, b_n$ realizes $p|A$ and $\{b_1, \ldots, b_n\}$ remains A-independent. Moreover, by Corollary 1.20, $\mathrm{stp}(b_1, \ldots, b_n, a/A)$ is p-semi-regular with p-weight n. By Proposition 5.6, $\mathrm{stp}(b_1, \ldots, b_n, a/A)$ is the generic type of a coset X of some p-semi-regular subgroup S of G^{n+1}, with $w_p(S) = n$, and both X, S defined over $\mathrm{acl}(A)$. In fact S is also defined over $\mathrm{cl}_p(\varnothing)$ by 5.7.) The projection of S on the first n coordinates must also have p-weight n (as $\mathrm{stp}(b_1, \ldots, b_n/A)$ does), so must be G^n (as G^n is p-semi-regular of p-weight n). So the cokernel of S has p-weight 0 and is contained in G_0. Thus $X = S + (0, \ldots, 0, d)$ for some $d \in G_0$.

We have $(b_1, \ldots, b_n, a - d) \in S$. As $\{b_1, \ldots, b_n\}$ are generic independent in G over A, there is for each $i = 1, \ldots, n$, some c_i such that $(0, \ldots, b_i, \ldots, 0, c_i) \in S$. As the cokernel of S has p-weight 0, $c_i \in \mathrm{cl}_p(b_i, A)$. So $w_p(b_i, c_i/A) = 1$. Augmenting A again by some elements of p-weight 0, we have some semi-regular $S_i \subseteq G \times G$ of p-weight 1, such that $\mathrm{stp}(b_i, c_i/A)$ is the generic type of some coset of S_i (in fact S_i is just $\{(x_i, y): (0, \ldots, x_i, \ldots, 0, y) \in S\}$). In any case, as above there is $d_i \in \mathrm{cl}_p(A)$ such that $(b_i, c_i - d_i) \in S_i$. Note that the d_i are in G_0. Let σ_i be the endomorphism of G/G_0 corresponding to S_i. Then $\sum \sigma_i(b_i/G_0) = (\sum c_i)/G_0$. But $(b_1, \ldots, b_n, \sum c_i) \in S$. So $\sum c_i - (a - d)$ is in cokernel of S, so in G_0. As $d \in G_0$, $(\sum c_i)/G_0 = a/G_0)$. Thus $a/G_0 = \sum \sigma_i(b_i/G_0)$, completing the proof. ∎

Note that if we name some parameters A and consider now $p|A$ the generic type of G over A, the only change to Proposition 5.10 will be to G_0 which is now replaced by $\mathrm{cl}_p(A) \cap G$. Proposition 5.10 shows that the geometry attached to p is precisely the geometry of an infinite-dimensional projective space over D.

6 Geometries

In this final section we characterize the geometry of a locally modular regular type $p \in S(A)$ as degenerate, or affine or projective geometry, over some division ring. This is exactly what was done in Chapter 5, section 2, for minimal types, and the proof there can be copied into the regular type context. However, for the sake of variety and entertainment, we give a slightly different proof, in which the use of the 'binding group' (Proposition 5.2.3) is replaced by a slightly more elaborate version of the construction of a modular type (Proposition 5.2.1). We will be rather informal and the reader is invited to fill in additional details.

Let us fix a stationary locally modular regular type $p \in S(\emptyset)$. We are interested in the geometry attached to p, that is the geometry corresponding to the pregeometry $(p^{\mathfrak{C}}, \mathrm{cl}_p(\text{-}))$. If p is trivial, we obtain the trivial geometry. So assume p non-trivial. We immediately use Theorem 3.3, which gives us a set E, with $w_p(E) = 1$, a connected ∞-definable group G, defined over E, such that the generic type of G is regular (and non-orthogonal to p) and G is a group of germs of invertible functions on $q \in S(E)$, where q is regular and not weakly orthogonal to $p|E$. We have two cases.

Case I

q is modular. In this case, as the generic type of G is also modular (by Proposition 5.10), by Corollary 2.9, q is not weakly orthogonal to the generic type of G. Thus the generic type of G (over E) is not weakly orthogonal to $p|E$. We will obtain a situation like that in the proof of Proposition 5.2.1. We work over $\mathrm{acl}(\emptyset)$ and sometimes may add parameters from $\mathrm{cl}_p(\emptyset)$.

Write G as $G(E)$, and let $G_0(E)$ denote $\{x \in G: w_p(x/E) = 0\}$. Let d realize $p|E$ and let $a \in G(E) - G_0(E)$ be such that a forks with d over E. By Corollary 1.20 we may assume (by augmenting E) that $\mathrm{tp}(a, d/E)$ is stationary and p-semi-regular of p-weight 1 (i.e. regular). Let F_E denote the division ring of 'definable' endomorphisms of $G(E)/G_0(E)$ given in section 5.

Let $\mathrm{tp}(E') = \mathrm{tp}(E)$ with E' independent from Ead over $\emptyset$. In particular $\mathrm{tp}(E, d) = \mathrm{tp}(E', d)$, so there is $a' \in G(E')$ such that $\mathrm{tp}(E, a, d) = \mathrm{tp}(E', a', d)$ $(= q_1$, say). (For future purposes, let $A' \subseteq \mathrm{dcl}(E') \cap \mathrm{cl}_p(\emptyset)$ be such that $\mathrm{tp}(E'/A')$ is stationary and p-semi-regular (of p-weight 1).)

A p-weight computation shows that $w_p(a, a'/E, E') = 1$. By Corollary 1.20 there is algebraically closed $D \supseteq E \cup E'$, with $D \subseteq \mathrm{cl}_p(E \cup E')$ such that $\mathrm{stp}(a, a'/D)$ is p-semi-regular of p-weight 1. By Proposition 5.6, $\mathrm{stp}(a, a'/D)$ is the generic type of a coset $S = S_{E,E'}$. That is, $S = H_{E,E'} + (a, a')$ for some regular subgroup $H_{E,E'}$ of $G(E) \times G(E')$, where S and H are D-definable. It is, moreover, easy to see that $H_{E,E'} \neq G(E) \times \{0\}$ or $\{0\} \times G(E')$. $H_{E,E'}$ also defines an isomorphism $f_{E,E'}$ between the division rings F_E and $F_{E'}$, by the following: if K is a semi-regular p-weight 1 subgroup of $G(E) \times G(E)$ (defining thus a member of F_E) then $f_{E,E'}(K)$ is the p-connected component of $H_{E,E'} \otimes K \otimes (H_{E,E'})^{-1}$ (where as above $\otimes$ denotes composition of relations).

Let M be a very big saturated model containing d such that M is independent with E over $\emptyset$. Let P be the set of realizations of p in $\mathrm{cl}_p(M)$. We will characterize $(P, \mathrm{cl}_p(\text{-}))$ which is enough.

Let $X(E) = \{a' \in G(E) - G_0(E)$: there is $d' \in P$ such that $\mathrm{tp}(E, a', d') = q_1\}$.

Lemma 6.1. *Either $(X(E)/G_0(E)) \cup \{0\}$ is a linear subspace of the F_E-vector space $G(E)/G_0(E)$, or $X(E)/G_0(E)$ is a non-linear affine subspace of $G(E)/G_0(E)$.*

Proof. If S is actually a subgroup of $G(E) \times G(E')$ (i.e. $S = H_{E,E'}$) then as in the proof of Proposition 5.2.1 (specifically the proofs of Claims I and II therein) $\mathrm{cl}_p(X(E) \cup E) \cap (G(E) - G_0(E)) = X(E)$, so by Proposition 5.10, $X(E)/G_0(E) \cup \{0\}$ is a linear subspace of $G(E)/G_0(E)$. So we may assume that $S \neq H_{E,E'}$.

We will now show

(i) if $a_1, a_2, a_3 \in X(E) - G_0(E)$ and $\{a_1, a_2, a_3\}$ is E-independent, then $a_1 + a_2 - a_3 \in X(E) - G_0(E)$.

(ii) if $a_1, a_2 \in X(E) - G_0(E)$, $\{a_1, a_2\}$ is E-independent and $\alpha \in F(E)$ then for some $a_3 \in X(E) - G_0(E)$, $a_3/G_0(E) = \alpha(a_1/G_0(E)) + (1 - \alpha)(a_2/G_0(E))$.

Proof of (i). Let $d_i \in P$ for $i = 1, 2, 3$ be such that $\mathrm{tp}(E, a_i, d_i) = q_1$. Let A'' be chosen in M such that $\mathrm{tp}(A'') = \mathrm{tp}(A')$ and A'' is independent from $\{d_1, d_2, d_3\}$ over $\emptyset$. Let E'' be chosen in M such that $\mathrm{tp}(E'', A'') = \mathrm{tp}(E', A')$ and $E''A''$ is independent from $\{d_1, d_2, d_3\}$ over $\emptyset$. The reader should then check that

(*) E'' is independent from $E \cup \{a_1, a_2, a_3, d_1, d_2, d_3\}$ over $\emptyset$; and

(**) $\mathrm{tp}(E, E'', a_i, d_i/\emptyset) = \mathrm{tp}(E, E', a, d/\emptyset)$ for $i = 1, 2, 3$.

Fix D' such that $\mathrm{tp}(E, E'', D'/\emptyset) = \mathrm{tp}(E, E', D/\emptyset)$.

As $\mathrm{tp}(a, d/E)$ is p-semi-regular, it follows from (**) that

$$\mathrm{tp}(E, E'', D', a_i, d_i/\emptyset) = \mathrm{tp}(E, E', D, a, d/\emptyset)$$

for each i. So we can find b_i such that

$$\mathrm{tp}(E, E'', D', a_i, b_i, d_i) = \mathrm{tp}(E, E', D, a, a', d).$$

Let $H_{E,E''}$ and S' be the images of $H_{E,E'}$ and S under an automorphism taking

E, E', D to E, E'', D'. Then $\mathrm{tp}(a_i, b_i/D')$ is a generic type of S', and $\{(a_i, b_i): i = 1, 2, 3\}$ is D'-independent. Also $b_i \in \mathrm{cl}_p(M)$.

Thus, as S is a coset, $(a_1, b_1) + (a_2, b_2) - (a_3, b_3) = (a_1 + a_2 - a_3, b_1 + b_2 - b_3) = (a_4, b_4)$, say, realizes the generic type of S over D', and $b_4 \in \mathrm{cl}_p(M)$. We can find d_4 such that $\mathrm{tp}(E, E', D, a, a', d) = \mathrm{tp}(E, E'', D', a_4, b_4, d_4)$. Then in particular $\mathrm{tp}(E, a_4, d_4) = q_1$ and $d_4 \in \mathrm{cl}_p(M)$. Thus $a_3 \in X(E) - G_0(E)$, as required.

Proof of (ii). First, as in (i) find E'', d_1 d_2, b_1, b_2 all in $\mathrm{cl}_p(M)$, and D' such that $\mathrm{tp}(E, E'', D', a_i, b_i, d_i) = \mathrm{tp}(E, E', D, a, a', d)$ for $i = 1, 2$. Let $H_{E,E''}$ and S' be as in (i). Then $\{(a_1, b_1), (a_2, b_2)\}$ is D'-independent and each (a_i, b_i) is a generic of S' over D', so $(a_1 - a_2, b_1 - b_2)$ is generic in $H_{E,E''}$ over D'. Let $\beta = f_{E,E''}(\alpha)$. (So $\beta \in F_{E''}$.) We can then easily find generic (a_3, b_3) in $H_{E,E''}$ over D' such that $a_3/G_0(E) = \alpha((a_1 - a_2)/G_0(E))$ and $b_3/G_0(E'') = \beta((b_1 - b_2)/G_0(E))$. Then $b_3 \in \mathrm{cl}_p(M)$. Let $(a_4, b_4) = (a_3, b_3) + (a_2, b_2)$. So easily (a_4, b_4) is generic in S' over D', whereby there is d_4 such that $\mathrm{tp}(E, E'', D', a_4, b_4, d_4) = \mathrm{tp}(E, E', D, a, a', d)$. As $b_4 \in \mathrm{cl}_p(M)$, also $d_4 \in \mathrm{cl}_p(M)$. Thus $a_4 \in X(E) - G_0(E)$, as required.

Now Lemma 6.1 follows from (i) and (ii). If $X(E)/G_0(E) \cup \{0\}$ is a linear subspace of $G(E)/G_0(E)$, then by Proposition 5.10 the geometry attached to $(X(E), \mathrm{cl}_p(\text{-}))$ (over E) is projective geometry over F_E. The same is thus true of the geometry attached to $(P, \mathrm{cl}_p(\text{-}))$ over E. As P is independent with E over $\varnothing$, also $(P, \mathrm{cl}_p(\text{-}))$ has the same geometry. If on the other hand $X(E)/G_0(E)$ is an affine non-linear subspace of $G(E)/G_0(E)$, then as can be checked, F_E-linear closure on $X(E)/G_0(E)$ equals affine closure (i.e. for W a subset of $X(E)/G_0(E)$ and for $v \in X(E)/G_0(E)$, v is in the F_E-vector space generated by W iff v is in the F_E-affine space generated by W). Thus $\mathrm{cl}_p(\text{-})$ on $X(E)$ gives rise to affine geometry over F_E. As above, this is also true of the geometry on P.

Case II

q is non-modular. In this case by Lemma 2.8 if a realizes q, then a is independent with $\mathbf{b}$ over E for any tuple $\mathbf{b}$ in $G(E)$.

Let Q be the set of realizations of q. It is then easy to see that G acts transitively on Q, and, moreover, as G is abelian, G acts regularly on Q. As in the proof of Proposition 5.2.4, it follows that the geometry induced by p-closure over E on Q is precisely affine geometry over F, where F is the division ring corresponding to G: for $a_0, \ldots, a_n \in Q$, $\{a_0, \ldots, a_n\}$ is dependent over E iff $\{(a_1 - a_0)/G_0, \ldots, (a_n - a_0)/G_0\}$ is linearly dependent over F (where $G_0 = \mathrm{cl}_p(E) \cap G$) and $a_i - a_0$ denotes the unique $g \in G$ such that

$g.a_0 = a_i$). Also if we define $a_1 \approx a_2$ to mean a_1 forks with a_2 (for $a_i \in Q$), then G/G_0 acts regularly on $Q' = Q/\approx$.

Again we write $G = G(E)$, $Q = Q(E)$, and let $a \in Q$ and d realize $p|E$ such that d forks with a over E. Let E' be a copy of E, independent with E, a, d over $\varnothing$ and let a' be such that $\mathrm{tp}(E, a, d) = \mathrm{tp}(E', a', d)$. Write $G(E')$, $Q(E')$ for the copies of $G(E)$ and $Q(E)$ over E'. Then $G(E) \times G(E')$ acts regularly on $Q(E) \times Q(E')$, and it is not difficult to conclude from Proposition 5.6 (or rather the homogeneous space version of it) that for some, say, algebraically closed set D, containing $E \cup E'$ and contained in $\mathrm{cl}_p(E \cup E')$, $\mathrm{tp}(a, a'/D)$ is the generic type of some orbit (or if you want, translate) $S \subseteq Q(E) \times Q(E')$ of a regular subgroup $H_{E,E'}$ of $G(E) \times G(E')$. Let again M be a big model containing d such that M is independent with E over $\varnothing$. Let $P = p^{\mathbf{C}} \cap \mathrm{cl}_p(M)$, and let $X = X(E) = \{a_1 \in Q$: there is $d_1 \in P$ with $\mathrm{tp}(E, a_1, d_1) = \mathrm{tp}(E, a, d)\}$. As in the proof of case I above, one shows that X is an 'affine subspace' of Q, that is $X' = X/\approx$ is an orbit of an F-subspace of G/G_0. But then p-closure over E on X gives us an affine geometry, isomorphic to the geometry arising from p-closure over E on P. As P is independent with E over $\varnothing$, $(P, \mathrm{cl}_p(\text{-}))$ gives us the same affine geometry over F.

This completes the proof of

Theorem 6.2. *Let p be a stationary, regular, locally modular type over $\varnothing$. Then the geometry of $(p^{\mathbf{C}}, \mathrm{cl}_p(\text{-}))$ is either trivial, or affine or projective geometry over some division ring.*

We should mention that the other results proved in Chapter 5 are also valid in the regular type context: (assuming p is a regular, locally modular type over $\varnothing$) there is some modular p_1 over $\mathrm{cl}_p(\varnothing)$ which is non-orthogonal to p, and if p itself is non-modular, then there is a regular group G defined over $\mathrm{cl}_p(\varnothing)$ such that for a realizing p there is $a' \in \mathrm{dcl}(a, \mathrm{cl}_p(\varnothing))$ such that G acts regularly on the set of realizations of $\mathrm{tp}(a')$. Also:

Proposition 6.3. *Let $p \in S(\varnothing)$ be stationary, regular, locally modular but not modular. Then for any A, $p|A$ is modular iff p is realized in $\mathrm{cl}_p(A)$.*

Proof. Just like Corollary 5.2.5. ■

8

Superstable theories

In this final chapter we give some scattered results on superstable theories. One of the main lines here is the development of some technology which helps towards generalizing results from the finite rank context to the general superstable context. A key point is the interaction between the local theory developed in Chapter 7 and Shelah's 'infinity rank', R^∞. This will yield in some situations 'continuity and definability' of p-weight. In any case one of the goals of this chapter is to suitably generalize Buechler's dichotomy theorem to regular types in superstable theories. This is done in section 5, under the assumption that T is countable and has NDOP and NOTOP (two structure properties); this result has great importance for some fine questions concerning the spectrum function. Note we also give some other versions of the dichotomy theorem, which do not depend on assuming NDOP and NOTOP, and which are relevant to Vaught's conjecture for superstable theories. In the process of developing the technology, we mention various related 'non-geometric' results, such as the existence of regular types between arbitrary models, and definability questions in non-multidimensional theories.

1 Analysability

We first take the opportunity to introduce a notion generalizing internality, namely analysability. For now we only assume stability of T. We use the notation of section 4 of Chapter 7. **P** will denote a $\emptyset$-invariant family of partial types, unless otherwise stated.

Definition 1.1. (i) *Let* $q = \mathrm{stp}(a/A)$. *q is said to be* **P**-*analysable if there is an ordinal* λ *and elements* a_α *for* $\alpha \leq \lambda$ *such that* $a = a_\lambda$ *and* $\mathrm{stp}(a_\alpha/A \cup \{a_\beta\colon \beta < \alpha\})$ *is* **P**-*internal for all* α. *Such a sequence* $(a_\alpha\colon \alpha \leq \lambda)$ *is called a* **P**-*analysis of a over A of length* $\lambda + 1$.

(ii) *A formula* θ *is said to be* **P**-*internal* (**P**-*analysable*) *if every strong type containing* θ *is.*

We summarize some basic properties (possibly repeating observations made in Chapter 7).

LEMMA 1.2. (i) *If* stp(a/A) *is* **P**-*internal* (**P**-*analysable*) *then for any* $B \supseteq A$ stp(a/B) *is* **P**-*internal* (**P**-*analysable*).

(ii) *If* stp(a/A), stp(b/A) *are* **P**-*internal* (**P**-*analysable*) *then so is* stp$(a, b/A)$.

(iii) *If* stp(a/bA) *and* stp(b/A) *are* **P**-*analysable then so is* stp(a/A).

(iv) *If* stp(a/A) *is foreign to* **P** *then* stp(a/A) *is orthogonal to any* **P**-*internal* (**P**-*analysable*) *type.*

Proof. The **P**-internal version of (i) follows from Lemma 7.4.2, and the **P**-analysable version then follows immediately. The same is true for (ii). For (iii) let $\{a_\alpha: \alpha \leq \lambda\}$ witness the **P**-analysability of stp(a/bA) and let $\{b_\beta: \beta \leq \mu\}$ witness the **P**-analysability of stp(b/A). By (i) stp$(a_\alpha/A \cup \{a_\gamma: \gamma < \alpha\} \cup \{b_\beta: \beta \leq \mu\})$ is **P**-internal for all α, so clearly stp(a/A) is **P**-analysable.

(iv) Let $p = $ stp(a/A). Let q be a **P**-internal strong type over a set B. Let M be a saturated model containing $A \cup B$, and let a realize $p|M$ and b realize $q|M$. Clearly there are realizations $c_1, \ldots, c_n$ of $\mathbf{P}|M$ such that $b \in \mathrm{dcl}(c_1, \ldots, c_n, M)$. But p is foreign to **P**, so in particular a is independent with **c** over M; thus a is independent with b over M. So p is orthogonal to q. It easily follows that p is orthogonal to any **P**-analysable type. ■

LEMMA 1.3. (i) stp(a/A) *is* **P**-*analysable iff for all* $B \supseteq A$, stp(a/B) *is algebraic or not foreign to* **P**.

(ii) *If* stp(a/A) *is* **P**-*analysable then there is some* **P**-*analysis of* a *over* A *which is contained in* dcl(aA).

Proof. (i) Using Lemma 1.2, if stp(a/A) is **P**-analysable then stp(a/B) is also **P**-analysable for $B \supseteq A$, and is either algebraic or not foreign to **P**. Now suppose the right-hand side. By 7.4.6 we can find inductively a_α in $\mathrm{dcl}(aA) - \mathrm{acl}(A \cup \{a_\beta: \beta < \alpha\})$ such that as long as stp$(a/A \cup \{a_\beta: \beta < \alpha\})$ is non-algebraic, stp$(a_\alpha/A \cup \{a_\beta: \beta < \alpha\})$ is **P**-internal. So for some λ, stp$(a/A \cup \{a_\alpha: \alpha < \lambda\})$ is algebraic, so **P**-internal. Put $a_\lambda = a$, and we see that $\{a_\alpha: \alpha \leq \lambda\}$ is a **P**-analysis of a over A.

(ii) follows from (i) and its proof. ■

COROLLARY 1.4. *Suppose* stp(a/B) *is* **P**-*analysable and does not fork over* A; *then* stp(a/A) *is* **P**-*analysable.*

Proof. Let $C \supseteq A$. We want to see that stp(a/C) is algebraic or not foreign to **P**. As **P** is $\emptyset$-invariant we may assume that B is independent with C over aA. Thus stp(a/BC) does not fork over C. Assuming stp(a/C) non-algebraic, we obtain that $\operatorname{stp}(a/BC)$ is non-algebraic, and by Lemma 1.3(i), $\operatorname{stp}(a/BC)$ is not foreign to $\mathbf{P}$. So as a is independent from B over C, $\operatorname{stp}(a/C)$ is not foreign to $\mathbf{P}$. By Lemma 1.3 again, $\operatorname{stp}(a/A)$ is $\mathbf{P}$-analysable. ■

LEMMA **1.5.** *Suppose* $\mathbf{P}$ *to be a family of* formulae.

(i) *If* $\operatorname{stp}(a/A)$ *is* $\mathbf{P}$*-internal then there is a formula* $\varphi(x)$ *in* $\operatorname{tp}(a/A)$ *which is* $\mathbf{P}$*-internal.*

(ii) *Suppose* $\operatorname{stp}(a/A)$ *is* $\mathbf{P}$*-analysable. Then there is a finite subset* A_0 *of* A *and a* finite $\mathbf{P}$*-analysis* $\{a_i: i \leq n\}$ *of* a *over* A_0 *which is contained in* $\operatorname{dcl}(aA_0)$.

Proof. (i) Let $p(x) = \operatorname{tp}(a/A)$. By Lemma 7.4.2(i) (and our assumption that $\mathbf{P}$ is $\varnothing$-invariant) and compactness there is a definable function $f(\mathbf{y})$ and formulae $\psi_i(y_i)$ in $\mathbf{P}$ for $i = 1, \ldots, m$ such that

$$p(x) \models (\exists y_1 \ldots y_m)(\textstyle\bigwedge_i \psi_i(y_i) \wedge f(\mathbf{y}) = x).$$

($f(\mathbf{y})$ and the $\psi_i(y_i)$ may have parameters outside A.)

By compactness again there is a formula $\varphi(x)$ in $p(x)$ such that

$$\varphi(x) \models (\exists y_1 \ldots y_m)(\textstyle\bigwedge_i \psi_i(y_i) \wedge f(\mathbf{y}) = x).$$

Clearly $\varphi(x)$ is $\mathbf{P}$-internal.

(ii) is a simple consequence of (i) and the fact that there is no infinite descending sequence of ordinals. ■

REMARK **1.6.** (i) *We cannot in general conclude from the assumptions of* 1.5(ii) *that some formula in* $\operatorname{tp}(a/A)$ *is* $\mathbf{P}$*-analysable. For example, if we have unary predicates* P, Q, *a surjective infinite-to-one function* $f: P \to Q$, *and an infinite set* $\{c_i: i < \omega\}$ *of constants in* Q, *let* $q(x)$ *be the complete type over* $\varnothing$ *axiomatized by* $\{Q(x)\} \cup \{x \neq c_i: i < \omega\}$. *Let* $\mathbf{P} = \{Q(x)\} \cup \{f(y) = b: b$ *realizes* $q\}$. *Then* $\mathbf{P}$ *is a* $\varnothing$*-invariant family of formulae. Let* $p(y)$ *be axiomatized by* $\{P(y)\} \cup \{f(y) \neq c_i: i < \omega\}$. *Then* p *is* $\mathbf{P}$*-analysable, but no formula in* p *is* $\mathbf{P}$*-analysable.*

(ii) *If* T *is superstable,* $\mathbf{P}$ *a family of partial types, and* $\{a_\alpha: \alpha \leq \lambda\}$ *a* $\mathbf{P}$*-analysis of* a *over* A *such that* $a_\alpha \notin \operatorname{acl}(A \cup \{a_\beta: \beta < \alpha\})$ *for all* α, *then clearly* λ *is finite.*

It is convenient to introduce the following notion:

DEFINITION **1.7.** *Let* P *be some property of elements (or tuples) from* $\mathbf{C}^{\mathrm{eq}}$. *We say that* P *holds of* $\mathbf{a}$ provably over A, *if there is some formula* $\varphi(\mathbf{x})$ *in* $\operatorname{tp}(\mathbf{a}/A)$ *such that whenever* $\mathbf{b}$ *satisfies* $\varphi(\mathbf{x})$ *then* $\mathbf{b}$ *has property* P. *We may also use this notation when* $\mathbf{a}$ *is a possibly infinite tuple, but note in this case* *that the formula $\varphi(\mathbf{x})$ can only talk about some finite subsequence of* $\mathbf{a}$. *So for example we may speak of some property P holding of* $\mathrm{stp}(a/B)$ *provably over* $\mathbf{A}$, *meaning that there is a formula $\varphi(x, \mathbf{y})$ over A and some* $\mathbf{b}$ *from B such that $\models \varphi(a, \mathbf{b})$ and whenever $(a', \mathbf{b}')$ satisfies φ then P holds of* $\mathrm{stp}(a'/\mathbf{b}')$.

We will say that $\mathbf{P}$ is a *union of $\varnothing$-definable families of formulae*, if there are formulae $\chi_i(x_i, y_i)$ and $\delta_i(y_i)$ without parameters, for i in some index set I, such that $\mathbf{P} = \{\chi_i(x_i, c) : i \in I, c \text{ in } \mathbf{C}, \text{ and } \models \delta_i(c)\}$.

Lemma 1.8. *Let* $\mathbf{P}$ *be a union of $\varnothing$-definable families of formulae. Suppose* $\mathrm{stp}(a/A)$ *is* $\mathbf{P}$*-internal* ($\mathbf{P}$*-analysable*). *Then* $\mathrm{stp}(a/A)$ *is* $\mathbf{P}$*-internal* ($\mathbf{P}$*-analysable*), *provably over* $\varnothing$.

Proof. Suppose first $\mathrm{stp}(a/A)$ to be $\mathbf{P}$-internal. By Lemma 1.5(i) there is a formula $\varphi(x)$ in $\mathrm{tp}(a/A)$, formulae $\psi_i(y_i)$ in $\mathbf{P}$, and definable function $f(\mathbf{y})$ such that

$$\models \varphi(x) \to (\exists y_1 \ldots y_m)(\wedge \psi_i(y_i) \wedge f(\mathbf{y}) = x).$$

Let $\mathbf{c}$ list the parameters (from A) in the formula $\varphi(x)$ and let $\mathbf{d} \supseteq \mathbf{c}$ contain the parameters from $f(\mathbf{y})$ and the $\psi_i(y_i)$. Rewrite φ as $\varphi(x, \mathbf{c})$, f as $f(\mathbf{y}, \mathbf{d})$, and ψ_i as $\psi_i(y_i, \mathbf{d})$. By our assumptions on $\mathbf{P}$ there is a formula $\delta(\mathbf{z})$ in $\mathrm{tp}(\mathbf{d}/\varnothing)$ such that whenever $\mathbf{e}$ satisfies $\delta(\mathbf{z})$ then $f(\mathbf{y}, \mathbf{e})$ is a function and the formulae $\psi_i(y_i, \mathbf{e})$ are in $\mathbf{P}$.

Let $\xi(x, \mathbf{w})$ be the following formula:

$$\varphi(x, \mathbf{w}) \wedge (\exists \mathbf{z} \supseteq \mathbf{w})(\delta(\mathbf{z}) \wedge (\forall x)(\varphi(x, \mathbf{w}) \to (\exists y_1 \ldots y_m)(\wedge \psi_i(y_i, \mathbf{z}) \wedge f(\mathbf{y}, \mathbf{z}) = x))).$$

Then clearly $\xi(a, \mathbf{c})$ holds, and, moreover, whenever $\xi(a', \mathbf{c}')$ holds and C contains $\mathbf{c}'$ then $\mathrm{stp}(a'/C)$ is $\mathbf{P}$-internal. Thus ξ witnesses that $\mathrm{stp}(a/A)$ is $\mathbf{P}$-internal, provably over $\varnothing$.

It easily follows from this and Lemma 1.5(ii) that whenever $\mathrm{stp}(a/A)$ is $\mathbf{P}$-analysable, then this also holds provably over $\varnothing$. ■

Examples of families $\mathbf{P}$ in which we may be interested are $\mathbf{RM}$, the family of formulae θ such that $RM(\theta) < \infty$, $\mathbf{R}^\infty$, the family of formulae θ such that $R^\infty(\theta) < \infty$, and $\mathbf{R}^\infty < \alpha$, the family of formulae θ such that $R^\infty(\theta) < \alpha$. These are all clearly $\varnothing$-invariant families. Note that if $\mathbf{P}$ is the empty family then the $\mathbf{P}$-analysable types are precisely the algebraic types. The following examples give cases where $\mathbf{P}$ is 'analysable-closed'.

Example 1.9. (i) Suppose the formula θ is $\mathbf{R}^\infty$-analysable. Then $R^\infty(\theta) < \infty$.

(ii) Suppose T is countable and the formula θ is $\mathbf{RM}$-analysable. Then $RM(\theta) < \infty$.

Proof. We will use the following basic facts (which either appear in or can be proved using results from section 3 of Chapter 1):

(a) For $R = RM$ or R^∞, if $\text{stp}(a/A)$ is $\mathbf{R}$-internal then $R(a/A) < \infty$.

(b) $U(p) \leq R^\infty(p)$ for any complete type.

(c) If T is countable and θ is a formula, then $\theta \in \mathbf{RM}$ iff for any countable set A containing the parameters in θ, there are at most countably many complete types over A containing θ.

(d) For any formula θ, $\theta \in \mathbf{R}^\infty$ iff $U(p) < \infty$ for every complete type containing θ.

Now we prove (i). Suppose $\text{stp}(a/A)$ contains θ. By Fact 1.5(ii) there are $a_0, a_1, \ldots, a_n = a$ such that $\text{stp}(a_i/a_0 a_1 \ldots a_{i-1}A)$ is $\mathbf{R}^\infty$-internal for $i = 1, \ldots, n$. By (a) and (b) above, $U(a_i/a_0, \ldots, a_{i-1}A) < \infty$, for each i. The U-rank inequalities 1.3.26 or 4.3.15 now imply that $U(a/A) < \infty$. By (d) above, $R^\infty(\theta) < \infty$.

(ii) Let A be a countable set containing the parameters from θ. By Fact 1.5(ii) and (a) above, for any a satisfying θ there are $a_1, \ldots, a_m = a$ such that $RM(a_i/a_0 \ldots a_{i-1}A) < \infty$. It easily follows from (c) above that there are only countable many complete types over A containing θ, so by (c) again $RM(\theta) < \infty$. ■

We now show how we can produce regular types using the notion of analysability. First, let us call a partial type $\Psi(x)$ $\mathbf{P}$-internal ($\mathbf{P}$-analysable) if every stationary extension of Ψ is $\mathbf{P}$-internal ($\mathbf{P}$-analysable). This agrees with Definition 1.1(i) if Ψ is a single formula. Also, by virtue of Lemma 1.2(i), this notion is unambiguous in the case where Ψ happens to be a stationary complete type over a set A. We will say that $\Psi' \supseteq \Psi$ is a forking extension of Ψ if Ψ' forks over the set $[\Psi]$ of canonical parameters for the formulae in Ψ. (Where the partial type Ψ is said to fork over A if some finite conjunction of members of Ψ forks over A.) The reader should also note that this agrees with the usual notion when Ψ is a complete type over some set A. In any case we will be typically interested in two cases, when Ψ is a formula and when Ψ is a complete stationary type.

$\mathbf{P}$ remains a $\varnothing$-invariant family of partial types.

Definition 1.10. (i) *Let $\Psi(x)$ be a partial type. We say that Ψ is weakly* $\mathbf{P}$*-minimal, if Ψ is not* $\mathbf{P}$*-analysable, but every forking extension of Ψ is* $\mathbf{P}$*-analysable.*

(ii) *The partial type Ψ is said to be* $\mathbf{P}$*-minimal if Ψ is weakly* $\mathbf{P}$*-minimal and for any set A containing the parameters in Ψ, there is a unique non-*$\mathbf{P}$*-analysable strong type over A extending Ψ.*

REMARK **1.11.** (i) *Any stationary complete weakly* **P**-*minimal type is* **P**-*minimal.*

(ii) *If* Ψ *is weakly* **P**-*minimal, then every stationary complete non*-**P**-*analysable extension of* Ψ *is* **P**-*minimal.*

(iii) *A formula* θ *is weakly* $\emptyset$-*minimal iff it is weakly minimal, and a complete stationary type is* $\emptyset$-*minimal iff it is minimal.*

(iv) *If* Ψ *is weakly* **P**-*minimal then there are, up to parallelism, only a bounded number of stationary complete non*-**P**-*analysable extensions of* Ψ.

(v) *If* p *is a strong type over* A *and is* **P**-*minimal then* p *is regular and is, moreover, orthogonal to all* **P**-*analysable strong types.*

(vi) *If* p *is a strong type over* A *which is* **P**-*minimal, then* p *is foreign to* **P**.

(vii) Ψ *is weakly* **P**-*minimal if and only* [Ψ *is not* **P**-*analysable and for some (any)* A *containing* $[\Psi]$, *for all* $B \supseteq A$, *if* $q(x) \in S(B)$ *contains* Ψ *and forks over* A *then* q *is* **P**-*analysable*].

Proof. (i) Let p be a complete stationary type over A which is weakly **P**-minimal. Let $B \supseteq A$. Any non-**P**-analysable strong type over B which extends p must be a non-forking extension of p and thus is unique.

(ii) follows from (i).

(iii) is true because $\emptyset$-analysable means algebraic.

(iv) Let A be the set of canonical parameters for Ψ. Assuming Ψ to be weakly **P**-minimal, then every stationary complete non-**P**-analysable extension of Ψ does not fork over A. So there are a bounded number of such types (up to parallelism).

(v) Suppose p is a strong type over A and p is **P**-minimal. Suppose q is a **P**-analysable strong type, and suppose by way of contradiction that p is non-orthogonal to q. Then there are $B \supseteq A$ and a realizing $p|B$, b realizing $q|B$ such that a forks with b over B. Then $\text{stp}(b/B)$ is **P**-analysable, and $\text{stp}(a/Bb)$ is **P**-analysable, so by Lemma 1.2(ii) $\text{stp}(a/B)$ is **P**-analysable. By Corollary 1.4, p is **P**-analysable, a contradiction. Thus p is orthogonal to all **P**-analysable types. In particular, as any forking extension of p is **P**-analysable, p must be regular.

(vi) The proof is contained in that of (v).

(vii) Clear. ∎

LEMMA **1.12.** (i) *Suppose* **P** *is a family of formulae and that the universe '*$x = x$*' is not* **P**-*analysable. Then there is some* **P**-*minimal type.*

(ii) *Suppose* T *is superstable and the universe is not* **P**-*analysable. Then there is a weakly* **P**-*minimal formula.*

(iii) *Suppose T superstable. Let q be a stationary complete type over some set. Let θ be a formula of least R^∞, say α, such that q is not foreign to θ. Then θ is weakly $\mathbf{R}^\infty < \alpha$-minimal (and q is non-orthogonal to some $\mathbf{R}^\infty < \alpha$-minimal stationary type containing θ).*

Proof. (i) As the universe is not **P**-analysable, there is some complete stationary non-**P**-analysable type $p(x)$. By Lemma 1.5(ii) it is easy to see that if $(p_i(x))_{i<\beta}$ is an increasing chain of strong types, none of which is **P**-analysable, then the union of the chain is not **P**-analysable. So stability enables us to find a strong type p which is not **P**-analysable, although every forking extension of p is **P**-analysable. That is, p is **P**-minimal.

(ii) Choose $\theta(x)$ to be a formula which is not **P**-analysable and such that $R^\infty(\theta) = \alpha$ is least possible. If $\Psi(x)$ contains $\theta(x)$ and forks over the canonical parameter for θ, then clearly $R^\infty(\Psi) < \alpha$, so there is a formula ψ in (the logical closure of) Ψ with $R^\infty(\psi) < \alpha$. But then ψ is **P**-analysable, so Ψ is **P**-analysable. Thus θ is weakly **P**-minimal.

(iii) Let q and θ be as given. Let $\mathbf{P} = \mathbf{R}^\infty < \alpha$. Then q is foreign to **P**. By Lemma 1.2(iv) q is orthogonal to any **P**-analysable stationary type. Thus θ is not **P**-analysable. But clearly any forking extension of θ is **P**-analysable, so θ is weakly **P**-minimal. If p is a stationary complete type which contains θ and is non-orthogonal to q, then p is not **P**-analysable, so by Remark 1.11(ii), p is **P**-minimal. ■

We finish this section with two applications of the notion of analysability. The first is somewhat technical but provides a useful criterion for a countable theory to be ω-stable.

Lemma 1.13. *Suppose T is countable, and that for every non-algebraic strong type q, there is an ordinal α and some $\mathbf{R}^\infty < \alpha$-minimal formula θ such that q is non-orthogonal to the (unique) $\mathbf{R}^\infty < \alpha$-minimal strong type containing θ. Then T is ω-stable.*

Proof. We show by induction on α that

(*) every $\mathbf{R}^\infty < \alpha$-minimal formula θ is in **RM** (i.e. has ordinal-valued Morley rank).

When $\alpha = 1$, $\mathbf{R}^\infty < \alpha$ is simply the set of algebraic formulae, so clearly θ is strongly minimal. Suppose (*) is proved for all $\alpha < \beta$ and that θ is $\mathbf{R}^\infty < \beta$-minimal. Let $p(x)$ be the unique non-$\mathbf{R}^\infty < \beta$-analysable strong type (over the parameters of θ) which contains θ.

CLAIM. *Suppose* $\models \psi(x) \rightarrow \theta(x)$, *and* $\psi(x)$ *is not in any non-forking extension of* p. *Then* $\psi \in \mathbf{RM}$.

Proof. By uniqueness of p, ψ must be $\mathbf{R}^{\infty} < \beta$-analysable. We may assume ψ is non-algebraic. Let r be any (non-algebraic) stationary type containing ψ. By assumption, there is α and $\mathbf{R}^{\infty} < \alpha$-minimal φ such that r is non-orthogonal to the unique $\mathbf{R}^{\infty} < \alpha$-minimal strong type, say t, containing φ. But r is $\mathbf{R}^{\infty} < \beta$-analysable, and by 1.11(v) t is orthogonal to all $\mathbf{R}^{\infty} < \alpha$-analysable types. Thus $\alpha < \beta$. By the induction hypothesis, $\varphi \in \mathbf{RM}$. We have shown that every non-algebraic strong type containing ψ is not foreign to $\mathbf{RM}$. By Lemma 1.3(i), ψ is $\mathbf{RM}$-analysable and by Example 1.9(ii), $\psi \in \mathbf{RM}$.

The claim is proved. ■

It follows from the claim that $RM(\theta(x)) < \infty$. (Let $\gamma = \min\{RM(\psi)$: ψ implies θ and ψ is not in any non-forking extension of $p\}$. Then for any M containing the parameters of θ, $p|M$ is the unique extension of θ over M which has Morley rank $\geq \gamma$. Hence $RM(\theta) = \gamma$.) Thus, by induction $(*)$ is proved for all α. Our assumptions, together with Lemma 1.3(i), imply that the formula '$x = x$' is $\mathbf{RM}$-analysable. Thus by Example 1.9(ii) again $RM(x = x) < \infty$. So T is ω-stable. ■

REMARK **1.14.** *Lemma* 1.13 *fails without the countability assumption, basically because Example* 1.9(*ii*) *needs countability of* T. *We describe briefly a counterexample. The theory* T *will have unary predicates* P_η *for* $\eta \in {}^{\omega>}2$, *and equivalence relations* E_σ *for* $\sigma \in {}^{\omega}2$. $P_\varnothing$ *is the universe and for each* η, P_η *is the disjoint union of* $P_{\eta^\wedge\langle 0\rangle}$ *and* $P_{\eta^\wedge\langle 1\rangle}$, *each of which is infinite. For each* $\sigma \in {}^{\omega}2$, $n < \omega$, *and* $i = 0, 1$, *if* $\sigma(n) \neq i$, *then* $P_{(\sigma|n)^\wedge\langle i\rangle}$ *is an* E_σ-*class. This describes a complete theory which is superstable.* $R^{\infty}(x = x) = 2$. *For* $\sigma \in {}^{\omega}2$, *let* $p_\sigma(x)$ *be the complete type* $\{P_{\sigma|n}(x)$: $n < \omega\}$. *On the set of realizations of* p_σ, E_σ *defines an equivalence relation with infinitely many classes, each class being a (definable) indiscernible set. In particular, for* a *realizing* p_σ, $E_\sigma(x, a)$ *is a strongly minimal set. Also, for any* σ, *the set of* E_σ-*classes is a strongly minimal set, say* X_σ (*in* $\mathbf{C}^{\mathrm{eq}}$). *The universe is clearly analysable over these strongly minimal formulae. However,* T *is not totally transcendental.*

It is also worth pointing out that we cannot replace the assumptions of 1.13 *by '*T *is countable and every strong type is non-orthogonal to some strongly regular type'. For example, let* T *be the theory of unary predicates* $P_i(i < \omega)$, *every finite Boolean combination of which is consistent. For* S *a subset of* ω, *let* $p_S = \{P_i(x)$: $i \in S\} \cup \{\neg P_i(x)$: $i \notin S\}$. *Then for each* S, p_S *is strongly regular (via the formula '*$x = x$*') and these account for all regular types. But* T *is not* ω-*stable.*

Proposition 1.15. *If T is unidimensional then T is superstable.*

Proof. Recall that the (stable) theory T is said to be unidimensional if any two non-algebraic types are non-orthogonal.

Let us assume T to be unidimensional.

Suppose first that there is some non-algebraic formula θ, such that $R^\infty(\theta) < \infty$. Without loss θ is over $\emptyset$. So every non-algebraic strong type is not foreign to $\{\theta\}$. By Lemma 1.3(i), every non-algebraic strong type is $\{\theta\}$-analysable. By Example 1.9(i), R^∞('$x = x$') $< \infty$. So T is superstable.

Now we do not know as yet whether there is such a formula θ, but we *do* know (by for example Lemma 1.12(i)) that there *is* a minimal type $p(x)$, that is $U(p) = 1$. Again we may assume p is over $\emptyset$. Again by unidimensionality, every non-algebraic strong type is $\{p\}$-analysable.

Case (i). Whenever $\mathrm{stp}(a/A)$ is $\{p\}$-internal then there are realizations $b_1, \ldots, b_n$ of p such that $a \in \mathrm{acl}(b_1, \ldots, b_n, A)$. In this case, for any element a, by looking at a $\{p\}$-analysis of a over $\emptyset$, we see that a is in the algebraic closure of some (finite) set of realizations of p. It easily follows that $U(\mathrm{tp}(a/\emptyset)) < \omega$. In particular (as a was arbitrary) T is superstable.

Case (ii). Not case (i). Then, without loss of generality we have a strong type q over $\emptyset$, such that q is $\{p\}$-internal but q is not realized in $\mathrm{acl}(A)$ for any set A of realizations of p. Let P be the set of realizations of p, and Q the set of realizations of q. Let $G = \mathrm{Aut}(Q/P)$ be the class of permutations of Q which are induced by elementary maps which fix $P \cup \mathrm{acl}(\emptyset)$ pointwise. *Then G is infinite,* and by Lemma 7.4.8, G can be identified with an ∞-definable group. By Lemma 1.6.18, G is contained in a *definable group H*.

By unidimensionality of T, the generic types of H are non-orthogonal to p. By Lemma 7.4.7 (or rather the remark following it), H has a definable normal subgroup N of infinite index such that the generic types of H/N are $\{p\}$-internal. The U-rank inequalities imply that the generic types of H/N have ordinal-valued U-rank. But every element of H/N is a product of elements whose type is generic (over a set of definition of H/N). Thus every type in H/N has ordinal-valued U-rank. By fact (d) in the proof of Example 1.9, $R^\infty(H/N) < \infty$. So we have found a non-algebraic *formula* in $\mathbf{R}^\infty$. By the remarks at the beginning of this proof, we can conclude that T is superstable. ■

2 Definability of analysability and weight

We saw in section 5 of Chapter 1 that if, for example, D is a weakly minimal formula over $\emptyset$, $a \in D^{\mathrm{eq}}$ and $U(a/A) = n$, then there is a formula $\varphi(x, y)$ and some b in A such that $\varphi(a, b)$ holds and, moreover, for all b', $\varphi(x, b')$, if consistent, has U-rank n. Also in this context U-rank $= R^\infty$.

In this section, among other things we generalize this observation to weakly **P**-minimal formulae, under various assumptions on the provability of **P**-analysability. U-rank can be replaced by p-weight or Π-weight for a suitable family Π of regular types. In fact the weakly minimal theory generalizes immediately after replacing algebraic dependence by '**P**-dependence' and making some assumption on the provability of **P**-analysability. Nevertheless, below we go through some details of this generalization (even though it is completely routine).

P remains some $\varnothing$-invariant family of partial types.

DEFINITION **2.1.** (i) *Let $\theta(x)$ be a formula, whose parameters are in a set A. By $(\theta_A)^{\mathrm{eq}}$ we mean $\{b \in \mathbf{C}^{\mathrm{eq}}$: for some $c_1, \ldots, c_n$ satisfying $\theta(x)$, $b \in \mathrm{dcl}(c_1, \ldots, c_n, A)\}$.*

(ii) *We say that a is **P**-dependent on A if $\mathrm{stp}(a/A)$ is **P**-analysable. We say that a set X of elements is **P**-independent over A if for each $a \in X$, a is not **P**-dependent on $A \cup \{X - \{a\}\}$.*

LEMMA **2.2.** (i) ***P**-dependence is transitive; that is, if each $a \in X$ is **P**-dependent on Y and b is **P**-dependent on X then b is **P**-dependent on Y.*

(ii) *Suppose $\theta(x)$ is weakly **P**-minimal, and A contains the parameters of θ. Then the relation 'a is **P**-dependent on $B \cup A$' for $\{a\} \cup B$ included in $\theta^{\mathbf{C}}$ defines a pregeometry on θ.*

Proof. (i) is simply a translation of Lemma 1.2(ii) and (iii).

(ii) Note first that if $\theta(a)$ holds and $\mathrm{stp}(a/A)$ is not **P**-analysable, then for any B, a is **P**-dependent on $B \cup A$ iff a forks with B over A. (ii) now follows from forking symmetry and (i). ■

By Lemma 2.2 (ii) we can define **P**-dimension inside a weakly **P**-minimal formula (where our notation may clash with earlier notation):

DEFINITION **2.3.** *Let $\theta(x)$ be a weakly **P**-minimal formula defined over a set A.*

(i) *Let X be a set of realizations of $\theta(x)$. We define $\dim_{\mathbf{P}}(X/A)$ to be the cardinality of a maximal subset Y of X which is **P**-independent over A.*

(ii) *Let $a \in (\theta_A)^{\mathrm{eq}}$. We define $\dim_{\mathbf{P}}(a/A)$ to be $\min\{\dim_{\mathbf{P}}(\mathbf{c}/\mathbf{b}A)$: $\mathbf{bc}$ is a sequence of realizations of θ, a is independent with $\mathbf{b}$ over A, and $a \in \mathrm{dcl}(\mathbf{bc}A)\}$.*

(iii) *If $\varphi(z)$ is over A and implies that $z \in (\theta_A)^{\mathrm{eq}}$, then $\dim_{\mathbf{P}}(\varphi) = \max\{\dim_{\mathbf{P}}(a/A)$: a satisfies $\varphi\}$.*

Lemma 2.4. (*With notation of* 2.3) *Let* $a \in (\theta_A)^{\mathrm{eq}}$.

(i) *Suppose* X *is a finite set of realizations of* θ *such that* $a \in \mathrm{dcl}(XA)$. *Let* Y *be a maximal subset of* X *which is* **P**-*independent over* $A \cup \{a\}$. *Then* $\dim_{\mathbf{P}}(a/A) = \dim_{\mathbf{P}}(X/Y \cup A)$.

(ii) $\dim_{\mathbf{P}}(a/A) = \min\{\dim_{\mathbf{P}}(\mathbf{c}/B)\colon B \supseteq A$, *a is independent with* B *over* A, **c** *is a tuple of realizations of* θ, *and* $a \in \mathrm{dcl}(\mathbf{c}B)\}$.

Proof. Left to the reader. ■

Lemma 2.5. *Suppose* $\theta(x)$ *is weakly* **P**-*minimal and defined over* A. *Suppose also that whenever* a *satisfies* θ *and* $\mathrm{stp}(a/B \cup A)$ *is* **P**-*analysable, then this (considered as a property of aB) holds provably over* A.

(i) *Let* $\psi(x, y)$ *be a formula over* A *which implies* $\theta(x)$. *Then there is a formula* $\delta(y)$ *over* A *such that for any* b, $\models\delta(b)$ *iff* $\psi(x, b)$ *is* **P**-*analysable.*

(ii) *Let* $\psi(x_1, \ldots, x_n, y)$ *be a formula over* A *which implies* $\bigwedge_i \theta(x_i)$. *Then there is a formula* $\delta(y)$ *over* A *such that for any* c, $\models\delta(c)$ *iff* $\dim_{\mathbf{P}}(\psi(x_1, \ldots, x_n, c)) = n$ (*iff for some type* $(a_1, \ldots, a_n)$ *satisfying* $\psi(x_1, \ldots, x_n, c)$, $\{a_1, \ldots, a_n\}$ *is* **P**-*independent over Ac*).

(iii) *Let* $\psi(x_1, \ldots, x_n, y)$ *be as in* (ii). *Let* $m \le n$. *Then there is a formula* $\delta(y)$ *over* A *such that for any* c, $\models\delta(c)$ *iff* $\dim_{\mathbf{P}}(\psi(x_1, \ldots, x_n, c)) = m$.

Proof. (i). Let **R** be the family of **P**-minimal types containing $\theta(x)$. Each such type is definable over $\mathrm{acl}(A)$. For each $r \in \mathbf{R}$, let $\chi_r(y)$ be the $\psi(x, y)$-definition of r. So χ_r is over $\mathrm{acl}(A)$. So for any b, $\psi(x, b)$ is **P**-analysable iff for all $r \in \mathbf{R}$ $\psi(x, b) \notin r|Ab$ iff for all $r \in \mathbf{R} \models \neg\chi_r(b)$. On the other hand our assumptions yield a set $\Phi(y)$ of formulae over A such that for any b, $\psi(x, b)$ is **P**-analysable iff $\models \phi(b)$ for some $\phi \in \Phi$. (Let Φ be the set of all formulae $\varphi(x, y)$ such that $\models \varphi(x, y) \to \psi(x, y)$ and for all a', b', $\models \varphi(a', b')$ implies $\mathrm{stp}(a'/Ab')$ is **P**-analysable.) Compactness now yields a single formula $\delta(y)$ over A as required.

(ii) This is standard and follows by iterating (i). For example, suppose $n = 2$. By (i), let $\delta_1(x_1, y)$ be a formula over A such that $\models \delta(a_1, c)$ iff $\psi(a_1, x_2, c)$ is not **P**-analysable. Let $\delta_2(y)$ be a formula over A such that $\models \delta_2(c)$ iff $\delta_1(x_1, c) \wedge \theta(x_1)$ is not **P**-analysable. Then $\delta_2(y)$ is the required formula.

(iii) is like (ii) and left to the reader. ■

Lemma 2.6. *Assume the hypotheses of Lemma* 2.5. *Then*

(i) If $a \in (\theta_A)^{\mathrm{eq}}$ *and* $\dim_{\mathbf{P}}(a/A) = n$, *then there is a formula* $\varphi(z)$ *in* $\mathrm{tp}(a/A)$ *such that* $\dim_{\mathbf{P}}(\varphi(z)) = n$.

(ii) *Suppose $\psi(z, y)$ is a formula over A such that for all b, $\psi(z, b)^{\mathbf{C}}$ is contained in $(\theta_{Ab})^{\text{eq}}$. For any n there is a formula $\delta(y)$ over A such that for all c, $\models \delta(c)$ iff $\dim_{\mathbf{P}}(\psi(z, c)) = n$.*

Proof. (i). By Lemma 2.4(i) there are $c_1, \ldots, c_k, c_{k+1}, \ldots, c_m$ satisfying θ such that $\{c_1, \ldots, c_k\}$ is **P**-independent over Aa, c_i is **P**-dependent on $Ac_1 \ldots c_k a$ for $i > k$, $a = f(c_1, \ldots, c_m)$ for some A-definable function f, and $\dim_{\mathbf{P}}(c_{k+1}, \ldots, c_m/Ac_1 \ldots c_k) = n$. By our assumptions on the provability of **P**-analysability (in θ), there is a formula $\chi(z, x_1, \ldots, x_m)$ in $\text{tp}(a, c_1, \ldots, c_m/A)$ which implies that $\dim_{\mathbf{P}}(x_{k+1}, \ldots, x_m/Ax_1 \ldots x_k) \leq n$, x_i is **P**-dependent on $Ax_1 \ldots x_k z$ for $i > k$, and $z = f(x_1, \ldots, x_m)$. Let $\xi(z, x_1, \ldots, x_k)$ be the formula $(\exists x_{k+1}, \ldots, x_m)(\chi(z, x_1, \ldots, x_k, \ldots, x_m))$, and by Lemma 2.5(ii) let $\varphi(z)$ be the formula '$\dim_{\mathbf{P}}(\xi(z, x_1, \ldots, x_k)) = k$'.

(ii) We may assume that there is some A-definable function f such that $\models \psi(z, y) \rightarrow (\exists x_1 \ldots x_m)(\bigwedge_i \theta(x_i) \wedge f(x_1, \ldots, x_m, y) = z)$.

Let $\sigma(z, x_1, \ldots, x_m, y)$ be the formula '$\bigwedge_i \theta(x_i) \wedge \psi(z, y) \wedge z = f(x_1, \ldots, x_m, y)$'. We may assume, by Lemma 2.5 and partitioning $\psi(z, y)$ suitably, that for some fixed $k \leq m$,

(a) whenever $\models \psi(b, c)$ then

$$\dim_{\mathbf{P}}((\exists x_{k+1} \ldots x_m)(\sigma(b, x_1, \ldots, x_k, x_{k+1}, \ldots, x_m, c))) = k; \text{ and}$$

(b) $\sigma(z, x_1, \ldots, x_m, y) \rightarrow x_i$ is **P**-dependent on $Azx_1 \ldots x_k y$ for $i = k + 1, \ldots, m$.

It is then easy to check that for any c, $\dim_{\mathbf{P}}(\psi(z, c)) = n$ iff 'there are $a_1, \ldots, a_k$, such that $\exists z(\sigma(z, a_1, \ldots, a_k, x_{k+1}, \ldots, x_m, c))$ is consistent, and, moreover, for any such $a_1, \ldots, a_k$, $\dim_{\mathbf{P}}((\exists z)(\sigma(z, a_1, \ldots, a_k, x_{k+1}, \ldots, x_m, c))) = n$'.

Again by Lemma 2.5, there is a formula $\delta(y)$ over A such that the latter condition is equivalent to $\models \delta(c)$. ■

We briefly connect the notion of $\dim_{\mathbf{P}}$(-) above with '**R**-weight'.

Definition 2.7. *Let **R** be a family of (stationary) regular types, closed under parallelism. **R** is said to be a* regular family *if whenever $p, q \in \mathbf{R}$, and p_1 is a forking extension of p, then p_1 is orthogonal to q.*

Example 2.8. If **R** is the family of **P**-minimal types, then **R** is a regular family.

Remark 2.9. *Suppose **R** is a regular family. Then the notions '**R**-simple type', '**R**-weight', '**R**-semi-regular' can be defined just as in Chapter 7, section 1. To be more precise:*

(i) *The strong type p is said to be orthogonal to* **R** *if p is orthogonal to q for all q in* **R**. *Similarly for p being hereditarily orthogonal to* **R**.

(ii) *Suppose p is a strong type over* $A = \mathrm{acl}(A)$ *and every* $q \in \mathbf{R}$ *is non-orthogonal to A. We say p is* **R**-*simple if for some* $B \supseteq A$ *there is a realizing* $p|B$ *and some B-independent set I of realizations of types* $r|B$ *for* $r \in \mathbf{R}$, *such that* $\mathrm{stp}(a/BI)$ *is hereditarily orthogonal to* **R**. (*So in particular* $B \supseteq \mathrm{Cb}(q)$ *for every* $r \in \mathbf{R}$ *realized by an element of I.*)

(iii) *If p is* **R**-*simple then* $w_{\mathbf{R}}(p)$, *the* **R**-*weight of p, is the minimum possible cardinality of a set I as in* (ii).

(iv) *The strong type p is said to be* **R**-*semi-regular if p is* **R**-*simple and domination equivalent to a finite product of types in* **R**.

Analogues of the results on *p*-simple types, *p*-weight, and *p*-semi-regularity which were proved in Chapter 7 hold for **R**-simple types etc. Details are left to the reader. The obvious connection with weakly **P**-minimal types and **P**-dimension is:

LEMMA 2.10. *Let* $\theta(x)$ *be a weakly* **P**-*minimal formula, defined over A. Let* **R** *be the family of* **P**-*minimal strong types containing* θ. *Let* $a \in (\theta_A)^{\mathrm{eq}}$. *Then* $\mathrm{stp}(a/A)$ *is* **R**-*simple, and, moreover,* $w_{\mathbf{R}}(\mathrm{stp}(a/A)) = \dim_{\mathbf{P}}(a/A)$.

Proof. This is a straightforward translation and is left as a good exercise for the reader. ■

The conclusion of Lemma 2.6 can now be expressed as giving the 'definability' of **R**-weight in θ^{eq} (given the assumptions on provability of **P**-analysability).

We are therefore interested in situations where **P**-analysability is provable. *T* is said to be *non-multidimensional* if every strong type is non-orthogonal to $\varnothing$. The following takes care of the non-multidimensional case:

PROPOSITION 2.11. *Suppose* **P** *is a* ($\varnothing$-*invariant*) *family of formulae, and that T is non-multidimensional. Suppose* $\mathrm{stp}(a/A)$ *is* **P**-*analysable. Then* $\mathrm{stp}(a/A)$ *is* **P**-*analysable, provably over* $\varnothing$.

Proof. Let **Q** be the family of formulae over $\varnothing$ which are **P**-analysable. We first prove:

CLAIM. *For any strong type p, p is* **P**-*analysable* iff *p is* **Q**-*analysable.*

Proof of claim. If *p* is **Q**-analysable, then by 1.3(i) every non-algebraic extension of *p* is not foreign to **Q**, and so by 1.2(iv) every non-algebraic extension of *p* is not foreign to **P**. By 1.3(i) again, *p* is **P**-analysable.

In order to prove the converse (the left to right direction), it suffices, by the same argument, to show that

SUBCLAIM. *If p is not foreign to* **P** *then p is not foreign to* **Q**.

Proof of subclaim. Let $p = \text{stp}(a/A)$. If p is not foreign to **P** then by Corollary 7.4.6 there is $a_1 \in \text{dcl}(aA) - \text{acl}(A)$ such that $\text{stp}(a_1/A)$ is **P**-internal. By non-multidimensionality, $\text{stp}(a_1/A)$ is non-orthogonal to $\varnothing$, that is there is $B \supseteq A$ independent with a_1 over A, and some element d independent with B over $\varnothing$ such that a_1 forks with d over B. In particular a_1B forks with d over $\varnothing$. Let $e \in \text{Cb}(\text{stp}(a_1B/d)) - \text{acl}(\varnothing)$. Then $e \in \text{acl}(d)$. Also $e \in \text{dcl}(a_1B_1, \ldots, a_nB_n)$, where $B_1 = B$ and a_iB_i is a Morley sequence in $\text{stp}(a_1B/d)$. Note that d is independent with $B_1 \ldots B_n$ over $\varnothing$; thus also e is independent with $B_1 \ldots B_n$ over $\varnothing$. By $\varnothing$-invariance of **P**, $\text{stp}(a_i/B_i)$ is **P**-internal for each i. By Lemma 1.2(i) and (ii), $\text{stp}(a_1, \ldots, a_n/B_1 \ldots B_n)$ is **P**-internal. Thus also $\text{stp}(e/B_1 \ldots B_n)$ is **P**-internal. Thus (as e is independent with $B_1 \ldots B_n$ over $\varnothing$) $\text{stp}(e/\varnothing)$ is **P**-internal. By Lemma 1.5 there is a formula $\varphi(x)$ in $\text{tp}(e/\varnothing)$ which is **P**-internal. So $\varphi \in$ **Q**. On the other hand, clearly e forks with a_1 over B. Thus $\text{stp}(a_1/B)$ is not foreign to φ. Clearly then $\text{stp}(a/A)$ is not foreign to φ. The subclaim is proved, and so also the remaining part of the claim. ■

Now by Lemma 1.8, whenever $\text{stp}(a/A)$ is **Q**-analysable, this holds provably over $\varnothing$. By the claim this is also true of **P**-analysability. The proposition is proved. ■

The proposition allows us to prove a version of Example 1.9(ii) for (not necessarily countable) non-multidimensional theories:

REMARK **2.12.** *Suppose T is non-multidimensional. Let R be R^∞ or RM. Suppose* $\text{stp}(a/A)$ *is* **R***-analysable. Then* $R(\text{stp}(a/A)) < \infty$, *and this holds provably over* $\varnothing$.

Proof. Let $\mathbf{R}_\varnothing$ be the set of formulae over $\varnothing$ which are **R**-internal. By fact (a) in Example 1.9, if $\varphi \in \mathbf{R}_\varnothing$, then $R(\varphi) < \infty$. Suppose now $\text{stp}(a/A)$ is **R**-analysable. By (the claim in) Proposition 2.11, $\text{stp}(a/A)$ is $\mathbf{R}_\varnothing$-analysable. By the proof of Lemma 1.8 there is a formula $\psi(x)$ in $\text{tp}(a/A)$ and formulae $\psi_i(x_1, \ldots, x_i)$ over A for $i = 1, \ldots, n$ such that

(i) $\psi(x) \to (\exists x_1, \ldots, x_{n-1})(\psi_1(x_1) \wedge \cdots \wedge \psi_n(x_1, \ldots, x_{n-1}, x))$; and

(ii) for each i, and $b_1, \ldots, b_{i-1}$, the formula $\psi_i(b_1, \ldots, b_{i-1}, x_i)$ is $\mathbf{R}_\varnothing$-internal.

We have just seen that every formula in $\mathbf{R}_{\varnothing}$ is in $\mathbf{R}$. Thus by fact (a) in Example 1.9, for each i and $b_1, \ldots, b_{i-1}$, $R(\psi(b_1, \ldots, b_{i-1}, x_i)) < \infty$.

It easily follows (by for example Theorem V.7.8 in [Sh1]) that $R(\psi(x)) < \infty$. So $R(\mathrm{stp}(a/A)) < \infty$. The fact that this holds provably over $\varnothing$ follows again from Proposition 2.11. ■

While we are on the topic of non-multidimensional theories, let us give a quick proof of

REMARK **2.13.** *Let T be non-multidimensional. Let $\varphi(x, y)$ be any formula over $\varnothing$. Then there is a formula $\delta(y)$ over $\varnothing$ such that for all b, $\models \delta(\mathrm{b})$ iff $\varphi(x, b)$ is algebraic.*

Proof. Suppose $\varphi(x, b)$ is not algebraic. Let M be an a-model containing b. Let a satisfy $\varphi(x, b)$ with $a \notin M$. By non-multidimensionality there is some element c, independent with M over $\varnothing$, such that a forks with c over M. Let $p = \mathrm{stp}(c/\varnothing)$, and let d be an element in M^{eq} such that c forks with ad over $\varnothing$. Let $\chi(x, z, c)$ be a formula true of (a, d) which forks over $\varnothing$. Let $\xi(y)$ be the formula $(\exists z)$ (for c' realizing $p|yz$, $(\exists x)(\varphi(x, y) \wedge \chi(x, z, c')))$. Then $\xi(y)$ is true of b, and clearly if $\xi(b')$ holds then $\varphi(x, b')$ is non-algebraic. Let $\Xi(y)$ be the collection of all formulae $\xi(y)$ obtained in this way (as b varies). Then for any b, $\varphi(x, b)$ is non-algebraic iff $\models \zeta(b)$ for some $\xi \in \Xi$. Compactness yields the desired formula $\delta(y)$. ■

We return now to the general question of provability of **P**-analysability. If we are concerned with the geometry of a **P**-minimal type p, we can of course assume p to be non-trivial. Somewhat surprisingly, this is enough to obtain provability of **P**-analysability inside a suitable weakly **P**-minimal formula.

LEMMA **2.14.** *Suppose θ is weakly* **P***-minimal, defined over the finite set A, and p is a* **P***-minimal strong type containing θ, which is non-trivial. Then there is a weakly* **P***-minimal formula $\varphi(x)$ over some finite set $B \supseteq A$, such that*

(i) *Whenever b satisfies $\varphi(x)$, $C \supseteq B$, and* $\mathrm{stp}(b/C)$ *is* **P***-analysable, then this holds provably over B.*

(ii) *Every* **P***-minimal strong type containing $\varphi(x)$ is non-orthogonal to p.*

(iii) *For any $C \supseteq B$ and $a \in (\varphi_C)^{\mathrm{eq}}$,* $\mathrm{stp}(a/C)$ *is p-simple and $w_p(a/C) =$* $\dim_{\mathbf{P}}(a/C)$.

Proof. By non-triviality of p we can assume after augmenting A that there are realizations a_1, a_2, a_3 of $p|\mathrm{acl}(A)$ which are pairwise A-independent, but A-dependent. Let $c \in \mathrm{Cb}(\mathrm{stp}(a_1, a_2/Aa_3))$ be such that (a_1, a_2) forks with c

over A. Then $c \in (\theta_A)^{\mathrm{eq}}$. Let $\delta(z, w)$ be a formula over A true of (c, a_3) and implying that $z \in \mathrm{acl}(A, w)$. Let $\theta_0(z)$ be a formula in $\mathrm{tp}(c/A)$ which implies that $z \in (\theta_A)^{\mathrm{eq}}$ and that $(\exists w)(\theta(w) \wedge \delta(z, w))$. It is clear that $\theta_0(z)$ is weakly **P**-minimal and that $\mathrm{stp}(c/A)$ is **P**-minimal (see 1.11(vii)).

Now each of a_1, a_2 is independent with c over A, but a_1 forks with (a_2, c) over A. Let $\chi(x, y, z)$ be a formula in $\mathrm{tp}(a_1, a_2, c/A)$ such that whenever a realizes $p|A$, then $\chi(a, y, z)$ forks over A. We may assume that $\models \chi(x, y, z) \rightarrow \theta(x) \wedge \theta(y) \wedge \theta_0(z)$.

Let $(d_1, e_1), \ldots, (d_n, e_n), \ldots$ be a Morley sequence in $\mathrm{stp}(a_1, a_2/Ac)$. Then there is an A-definable function f, such that $c = f(d_1, \ldots, d_n, e_1, \ldots, e_n)$ for some n. Note also that $\models \chi(d_i, e_i, c)$ for each $i = 1, \ldots, n$.

Note finally that $d_1, d_2, \ldots$ is a Morley sequence in $p|Ac$ (as a_1 is independent with c over A).

Let $\varphi(z)$ be the following formula (over $\mathrm{acl}(A)$): for $(x_1, \ldots, x_n)$ realizing $p^{(n)}|(Az)\ ((\exists y_1 \ldots y_n)\ (z = f(x_1, \ldots, x_n, y_1, \ldots, y_n) \wedge \bigwedge_i \chi(x_i, y_i, z))$.

By adding to A some elements in $\mathrm{acl}(A)$ we may assume φ is over A. By the above remarks φ holds of c and is weakly **P**-minimal. Note that in the proof B is formed implicitly by augmentation of A.

Proof of (i). So suppose $C \supseteq A$, $\models \varphi(c')$, and $\mathrm{stp}(c'/C)$ is **P**-analysable. Let $(d'_1, \ldots, d'_n)$ realize $p^{(n)}|(Cc')$. So there are $e'_1, \ldots, e'_n$ such that

$$c' = f(d'_1, \ldots, d'_n, e'_1, \ldots, e'_n)$$

and $\models \chi(d'_1, e'_i, c')$ for each $i = 1, \ldots, n$.

CLAIM. *For each i, d'_i forks with e'_iC over A.*

Proof. By choice of χ, d'_i forks with $e'_ic'C$ over A. Thus $\mathrm{stp}(d'_i/e'_ic'C)$ is **P**-analysable. But $\mathrm{stp}(c'/e'_iC)$ is **P**-analysable. So by transitivity, $\mathrm{stp}(d'_i/e'_iC)$ is **P**-analysable. However, d'_i realizes $p|C$ and so in particular $\mathrm{stp}(d'_i/A)$ is not **P**-analysable. Thus d'_i forks with e'_iC over A as required. ■

By the claim, choose b in C such that d'_i forks with e'_ib over A for each i. For each i, let $\psi_i(x, y, w) \in \mathrm{tp}(d'_i, e'_i, b/A)$ be such that $\psi_i(d'_i, y, w)$ forks over A.

Let $\psi(z, w)$ be the formula (over $\mathrm{acl}(A)$): $\varphi(z) \wedge$ (for $\mathbf{x}$ realizing

$$p^{(n)}|(Azw)(\exists \mathbf{y}(\textstyle\bigwedge_i \theta(y_i) \wedge \bigwedge_i \psi_i(x_i, y_i, w) \wedge z = f(\mathbf{x}, \mathbf{y})))).$$

Then $\models \psi(c', b)$. It remains to be seen that whenever $\models \psi(c'', b')$ then $\mathrm{stp}(c''/Ab')$ is **P**-analysable. Choose again $\mathbf{d}''$ realizing $p^{(n)}|Ab'c''$, and $\mathbf{e}''$ such that $c'' = f(\mathbf{d}'', \mathbf{e}'')$ and $\psi_i(d''_i, e''_i, b')$ for each i. So for each i, d''_i forks with e''_i over Ab', and thus $\mathrm{stp}(e''_i/Ab'd''_i)$ is **P**-analysable. By Lemma 1.2, $\mathrm{stp}(\mathbf{e}''/Ab'\mathbf{d}'')$ is **P**-analysable, and thus $\mathrm{stp}(c''/Ab'\mathbf{d}'')$ is **P**-analysable. But c'' is independent with $\mathbf{d}''$ over Ab'. Thus (by Corollary 1.4) stp(c''/Ab') is **P**-analysable. (i) is proved.

Proof of (ii). This is similar to (i). Suppose $\models \varphi(c')$, $C \supseteq A$, and stp(c'/C) is not **P**-analysable. Let $\mathbf{d}'$ realize $p^{(n)}|(Cc')$ and let $\mathbf{e}'$ be such that $c' = f(\mathbf{d}', \mathbf{e}')$ and $\chi(d'_i, e'_i, c')$ holds for each i. As c' is independent with $\mathbf{d}'$ over C, and $c' = f(\mathbf{d}', \mathbf{e}')$, it follows that for some i, stp(e'_i/d'_iC) is not **P**-analysable. So e'_i is independent with d'_i over C, and thus d'_i realizes $p|(Ce'_i)$. But stp$(d'_i/e'_ic'C)$ forks over A; thus d'_i forks with c' over Ce'_i. As stp(d'_i/Ce'_i) and stp(c'/C) are **P**-minimal (and a **P**-minimal type is orthogonal to any **P**-analysable type) it follows that c' is independent with e'_i over C. Thus we see that stp(c'/C) is non-orthogonal to p, as required.

Proof of (iii). Let **R** be the family of **P**-minimal strong types containing the formula $\varphi(x)$. By (ii) every $r \in \mathbf{R}$ is non-orthogonal to p. It is then clear that the notions **R**-simple, **R**-semi-regular, **R**-weight are equivalent to (or the same as) the notions p-simple, p-semi-regular, p-weight. So (iii) follows from 2.10. ■

DEFINITION **2.15.** *Let p be a regular type, and let A be a set such that any A-conjugate of p is non-orthogonal to p (so this is stronger than saying that p is non-orthogonal to A). Let θ be a formula over A. We will say that p-weight is* definable inside θ over A, if

(i) *θ is p-simple (i.e. for all a satisfying θ,* stp(a/A) *is p-simple); and*

(ii) *whenever $B \supseteq A$, $a \in (\theta_B)^{\mathrm{eq}}$ and $w_p(a/B) = n$, then there is a formula $\psi(x, y)$ over A such that $\models \psi(a, b)$ for some b in B and such that for all b', the formula $\psi(x, b')$ is either inconsistent, or p-simple with p-weight n (where the p-weight of a formula is the maximum of the p-weights of strong types containing the formula).*

REMARK **2.16.** (i) *One can make an analogous definition for* **R**-*weight being definable inside an* **R**-*simple formula, where the condition on A is replaced by '**R** is A-invariant.'*

(ii) *Suppose θ is over A and is weakly* **P**-*minimal. Suppose also whenever a realizes θ, $B \supseteq A$, and* stp(a/B) *is* **P**-*analysable, then this holds provably over A. Let* **R** *be the family of* **P**-*minimal strong types containing θ. Then* **R**-*weight is definable inside θ over A.*

(iii) *Let p, φ, and B be as given by Lemma* 2.14. *Then p-weight is definable inside φ over B.*

Proof. (ii) By Lemmas 2.6 and 2.10. Note that **R** is A-invariant.

(iii) By 2.6, and 2.14. Note that (ii) in 2.14 will ensure that any B-conjugate of p is non-orthogonal to p. ■

Here we make out first genuine assumption of superstability.

Proposition 2.17. *Suppose T is superstable, and let p be a regular non-trivial type. Let α be the least ordinal such that p is not foreign to some formula φ with $R^\infty(\varphi) = \alpha$. Let $\mathrm{stp}(a/A)$ be p-simple (necessarily of finite p-weight as T is superstable), where A is algebraically closed. Then there is some $a_1 \in \mathrm{dcl}(a, A)$, and a formula $\theta(x) \in \mathrm{stp}(a_1/A)$ over some finite A_0 contained in A, such that*

(a) *$w_p(a/a_1A) = 0$ and p-weight is definable inside $\theta(x)$ over A_0.*

(b) *For all $B \supseteq A_0$ and $a' \in (\theta_B)^{\mathrm{eq}}$, $w_p(a'/B) = 0$ iff $\mathrm{stp}(a'/B)$ is $\mathbf{R}^\infty{<}\alpha$-analysable.*

Proof. By Lemma 1.12(iii) we find a formula $\varphi(x)$ over some finite set b with $R^\infty(\varphi) = \alpha$, such that φ is weakly $\mathbf{R}^\infty{<}\alpha$:minimal, and some $\mathbf{R}^\infty{<}\alpha$-minimal strong type q contains φ and is non-orthogonal to p. Without loss of generality (as q is regular and non-triviality is preserved by non-orthogonality between regular types) $q = p$. Let $\mathbf{P}$ denote $\mathbf{R}^\infty{<}\alpha$. By Lemma 2.14 (as p is non-trivial) we may assume that φ over a finite set b satisfies the conclusions of Lemma 2.14. In particular.

(i) For any C containing b, if d satisfies φ and $\mathrm{stp}(d/C)$ is $\mathbf{P}$-analysable, then this holds provably over b; and

(ii) every $\mathbf{P}$-minimal strong type containing φ is non-orthogonal to p.

We may assume that a is independent with b over A (as p is non-orthogonal to $A = \mathrm{acl}(A)$). As $\mathrm{stp}(a/A)$ is p-simple, there is $C \supseteq Ab$ such that a is independent with C over A, and there is some set $\{c_1, \ldots, c_k\}$ of realizations of $p|C$ such that $w_p(a/Cc_1 \ldots c_k) = 0$. By augmenting b and superstability we may assume that $C = Ab$. Let c denote the tuple $(c_1, \ldots, c_k)$. Recall the version Cb_1 of canonical bases given in Remark 1.16 of Chapter 7. Let $a_1 \in \mathrm{Cb}_1(\mathrm{tp}(bc/Aa))$ be such that bc is independent with Aa over Aa_1 (using superstability). Then $a_1 \in \mathrm{dcl}(Aa)$, and also $a_1 \in \mathrm{dcl}(b^1c^1, \ldots, b^nc^n)$ where $\{b^ic^i \colon i = 1, \ldots, n\}$ is an Aa-independent set of realizations of $\mathrm{tp}(bc/Aa)$ and $b^1c^1 = bc$. Let us rewrite $\varphi(x)$ as $\varphi(x, b)$ (so we exhibit the parameters in φ). As b is independent with a over A, we see that $\{b^1, \ldots, b^n\}$ is A-independent. Let p^i be the image of p under some A-automorphism which takes b^1 to b^i. As A is algebraically closed and p is non-orthogonal to A, it follows that the p^i are pairwise non-orthogonal. Note also that each formula $\varphi(x, b^i)$ is still weakly $\mathbf{P}$-minimal for each $i = 1, \ldots, n$. Let $\psi(x) = \varphi(x, b^1) \vee \cdots \vee \varphi(x, b^n)$. Let $\mathbf{b} = b^1 \ldots b^n$. Then $\psi(x)$ is weakly $\mathbf{P}$-minimal and clearly

(iii) for any C containing $\mathbf{b}$, if d realizes $\psi(x)$ and $\mathrm{stp}(d/C)$ is $\mathbf{P}$-analysable, then this holds provably over $\mathbf{b}$;

(iv) every **P**-minimal strong type containing $\psi(x)$ is non-orthogonal to p.

As in Lemma 2.14 we also obtain

(v) for any C containing $\mathbf{b}$, and $d \in (\psi_C)^{eq}$, $\text{stp}(d/C)$ is p-simple and $w_p(d/C) = \dim_{\mathbf{P}}(d/C)$.

As in 2.16(iii) we conclude that

(vi) p-weight is definable inside ψ over $\mathbf{b}$.

Now $a_1 = f(\mathbf{b}, \mathbf{c})$ (where $\mathbf{c} = c^1 \dots c^n$), and a_1 is independent with $\mathbf{b}$ over A. Let $r = \text{tp}(\mathbf{b}/A)$ (a stationary type, as A is algebraically closed). Let $\theta(y)$ be the formula: for $\mathbf{b}'$ realizing $r|Ay$ $(\exists \mathbf{z}(\bigwedge_{i,j} \psi(z^i_j, \mathbf{b}') \wedge y = f(\mathbf{b}', \mathbf{z})))$. Then $\theta(y) \in \text{tp}(a_1/A)$. We claim that θ satisfies the requirements of the proposition. Let M be a very saturated model containing Aa_1, and such that $\mathbf{b}$ is independent with M over A. It is enough to show that p-weight is definable inside θ over A, working in the model M.

CLAIM I. *For all d in M satisfying θ,* $\text{stp}(d/A)$ *is p-simple.*

Proof. As $\mathbf{b}$ is independent with d over A, $d = f(\mathbf{b}, \mathbf{c}')$ for suitable $\mathbf{c}'$. So $d \in (\psi_{\mathbf{b}A})^{eq}$. Thus $\text{stp}(d/A\mathbf{b})$ is p-simple. As d is independent with $\mathbf{b}$ over A, $\text{stp}(d/A)$ is p-simple. ■

CLAIM II. *Suppose C, d are in M, $C \supseteq A$, $d \in (\theta_C)^{eq}$,* and $w_p(d/C) = m$. *Then there are formulae $\chi(y, w)$ and $\delta(w)$ over A such that $\chi(d, e)$ holds for some e in C satisfying δ, and such that whenever e' is in M and $\delta(e')$ holds, then $w_p(\chi(y, e')) = m$.*

Proof. Again $d \in (\psi_{C\mathbf{b}})^{eq}$, and as $\mathbf{b}$ is independent with dC over A, we have that $w_p(d/C) = w_p(d/C\mathbf{b}) = m$. By (vi) above there are formulae $\chi_1(y, w, \mathbf{b})$ and $\delta_1(w, \mathbf{b})$ over A, such that for some e in C, $\models \chi_1(d, e, \mathbf{b}) \wedge \delta_1(e, \mathbf{b})$ and such that whenever e' satisfies $\delta_1(w, \mathbf{b})$ then $w_p(\chi_1(y, e', \mathbf{b})) = m$.

Let $\chi(y, w)$ say: 'for $\mathbf{b}'$ realizing $r|Ayw$ $(\chi_1(y, w, \mathbf{b}'))$', and similarly for $\delta(w)$. Then $\models \chi(d, e) \wedge \delta(e)$, and clearly these formulae satisfy the claim. ■

Claims I and II show that p-weight is definable inside θ over A. From (v) above it easily follows by similar arguments that

CLAIM III. *Whenever C, d are in M, $C \supseteq A$, and $d \in (\theta_C)^{eq}$, then $w_p(d/C) = 0$ iff* $\text{stp}(d/C)$ *is* **P**-*analysable.*

As a_1 satisfies θ and $w_p(a/a_1A) = 0$ (since $w_p(a/Abc) = 0$ and a is independent from bc over Aa_1), we have proved (a) and (b) of the proposition with A instead of A_0. It is left as an exercise for the reader to find a suitable finite subset A_0 of A. ■

COROLLARY **2.18.** *Suppose T is superstable, p is a non-trivial regular type, and p is non-orthogonal to $A = \mathrm{acl}(A)$. Suppose again that α is least such that p is not foreign to some formula with ∞-rank α. Then there is a formula $\theta(x)$ over a finite subset A_0 of A, which is p-simple of non-zero p-weight, such that*

(a) *p-weight is definable inside θ over A_0.*

(b) *For any $B \supseteq A_0$ and $c \in (\theta_B)^{\mathrm{eq}}$, $w_p(c/B) = 0$ iff $\mathrm{stp}(c/B)$ is $\mathbf{R}^\infty{<}\alpha$-analysable.*

Proof. Let $\mathrm{stp}(a/A)$ be non-orthogonal to p. By Lemma 7.1.17 there is $a' \in \mathrm{dcl}(aA)$ such that $\mathrm{stp}(a'/A)$ is p-simple of non-zero p-weight. Now apply Proposition 2.17 to a' and A to obtain a_1 and formula θ. Note that $w_p(a_1/A) > 0$, so θ has non-zero p-weight. ■

We now give an easy result valid for even trivial regular types, which is implicit in the proof of 2.17.

LEMMA **2.19.** *Suppose T superstable. Let p be a regular type, and suppose $\mathrm{stp}(a/A)$ is non-orthogonal to p where $A = \mathrm{acl}(A)$. Then there is $a_1 \in \mathrm{dcl}(aA)$ and a p-simple formula $\varphi(z)$ in $\mathrm{tp}(a_1/A)$ such that $w_p(a_1/A) \neq 0$ (so $\mathrm{stp}(a_1/A)$ is non-orthogonal to p).*

Proof. We may assume (by say 1.12(iii)) that for some ordinal α, p is $\mathbf{R}^\infty{<}\alpha$-minimal and contains a weakly $\mathbf{R}^\infty{<}\alpha$-minimal formula $\psi(x, b)$. Let $\mathbf{P} = \mathbf{R}^\infty{<}\alpha$. As p is non-orthogonal to A we may assume that b is independent with a over A. We may choose B containing A such that $b \in B$, a is independent with B over A, and for some c realizing $p|B$, c forks with a over B. Without loss of generality $B = A \cup \{b\}$. Let $a_1 \in \mathrm{Cb}_1(\mathrm{tp}(cb/aA)) - \mathrm{acl}(A)$. Then (as in 7.1.17) $a_1 \in \mathrm{dcl}(c_0b_0, \ldots, c_nb_n)$ where $c_0b_0 = cb$ and $\{c_ib_i\}$ is an Aa-independent set of realizations of $\mathrm{tp}(cb/aA)$. Let f be a $\emptyset$-definable function such that $a_1 = f(c_0b_0, \ldots, c_nb_n)$. Note that $(b_0, \ldots, b_n)$ is independent with a over A. Let $q = \mathrm{tp}(b_0, \ldots, b_n/A)$. Let $\varphi(z)$ say: 'for $(b'_0, \ldots, b'_n)$ realizing some non-forking extension of q over Az, $(\exists c'_0 \ldots c'_n)(\bigwedge \psi(c'_i, b'_i) \wedge z = f(c'_0b'_0, \ldots, c'_nb'_n))$'.

Note that $\varphi(z)$ is satisfied by a_1, and, moreover, that if $(b'_0, \ldots, b'_n)$ realizes q and $\models \psi(c'_i, b'_i)$ for each i, then for each i $\mathrm{stp}(c'_i/\mathbf{b}'A)$ is either regular and non-orthogonal to p or herditarily orthogonal to p. It clearly follows that whenever a'_1 satisfies $\varphi(z)$, then $\mathrm{stp}(a'_1/A)$ is p-simple.

So φ is as required. ■

Finally we show the existence of elements with p-weight 1, in between models. This will be used in the next section.

Lemma 2.20. *Assume T is superstable. Let $M < N$ be models with $M \neq N$, Let p be some non-trivial regular type which is non-orthogonal to* $\mathrm{tp}(N/M)$. *Then there is a formula θ over M such that p-weight is definable inside θ over M, $w_p(\theta) = 1$, and there is some $c \in N$* such that $N \models \theta(c)$, and $w_p(c/M) = 1$. *Moreover, if α is least such that p is not foreign to a formula with ∞-rank α, we may choose θ such that inside $(\theta_M)^{\mathrm{eq}}$, having p-weight 0 means being $\mathbf{R}^\infty < \alpha$-analysable.*

Proof. Let α be as in the moreover clause.

By Lemma 7.1.17 there is $a_1 \in N$, such that $\mathrm{tp}(a_1/M)$ is p-simple with non-zero p-weight. By 2.17 above, we find $a_2 \in \mathrm{dcl}(a_1, M)$ (so $a_2 \in N$), and p-simple formula θ over M, such that θ satisfies the moreover clause, $N \models \theta(a_2)$, and $w_p(a_2/M) > 0$. The only thing necessary is to find such an element with p-weight 1 over M. Let M' be a saturated model containing M which is independent from a_2 over M. It is easy to find a *regular* type $q \in S(M')$ which contains θ and is non-orthogonal to p. Hence $w_p(q) = 1$. Then q is non-orthogonal to $\mathrm{tp}(a_2/M')$. As M' is saturated, there is (by 1.4.3.1), some c_1 realizing q such that c_1 forks with a_2 over M'. Note that $w_p(c_1/M') = 1$, $w_p(c_1/a_2, M') = 0$, and hence $w_p(a_2/c_1, M') < w_p(a_2/M')$ ($=n$, say). As p-weight is definable inside θ over M, we can find a formula $\chi(x, y, z)$ over M, and some $b \in M'$, such that $\models \chi(a_2, c_1, b)$, such that

($*$) for any a'_2, c'_1, b', if $\models \chi(a'_2, c'_1, b')$ *then* $w_p(c'_1/M) \leq 1$, and

$$w_p(a'_2/c'_1, b', M) < n.$$

Now we have $\models \exists y(\chi(a_2, y, b))$. As a_2 is independent from b over M, and M is a model, $\mathrm{tp}(b/Ma_2)$ is finitely satisfiable in M, so we have some $b' \in M$ such that $\models \exists y(\chi(a_2, y, b'))$. As N is a model (and $a_2 \in N$) there is $c \in N$, such that $\models \chi(a_2, c, b')$.

By ($*$), $w_p(a_2/c, M) < n = w_p(a_2/M)$. The p-weight equality (7.1.11) implies that $w_p(c/a_2, M) < w_p(c/M)$. As $w_p(c/M) \leq 1$ (by (*), it follows that $w_p(c/M) = 1$.

Now the original θ need not have p-weight 1. But as c satisfies θ, $w_p(c/M) = 1$, and p-weight is definable inside θ over M, we may choose $\theta' \in \mathrm{tp}(c/M)$ such that $\models \theta' \to \theta$, and $w_p(\theta') = 1$. So also p-weight is definable inside θ' over M.

3 Existence of regular types

The aim in this section is to prove results of the form: given $M < N$ models of the superstable theory T, there is $a \in N - M$ such that $\mathrm{tp}(a/M)$ is regular. Lemma 1.4.5.6 gives this if M and N are a-models. The definability results of section 2 will be used.

We first begin with an easy remark concerning *trivial* regular types, and not requiring superstability.

LEMMA 3.1. *Let p be a trivial regular type. Suppose* $\mathrm{stp}(a/B)$ *is non-orthogonal to p, but orthogonal to all forking extensions of p. Then there is $a_1 \in \mathrm{acl}(aB)$ such that* $\mathrm{stp}(a_1/B)$ *is regular and non-orthogonal to p.*

Proof. We begin by 'redoing' the proof of Lemma 7.1.17 in a rather more straightforward way. Let $C \supseteq B$ be such that a is independent with C over B, p is definable over C, and $p|C$ is not weakly orthogonal to $\mathrm{stp}(a/C)$. Let c realize $p|C$ such that a forks with c over C. Let $a_1 \in \mathrm{Cb}(\mathrm{stp}(cC/aB))$ be such that a_1 forks with c over C. $a_1 \in \mathrm{acl}(aB)$. Also $a_1 \in \mathrm{dcl}(c_1C_1, \ldots, c_nC_n)$ where $c_1C_1 = cC$ and $(c_iC_i: i < \omega)$ is a Morley sequence in $\mathrm{stp}(cC/aB)$. The proof of 7.1.19 already shows that $\mathrm{stp}(a_1/B)$ is p-semi-regular. (This is where we use the hypothesis that $\mathrm{stp}(a/B)$ is orthogonal to all forking extensions of p.) Note also that $\mathrm{stp}(a_1/B)$ has p-weight > 0. So it suffices, by Corollary 7.1.15, to show that $w_p(a_1/B) \leq 1$. Let $D = \cup\{C_i: i < \omega\}$.

Now as C is independent with a over B it follows that

(i) $\{C_i: i < \omega\}$ is B-independent, and a_1 is independent with D over B.

(ii) For each m, $c_1. \ldots, c_m$ is independent with D over $C_1 \ldots C_m$.

Also, as p is non-orthogonal to B and $\mathrm{stp}(C_i/B) = \mathrm{stp}(C/B)$ for each i, it follows that

(iii) For each i, $\mathrm{stp}(c_i/C_i)$ and thus also $\mathrm{stp}(c_i/D)$ is regular and non-orthogonal to p.

CLAIM. *For each $i = 1, \ldots, n$, c_i forks with c_1 over $C_1 \ldots C_n$.*

Proof of claim. As cC forks with aB over B, and (c_iC_i) is a Morley sequence in $\mathrm{stp}(cC/aB)$, $\{c_iC_i: i < \omega\}$ cannot be B-independent (note that this does *not* need superstability). So by (i) and (ii) above $\{c_i: i < \omega\}$ is not D-independent. By (iii) and Corollary 7.2.9, for each i there is b_i realizing $p|D$ such that b_i forks with c_i over D. So (by regularity of the types of b_i and c_i over D), we conclude that $\{b_i: i < \omega\}$ is not D-independent. Triviality of p then implies that $\{b_i: i < \omega\}$ is pairwise D-dependent: say b_1 forks with b_2 over D. Regularity of the b_i and c_i again implies that c_1 forks with c_2 over D. By (ii) above, then c_1 forks with c_2 over C_1C_2. Indiscernibility of the sequence (c_iC_i) implies that c_i forks with c_k over C_iC_k for all $i \neq k$. Thus c_i forks with c_1 over $C_1 \dots C_n$ for all $i = 1, \dots, n$. The claim is proved. ■

By the claim and (iii) above, $w_p(c_1, \dots, c_n/C_1 \dots C_n) = 1$. But

$$a_1 \in \mathrm{dcl}(c_1C_1 \dots c_nC_n).$$

Thus $w_p(a_1/C_1 \dots C_n) \leq 1$. As a_1 is independent with $\cup C_i$ over B (by (i) above), $w_p(a_1/B) \leq 1$. As remarked already, this suffices to prove that $\mathrm{stp}(a_1/B)$ is regular and non-orthogonal to p. ■

REMARK. *In Lemma* 3.1, *we cannot demand that* $a_1 \in \mathrm{dcl}(aB)$, *even if* B *is algebraically closed. For example, consider the theory of an infinite set with no additional structure.* $\emptyset$ *is algebraically closed in* $\mathbf{C}^{\mathrm{eq}}$. *Let* a_1, a_2 *be distinct elements of the home sort. Let* $a = \{a_1, a_2\}$ *(as an object in* $\mathbf{C}^{\mathrm{eq}}$*). Then* $a_1 \in \mathrm{acl}(a)$, *but there is no element of rank* 1 *in* $\mathrm{dcl}(a)$.

PROPOSITION **3.2.** *Suppose* T *to be superstable. Let* $M < N$ *be models of* T $(M \neq N)$. *Then there is* $e \in N$ *such that* $\mathrm{tp}(e/M)$ *is regular.*

Proof. Note first that Lemma 1.12(iii) is still valid if q is a complete stationary type in infinitely many variables. Thus Lemma 1.12(iii) yields some regular type p, with $R^\infty(p) = \alpha$, such that $\mathrm{tp}(N/M)$ is non-orthogonal to p, but $\mathrm{tp}(N/M)$ is foreign to $\mathbf{R}^\infty < \alpha$. In particular $\mathrm{tp}(N/M)$ is orthogonal to all forking extensions of p. If p is trivial, then Lemma 3.1 yields some a_1 in N such that $\mathrm{tp}(a_1/M)$ is regular (and non-orthogonal to p). So we may assume p to be non-trivial. Note that p also is foreign to $\mathbf{R}^\infty < \alpha$.

Let $\theta(x)$ and $c \in N$ be as given (for this choice of p and α) by Lemma 2.20.

CLAIM. $\mathrm{tp}(c/M)$ *is regular (and hence non-orthogonal to* p*).*

Proof of claim. If $e \in \mathrm{dcl}(cM) - M$ then $e \in (\theta_M)^{\mathrm{eq}}$ and $e \in N - M$, so by the properties of θ (see 2.18), $w_p(e/M) = 0$ iff $\mathrm{stp}(e/M)$ is $\mathbf{R}^\infty < \alpha$-analysable. But $\mathrm{stp}(N/M)$ is foreign to $\mathbf{R}^\infty < \alpha$, and so by Lemma 1.2(iv), $w_p(e/M) \geq 1$. By Lemma 7.1.18, $\mathrm{tp}(c/M)$ is p-semi-regular. The claim then follows from Corollary 7.1.15. ■

The proposition is proved.

We now give some slight 'extensions' of Proposition 3.2. The usual inclusion of models will be replaced by a stronger relation: '*na*-inclusion'. The benefit of doing this is that one can show (in the superstable case) that whenever M *is na*-included in N, and p is *any* regular type non-orthogonal to $\mathrm{tp}(N/M)$ then there is a regular type $q \in S(M)$ which is non-orthogonal to p and realized in N. This has various consequences, among them being that if T is superstable with NDOP then any model of T is minimal over a non-forking tree of small models, where the inclusions are *na*-inclusions. We will not, however, pursue such things here.

DEFINITION 3.3. *We say that M is na-included in N, written $M <_{na} N$, if $M < N$ and whenever θ is a formula over a finite subset A of M and θ is realized in $N - M$ then θ^M is not included in* $\mathrm{acl}(A)$.

The following is valid in any theory (stable or not) and is proved by a routine downward Lowenheim–Skolem theorem type argument.

REMARK 3.4. *If $M < N$, then there is M' such that $M < M' <_{na} N$ and $|M'| \leq |M| + |T|$.*

PROPOSITION 3.5. *Suppose T superstable and $M <_{na} N$. Let p be any regular type such that* $\mathrm{tp}(N/M)$ *is non-orthogonal to p. Then there is $b \in N - M$ such that* $\mathrm{tp}(b/M)$ *is regular and non-orthogonal to p. Moreover, there is a (p-simple) formula φ in* $\mathrm{tp}(b/M)$ *such that whenever $b' \in N - M$ realizes φ and* $\mathrm{tp}(b'/M)$ *is non-orthogonal to p then* $\mathrm{tp}(b'/M)$ *is regular.*

Proof. We separate into cases.

Case (i). p is trivial. The first step is to find a p-simple formula φ over M such that

$(*)$ for some a_1 in N, $\models \varphi(a_1)$, and $\mathrm{tp}(a_1/M)$ is non-orthogonal to p.

This is accomplished by Lemma 2.19.

The next step is to choose such a formula φ satisfying $(*)$ and with $R^\infty(\varphi)$ minimized. We will show that φ satisfies the requirements. Let $\alpha = R^\infty(\varphi)$. Let $a \in N$ satisfy φ, and suppose $\mathrm{tp}(a/M)$ is non-orthogonal to p. Thus $w_p(a/M) = n$ for some $n > 0$. We have to show that $\mathrm{tp}(a/M)$ is regular.

CLAIM I. *There is $c \in \mathrm{acl}(aM)$ such that* $\mathrm{stp}(a/cM)$ *is regular and non-orthogonal to p.*

Proof of Claim I. Suppose $w_p(a/M) = n > 0$. By Corollary 7.1.20 and superstability, there is $c_1 \in \mathrm{dcl}(aM)$ such that $\mathrm{stp}(a/c_1M)$ is p-semi-regular with p-weight n. If $n = 1$ we are finished. Otherwise, by Lemma 3.1, there is $a_1 \in \mathrm{acl}(ac_1M)$ such that $w_p(a_1/c_1M) = 1$. So by additivity of p-weight, $w_p(a/a_1c_1) = n - 1 > 0$. Iterating this procedure gives us the required c. ■

CLAIM **II.** *c is in M (c given by Claim I).*

Proof of Claim II. Suppose not. Let q_c denote $\mathrm{stp}(a/cM)$. Let d realize $\mathrm{tp}(c/M)$ with d independent with ac over M. Let f be an M-automorphism taking c to d and let $q_d = \mathrm{stp}(f(a)/f(c), M)$. Then q_d is a trivial regular type. As q_c is non-orthogonal to M, q_c is non-orthogonal to q_d. By Corollary 7.2.9, $q_c|Mcd$ is non-weakly orthogonal to $q_d|Mcd$. Thus there is b realizing $q_d|Mcd$ such that a forks with b over Mcd. Note that then a forks with bd over Mc. Let $e \in M$ be such that φ is over e and $c \in \mathrm{acl}(ae)$. Let $\chi(w, z, x, y)$ be a formula over M such that

(a) $\models \chi(w, z, x, y) \to (\varphi(x) \wedge$ '$y \in \mathrm{acl}(xe)$');

(b) $\chi(a, c, x, y)$ forks over Mc; and

(c) $\models \chi(a, c, b, d)$.

Let $\delta(y)$ be a formula over M saying 'for (a', c') realizing $\mathrm{tp}(a, c/M)|My$, $(\exists x)(\chi(a', c', x, y))$'.

So $\delta(y)$ is satisfied by d, and thus also by c. As we are supposing $c \notin M$, the fact that $M <_{na} N$ implies that there is $d' \in M - \mathrm{acl}(e)$ such that $\models \delta(d')$. Now trivially (a, c) realizes $\mathrm{tp}(a, c/M)|Md'$. Thus there is b' in N such that $\models \chi(a, c, b', d')$. So $d' \in \mathrm{acl}(b'e) - \mathrm{acl}(e)$. In particular b' forks with M over e, whereby $R^\infty(\mathrm{tp}(b'/M)) < R^\infty(\mathrm{tp}(b'/e)) \leq R^\infty(\varphi) = \alpha$. So there is a formula ψ over M true of b' such that $R^\infty(\psi) < \alpha$ and $\models \psi \to \varphi$. In particular ψ is also p-simple.

On the other hand (b) above shows that b' forks with a over Mc. As $\mathrm{stp}(a/cM)$ is regular, non-orthogonal to p, it follows that $w_p(b'/M) \neq 0$. ψ and b' then give a contradiction to the choice of φ minimizing α. This contradiction proves Claim II. ■

Thus from Claim I, $\mathrm{tp}(a/M)$ is regular, non-orthogonal to p. The proposition is proved in the case where p is trivial.

Case (ii). p is non-trivial. The strategy is first to find a p-simple formula φ over M which has p-weight 1 and is realized in N by some a with $w_p(a/M) \neq 0$, and then to minimize $R^\infty(\varphi)$ and show by an argument like that of Case(i) that φ works.

The first part is given by Lemma 2.20.

Now minimize $R^\infty(\varphi)$. That is, choose formula φ over M with least possible ∞-rank subject to: φ is p-simple of p-weight 1 and there is $a \in N$ satisfying φ with $w_p(a/M) = 1$. Suppose $R^\infty(\varphi) = \alpha$.

We will show that φ works.

Let a be any realization of φ in N such that $w_p(a/M) \neq 0$. So $w_p(a/M) = 1$, and by Corollary 7.1.20 and superstability there is $c \in \mathrm{dcl}(aM)$ such that $\mathrm{stp}(a/cM)$ is p-semi-regular of p-weight 1 (and so regular), and $w_p(c/M) = 0$.

CLAIM. *c is in M.*

Proof of claim. This is like the proof of Claim II in Case (i) above.

Suppose for a contradiction that $c \notin M$. Let $e \in M$ be such that φ is over e and $f(a, e) = c$ for some $\varnothing$-definable function. Let M' be a big saturated model containing M such that M' is independent with N over M. In particular

(i) ac is independent with M' over M.

We may assume p to be over M'. So as $w_p(ac/M') = 1$ there is e' realizing $p|M'$ such that e' forks with ac over M'. As $w_p(c/M) = 0$, we see from (i) that

(ii) c is independent with $e'M'$ over M.

Let d realize the non-forking extension of $\mathrm{tp}(c/M)$ over $M'e'ac$. So by (ii) $\mathrm{tp}(de'/M') = \mathrm{tp}(ce'/M')$. Thus we can find b such that $\mathrm{tp}(bde'/M') = \mathrm{tp}(ace'/M')$. A p-weight computation now shows that bd forks with ac over M'. On the other hand, by choice of d, and as $w_p(c/M') = 0$ and $\mathrm{stp}(b/dM')$ is p-semi-regular, c is independent with bd over M', and thus by (i), c is independent with bdM' over M. Thus we have

(iii) ac forks with bdM' over Mc.

So we can find a formula $\chi(w, z, x, y, u)$ over M such that

(a) $\models \chi(a, c, b, d, e'')$ for some e'' in M';

(b) $\models \chi(w, z, x, y, u) \rightarrow \varphi(x) \wedge y = f(x, e)$;

(c) $\chi(a, c, x, y, u)$ forks over Mc.

So $\models (\exists x)(\chi(a, c, x, d, e''))$.

By (i) and the choice of d, acd is independent with M' over M, and thus, for some e''' in M, $\models (\exists x)(\chi(a, c, x, d, e'''))$.

Now ac is independent with d over M, so as in the proof of Claim II in Case (i) above, we can find d' in $M - \mathrm{acl}(e)$ such that $\models (\exists x)(\chi(a, c, x, d', e'''))$.

Let b' in N be such that $\models \chi(a, c, b', d', e''')$. Then $d' = f(b', e)$ and $d' \notin \mathrm{acl}(e)$ so d' forks with b' over e. So there is a formula ψ in $\mathrm{tp}(b'/M)$ such that $\models \psi \to \varphi$ and $R^\infty(\psi) < \alpha$. On the other hand by (c) above b' forks with ac over Mc and thus easily $w_p(b'/M) \neq 0$. ψ and b' now contradict the minimal choice of φ. This contradiction proves the claim. ■

The claim shows that $\mathrm{tp}(a/M)$ is regular, non-orthogonal to p, as required. This completes the proof of the proposition. ■

We complete this section with two '3-model theorems' concerning *na*-extensions.

Proposition 3.6. *Suppose T superstable. Let $M_0 < M_1 < M_2$ be models such that $M_0 <_{na} M_2$. Suppose that $a_1 \in M_2$, and $p = \mathrm{tp}(a_1/M_1)$ is regular and non-orthogonal to M_0. Then there is a_2 in M_2 such that a_2 forks with a_1 over M_1, and $\mathrm{tp}(a_2/M_1)$ is regular and does not fork over M_0.*

Proof. We first claim that $\mathrm{tp}(a_1 M_1/M_0)$ is non-orthogonal to p. For this, let θ be a p-simple formula over M_0 of non-zero p-weight. (θ is given by 2.19, for example.) Now there is $M_3 \supseteq M_1$ and c' satisfying θ, with a_1 independent from M_3 over M_1, and a_1 forking with c' over M_3. Then a_1 forks with $M_3 c'$ over M_1 so (by symmetry) there are $\mathbf{d}$ in M_3 and formula $\chi(x, \mathbf{y}, z) \wedge \theta(x) \in \mathrm{tp}(c', \mathbf{d}, a_1/M_1)$ such that $\chi(x, \mathbf{y}, a_1) \wedge \theta(x)$ forks over M_1. But $\mathrm{tp}(\mathbf{d}/M_1 a_1)$ is finitely satisfiable in M_1, and hence there is $\mathbf{d}'$ in M_1, and c, such that $\chi(c, \mathbf{d}', a_1) \wedge \theta(c)$. So a_1 forks with c over M_1. But then, as $\mathrm{tp}(a_1/M_1)$ is regular $w_p(a_1/cM_1) = 0 < w_p(a_1/M_1)$. So $w_p(c/a_1 M_1) < w_p(c/M_1) \leq w_p(c/M_0)$. It is then easy to conclude that $\mathrm{tp}(a_1 M_1/M_0)$ is non-orthogonal to p. (The relevant observation here is that if $\mathrm{stp}(d/A)$ is p-simple of p-weight n and $\mathrm{stp}(e/A)$ is orthogonal to p, then $w_p(d/Ae) = n$ too.)

So by Proposition 3.5 there is a p-simple formula ψ over M_0 such that for some a_2 satisfying ψ in M_2, $\mathrm{tp}(a_2/M_0)$ is regular and non-orthogonal to p, and such that, moreover, whenever $a_2 \in M_2$ satisfies ψ and $\mathrm{stp}(a_2/M_0)$ is non-orthogonal to p, then $\mathrm{tp}(a_2/M_0)$ is regular. As in the previous paragraph we can find a_2 in M_2 realizing ψ such that a_2 forks with a_1 over M_1. But then $w_p(a_2/M_1) \geq 1$, whereby $w_p(a_2/M_0) \geq 1$. Thus $\mathrm{tp}(a_2/M_0)$ is regular, and so a_2 is independent with M_1 over M_0. ■

Corollary 3.7. *Suppose T superstable. Let $M_0 <_{na} M_1 <_{na} M_2$. Suppose $a \in M_2$ and $\mathrm{tp}(a/M_1)$ is non-orthogonal to M_0. Then there is $b \in M_2$ such that $\mathrm{tp}(b/M_1)$ is regular and does not fork over M_0, and is non-orthogonal to $\mathrm{tp}(a/M_1)$.*

Proof. By superstability there is some regular type p such that $\mathrm{tp}(a/M_1)$ is non-orthogonal to p, and p is non-orthogonal to M_0. Now apply Propositions 3.5 and 3.6. ■

4 NDOP and types of depth 0

In this section we give some additional results on provability and definability of **P**-analysability/p-weight under various assumptions. In section 2, we proved, among other things, that if T is non-multidimensional, **P** is a $\varnothing$-invariant family of formulae, and $\mathrm{stp}(a/B)$ is **P**-analysable, then this holds provably over $\varnothing$. We will first prove in Proposition 4.6 a 'local' version of this result, under weaker hypotheses. The result is not essential for subsequent results but has a nice proof.

In this section T is assumed to be superstable.

Definition 4.1. *T is said to have the dimensional order property* (DOP) *if there are a-models M_0, M_1, M_2, N and a type $p \in S(N)$ such that*

(i) *$M_0 < M_i$ for $i = 1, 2$ and M_1 is independent with M_2 over M_0.*

(ii) *N is a-prime over $M_1 \cup M_2$.*

(iii) *p is orthogonal to M_i for $i = 1, 2$.*

Remark 4.2. *The definition above would be equivalent if we required p to be regular.*

DOP is one of the important 'non-structure' properties: a superstable theory with DOP has the maximum possible number of models (a-models) in cardinalities $> |T|$.

Definition 4.3. (i) *Let M be an a-model and $p \in S(M)$ a regular type. We say p has* depth 0, *if there* do not *exist $N > M$ and non-algebraic $q \in S(N)$ such that*

(a) *N is a-prime over $M \cup \{a\}$ for some realization a of p;*

(b) *q is orthogonal to M.*

(ii) *An arbitrary regular type p is said to have depth* 0 *if some stationarization of p over some a-model has depth* 0.

Fact 4.4. *Suppose p is a regular type of depth* 0, *and* $\mathrm{stp}(b/B)$ *is domination equivalent to a power of p. Then any non-algebraic strong type which is non-orthogonal to bB is non-orthogonal to B.*

Proof. Note first that (by superstability) for any strong type r and any set C, r is non-orthogonal to C if and only if there is a regular type r_1 which is non-orthogonal to r and non-orthogonal to C. So it is enough to prove that any *regular* type r which is non-orthogonal to bB is non-orthogonal to B. Let r be such. Assume by way of contradiction that r is orthogonal to B. By 1.4.3.3 there is an a-model $M \supseteq B$ such that b is independent from M over B and *r is orthogonal to M*. We may assume p is over M. Let N be the a-prime model over $M \cup \{b\}$. By 1.4.5.4, there is an M-independent set $\{a_1, \ldots, a_m\}$ of realizations of $p|M$ such that N is a-prime over $M \cup \{a_1, \ldots, a_m\}$. Let $M_1 < \cdots < M_m < N$ be such that M_1 is a-prime over $M \cup \{a_1\}, \ldots, M_m$ is a-prime over $M_{m-1} \cup \{a_m\}$. Then M_m is also a-prime over $M \cup \{a_1, \ldots, a_m\}$, so by 1.4.2.4 is isomorphic to N over $M \cup \{a_1, \ldots, a_m\}$. We may thus assume that $M_m = N$. By 1.4.5.5 we may assume r is over N. Note that $\text{tp}(a_i/M_{i-1}) = p|M_{i-1}$ for $i = 1, \ldots, m$. By the fact that p has depth 0, we see from Definition 4.3 that r is non-orthogonal to M_{m-1}, so (as the latter is an a-model) we may (by replacing r by a regular type non-orthogonal to it), assume r is over M_{m-1}. Continuing this way we end up seeing that r is non-orthogonal to M, a contradiction. This proves the fact. ■

FACT **4.5.** *Suppose T has* NDOP (*i.e. does not have* DOP). *Then any regular non-trivial type has depth* 0.

Proof. Suppose that the regular type p is non-trivial but of depth > 0. We may assume $p \in S(M)$ where M is an a-model, and for a realizing p, there is non-algebraic $q \in S(M[a])$, such that q is orthogonal to M. As p is non-trivial (and M is an a-model) there are realizations a_1, a_2, a_3 of p which are pairwise M-independent, but M-dependent. Let M_1 be a-prime over $M \cup \{a_1\}$, M_2 be a-prime over $M \cup \{a_2\}$. So M_1 is independent from M_2 over M. Let N be a-prime over $M_1 \cup M_2$. Let $A \subseteq M$ be finite such that $\text{tp}(a_1, a_2, a_3/M)$ does not fork over A. Let $a_3' \in N$ realize $\text{stp}(a_3/a_1, a_2, A)$. Then $\text{tp}(a_3'/M) = p$, and $\{a_1, a_2, a_3'\}$ is pairwise M-independent, but M-dependent. Let $M_3 < N$ be an a-prime model over $M \cup \{a_3'\}$. Let q' be a non-algebraic type over M_3 which is orthogonal to M. Note that each of M_1, M_2 is independent from M_3 over M. By 1.4.3.3, q' is orthogonal to M_1 and to M_2. The same is true for q'', the non-forking extension of q' over N. But then T has the DOP. ■

PROPOSITION **4.6.** *Suppose T has NDOP (i.e. does not have DOP). Let* **P** *be some $\varnothing$-invariant family of partial types, θ a weakly* **P**-*minimal formula, and $p \in S(A)$ a* **P**-*minimal strong type containing θ, where A is algebraically closed. Suppose p has depth* 0. *Then any forking extension of p is* **P**-*analysable, provably over A.*

Proof. We will first prove the proposition in the case where A is an a-model M. Let Θ be the family of formulae over M which are $\mathbf{P}$-analysable. We start with

LEMMA. *Let* $\mathrm{stp}(a/bM)$ *be a non-algebraic forking extension of* p. *Then* $\mathrm{stp}(a/bM)$ *is not foreign to* Θ.

Proof of lemma. Let N be an a-model containing Mb such that a is independent with N over Mb. Let A_0 be a finite subset of M such that p is definable over A_0 (A_0 can be found as M is an a-model). We can assume that θ is over A_0. Let $p_0 = p|A_0$. Let I be a basis for p in N, that is I is a maximal M-independent set of realizations of p in N. Let $C = \{c \in N\colon \mathrm{tp}(c/M)$ is hereditarily orthogonal to $p\}$. Note that I is independent with C over M. Finally let $M_1 < N$ be a-prime over $M \cup I$ and let $M_2 < N$ be a-prime over $M \cup C$. Let $M_3 < N$ be a-prime over $M_1 \cup M_2$. Note that M_1 is independent with M_2 over M.

CLAIM I. *Every realization of* p_0 *in* N *is contained in* M_3.

Proof of Claim I. Note first

(*) if $a_1 \in N$ realizes p_0, and $\mathrm{tp}(a_1/M) \neq p$, then $a_1 \in C$.

Suppose by way of contradiction that $a_1 \in N$ realizes p_0 and $a_1 \notin M_3$. From NDOP and superstability we get $a_2' \in N$ such that a_2' forks with a_1 over M_3, $\mathrm{tp}(a_2'/M_3)$ is regular, and for some $i = 1, 2$, $\mathrm{tp}(a_2'/M_3)$ does not fork over M_i. By a standard 'a-model' argument there is $\beta \in M_i[a_2'] - M_i$ realizing p_0. Using the proof of 1.4.5.6 we can assume that $\mathrm{tp}(\beta/M_i)$ is regular. Note then that $\beta \in N$, β realizes p_0, β forks with a_1 over M_3, and β is independent with M_3 over M_i. Summarizing, we have found $a_2 \in N$ such that a_2 forks with a_1 over M_3, a_2 realizes p_0, $\mathrm{tp}(a_2/M_3)$ is regular, and either

(a) a_2 is independent with M_3 over M_1, or

(b) a_2 is independent with M_3 over M_2.

By (*) $\mathrm{tp}(a_2/M) = p$. Thus a_2 is independent with C and so also with M_2 over M. So if a_2 were independent with M_3 over M_2, then a_2 would be independent with I over M, contradicting maximality of I. Thus (a) above holds, that is a_2 is independent with M_3 over M_1. Now $\mathrm{tp}(M_1/M)$ is domination equivalent to an (infinite) power of p. So by Fact 4.4, $\mathrm{tp}(a_2/M_1)$ (and thus also $\mathrm{tp}(a_2/M_3)$) is non-orthogonal to M. Thus, as before we can find a_3 in N such that a_3 forks with a_2 over M_3, a_3 realizes p_0, and a_3 is independent with M_3 over M. By $(*)$ again $\text{tp}(a_3/M) = p$, and we again contradict the maximality of I. This proves Claim I. ■

By Claim I, the fact that N is an a-model (and 1.4.5.9(ii)), we conclude that $\text{tp}(a/N)$ does not fork over M_3. So to prove the lemma it suffices to prove that $\text{tp}(a/M_3)$ is not foreign to Θ.

Let N_1 be a-prime over $M_3 \cup \{a\}$.

CLAIM **II**. *There is a_1 in N_1 such that a_1 forks with a over M_3, a_1 realizes p_0, and* $\text{tp}(a_1/M) \neq p$.

Proof. Note first that as $\text{tp}(a/M_3)$ is orthogonal to p, $p|M_3$ is not realized in N_1 and so I remains a basis of p in N_1. Using this remark, Claim II is proved just as Claim I. Details are left to the reader. ■

We now complete the proof of the lemma. As $\text{tp}(a_1/M)$ is a forking extension of p_0, there is a formula $\psi(x)$ in $\text{tp}(a_1/M)$ which forks over A_0. Then clearly $\psi(x) \wedge \theta(x)$ must be **P**-analysable. So $\text{tp}(a_1/M)$ contains a formula in Θ. As a forks with a_1 over M_3, $\text{tp}(a/M_3)$ is not foreign to Θ. But as remarked above, $\text{stp}(a/bM)$ is parallel to $\text{tp}(a/M_3)$. Thus $\text{stp}(a/bM)$ is not foreign to Θ. The lemma is proved. ■

By Lemma 1.3, (working over M) we see that every forking extension of p is Θ-analysable. By Lemma 1.8, in fact every forking extension of p is Θ-analysable, provably over M. But clearly Θ-analysability implies **P**-analysability. Thus every forking extension of p is **P**-analysable, provably over M. We have proved the proposition in the case where A is an a-model.

For the general case, let $\text{stp}(a/bA)$ be a forking extension of p. Let M be an a-model containing A and independent from ab over A. Then $\text{stp}(a/bM)$ is a forking extension of $p|M$. By what we have just seen there is a formula $\psi(x, y, z)$ over A and some m in M, such that $\models \psi(a, b, m)$ and such that for all b', the formula $\psi(x, b', m)$ is **P**-analysable. Let $r = \text{stp}(m/A)$, and let $\delta(x, y)$ be the formula: 'for m' realizing $r|Axy$, $\psi(x, y, m')$'. δ is over A. As **P** is $\varnothing$-invariant, we see (using Corollary 1.4) that for all b', $\delta(x, b')$ is **P**-analysable. Thus $\text{stp}(a/bA)$ is **P**-analysable, provably over A. The proposition is proved. ■

We now aim for a result (4.9) substantially more powerful than 4.6 (and in which the NDOP hypothesis is dropped). Here definable automorphism groups will intervene. We will first need some additional remarks concerning internality.

LEMMA 4.7. *Let a, b, B, C, and π be such that: $C \supseteq B$, π is a partial type over C,* stp(a/bB) *is not foreign to π, and ab is independent with C over B. Then*

(i) *There are copies $C_0 = C, C_1, \ldots, C_n$ of C over* acl(B) *such that $C_0 \cdots C_n$ is independent with ab over B, and there is $a' \in$* acl(abB) – acl(bB), such that stp(a'/bB) *is $\{\pi_0, \ldots, \pi_n\}$-internal, where π_i is the copy of π over C_i.*

(ii) *If, moreover, π is a single formula θ, say, and Θ is the family of* acl(B)*-conjugates of θ, then there is $a' \in$* acl(abB) – acl(bB) *such that* stp(a'/bB) *is Θ-internal, provably over* acl(B).

Proof. (i). Let M be a big model containing bC, such that a is independent with M over bC (and so also over bB), and let e realize $\pi|M$ such that a forks with e over M. Let $a' \in \mathrm{Cb}(\mathrm{stp}(eM/abB)) - \mathrm{acl}(bB)$. Then $a' \in \mathrm{acl}(abB)$. Also $a' \in \mathrm{dcl}(e_0M_0, \ldots, e_nM_n)$ where $e_0M_0 = eM$ and $\{e_iM_i\}_i$ is a Morley sequence in stp(eM/abB). Let C_i be the 'copy' of C in M_i. Then $\{C_i\}_i$ is a Morley sequence in stp(C/abB), so by our assumptions, $C_0 \ldots C_n$ is independent from ab over B. Also e_i realizes π_i, the copy of π over C_i. Clearly a' is independent from $M_0 \ldots M_n$ over bB. Thus stp(a'/bB) is $\{\pi_0, \ldots, \pi_n\}$-internal.

(ii). In the case where π is a formula θ, let θ_i denote π_i in (i) above. Then each θ_i is in Θ. Let a' be as given by the proof of (i). Then a' is independent from $C_0 \ldots C_n$ over bB, so clearly stp(a'/bB) is Θ-internal. Let $\psi = \vee\{\theta_i: i = 0, \ldots, n\}$. Let $\mathbf{c} \in C_0 \ldots C_n$ be such that ψ is over $\mathbf{c}$. Then stp$(a'/bB\mathbf{c})$ is ψ-internal. At this point 1.8 or 1.5 could be applied. We use 1.5. So by Lemma 1.5 there is a formula $\varphi(x, b, \mathbf{c})$ in tp$(a'/b\mathbf{c}(\mathrm{acl}(B)))$ which is ψ-internal. Let $r = \mathrm{stp}(\mathbf{c}/B)$. Note that $a'b$ is independent with $\mathbf{c}$ over B. Let $\chi(x, y)$ be the formula over acl(B) saying: for $\mathbf{c}'$ realizing $r|(x, y, B)(\varphi(x, y, \mathbf{c}'))$. Then $\chi(x, y)$ clearly witnesses that stp(a'/bB) is Θ-internal, provably over acl(B). ∎

PROPOSITION 4.8. *Let p be a regular type of depth* 0. *Suppose* stp(a/B) *is p-semi-regular, $b \in$* acl(aB), *and $w_p(a/bB) = 0$. Then $w_p(a/bB) = 0$, provably over* acl(B).

Proof. We will first prove.

LEMMA. *Let p, a, b, B be as in the proposition. We may assume that $a \notin$* acl(Bb). *Then there is $a' \in$* acl(aB) – acl(bB) *such that $w_p(a'/Bb) = 0$ provably over* acl(B).

Proof of lemma. a' will be obtained by a number of approximations. Note that (by 7.1.18) stp(b/B) is also p-semi-regular. By Fact 4.4, every regular type r non-orthogonal to stp(a/bB) is non-orthogonal to B. This is used immediately to allow us to apply 7.1.19. Choose regular r non-orthogonal to stp(a/bB) with $R^{\infty}(r)$ minimized. By Corollary 7.1.19, there is $a_1 \in \mathrm{acl}(abB)$ ($= \mathrm{acl}(aB)$) such that stp(a_1/bB) is r-semi-regular of non-zero r-weight. As r is non-orthogonal to B, we may assume that $r \in S(C)$ for some $C \supseteq B$ with C independent from ab over B. By Lemma 4.7 there are acl(B)-conjugates $C_0 = C, C_1, \ldots, C_n$ of C such that $C_0C_1 \ldots C_n$ is independent with a_1b over B, and there is $a_2 \in \mathrm{acl}(a_1bB) - \mathrm{acl}(bB)$ such that stp($a_2/bBC_0 \ldots C_n$) is $\{r_0, \ldots, r_n\}$-internal (where r_i is the conjugate of r over C_i and the r_i are now thought of as partial types). Note that for each i the stationary type r_i is non-orthogonal to r. Clearly then stp($a_2/bBC_0 \ldots C_n$) is also r_i-semi-regular, and hence (by 7.1.14) orthogonal to any forking extension of r_i. Let C' denote $C_0 \ldots C_n$.

Claim. *Let* **c** *be any tuple of realizations of the* r_i. *Then* a_2 *is independent from* **c** *over* bBC'.

Proof of claim. We may write **c** as **ed**, where $\mathbf{e} = (e_1, \ldots, e_m)$, $\mathbf{d} = (d_1, \ldots, d_k)$, and $\{e_1, \ldots, e_m\}$ is a bBC'-independent set of realizations of $r_i | bBC'$ for various i, and each d_s realizes over $\mathbf{e}bBC$ a forking extension of some r_i. Now a_2b is independent from BC' over B, and clearly stp(a_2b/B) is p-semi-regular. As p is orthogonal to each r_i it follows that a_2b is independent from **e** over BC', and thus a_2 is independent from **e** over bBC'. Thus stp($a_2/\mathbf{e}bBC'$) is still r-semi-regular, and so a_2 is independent from **d** over $\mathbf{e}bBC'$. Putting this together we conclude that a_2 is independent from **c** over bBC', yielding the claim. ■

As stp(a_2/bBC') is $\{r_0, \ldots, r_n\}$-internal, we can apply Theorem 7.4.8. So, by the claim, stp(a_2/bBC') is the generic type of some ∞-acl(bBC')-definable homogeneous space X (acted on by the ∞-acl(bBC')-definable (connected) group G). By superstability, 1.6.20, and 1.6.22(ii), there is an acl(bBC')-*definable* homogeneous space (G_1, X_1) such that $G < G_1$, $X \subseteq X_1$, the action of G_1 on X_1 extends the action of G on X, and $U(G_1) = U(G)$, $U(X_1) = U(X)$. Now $r_2 = \mathrm{stp}(a_2/bBC')$ is *the* generic type of X, so is *a* generic type of X_1. The theory of homogeneous spaces from section 6 of Chapter 1 tells us that *every* type in X_1 is an extension of a generic type of X_1. But G_1 acts transitively on the generic types of X_1. Thus every type in X_1 is an extension of a translate of r_2. Thus X_1 is an acl(bBC')-definable set of p-weight 0. Now X_1 is defined by a formula $\delta(x, d)$ say, where $d \in \mathrm{acl}(bBC')$. Let $d \in \mathrm{acl}(bB, c)$ for some $c \in C'$, and let $\chi(z, y)$ be a formula over acl(bB) which isolates tp(d/acl(bB), c). Note that for each d' realizing tp(d/acl(bB), c), $\delta(x, d')$ also has p-weight 0. Now a_2 is independent with C' over bB. Thus a_2 satisfies the formula $\delta'(x)$

over $\mathrm{acl}(bB)$: 'for c' realizing $\mathrm{stp}(c/bB)|(xbB)$, there is d' such that $\chi(d', c')$ and $\delta(x, d')$'. Clearly $\delta'(x)$ has p-weight 0. *We have found a formula (i.e.* $\delta'(x)$*) in* $\mathrm{stp}(a_2/bB)$ *which has p-weight* 0.

The final step is again to use the fact (Fact 4.4) that $\mathrm{stp}(a_2/bB)$ is non-orthogonal to B. Let b' realize $\mathrm{stp}(b/B)$ such that b' is independent with b over B, and let a_2' realize a copy of $\mathrm{stp}(a_2/Bb)$ over $\mathrm{acl}(Bb')$, such that a_2' is independent from ba_2' over Bb'. So $\mathrm{stp}(a_2'/b'B)$ is non-orthogonal to $\mathrm{stp}(a_2/bB)$. Let $\delta''(x)$ be the copy of $\delta'(x)$ under an $\mathrm{acl}(B)$-automorphism taking b to b'. It is clear that the formula $\delta''(x)$ has p-weight 0 (as p is non-orthogonal to B), and that $\mathrm{stp}(a_2/bB)$ is not foreign to $\delta''(x)$. Let Δ be the family of $\mathrm{acl}(B)$-conjugates of δ''. Again each formula in Δ has p-weight 0. As the parameters in δ'' are independent with a_2b over B, it follows from Lemma 4.7(ii) that there is $a' \in \mathrm{acl}(a_2bB) - \mathrm{acl}(bB)$ such that $\mathrm{stp}(a'/bB)$ is Δ-internal provably over $\mathrm{acl}(B)$. Thus clearly $\mathrm{stp}(a'/bB)$ has p-weight 0, provably over $\mathrm{acl}(B)$. This completes the proof of the lemma. ■

Let a' be as given by the lemma. Clearly then still $w_p(a/a'bB) = 0$. Thus we can iterate, and by superstability we find $a_1, \ldots, a_n$ in $\mathrm{acl}(aB)$ such that $w_p(a_i/ba_1 \ldots a_{i-1}B) = 0$ provably over $\mathrm{acl}(B)$ for each i, and $a \in \mathrm{acl}(ba_1 \ldots a_nB)$. Then clearly $w_p(a/bB) = 0$, provably over $\mathrm{acl}(B)$, proving the proposition. ■

COROLLARY 4.9. *Suppose* $\mathrm{stp}(a/B)$ *is p-semi-regular, where p is a non-trivial regular type of depth* 0. *Then there is some formula* $\varphi(x, b)$ *in* $\mathrm{stp}(a/B)$ *such that p-weight is definable inside* φ *over b.*

Proof. First, by 2.17, find some $a_1 \in \mathrm{dcl}(a, \mathrm{acl}(B))$ and $\theta(y, b_1) \in \mathrm{stp}(a_1/B)$ such that $\theta(y, b_1)$ is p-simple, p-weight is definable inside $\theta(y, b_1)$ over b_1, and $w_p(a/a_1B) = 0$. By Proposition 4.8, $w_p(a/a_1B) = 0$, provably over $\mathrm{acl}(B)$. Thus for some $\mathrm{acl}(B)$-definable function f, and formula $\psi(x, y)$ over $\mathrm{acl}(B)$, $f(a) = a_1$, $\models \psi(a, a_1)$, and for any a_1', the formula $\psi(x, a_1')$ has p-weight 0. Let $\varphi(x)$ be the formula $\theta(f(x), b_1) \wedge \psi(x, f(x))$. Then $\varphi(x)$ is over some parameter b in $\mathrm{acl}(B)$ which contains b_1 and the parameters in f and ψ. We need to see that p-weight is definable inside $\varphi(x, b)$ over b. Clearly $\varphi(x, b)$ is p-simple of p-weight > 0. Suppose for example that $a^1, \ldots, a^m$ satisfy $\varphi(x, b)$ and $w_p(a^1, \ldots, a^m/C) = k$ where C contains b. Then clearly $w_p(f(a^1), \ldots, f(a^m)/C) = k$, and thus there is some formula $\chi(y_1, \ldots, y_m, z)$ over b_1 such that χ holds of $(f(a^1), \ldots, f(a^m), c)$ for some c in C and such that for all c', $\chi(y_1, \ldots, y_m, c')$ is inconsistent or has p-weight k. Let $\xi(x_1, \ldots, x_m, z)$ be the formula $\bigwedge_i \psi(x_i, f(x_i)) \wedge \chi(f(x_1), \ldots, f(x_m), z)$. Clearly ξ holds of $(a^1, \ldots, a^m, c)$ and for all c', $\xi(x_1, \ldots, x_m, c')$ has p-weight k or is inconsistent.

Checking the general case where $a^1 \in (\varphi_C)^{\mathrm{eq}}$ for some C containing b is left to the reader. ■

COROLLARY **4.10.** *Suppose T has NDOP, $\mathbf{P}$ is a $\varnothing$-invariant family of partial types, θ is a $\mathbf{P}$-weakly minimal formula, and p is a non-trivial $\mathbf{P}$-minimal type over $A = \mathrm{acl}(A)$ which contains θ. Then there is a weakly $\mathbf{P}$-minimal formula $\delta(x, c)$ in p such that*

(i) *inside $\delta(x, c)$ having p-weight* 0 *is equivalent to being $\mathbf{P}$-analysable; and*

(ii) *p-weight is definable inside $\delta(x, b)$ over b.*

Proof. Note that by Fact 4.5, p has depth 0. So bearing in mind Corollary 4.9 we only have to find some weakly $\mathbf{P}$-minimal formula $\delta(x, b)$ in p which satisfies (i).

As p is non-trivial, there is some b and there are a_1, a_2, a_3 realizations of $p|Ab$ such that $\{a_1, a_2, a_3\}$ is pairwise Ab-independent but Ab-dependent. By Proposition 4.6 there is a formula $\chi(x_1, x_2, x_3, w)$ over A true of (a_1, a_2, a_3, b) such that whenever (a_1', a_2', a_3', b') satisfies χ then the strong type of each a_i' over the other two together with Ab' is $\mathbf{P}$-analysable. We may also assume that $\models \chi(x_1, x_2, x_3, w) \rightarrow \bigwedge_i \theta(x_i)$. Let $r(x_3, z) = \mathrm{tp}(a_3, b/A)$. Let $\delta(x_1)$ be the formula: 'for (x_3, w) realizing $r|Ax_1$, $(\exists x_2)(\chi(x_1, x_2, x_3, w))$'. Then $\delta(x_1)$ is over some $b \in A$, and is true of a_1. Clearly $\delta(x, b)$ is weakly $\mathbf{P}$-minimal. Note that any $\mathbf{P}$-minimal strong type containing $\delta(x, b)$ is non-orthogonal to p, and thus inside $\delta(x, b)$, $\mathbf{P}$-analysability is the same as having p-weight 0. We leave it to the reader to check that $\delta(x_1)$ is weakly $\mathbf{P}$-minimal and, moreover, that any $\mathbf{P}$-minimal strong type containing δ is non-orthogonal to p.

Let δ be over the finite set A_0. As depth 0-ness is invariant under non-orthogonality of regular types, we see that every $\mathbf{P}$-minimal strong type containing δ also has depth 0. ■

5 Generalized existence and dichotomy theorems

In this final section we are concerned with generalizing the dictotomy theorem of Chapter 2 in various ways. In this section T will be superstable. We begin by more or less repeating part of Proposition 7.2.4.

LEMMA **5.1.** *Suppose $p = \mathrm{stp}(a/A)$ is regular and not locally modular. Then there exist $C \supseteq A$, and a_1, a_2, b such that*

(i) *a_1 and a_2 both realize $p|C$ and they are independent over C;*

(ii) *$\mathrm{stp}(b/C)$ is p-semi-regular of p-weight* 2;

(iii) $\mathrm{stp}(a_1a_2/bC)$ *is p-semi-regular of p-weight* 1;

(iv) b is (interalgebraic with) $\mathrm{Cb}(\mathrm{stp}(a_1a_2/bC))$;

(v) $\mathrm{w}_p(b/a_1a_2C) = 1$.

Proof. Proposition 7.2.4 gives us a_1, a_2 independent realizations of p, and (using superstability) b such that $\mathrm{stp}(a_1a_2/bA)$ is p-semi-regular of p-weight 1, and b is interalgebraic with $\mathrm{Cb}(\mathrm{stp}(a_1a_2/bA))$, and $w_p(\mathrm{stp}(b/A)) > 1$. We can easily find algebraically closed $C \supseteq A$ such that $\mathrm{stp}(b/C)$ is p-semi-regular with p-weight 2. Moreover, C can be chosen to be independent with a_1a_2 over Ab. Thus, in particular, $\mathrm{stp}(a_1a_2/bC)$ is semi-regular of p-weight 1, and b is its canonical base. It remains to be seen that a_1, a_2 are C-independent realizations of $p|C$, that is that $w_p(a_1a_2/C) = 2$. If not, then $w_p(a_1a_2/C) = 1 = w_p(a_1a_2/bC)$. But then, by Lemma 7.2.1, $b \in \mathrm{cl}_p(C)$, contradicting $w_p(b/C) = 2$. So (i) to (iv) have been shown, (v) follows from computing $w_p(a_1a_2b/C)$ two ways (and noting the value is 3). ■

Note that in 5.1, a_1, a_2, b are in p^{eq} (and we can choose C in p^{eq} too).

Note also that in 5.1, $\mathrm{stp}(b/a_1a_2C)$ may not be p-semi-regular. For the purposes of Lemma 5.3 below and applications to **RM**-minimal types, this does not present a problem. However for further results, semi-regularity of this type would be convenient. In fact a slight modification of the pair a_1a_2 and of C will provide this. First replace C by $C_1 = \mathrm{cl}_p(C) \cap \mathrm{dcl}(a_1a_2, b, C)$. The conclusion of 5.1 remains true with C_1 replacing C, except that now $\mathrm{tp}(a_1, a_2, b/C_1)$ is p-semi-regular. Now let $A' = \mathrm{dcl}(a_1a_2, b, C_1) \cap \mathrm{cl}_p(a_1a_2, C_1)$. By 7.1.20, $\mathrm{stp}(b/A')$ is semi-regular (of p-weight 1). Let a' be a finite part of A' extending (a_1, a_2) such that $\mathrm{stp}(b/a'C_1)$ is (semi-)regular. Replace a_1a_2 by a'. Now (ii) to (v) of 5.1 are true with a' replacing a_1a_2 and with C_1 replacing C. We also have that $\mathrm{stp}(a'/C_1)$ is p-semi-regular of p-weight 2 (by choice of C_1). We sum this up in

COROLLARY **5.2.** *Suppose p is a regular non-locally modular strong type over A. Then there are $C \supseteq A$, and elements a and b (in p^{eq}) such that*

(i) $\mathrm{stp}(a/C)$ *is p-semi-regular of p-weight* 2, *and* $p|Cb$ *is realized in* $\mathrm{dcl}(aC)$.

(ii) $\mathrm{stp}(b/C)$ *is p-semi-regular of p-weight* 2.

(iii) $\mathrm{stp}(a/bC)$ *is p-semi-regular with p-weight* 1.

(iv) $w_p((\mathrm{Cb}(\mathrm{stp}(a/bC))/C)) = 2$ (*in fact* $\mathrm{Cb}(\mathrm{stp}(a/bC))$ *is interalgebraic with b over C*).

(v) $\mathrm{stp}(b/aC)$ *is p-semi-regular of p-weight* 1.

The following lemma represents a rather straightforward case of the basic argument, and requires only Lemma 5.1.

LEMMA 5.3. *Suppose* $p = \mathrm{stp}(a/A)$ *is non-locally modular. Suppose also that* $D(x)$ *is a* p*-simple formula over* A *such that* a *satisfies* D *and* p*-weight is definable inside* D *over* A*. Then there is some* p*-simple formula* $\psi_1(x)$ *in* p *and some formula* $\chi_1(x)$ *over a set* B *containing* A*, such that*

(i) $w_p(\chi_1) = 0$; *and*

(ii) *whenever* q *is a strong type over* A *which contains* ψ_1 *and* $w_p(q) \neq 0$, *then* q *is realized by an element satisfying* χ_1.

Proof. Let a_1, a_2, b, C be as given by Lemma 5.1. We can assume that a_1, a_2, b are all inside $D_A{}^{\mathrm{eq}}$ and that C is algebraically closed. Let $\mathbf{a}$ denote a_1a_2.

CLAIM I. *Let* b' *be such that* $\mathrm{stp}(b'/\mathbf{a}C) = \mathrm{stp}(b/\mathbf{a}C)$ *and* $w_p(b'/b\mathbf{a}C) \neq 0$. *Then*

(i) $w_p(\mathbf{a}/bb'C) = 0$;

(ii) b *is independent from* b' *over* C.

Proof of Claim I. As $w_p(b'/b\mathbf{a}C) \neq 0$, $b' \notin \mathrm{acl}(bC)$. Thus $\mathrm{stp}(\mathbf{a}/bC)$ is not parallel to $\mathrm{stp}(\mathbf{a}/b'C)$. So $\mathbf{a}$ forks with b' over bC. As $\mathrm{stp}(\mathbf{a}/bC)$ is p-regular, we can conclude (i).

Now $w_p(b'/b\mathbf{a}C) = 1$. So by (i) and (vi) of 5.1, together with additivity of p-weight, we conclude that $w_p(\mathbf{a}bb'/C) = 4$. By (i) of the claim and additivity again we see that $w_p(bb'/C) = 4$. By Lemma 7.1.14 and additivity (and the fact that $\mathrm{stp}(b/C)$ is semi-regular of p-weight 2), we conclude that b is independent from b' over C. This completes the proof of Claim I. ■

Let $r(\mathbf{x}, y) = \mathrm{tp}(\mathbf{a}, b/C)$. So from Claim I we see

($*$) *if* $r(\mathbf{x}, y)$ and $r(\mathbf{x}, y')$ and $w_p(y'/y\mathbf{x}C) \neq 0$ and $\mathrm{stp}(y'/\mathbf{x}C) = \mathrm{stp}(y/\mathbf{x}C)$, *then* $w_p(\mathbf{x}/yy'C) = 0$.

The point now is that compactness can be applied to the implication ($*$). This is because our assumptions on the definability and continuity of p-weight inside $D_C{}^{\mathrm{eq}}$ mean that the expression $w_p(w/zC) \neq 0$ is a partial type in x, z over C. Thus we obtain a formula $\psi(\mathbf{x}, y)$ in $r(\mathbf{x}, y)$ (such that $\models \psi(\mathbf{x}, y) \to D(x_i)$ for $i = 1, 2$, and $\models \psi(\mathbf{x}, y) \to$ '$y \in D_C{}^{\mathrm{eq}}$') and a finite collection $\chi_i(\mathbf{x}, y, y')$ of formulae over C for $i = 1, \ldots, m$, such that for all d, d', and i, $\chi_i(\mathbf{x}, d, d')$, if consistent, has p-weight 0 and such that

($**$) *if* $\psi(\mathbf{x}, y)$ and $\psi(\mathbf{x}, y')$ hold, $w_p(y'/y\mathbf{x}C) \neq 0$, and y and y' have the same strong type over $C\mathbf{x}$, *then* for some $i = 1, \ldots, n$, $\chi_i(\mathbf{x}, y, y')$ holds.

Let $\chi(\mathbf{x}, y, y')$ be the disjunction of the χ_i. Then for any d, d', $\chi(\mathbf{x}, d, d')$ has p-weight 0 whenever it is consistent. We may also, by virtue of our definability assumptions, assume

(***) *if* $\psi(\mathbf{x}, y)$ *then* $w_p(\mathbf{x}/C)$, $w_p(y/C) \leq 2$, and $w_p(\mathbf{x}/yC)$, $w_p(y/\mathbf{x}C) \leq 1$.

Let us now fix b' such that b' is independent from b over C and realizes the same strong type over C as b.

CLAIM **II.** *If $q(\mathbf{x})$ is a strong type over Cb which contains $\psi(\mathbf{x}, b)$ and has non-zero p-weight, then q is realized by some element satisfying $\chi(\mathbf{x}, b, b')$.*

Proof. Let $\mathbf{a}'$ realize q. So $w_p(\mathbf{a}'b/C) = 3$. But by (***) $w_p(\mathbf{a}'/C) \leq 2$ and $w_p(b/\mathbf{a}'C) \leq 1$. So we obtain

$$w_p(\mathbf{a}'/C) = 2 \text{ and } w_p(b/\mathbf{a}'C) = 1. \tag{1}$$

Let b'' realize $\text{stp}(b/\mathbf{a}'C)$ such that b'' is independent with $b\mathbf{a}'C$ over $\mathbf{a}'C$. In particular $w_p(b''/b\mathbf{a}'C) = 1$, and $\psi(\mathbf{a}', b'')$ holds. By (**)

$$\chi(\mathbf{a}', b, b'') \text{ holds.} \tag{2}$$

Thus we have

$$w_p(b''/b\mathbf{a}'C) = 1, \text{ and } w_p(\mathbf{a}'/bb''C) = 0. \tag{3}$$

Now by (1) and (3) and the same computation as in the proof of Claim I(ii) we see that b is independent with b'' over C. In particular $\text{tp}(b'/Cb) = \text{tp}(b''/Cb)$. By (2) we conclude that there is $\mathbf{a}''$ realizing q such that $\chi(\mathbf{a}'', b, b')$ holds, as required. This establishes Claim II. ∎

To complete the proof of the lemma is quite easy now using Claim II. Let $\chi_1(x_1)$ be the formula $(\exists x_2)(\chi(x_1, x_2, b, b'))$. So clearly $w_p(\chi_1(x_1)) = 0$. By the open map theorem, let $\psi_1(x_1)$ be a formula over $\text{acl}(A)$ such that for any strong type $q(x_1)$ over A, $\psi_1(x_1) \in q$ iff $(\exists x_2)(\psi(x_1, x_2, b)) \in q|bC$. It is easy to see that ψ_1 and χ_1 satisfy the lemma (using the fact that a_1 realizes $p|bC$). ∎

COROLLARY **5.4.** *Suppose that T is either countable or non-multidimensional. Let θ be a weakly* **RM**-*minimal formula and p a stationary* **RM**-*minimal type containing θ. Then p is locally modular.*

Proof. The hypothesis that T is countable or non-multidimensional tells us, by Example 1.9 and Remark 2.12, that any **RM**-analysable type (or formula) has Morley rank.

We may assume p to be non-trivial. By Remark 2.16(iii), we may assume that p-weight is definable inside θ (over some set B over which θ is defined) and that every non-**RM**-analysable strong type over B containing θ has non-zero p-weight. Also, by Lemma 2.14, we may assume that inside θ^{eq} having p-weight 0 is equivalent to being **RM**-analysable (and thus having Morley rank). We may also assume that p is a strong type over B which contains θ. We will work inside θ. By Lemma 5.3, if p is not locally modular we obtain $\psi(x)$ in p (note that in 5.3 we may assume that $\models \chi_1(x) \rightarrow D(x)$) and some **RM**-analysable formula $\chi(x)$ (over a set containing B) such that any non-**RM**-analysable type over B which contains ψ is realized by some element satisfying χ. By the above remarks $RM(\chi) < \infty$.

Choose now ψ and χ such that $\psi(x)$ is over $\mathrm{acl}(B)$, $\chi(x)$ is over some set containing B, ψ is not **RM**-analysable, every non-**RM**-analysable strong type over B which contains ψ is realized in $\chi(x)$, and (RM, Morley degree)(χ) is least possible.

CLAIM. *There is a unique non-***RM***-analysable strong type over B which contains $\psi(x)$.*

Proof. If not then we can clearly find a formula $\xi(x)$ over $\mathrm{acl}(B)$ such that both $\psi(x) \wedge \xi(x)$, $\psi(x) \wedge \neg\xi(x)$ are non-**RM**-analysable. But then one of $\chi(x) \wedge \xi(x)$, $\chi(x) \wedge \neg\xi(x)$ has lower (RM, Morley degree) than $\chi(x)$, and we contradict minimal choice of χ. ■

Let $q(x)$ be the strong type over B given by the claim. Let the ordinal α be such that any type with ordinal-valued Morley rank has Morley rank $< \alpha$. Then by the claim, for any $C \supseteq B$, $q|C$ is the unique strong type over C which contains $\psi(x)$ and with $RM(q) \geq \alpha$. But then $RM(q) = \alpha$, a contradiction. This proves Corollary 5.4. ■

COROLLARY **5.5.** *Let T be (superstable), non-totally transcendental, and either countable or non-multidimensional. Then some regular type is locally modular.*

Proof. If $\theta(x)$ is a formula of least ∞-rank without Morley rank, then by 1.9(ii) and 2.12, θ is **RM**-weakly minimal and so we can apply Corollary 5.4. ■

Finally we consider (countable) theories with NOTOP and NDOP.

DEFINITION **5.6.** *T is said to have OTOP if there is a complete type $p(x, y, z)$ over $\emptyset$, such that for any cardinal λ and binary relation R on λ, there is a model M_R of T and a_α in M_R for $\alpha < \lambda$ such that for each $\alpha < \beta < \lambda$, $(\alpha, \beta) \in R$ iff $p(x, a_\alpha, a_\beta)$ is realized in M_R.*

It is pointed out in [Sh1, XII] that if T has OTOP then for any uncountable $\lambda \geq |T|$, T has 2^λ models of cardinality λ. On the other hand, according to our current understanding, we require *countability* of T to deduce structure theorems from NOTOP (and NDOP) (see 5.7 and 5.8 below).

So as to avoid a lengthy exposition of OTOP we will make use of a certain consequence of NOTOP (and NDOP) which is proved in [Sh1]. In some sense our resulting treatment will suffer from overkill.

Definition 5.7. *T has PMOP (Prime Models Over Pairs, also called the $\infty - 2$ existence property in [Sh1]) if whenever M_0, M_1, M_2 are models with $M_0 < M_i$ for $i = 1, 2$, and such that M_1 is independent with M_2 over M_0, then there is a* constructible *(and thus atomic) model M over $M_1 \cup M_2$.*

Fact 5.8. [Sh1, XII.5.14] *Suppose T is countable. If T has NOTOP and NDOP then T has PMOP.*

Remark 5.9. *If T (of any cardinality) has PMOP and NDOP then the results of section 3 can be used to show that any model of T is prime over a non-forking tree of models of cardinality $|T|$. Thus the OTOP/NOTOP dichotomy is of interest for* countable theories.

Lemma 5.10. *Suppose T has PMOP and NDOP. Suppose A_0, A_1, A_2, and c satisfy*

(i) *$A_i \supseteq A_0$ for $i = 1, 2$ and A_1 is independent with A_2 over A_0;*

(ii) whenever *$B_i \supseteq A_i$ for $i = 0, 1, 2$, B_0 is independent with $\cup A_i \cup c$ over A_0, and B_1 is independent with B_2 over B_0,* then *$\text{stp}(c/\cup B_i)$ does not fork over $\cup A_i$.*

Then *$\text{tp}(c/\cup A_i)$ is isolated.*

Proof. Let M_0 be an a-model containing A_0 such that M_0 is independent with $\cup A_i \cup c$ over A_0. Let M_i be a-prime over $M_0 \cup A_i$, for $i = 1, 2$. Then M_1 is independent with M_2 over M_0.

Claim 1. *$M_1 \cup M_2$ dominates M_1M_2c over M_0.*

Proof. Let $C \supseteq M_0$ be independent with M_1M_2 over M_0. Then $(M_0, C \cup M_1, M_2)$ satisfy the assumptions on (B_0, B_1, B_2) in (ii) above. Thus $\text{stp}(c/C \cup M_1 \cup M_2)$ does not fork over $\cup A_i$. In particular $\text{stp}(c/C \cup M_1 \cup M_2)$ does not fork over $\cup M_i$, so C is independent with $M_1 \cup M_2 \cup c$ over $\cup M_i$.

But C is independent with $\cup M_i$ over M_0, and thus C is independent with M_1M_2c over M_0, as required. ■

By the claim there is an a-prime model M over M_1M_2 which contains c. (See 1.4.3.4(iii).) By our PMOP assumption there is is a model M' such that $M_1 \cup M_2 < M' < M$, and M' is atomic over M_1M_2.

CLAIM **2.** $M' = M$.

Proof. Suppose $M' \neq M$. By Proposition 3.2 there is $e \in M - M'$ such that $\mathrm{tp}(e/M')$ is regular. By NDOP (considering the non-forking extension of $\mathrm{stp}(e/M')$ over M), $\mathrm{stp}(e/M')$ is without loss of generality non-orthogonal to M_1. By Proposition 3.6 (noting that any extension of an a-model is an na-extension), there is d in $M - M'$ such that $\mathrm{stp}(d/M')$ does not fork over M_1. In particular d is independent with M_2 over M_1. This contradicts the fact that M_2 dominates M over M_1. ■

REMARK **5.11.** *It follows from Bradd Hart's article [Ha] on OTOP (which expounds a treatment due to Harrington) that Lemma* 5.10 *remains true when T is countable and the 'PMOP and NDOP' assumption is replaced by 'NOTOP'. In the notation of his paper, the assumptions on A_i and c in Lemma* 5.10 *translate into* '(A_0, A_1, A_2) *V-dominates* $\cup A_i \cup c$'.

The purpose now is to show that under the assumptions of PMOP and NDOP, non-locally modular regular types are close to being strongly regular. In fact strong regularity is in a sense not sharp enough a notion.

DEFINITION **5.12.** (i) *A stationary type $p \in S(A)$ is said to be* strongly regular *if p contains a formula $\varphi(x)$ such that if q is any stationary type over a set $B \supseteq A$ and $\varphi(x) \in q$, then either q is orthogonal to p or q is a non-forking extension of p.*

(ii) *Let* **P** *be some (say ∅-invariant) family of partial types. A stationary type $p \in S(A)$ is said to be* strongly **P**-minimal *if p is not* **P***-analysable and p contains some* **P***-minimal formula.*

REMARK **5.13.** *Any strongly* **P***-minimal type is strongly regular. The interest is in* $\mathbf{P} = \mathbf{R}^\infty < \alpha$, *for various* α. *The second part of Remark* 1.14 *gives a strongly regular type which is not strongly* **P***-minimal, for any such* **P**.

PROPOSITION **5.14.** *Suppose T has PMOP and NDOP. Suppose* $p = \mathrm{stp}(a/A)$ *is regular and non-locally modular. Suppose also that $D(x)$ is a formula over*

A such that $D(x) \in p$ and p-weight is definable inside D over A. Then p is strongly regular.

Proof. The proof is broken into several lemmas and claims. From now on we work in D_A^{eq} (which is clearly permissible for proving the proposition). Let C, a, b be as given by Corollary 5.2 (and our assumption that p is not locally modular), with C algebraically closed. We begin by restating a slight modification of Claim I of Lemma 5.3.

CLAIM I. *Let b' be such that* $\text{stp}(b'/aC) = \text{stp}(b/aC)$ *and* $w_p(b'/baC) \neq 0$. *Then*

(i) $w_p(a/bb'C) = 0$;

(ii) *b is independent with b' over C;*

(iii) *b is independent with b' over aC (and so* $\text{tp}(a, b, b'/C)$ *is uniquely determined).*

Proof. (i) and (ii) follow exactly as in Claim I in 5.3.(iii) follows using Lemma 7.1.14, because now $\text{stp}(b/aC)$ is (semi-)regular. ■

Let b' be as in Claim I and let $\alpha = R^\infty(a/bb'C)$. By (iii), $\text{tp}(a, b, b'/C)$ depends only on $\text{tp}(a, b/C)$, and thus α depends only on a, b, and C. Write $\alpha = f(a, b, C)$.

Now choose a, b, C to satisfy (i)–(v) of Corollary 5.2, and such that $f(a, b, C)$ is least possible, with value α, say. Let again b' be such that $\text{stp}(b'/aC) = \text{stp}(b/aC)$ and $w_p(b'/baC) \neq 0$ (so the conclusion of Claim I holds). With these assumptions, we prove:

LEMMA. $\text{tp}(a/bb'C)$ *is isolated.*

Proof. We shall use Lemma 5.10. We have to check that (C, Cb, Cb') and a satisfy the assumptions on (A_0, A_1, A_2) and c in 5.10. The fact that Cb and Cb' are independent over C is just (ii) of Claim I.

Now suppose that $C_1 \supseteq C$ is independent with $bb'a$ over C. Then (i)–(v) of 5.2 are still true of a, b, C_1, and $f(a, b, C_1)$ is still α. Also b and b' have the same strong type over C_1a and are independent over C_1a. So there is no harm in assuming $C_1 = C$. Now let $B \supseteq Cb$, $B' \supseteq Cb'$ such that B is independent with B' over C. Our aim is to prove that a is independent with BB' over Cbb'.

CLAIM II. *Let* $B_0 = \text{cl}_p(Cb) \cap \text{dcl}(B)$, $B_0' = \text{cl}_p(Cb') \cap \text{dcl}(B')$. *Then a is independent with* $B_0 \cup B_0'$ *over* Cbb'.

Proof. If not we have $R^\infty(a/B_0B_0') < \alpha$. Let b_0, b_0' be in B_0, B_0', respectively, such that $R^\infty(a/b_0b_0') < \alpha$. Now as $\mathrm{stp}(b/aC)$ is p-semi-regular, $\mathrm{stp}(b/\mathrm{cl}_p(aC)) = \mathrm{stp}(b'/\mathrm{cl}_p(aC))$. So let g be an automorphism which fixes $\mathrm{cl}_p(aC)$ and takes b to b' (this is possible by stability, even though $\mathrm{cl}_p(aC)$ could have the same cardinality as $\mathbf{C}$). Let $c_0 = g^{-1}(b_0')$ and $c_0' = g(b_0)$. Then $\mathrm{tp}(b, b_0, c_0/\mathrm{cl}_p(aC)) = \mathrm{tp}(b', c_0', b_0'/\mathrm{cl}_p(aC))$. Note that $b_0, c_0 \in \mathrm{cl}_p(Cb)$. Let $d = (b, b_0, c_0)$, and $d' = (b', c_0', b_0')$. Let $C_1 = \mathrm{cl}_p(C) \cap \mathrm{dcl}(a, d, C)$. Using 7.1.18 and 7.1.20 we have

(∗) $\mathrm{stp}(d/C_1)$ is p-semi-regular of p-weight 2, $\mathrm{stp}(d/aC_1)$ has p-weight 1, $\mathrm{stp}(a/dC_1)$ is p-semi-regular of p-weight 1, $w_p(d'/daC_1) = 1$, $\mathrm{tp}(d/\mathrm{cl}_p(aC_1)) = \mathrm{tp}(d'/\mathrm{cl}_p(aC_1))$, **and** $R^\infty(a/dd'C_1) < \alpha$.

Note also that $\mathrm{stp}(a/C_1)$ is still p-semi-regular of p-weight 2.

By 7.1.20 and superstability we can choose $a_1 \in \mathrm{dcl}(daC_1) \cap \mathrm{cl}_p(aC_1)$ such that $a \in \mathrm{dcl}(a_1)$ and $\mathrm{stp}(d/a_1C_1)$ is p-semi-regular (of p-weight 1), and $\mathrm{dcl}(a_1C_1)$ contains a realization of $p|C_1$. Then still $\mathrm{stp}(a_1/dC_1)$ is p-semi-regular of p-weight 1, and $\mathrm{stp}(a_1/C_1)$ is p-semi-regular of p-weight 2 (by choice of C_1 and 7.1.20 again). Also of course we have $R^\infty(a_1/dd'C_1) < \alpha$. Thus a_1, d, C_1 satisfy (i)–(v) of 5.2, but d' witnesses a contradiction to the least choice of $f(a, b, C)$. This contradiction proves Claim II. ■

Now by choice of B_0 and 7.1.20, $\mathrm{stp}(B/B_0)$ is p-semi-regular, as is $\mathrm{stp}(B'/B_0')$. Now the independence of B with B' over C guarantees that B is independent with B' over each of B_0, B_0' (e.g. $w_p(B/B_0B') \geq w_p(B/\mathrm{cl}_p(CbB')) = w_p(B/CbB') = w_p(B/Cb) =$ (by 7.1.2) $w_p(B/B_0)$, so by p-semi-regularity, B is independent from B' over B_0). Thus also $\mathrm{stp}(BB'/B_0 \cup B_0')$ is p-semi-regular. As $\mathrm{stp}(a/B_0B_0')$ has p-weight 0, we conclude that a is independent with BB' over B_0B_0', and so by claim II, a is independent with BB' over Cbb'. We have succeeded in showing that the assumptions of 5.10 hold for (C, Cb, Cb') and a. So by Lemma 5.10, we conclude that $\mathrm{tp}(a/Cbb')$ is isolated, proving the lemma. ■

Let $\chi(x, y, y')$ be a formula over C such that $\chi(x, b, b')$ isolates $\mathrm{tp}(a/bb'C)$. Let $r(x, y) = \mathrm{tp}(a, b/C)$. By Claim I we have

if $r(x, y)$ holds $\wedge$ $r(x, y')$ holds $\wedge$ $w_p(y'/xyC) \neq 0 \wedge \mathrm{stp}(y'/xC) = \mathrm{stp}(y/xC)$ **then** $\chi(x, y, y')$ holds.

As in the proof of Lemma 5.3, we can, using our assumptions on the definability and continuity of p-weight, apply compactness to the above implication. Exactly the same arguments as in Lemma 5.3 then yield

Claim III. *There is a formula* $\delta(x, b)$ *in* $\mathrm{tp}(a/bC)$ *such that whenever* $q(x)$ *is*

a strong type over Cb which contains $\delta(x, b)$ and has non-zero p-weight, then $q(x)$ is realized by some element satisfying $\chi(x, b, b')$.

Now 5.2(i) gives us $a_1 \in \mathrm{dcl}(aC)$ realizing $p|Cb$, say $a_1 = f_1(a)$ with f_1 C-definable. Let $\xi(x_1, b)$ be the formula $(\exists x)(\delta(x, b) \wedge f_1(x) = x_1)$. Then $\models \xi(a_1, b)$. Let a_1' realize $\xi(x_1, b)$ and suppose $\mathrm{stp}(a_1'/C')$ is non-orthogonal to p, where $C' \supseteq Cb$. In particular $\mathrm{stp}(a_1'/Cb)$ and $\mathrm{stp}(a_1'/C)$ both have non-zero p-weight. Let a' be such that $f(a') = a_1'$ and $\delta(a', b)$. Then $w_p(a'/Cb) \neq 0$, so by Claim III, we have $\chi(a', b, b')$ holding. In particular $\mathrm{stp}(a'/C) = \mathrm{stp}(a/C)$ (recall that C is algebraically closed), so also $\mathrm{stp}(a_1'/C) = p|C$. As p is regular, a_1' realizes $p|C'$.

We have shown that $p|Cb$ is strongly regular. The open map theorem clearly implies that p is also strongly regular.

This concludes the proof of Proposition 5.14. ■

COROLLARY **5.15.** *Suppose T has PMOP and NDOP. Let* **P** *be some $\varnothing$-invariant family of partial types. Let θ be a* **P**-*weakly minimal formula and p a* **P**-*minimal strong type over A containing θ. Suppose p is not locally modular. Then p is strongly* **P**-*minimal.*

Proof. Assume p non-locally modular. In particular p is non-trivial, so we may assume by Corollary 4.10 that p-weight is definable inside θ (over A) and that inside θ, having p-weight 0 is equivalent to being **P**-analysable. By Corollary 5.14 p is strongly regular. So we have some formula ψ in p, where we may assume $\models \psi \rightarrow \theta$, and every strong type q over $B \supseteq A$ which contains ψ and has non-zero p-weight is $p|B$. But then every non-**P**-analysable strong type over B which contains ψ is $p|B$, whereby ψ is **P**-minimal and p is strongly **P**-minimal. ■

COROLLARY **5.16.** *Suppose T has PMOP and NDOP. Let p be a non-locally modular regular type. Then*

(i) *p is strongly regular.*

(ii) *Let q be a strong type of least ∞-rank, say α, such that q is non-orthogonal to p. Then q is strongly $\mathbf{R}^\infty{<}\alpha$-minimal.*

Proof. (i) follows by Proposition 5.14 and Corollary 4.9. (Note that as p is non-trivial and T has NDOP, p must, by 4.5, have depth 0.)

For (ii) let $\mathbf{P} = \mathbf{R}^\infty{<}\alpha$, and note that if $\theta(x) \in q$ and $R^\infty(\theta) = \alpha$, then θ is weakly **P**-minimal, and q is **P**-minimal. As q is also non-locally modular, we may apply Corollary 5.15 to conclude that q is strongly **P**-minimal. ■

Notes on chapters

Chapter 1

The general theory of stability and forking is due to Shelah. All the results of sections 1 to 5 are due in one form or another to Shelah [Sh1], unless stated otherwise below. There are several other books treating this material in greater or lesser depth, for example [B1], [L1], [P1], and [Po1]. The general theory of stable groups is due to Poizat (see Chapter 5 of [Po2]), although the theory was developed earlier in more specifc contexts (finite Morley rank, superstable) by Zilber, Cherlin, and Shelah (see [Z1], [Ch], [Ch–Sh]).

Section 1

Shelah (in [Sh1]) took a 1-sorted approach to imaginary elements. We follow Makkai's many-sorted approach [Makk]. The importance of 'eliminating imaginaries' was recognized by Poizat [Po3], whose also proved that various algebraic structures (algebraically closed fields, differentially closed fields) admit elimination of imaginaries.

Section 2

The theory of local forking developed here is a mixture of the approaches of Hrushovski and myself (see [H–P2] and [P2]).

Section 3

The notions of Morley rank and totally transcendental theory are due to Morley [M]. The *U*-rank introduced by Lascar [L2].

Section 4

The '*a*-model' notation is due to Makkai [Makk]. The notion of domination and its basic properties are due to Lascar [L3].

Section 5

Strongly minimal formulae were introduced by Marsh [Ma]. Baldwin and Lachlan [B–L] highlighted the basic role of strongly minimal formulae in the study of ω_1-categoricity.

The countable case of Theorem 5.18 is due to Morley [M]. 'Morley's theorem' (5.19(i)) was the starting point of classification theory. The 'Baldwin–Lachlan theorem' (5.19(ii)) appears in [B–L], which has always been for me among the most beautiful papers in mathematical logic. Finiteness of Morley rank in ω_1-categorical theories (a consequence of 5.8, 5.12, and 5.18) is due to Baldwin [B2], and independently to Zilber [Z2].

Section 6

Although originally the study of stable groups was considered and treated as part of model-theoretic algebra, it is now considered as an essential part of general stability theory. What I mean is that the point of view shifted from the study of groups, rings, etc., on which we externally (and arbitrarily) impose some stability-theoretic hypothesis, to the realization that the nature of groups definable in a stable structure is an important measure of the 'complexity' of the structure. In fact this new point of view is fundamental to geometric stability theory. Among the earlier results along these lines was Zilber's observation that a model of an ω_1-categorical theory is built up from a strongly minimal set by a finite sequence of 'definable fibre bundles'. This is Zilber's ladder theorem [Z3], which we do not formally state or prove here.

Early work on stable groups concentrated on chain conditions. The ω-stable DCC (6.22) is due to Macintyre [Mac1], the superstable DCC (6.23) to Cherlin and Shelah [Ch–Sh], and the stable DCC (6.13) to Baldwin and Saxl [B–S]. The use of more sophisticated consequences of stablity (such as definability of types) begins with Zilber [Z1] (where stabilizers first appear), continues with [Ch], [Ch–Sh], and reaches its general form with Poizat's theory of generic types in stable groups [Po2], [Po5]. Our approach to this general theory is influenced by that in [H–P2]; 6.19 is due to Hrushovski [H1] (following ideas of Poizat [Po4]); 6.23 is due to Poizat [Po5], following early work of Reineke [R]; 6.24 is also due to Poizat [Po5], with earlier results appearing in [Mac2] and [Ch–Sh]. The references following 6.25 are to [B–Ch–M] and [Ch]; 6.26 is due to Hrushovski [H2].

Chapter 2

Fact 1.11 is from [D–H].

The study of the geometry of strongly minimal sets and the interaction with the general theory of categoricity was initiated by Zilber. A good general reference for his work and point of view is [Z4]. One of the forms of what became known as 'Zilber's conjecture' was that if D is a (strongly) minimal set which is not locally modular, than an algebraically closed field is definable in D.

Zilber's proof of Theorem 4.17 (the classification of ω-categorical strongly minimal sets) appeared in the series of papers [Z5], [Z6], [Z7]. Zilber's correct proof of Theorem 6.6 (non-finite axiomatizability of totally categorical theories) appears in [Z8]. (Earlier proofs contained some gaps.) Cherlin [C–H–L] proved Theorem 4.17 using the classification of simple groups. This was independently observed by Mills and by Peter Neumann. In [C–H–L], Theorem 6.6 was also proved, and the proof we present here is essentially theirs. In [Bu1], Buechler studies theories of finite U-rank all of whose U-rank 1 types are locally modular, and Proposition 5.8 appears there; 3.1 and 3.2 are due to Buechler [Bu2, Bu3].

We should say something about our proof of Theorem 4.17. The general line of argument is that of Zilber [Z6 and Z7]. That is, given an ω-categorical strongly minimal set D, first show D to be 2-pseudolinear, then find a rank 2 group acting transitively on a rank 1 set, then get a contradiction. Zilber achieved the first step by ingenious counting arguments on a pseudoplane. These arguments were simplified, conceptualized, and placed in a more general context (unimodularity) by Hrushovski [H3], and this is exactly our Theorem 4.15. In the second step, we follow Zilber, although the final contradiction is somewhat simplified.

Chapter 3

Proposition 1.4 appears in [A–Z1]. A general discussion and survey of ω-categorical structures and their automorphism groups appear in [Ho–Ho–M]. Corollary 1.9 is remarked in [C–H–L].

The main result of this chapter (quasifinite axiomatizability of ω-categorical, ω-stable theories) is due to Hrushovski [H4], and the proof we present is his. The fundamental idea (finding nice enumerations of the coordinatizing structures) is due to Ahlbrandt and Ziegler [A–Z1], who used it to prove the quasifinite axiomatizability of almost strong minimal ω-categorical structures. The 'easy case' of the main theorem which we prove in section 3 is a special case of their result.

Additional results concerned with pinning down the precise structure of ω-categorical, ω-stable structures appear in [H4], [A-Z2], and [Ho–P], for example. In the latter two papers, cohomology enters the picture. This has led to a growing body of work on 'covers' (finite or otherwise) of structures. See [E–H] for example.

Chapter 4

Section 1

The notion of a weakly normal definable set appears in [P3]. There we also defined a *weakly normal theory* to be one in which every definable set is a Boolean combination of weakly normal definable sets. We had earlier worked with the stronger notion of a *normal* definable set (a definable set X every conjugate of which is equal to or disjoint from X). Similar ideas were developed by Soviet model theorists (specifically Palyutin). See for example [Pa]. The expression '1-based' is due to Buechler [Bu1] (the intended meaning being that if I is a Morley sequence in the strong type p, then for any element $a \in I$, the global non-forking extension of p does not fork over a). The notion of a pseudoplane is due to Lachlan [Lach], who observed that the existence of a definable pseudoplane was an obstruction to proving the ω-stablity of stable ω-categorical theories. Pseudoplanes and quasi-designs played an important role in Zilber's work on the fine structure of ω_1-categorical theories. In fact in [Z9] he proves his 'Trichotomy theorem', that in any ω_1-categorical structure—exactly one of the following holds: (i) a (ω_1-categorical) pseudoplane is definable in M, (ii) any strongly minimal set in M is locally projective, (iii) any strongly minimal set in M is 'trivial'.

The main point of, say, [P3] was to understand the meaning of 'no definable pseudoplanes' in arbitrary stable (not necessarily ω-categorical) theories.

Proposition 1.7 and Lemma 1.9 appear in [H–P1].

Section 2

The material here is largely due to myself, and appears in [Ha–P–St1]. Lemma 2.7 is due to Poizat [G], where he makes a useful study of various notions of triviality. Example 2.15(ii) is analyzed in [Bou–Po]. This is a special case of the locally free algebras studied by Belegradek [Be]. Lemma 2.5 is due to Starchenko.

Section 3

The notion of 'finite coding' and its consquences appear in [H5], where it is also proved that a countable non-ω-categorical theory with finite coding has infinitely many countable models (with the basic line of argument being 3.1 to 3.7). The weaker result, replacing the finite coding assumption by 1-basedness, appears in [P3] (with a quite different proof). See [P4] or [H5] for the proof of 3.8; 3.9 is essentially due to Hyttinen [Hy].

In [Lo–He–P–T–W], an analysis is done of theories with no dense forking chains (developing in the process the notion of U_α-rank), and it is shown that in such a theory every type is domination equivalent to a finite product of regular types.

Section 4

The main result (Proposition 4.5) and its consequences are from [H–P1]. This built on earlier work of Zilber (described in [Z9]) and myself [P5].

Section 5

The material here is due to Hrushovski [H6], where he worked in the more general regular type context. Again earlier work of Zilber (in the strongly minimal context) was fundamental.

Section 6

Proposition 6.3 and 6.4 are from [E–P–Po].

Chapter 5

The whole problem of the geometry of (strongly) minimal sets and the interaction with definable groups comes from Zilber. On the other hand he only proves the existence of definable groups in an ω-categorical (strongly minimal) context. If M is strongly minimal ω-categorical with locally projective geometry, then he observes that the corresponding projective geometry (over a finite field) is definable in M, as well as the associated vector space (see [Z10]). Also, as part of his proof that ω-categorical strongly minimal sets are locally modular, he proves that if the geometry is 2-pseudolinear then an infinite group is definable ([Z7]). (This is exactly the group appearing in Lemma 4.21 of Chapter 2.) Hrushovski found various levels of generalization of these constructions outside the ω-categorical context, and these generalizations (going under the general name of 'group configuration theorems') are the main focus of this chapter.

Section 1

Theorem 1.1 is due to Hrushovski and appears (in the more general context of regular types) in [H6]; 1.11 is due to Starchenko.

Section 2

The main results (2.1, 2.3, 2.4) are due to Hrushovski [H6] (again in a more general context). Our proof of 2.1 is a slight simplification of Hrushovski's. A weaker version of 2.2 (where the division ring is a field) appears in [P6], using other methods. In any case 2.2 was known to Hrushovski.

Section 3

Proposition 3.2 (in the special case of strongly minimal sets) appears in [Bu4], although the proof there had considerable input from Hrushovski. Here we present Hrushovski's proof from [H3]. Note that from 3.2 and 4.15 of Chapter 2, one concludes that a unimodular minimal set is locally modular.

Section 4

Hrushovski noticed that the existence of a definable group in, say, a minimal set D is *equivalent* to the existence of a certain configuration of points in D^{eq} *vis-à-vis* algebraicity. This result is called 'the group configuration theorem', and (sometimes generalized to a regular type context and p-weight) appears in [Bou] and [H3]. The generalization to arbitrary stable theories presented here (Theorem 4.5) was recognized by me in 1992, after which I found out that it was already known to Hrushovski. Corollary 4.13 is due to Bouscaren and Hrushovski [Bou–H].

In [Ha–P–St2], 4.13 was used, together with ideas from section 2 of Chapter 4, to prove the 'Main Gap' for the class of sufficiently saturated models (a-models) of a 1-based theory.

Chapter 6

This chapter contains the main applications of the finite rank geometric theory to classification theory, at least as far as this book is concerned. Theorem 2.1 on uncountable theories categorical in a higher power is due to Laskowski [Lask]. Corollary 5.13 (Vaught's conjecture for countable weakly minimal theories) is the combined work of Buechler [Bu5] and Newelski [N1], building on (and correcting) the pioneering work of Saffe [Sa]. Our treatment of the material follows to some extent the essentially expository paper [P–R].

Section 1

The material is largely from [P–R] (although influenced by and borrowing from [Lask]).

Section 2

As mentioned above Theorem 2.1 is from [Lask]. The corollary that if T is categorical in some $\lambda > |T|$ then T is categorical in all $\lambda > |T|$ (the generalization of Morley's theorem to theories of arbitrary cardinality) was proved earlier by Shelah [Sh2], using 'classical' methods. Proposition 2.3 and Corollary 2.4 are from [P–R].

Section 3

Lemma 3.3 is from [Bu6]. A proof of Lemma 3.6 appears in [N1], although the result was known earlier to Buechler and Hrushovski. The rest of the material is from [N1].

Section 4

Theorem 4.1 is the main result of [N1]. An incorrect proof appeared much earlier in [Sa]. In any case our exposition follows Newelski [N1].

Section 5

In [Bu5], Buechler proved Vaught's conjecture for countable weakly minimal theories, *assuming* 4.1. We follow his guidelines here, with some minor simplifications due to our assumption that T is unidimensional. Theorem 5.14 is from [P–R].

Vaught's conjecture (that a countable complete theory has *either* at most countably many, *or* exactly continuum many, countable models) remains an open and important problem in model theory. New methods were required even for the case of (not necessarily weakly minimal) unidimensional superstable theories. Buechler, building on work of Newelski [N2], proved Vaught's conjecture for superstable theories of finite U-rank ([Bu7], [Bu8]). Deep problems remain for dealing with arbitrary superstable theories. In [P8], the machinery of sections 2, 4, and 5 of this chapter was adapted to prove Vaught's conjecture for the theory of any locally modular regular superstable group, whose non-generic types all have Morley rank. Newelski has, in a series of papers (such as [N3], [N4], [N5]), proved a number of deep results (such as general analogues of Theorem 4.1), and built up a considerable machinery, all of which should play a major role in future work on Vaught's conjecture for superstable theories. There have been also a number of other, maybe less difficult, results around Vaught's conjecture for stable theories. In [L–P] Vaught's conjecture was proved for superstable theories in which no infinite group is definable. In [P7], Vaught's conjecture was proved for trivial 1-based theories. This was generalized in [Lo–T] to stable 'finitely based' theories in which no infinite group is definable.

Similarly, deep problems remain concerning the other topic of this chapter: classification theory for uncountable unidimensional superstable theories (and many more generally for arbitrary uncountable superstable theories). One would expect the geometric theory to continue to play an important role, but in conjunction with the theory of covers.

Chapter 7

Everything in this chapter is due to Hrushovski (unless stated otherwise below) and forms a substantial part of his doctoral thesis [H8]. The material is also distributed among [H6], [H1], [H3], and [Bou]. Another attempt to generalize the finite rank geometric theory to superstable theories appears in [P9].

Section 1

Notions such as **P**-simple, **P**-semi-regular, **P**-weight, etc. (for P a suitable family of regular types), originate with Shelah [Sh1, Chapter 5]. The definitions we work with here, however, are due to Hrushovski, who recast Shelah's theory in a more 'manageable' form. It should be noted that p-semi-regularity as defined here is *not* identical with Shelah's notion. Nevertheless the important existence theorems for p-simple and p-semi-regular types (1.17 and 1.19) are due essentially to Shelah.

Section 2

The inclusion of (iv) in the equivalences in Proposition 2.4 was recognized by myself (especially the *semi-regularity* requirement). Example 2.11 is due, I believe, to Poizat. Example 2.13 is due to myself, and is a 'trivialized' version of an example of Hrushovski [H6].

Section 4

Internality and the connection with definable automorphism groups has roots in the work of Zilber (although the precise definitions here and results in the general context of stable theories are due to Hrushovski). Some useful observations (such as Lemma 4.2) were made by Poizat [Po2]. Poizat [Po3] also made the important observation that this general machinery provides a model-theoretic explication of Kolchin's differential Galois theory.

Section 6

The proof presented here of 6.2 is due to myself (although the result is again due to Hrushovski).

Chapter 8

Section 1

The notion of analysability (a natural extension of internality) and its basic properties given here are due to Hrushovski ([H7], [H8]), although various forms of it were implicit in earlier work such as [B–L].

The notion of **P**-minimality is due to myself, appearing also in [P10] and [Ch–P]. A stronger (and incorrect) statement than Lemma 1.13 appeared in [H–Sh] (proposition 1.1 there).

Proposition 1.15 is due to Hrushovski ([H7], [H1]), and was one of the first major results in stability theory obtained using the machinery of analysability and definable automorphism groups.

Section 2

The basic results on definablity of weight are due to Hrushovski [H–Sh]. Related results appear in [Bu–Sh]. In any case, many of these results have been recast here in the context of **P**-analysability and **P**-minimality.

Proposition 2.11 is due to myself and appears in [P10]. Lemma 2.20 is from [Bu–Sh].

Section 3

The main results appearing here are due to Buechler and Shelah [Bu–Sh].

Section 4

The notions DOP and depth 0, as well as 4.4 and 4.5, are due to Shelah [Sh1]; 4.6 is from [Bu–Sh]. The results in the remainder of the section are due to Hrushovski and Shelah [H–Sh].

Section 5

The main results of this section (Proposition 5.14, Corollary 5.16(i)) are due to Hrushovski and Shelah [H–Sh], these being the ‘aim’ of the definability of weight machinery set up in section 2; 5.3, 5.4, and 5.5 are from a paper of Chowdhury and myself [Ch–P]. The notions OTOP and PMOP are due to Shelah [Sh1], but the reader should also see Bradd Hart's treatment [Ha]. The crucial Lemma 5.10 which states the required technical consequence of NDOP and NOTOP was observed and simply stated as a fact in [H–Sh]. The notion of a strongly regular type is due to Shelah [Sh1]. The variant 'strong **P**-minimality' is due to myself, as are the easy variants (5.15 and 5.16(ii)) of the main results of this section.

A natural 'showcase' for the results of this chapter would have been a presentation of Hrushovski's precise computation of the possible spectrum functions for countable superstable theories with NDOP and NOTOP (unpublished). However, I have not absorbed this material as yet, and from what I understand, it would require a separate monograph for its exposition.

Corollary 5.4, together with the group existence results of Chapter 7, is used in [Ch–P] to show that any theory T which has *uncountable* cardinality λ has infinitely many models of cardinality λ (strengthening a result of Shelah [Sh1] stating that T could not have exactly one model of cardinality λ).

Lemma 5.3 is also observed by Newelski [N4] and used by him to show that a regular type on which forking is 'meagre' is locally modular. He obtains also many results relevant to Vaught's conjecture for superstable theories (in for example [N3], [N4], [N5]).

References

[A] Artin, E. (1957). *Geometric Algebra.* Interscience Publishers Inc., New York.

[A–Z1] Ahlbrandt, G., and Ziegler, M. (1986). Quasi-finitely axiomatizable totally categorical theories. *Annals of Pure and Applied Logic*, **30**, 63–82.

[A–Z2] Ahlbrandt, G., and Ziegler, M. (1991). What's so special about $(\mathbf{Z}/4\mathbf{Z})^{\omega}$?, *Archive for Mathematical Logic*, **31**, 115–32.

[B1] Baldwin, J. T. (1988). *Fundamentals of Stability Theory*. Springer-Verlag, Berlin.

[B2] Baldwin, J. T. (1973). α_T is finite in ω_1-categorical theories. *Transactions of the American Mathematical Society*, **181**, 37–51.

[B–L] Baldwin, J. T., and Lachlan, A. H. (1971). On strongly minimal sets. *Journal of Symbolic Logic*, **36**, 79–96.

[B–S] Baldwin, J. T., and Saxl, J. (1976). Logical stability in group theory. *Journal of the Australian Mathematical Society*, *Ser. A*, **21**, 267–76.

[B–Ch–M] Baur, W., Cherlin, G. L., and Macintyre, A. J. (1979). Totally categorical groups and rings, and rings. *Journal of Algebra*, **57**, 407–40.

[Be] Belegradek, O. V. (1988). The model theory of locally free algebras (in Russian). In *Theory of mdoels and its applications* (all in Russian). Nauka, Novosibirsk. (English summary: Some model theory of locally free algebras, in *6th Easter Conference on Model Theory* (ed. B. I. Dahn and H. Wolter), Humboldt-University, Berlin.)

[Bou] Bouscaren, E. (1989). The group configuration–after Hrushovski. In *Model Theory of Groups* (ed. A. Nesin and A. Pillay), pp. 199–209. Notre Dame University Press.

[Bou–H] Bouscaren, E., and Hrushovski, E. (1994). On one-based theories. *Journal of Symbolic Logic*, **59**, 579–95.

[Bou–Po] Bouscaren, E., and Poizat, B. (1988). Des belles paires aux belles uples. *Journal of Symbolic Logic*, **53**, 434–42.

[Bu1] Buechler, S. (1986). Locally modular theories of finite rank. *Annals of Pure and Applied Logic*, **30**, 83–95.

[Bu2] Buechler, S. (1987). On nontrivial types of U-rank 1. *Journal of Symbolic Logic*, **52**, 548–51.

[Bu3] Buechler, S. (1985). The geometry of weakly minimal types. *Journal of Symbolic Logic*, **50**, 1044–53.

[Bu4] Buechler, S. (1991). Pseudoprojective strongly minimal sets are locally projective. *Journal of Symbolic Logic*, **56**, 1184–94.

[Bu5] Buechler, S. (1987). Classification of small weakly minimal sets I. *Classification Theory, Proceedings, Chicago 1985* (ed. J. T. Baldwin), pp. 32–71.

[Bu6] Buechler, S. (1988). Classification of small weakly minimal sets II. *Journal of Symbolic Logic*, **53**, 625–35.

[Bu7] Buechler, S. (1992). Vaught's conjecture for unidimensional theories. Preprint.

[Bu8] Buechler, S. (1992). Vaught's conjecture for superstable theories of finite rank. Preprint.

[Bu–Sh] Buechler, S., and Shelah, S. (1989). On the existence of regular types. *Annals of Pure and Applied Logic*, **45**, 277–308.

[C–K] Chang, C. C., and Keisler, H. J. (1990). *Model Theory* (3rd ed. North-Holland, Amsterdam.

[Ca–K] Cameron, P. J., and Kantor, W. M. (1979). 2-transitive and antiflag transitive collineation groups of finite projective spaces. *Journal of Algebra*, **60**, 384–422.

[Ch] Cherlin, G. L. (1979). Groups of small Morley rank. *Annals of Mathematical Logic*, **17**, 1–28.

[C–H–L] Cherlin, G. L., Harrington, L., and Lachlan, A. H. (1986). ω-categorical, ω-stable structures. *Annals of Pure and Applied Logic*, **28**, 103–35.

[Ch–Sh] Cherlin, G. L., and Shelah, S. (1980). Superstable fields and groups. *Annals of Mathematical Logic*, **18**, 227–70.

[Cho–P] Chowdhury, A., and Pillay, A. (1994). On the number of models of uncountable theories. *Journal of Symbolic Logic*, **59**, 1285–300.

[D–H] Doyen, J., and Hubaut, X. (1971). Finite regular locally projective geometries. *Mathematische Zeitschrift*, **119**, 83–8.

[E–H] Evans, D., and Hrushovski, E. (1993). On the automorphism groups of finite covers. *Annals of Pure and Applied Logic*, **62**, 83–112.

[E–P–Po] Evans, D., Pillay, A., and Poizat, B. (1990). Le groupe dans le groupe. *Algebra i Logika*, **29**, 244–52.

[G] Goode, J. B. (1991). Some trival considerations. *Journal of Symbolic Logic*, **56**, 624–31.

[Ha] Hart, B. (1985). An exposition of OTOP. *Classification Theory, Proceedings, Chicago 1985*, (ed. J. T. Baldwin). Springer-Verlag, Berlin.

[Ha–P–St1] Hart, B., Pillay, A., and Starchenko, S. (1993). Triviality, NDOP and stable varieties. *Annals of Pure and Applied Logic*, **62**, 119–46.

[Ha–P–St2] Hart, B., Pillay, A., and Starchenko, S. (1995). 1-based theories–the main gap for *a*-models. *Archives for Mathematical Logic*, **34**, 285–300.

[Ho] Hodges, W. A. (1993). *Model Theory*. Cambridge University Press, Cambridge.

[Ho–Ho–M] Hodges, W. A., Hodkinson, I. M., and Macpherson, D. (1990). Omega categoricity, relative categoricity and coordinatization. *Annals of Pure and Applied Logic*, **46**, 169–99.

[Ho–P] Hodges, W. A., and Pillay, A. (1994). Cohomology of structures and some problems of Ahlbrandt and Ziegler. *Journal of the London Mathematical Society, Ser.* 2, **50**, 1–16.

[H1] Hrushovski, E. (1990). Unidimensional theories are superstable. *Annals of Pure and Applied Logic*, **50**, 117–38.

[H2] Hrushovski, E. (1989). Almost orthogonal regular types. *Annals of Pure and Applied Logic*, **45**, 139–55.

[H3] Hrushovski, E. (1992). Unimodular minimal theories. *Journal of the London Mathematical Society, Ser. 2*, **46**, 385–96.

[H4] Hrushovski, E. (1989). Totally categorical structures. *Transactions of the American Mathematical Society*, **313**, 131–59.

[H5] Hrushovski, E. (1989). Finitely based stable theories. *Journal of Symbolic Logic*, **54**, 221–5.

[H6] Hrushovski, E. (1985). Locally modular regular types. In *Classification Theory, Proceedings, Chicago 1985*, (ed. J. J. Baldwin. Springer-Verlag, Berlin.

[H7] Hrushovski, E. (1986). Contributions to stable model theory. Ph.D. Thesis, Berkeley.

[H8] Hrushovski, E. (1989). Kueker's conjecture for stable theories. *Journal of Symbolic Logic*, **54**, 207–20.

[H–P1] Hrushovski, E., and Pillay, A. (1987). Weakly normal groups. *In Logic Colloquium 85* (ed. Paris Logic Group). 233–44. North-Holland, Amsterdam.

[H–P2] Hrushovski, E., and Pillay, A. (1994). Groups definable in local fields and pseudofinite fields. *Israel Journal of Mathematics*, **85**, 203–62.

[H–Sh] Hrushovski, E., and Shelah, S. (1989). A dichotomy theorem for regular types. *Annals of Pure and Applied Logic*, **45**, 157–69.

[Hy] Hyttinen, T. (1995). Remarks on structure theorems for ω_1-saturated models. *Notre Dame Journal of Formal Logic*, **36**, 269–278.

[Lach] Lachlan, A. H. (1974). Two conjectures regarding the stability of ω-categorical theories. *Fundamenta Mathematicae*, **81**, 133–45.

[L1] Lascar, D. (1987). *Stability in Model Theory*, Longman, Harlow, and John Wiley, New York.

[L2] Lascar, D. (1976). Ranks and definability in superstable theories. *Israel Journal of Mathematics*, **23**, 53–87.

[L3] Lascar, D. (1982). Order de Rudin-Keisler et poids dans les theories ω-stables. *Zeitschrift für Mathematische Logik*, **28**, 413–30.

[Lask] Laskowski, M. C. (1988). Uncountable theories that are categorical in a higher power. *Journal of Symbolic Logic*, **53**, 512–30.

[Lo–He–P–T–W] Loveys, J., Herwig, B., Pillay, A., Tanovic, P., and Wagner, F. O. (1992). Stable theories with no dense forking chains. *Archive for Mathematical Logic*, **31**, 297–303.

[Lo–T] Loveys, J., and Tanovic, P. (1996). Countable models of trivial theories which admit finite coding. *Journal of Symbolic Logic*.

[L–P] Low, L. F., and Pillay, A. (1992). Superstable theories with few countable models. *Archive for Mathematical Logic*, **31**, 457–65.

[Mac1] Macintyre, A. J. (1971). On ω_1-categorical theories of abelian groups. *Fundamenta Mathematicae*, **70**, 253–70.

[Mac2] Macintyre, A. J. (1971). On ω_1-categorical theories of fields. *Fundamenta Mathematicae*, **71**, 1–25.

[Makk] Makkai, M. (1984). A survey of basic stability theory with emphasis on regularity and orthogonality. *Israel Journal of Mathematics*, **49**, 181–238.

[Ma] Marsh, W. E. (1966). On ω_1-categorical and not ω-categorical theories. Doctoral Dissertation, Dartmouth College.

[M] Morley, M. (1965). Categoricity in power. *Transactions of the American Mathematical Society*, **114**, 514–38.

[N1] Newelski, L. (1990). A proof of Saffe's conjecture. *Fundamenta Mathematicae*, **134**, 143–55.

[N2] Newelski, L. (1994). On U-rank 2 types. *Transactions of the American Mathematical Society*, **344**, 553–81.

[N3] Newelski, L. (1995). M-rank and meager types. *Fundamenta Mathematicae*, **146**, 121–39.

[N4] Newelski, L. (1994). Meager forking. *Annals of Pure and Applied Logic*, **70**, 141–75.

[N5] Newelski, L. (1995). M-rank and meagre groups. Preprint.

[Pa] Palyutin, E. A. (1980). Categorical Horn classes 1. *Algebra i Logika*, **19**(5), 582–614.

[P1] Pillay, A. (1983). *An introduction to stability theory*. Oxford University Press.

[P2] Pillay, A. (1986). Forking, normalization and canonical bases. *Annals of Pure and Applied Logic*, **32**, 61–81.

[P3] Pillay, A. (1989). Stable theories, pseudoplanes, and the number of countable models. *Annals of Pure and Applied Logic*, **43**, 147–60.

[P4] Pillay, A. (1983). Countable models of stable theories. *Proceedings of the American Mathematical Society*, **89**, 660–72.

[P5] Pillay, A. (1986). Superstable groups of finite rank without pseudoplanes. *Annals of Pure and Applied Logic*, **30**, 95–101.

[P6] Pillay, A. (1991). Some remarks on modular regular types. *Journal of Symbolic Logic*, **56**, 1003–11.

[P7] Pillay, A. (1992). Countable models of 1-based theories. *Archive for Mathematical Logic*, **31**, 163–69.

[P8] Pillay, A. (1992). On certain locally modular regular superstable groups. Preprint.

[P9] Pillay, A. (1985). Simple superstable theories. In *Classification Theory, Proceedings, Chicago 1985*, (ed. J. T. Baldwin). Springer-Verlag, Berlin.

[P10] Pillay, A. (1994). Some remarks on nonmultidimensional superstable theories. *Journal of Symbolic Logic*, **59**, 151–65.

[P–R] Pillay, A., and Rothmaler, Ph. (1990). Non totally transcendental unidimensional theories. *Archive for Mathematical Logic*, **30**, 93–111.

[Po1] Poizat, B. (1985). *Cours de theorie des modéles*. Nur al-Mantiq wal-Ma'rifah, Villeurbanne.

[Po2] Poizat, B. (1987). *Groupes stables*. Nur al-Mantiq wal-Ma'rifah, Villeurbanne.

[Po3] Poizat, B. (1983). Une théorie de Galois imaginaire. *Journal of Symbolic Logic*, **48**, 1151–70.

[Po4] Poizat, B. (1981). Sous-groupes définissables d'un groupe stable. *Journal of Symbolic Logic*, **46**, 137–46.

[Po5] Poizat, B. (1983). Groupes stable avec types génériques réguliers. *Journal of Symbolic Logic*, **48**, 339–55.

[R] Reineke, J. (1975). Minimale Gruppen. *Zeitschrift für Mathematische Logik*, **21**, 357–9.

[Sa] Saffe, J. (1980). On Vaught's conjecture for superstable theories. Preprint.

[Sh1] Shelah, S. (1990). *Classification theory* (revised). North-Holland, Amsterdam.

[Sh2] Shelah, S. (1974). Categoricity of uncountable theories. *Proceedings of Tarski Symposium*, 187–203. AMS, Providence, RI.

[Z1] Zilber, B. I. (1977). Groups and rings whose theory is categorical. *Fundamenta Mathematicae*, **95**, 173–88.

[Z2] Zilber, B. I. (1974). On the transcendence rank of formulas of an ω_1-categorical theory. *Matematika Zametki*, **15**, 321–9.

[Z3] Zilber, B. I. (1983). The structure of models of ω_1-categorical theories. *Proceedings of International Congress of Mathematicians* (*Warsaw 1983*), 359–68. PWN, Warsaw.

[Z4] Zilber, B. I. (1993). *Uncountably categorical theories*. AMS Translations of Mathematical Monographs, Vol. 117.

[Z5] Zilber, B. I. (1980). Strongly minimal countably categorical theories (in Russian). *Sibirsk Matematika Zhurnal*, **21**, No. 2, 98–112.

[Z6] Zilber, B. I. (1984). Strongly minimal countably categorical theories II (in Russian). *Sibirsk Matematika Zhurnal*, **25**, No. 3, 71–88.

[Z7] Zilber, B. I. (1984). Strongly minimal countably categorical theories III (in Russian). *Sibirsk Matematika Zhurnal*, **25**, No. 4, 63–77.

[Z8] Zilber, B. I. (1981). On the finite axiomatizability problem of theories categorical in all infinite powers (in Russian). *Investigations in Theoretical Programming*, 69–74. Kazakh. Gos. University, Alma-Ata.

[Z9] Zilber, B. I. (1986). Structural properties of models of ω_1-categorical theories. *Logic, Methodology and Philosophy of Science VII*, pp. 115–28. North-Holland, Amsterdam.

[Z10] Zilber, B. I. (1981). Some problems concerning categorical theories. *Bulletin de l'Academie Polonaise de Sciences*, **XXIX**, 47–9.

Index